Dear MyCopy Customer,

This Springer book is a monochrome print version of the eBook to which your library gives you access via SpringerLink. It is available to you at a subsidized price since your library subscribes to at least one Springer eBook subject collection.

Please note that MyCopy books are only offered to library patrons with access to at least one Springer eBook subject collection. MyCopy books are strictly for individual use only.

You may cite this book by referencing the bibliographic data and/or the DOI (Digital Object Identifier) found in the front matter. This book is an exact but monochrome copy of the print version of the eBook on SpringerLink.

INTERNATIONAL ENERGY AGENCY (IEA)
SMALL SOLAR POWER SYSTEMS PROJECT (SSPS)

THE IEA/SSPS SOLAR THERMAL POWER PLANTS
– Facts and Figures –

Final Report of the
International Test and Evaluation Team (ITET)

Editors:
P. Kesselring and C. S. Selvage

Volume 1:
Central Receiver System (CRS)

Springer-Verlag Berlin Heidelberg GmbH 1986

Dr. sc. nat. Paul Kesselring

Head, Prospective Studies Division,
Swiss Federal Institute for Reactor Research (EIR),
Head, SSPS Test and Operation Advisory Board.

Clifford S. Selvage, BS

Head, SSPS International Test and Evaluation Team.

DOI 10.1007/978-3-642-82678-8

CIP-Kurztitelaufnahme der Deutschen Bibliothek
International Energy Agency / Small Solar Power Systems Project:
The IEA, SSPS solar thermal power plants: facts and figures; final report of the Internat. Test and Evaluation Team (ITET) / Internat. Energy Agency (IEA), Small Solar Power Systems Project (SSPS). Ed.: P. Kesselring and C. S. Selvage. – Berlin; Heidelberg; New York; Tokyo: Springer

NE: Kesselring, Paul [Hrsg.] HST

Vol. 1. Central receiver system (CRS). – 1986.

MyCopy version of the original edition 1986
www.springer.com/mycopy

INTERNATIONAL ENERGY AGENCY (IEA)
SMALL SOLAR POWER SYSTEMS PROJECT (SSPS)

THE IEA/SSPS SOLAR THERMAL POWER PLANTS

– Facts and Figures –

Final Report of the International Test and Evaluation Team (ITET)

Editors:
P. Kesselring and C. S. Selvage

Volume 1: Central Receiver System (CRS)

Springer-Verlag Berlin Heidelberg GmbH 1986

Dr. sc. nat. Paul Kesselring

Head, Prospective Studies Division,
Swiss Federal Institute for Reactor Research (EIR),
Head, SSPS Test and Operation Advisory Board.

Clifford S. Selvage, BS

Head, SSPS International Test and Evaluation Team.

DOI 10.1007/978-3-642-82678-8

CIP-Kurztitelaufnahme der Deutschen Bibliothek
International Energy Agency / Small Solar Power Systems Project:
The IEA, SSPS solar thermal power plants: facts and figures; final report of the Internat. Test and Evaluation Team (ITET) / Internat. Energy Agency (IEA), Small Solar Power Systems Project (SSPS). Ed.: P. Kesselring and C. S. Selvage. – Berlin; Heidelberg; New York; Tokyo: Springer

NE: Kesselring, Paul [Hrsg.] HST

Vol. 1. Central receiver system (CRS). – 1986.

MyCopy version of the original edition 1986
www.springer.com/mycopy

EDITORS PREFACE

The Project's origin

As a consequence of the so-called "first oil crisis", the interest in solar electricity generation rose sharply after 1973. The solar thermal way of solving the problem was attractive because the main task was simply to replace the fossil fuel by a "solar fuel" in an otherwise conventional thermal power plant -that was at least what many thought at that time. Thus more than half a dozen of solar thermal plant projects were created in the mid-seventies. One of them is the Small Solar Power Systems (SSPS) Project of the International Energy Agency (IEA). It consists of the design, development, construction, operation, test and evaluation of two dissimilar small solar thermal electric power systems each at a nominal power of 500 kW_e.

ITET and TOAB

In order to assist the Operating Agent (DFVLR - Deutsche Forschungs- und Versuchsanstalt für Luft- und Raumfahrt e.V.) in managing the project, the Executive Committee (EC) created two bodies called the "International Test and Evaluation Team" (ITET) and the "Test and Operation Advisory Board" (TOAB). The latter consisted of a group of experts from the different participating countries, meeting three to four times a year to articulate i.a. the technical interests and expectations of the different parties in the project. It was the TOAB that formulated e.g., the test and evaluation program, which for the final project evaluation boiled down to a list of topics -the so-called "deliverables"- that would be covered by the final report.

The ITET was a group of engineers and scientists that did most of the final evaluation work, supported by cooperative organizations such as e.g., DFVLR and contractors. The final result of this major effort is the present series of books, containing facts and figures, discussed and interpreted to fulfill the task defined in the "deliverables".

Information pyramid

In order not to get lost in the vast amount of information provided by the SSPS Project, an efficient information management was thought to be vital. We adopted a solution which we call "information pyramid" and which will be explained in the introduction. Part of the pyramid is the "Book of Summaries", which contains the abstracted contents of all three volumes of the final ITET-Evaluation Report. Thus, it is possible to get a quick overview of the work performed by the ITET members and to find their conclusion.

General Conclusions

What are these conclusions after all?

We certainly know now that the naive picture of the solar-fired, but otherwise conventional, thermal power plant is wrong. Here we have learned much from our difficulties with the systems as a whole. Our main technical success has been the good performance of the solar specific components and subsystems, such as e.g., receivers, collectors, heliostat fields, etc. They fulfilled most of our high expectations.

Thus, generally speaking, although we did not demonstrate routine power production from a utility's point of view, we were able to contribute considerably to the technical advancement of the solar thermal technology. Above all we developed confidence in the technical soundness of the solar thermal approach. Most companies involved in the project would be ready to go on with a commercially sized plant, provided there was a customer.

Acknowledgements

The editors, being the heads of ITET and TOAB, would like to thank their colleagues for all the work accomplished under sometimes necessarily less than ideal conditions. Without the motivation and perseverance of the ITET-crew and without the positive, critical minds of the TOAB members, the present series of books would not exist.

The Operating Agent, DFVLR, provided much valuable support throughout the project. It is a pleasure to acknowledge this help as well as the good services of the Plant Operating Authority, Cīa. Sevillana de Electricidad S.A.

The preparation of the manuscript for publication has been another formidable task, shared by Sandia National Laboratories, Livermore, and DFVLR again. Here we would like to thank in particular Miss Melissa McCreery, secretary of the ITET, Mrs. Sallie Fadda from Sandia and Dr. H. Ellgering of the DFVLR and their collaborators.

Last but not least, our thanks go to the Executive Committee of the SSPS Project, whose full support -within the limits of a complex international cooperation between nine countries- is gratefully acknowledged.

Würenlingen and Livermore, August 1985

P. Kesselring and C.S. Selvage, Editors

TABLE OF CONTENTS

INTERNATIONAL ENERGY AGENCY / SMALL SOLAR POWER SYSTEMS (SSPS)

EVALUATION REPORTS

1. INTRODUCTION

This introduction to the final evaluation report of the SSPS International Test and Evaluation Team (ITET) is split into two parts: The first part -written by the head of the Test and Operation Advisory Board (TOAB)- gives a picture of the SSPS evaluation effort as seen from the point of view of an observer far away from the project site in a participating country. The second part -written by the head of the International Test and Evaluation Team (ITET)- gives the general project overview.

1.1 The SSPS Project evaluation, as seen by the head of the TOAB

a) Structure and interaction of ITET and TOAB

In retrospect, the most astonishing feature of the SSPS Project to me is that it was possible to integrate the quite different interests of nine countries to the extent that such a large common venture -worth approximately 90 Million DM- could be realized. This general aspect -i.e., the need to integrate different, sometimes conflicting interests- was also of importance when it came to the organization of the project evaluation. It is e.g., reflected in the structure of the ITET. Only its head and the two senior evaluators were direct "employees" of the Project. All other members were seconded by the different countries to the Project. Their selection, in the countries, was only restricted by relatively loose boundary conditions, set by the Project (minimal duration of stay, minimal number of members to be seconded by a country, preference for certain qualification profiles). As a result, the ITET was a frequently changing group of people, differing not only in nationality but also with respect to the background of education and interests. It was held together by the common task.

While ITET members during their stay in Almería worked full time and on site for the project, many members of the Test and Operation Advisory Board (TOAB) devoted a few days per year only to SSPS activities. The members of this board were designated by their countries in order to help the project with their professional expertise and at the same time to articulate the interest of their countries on a technical level. Thus, in making a main contribution to the definition of the evaluation program, the TOAB selected from the very large number of imaginable R&D subjects, a small fraction, lying within reach of the ITET and reflecting the technical priorities as well as the national interests.

The interaction between TOAB and ITET -whose head and senior evaluators as well as the OA, took part in TOAB meetings ex officio- led to very beneficial side effects: The ITET, struggling with the daily on site problems, could not forget about the needs of the far away home countries and the sometimes (too) high expectations of the TOAB were brought down to the reality of the hard facts in Almería.

b) Structure and character of the ITET Final Evaluation Report

Thus, the stage was set for the final evaluation. It was carried out in the following way: The evaluation topics defined by the "deliverables" were discussed within the ITET and subtasks assigned to the members of the group. The responsibles for each subtask then became the authors of a self-consistent paper, describing their work, results and conclusions. It is the collection of all these individual papers -written within the common framework explained before- that forms the main body of the Final Evaluation Report of the ITET.

The history of the report makes it clear that one should not expect a homogeneous document, covering every possible R&D aspect of the two plants in a comprehensive way. What we must expect and find, is consistency between the different contributions and their conclusions. A variation in the depth and quality of treatment is obvious and finds its natural explanation in the fact that the spectrum of authors begins with engineers, recently graduated from engineering schools, and ends with professors from technical universities. The reader, missing a paper on a topic of high interest to him, must be reminded that time and resources were limited and obviously any selection of priorities is debatable to some degree.

c) The Information Pyramid

Even considering these restrictions, the present volumes contain a very large amount of very valuable information concerning solar thermal power plants. In order to manage this information avalanche efficiently, we have introduced a hierarchy of publications, which we call the "information pyramid". It begins at the top with a book giving a synthesis of the SSPS work in the context of solar thermal power plant development in general. The book makes reference to the present collection of papers frequently. It also appears in a Springer edition and is written by an author hired by the project. The language is such that students and young engineers will be able to follow and the mature engineer gets a quick overview of the important aspects of the solar thermal technology.

The reader, willing to go into more detail, may then take the "Book of Summaries", containing the abstracts of all the papers included in the 3 volumes of the ITET Final Report. He thus has the possibility to decide quickly which of the references given in the book are most important to him and whether or not he should dig into the thick volumes in order to study the complete papers. Complete papers make reference to SSPS Technical Reports and/or to the lowest level, the SSPS Internal Reports. This information as well as raw data are available upon request via the Executive Committee members. Thus, there is a simple and efficient way to get down from the most general, highly aggregated information into more detail, step by step, to end up with the raw data, if necessary.

In parallel to the ITET's Report, the Operating Agent's point of view of the SSPS Project is given in his final report (SR-7: SSPS - Results of Test and Operation, 1981 - 1984).

d) Lessons learned

Concluding my part of the introduction, I would like to give a short, personal view of the lessons learned from the existing solar thermal power plants in general and the SSPS Project in particular. Such statements are necessarily simplifying and incomplete but nevertheless, useful in characterizing the status of a development at a given point in time [1]:

- The development of the solar specific components and subsystems of solar thermal power plants during the last 8 years has been a technical success. Receivers, collectors and heliostat fields perform to a large extent as expected.
- The problems arising from solar specific systems aspects have been underestimated. We mention in particular:
  - Start-up, shut-down time of plant (transient behavior)
  - Heat energy management in storage systems
  - Troubles with "from the shelf" components and subsystems.

  They have been the source for a great part of the difficulties encountered in the existing prototype plants.
- These problems are manageable and can be handled by good design including in particular
  - fast "first stages" (receiver + energy transport system to storage and e.g., steam generator)
  - higher solar multiples and storage
  - carefully chosen power conversion systems, matched to the solar specific requirements of the plant as a whole
  - larger plants (e.g., $\geqslant$30 MW(el))
  - minimizing plant internal consumption (10% of the annual gross output seems to be a feasible goal for larger plants).
- Site selection is very important. Local meteo conditions must be evaluated carefully. On site measurements of direct normal radiation are necessary before final site selection. Mean values are not sufficient, information concerning the intensity distribution in time is required.

In conclusion, we may say that when we started the design of the present generation of solar thermal plants in the mid-seventies, we thought that we would demonstrate commercial operation on a small scale. We were too optimistic. As a matter of fact, we have been one plant generation further away from commercial operation than we thought at the time. This is the reason why I call existing plants "prototype plants" or "experimental plants" and not "pilot plants" as it is usually done. However, if the lessons learned from the existing experimental plants are incorporated properly into future designs, a satisfactory performance of commercially sized future demonstration plants may be expected now.

1) Statements taken from a lecture given at the 2nd Igls Summer School on Solar Energy 1985, 31.7-9.8.1985 (Papers to be published by ESA)

1.2 Introduction to the SSPS Project

One objective of the International Energy Agency's (IEA's) energy research, development, and demonstration (R§D) program is to promote the development and application of new and improved energy technologies which could potentially make a significant contribution to our energy needs. Towards this objective, the IEA has established and conducted energy research, development, and demonstration projects, one of which is the Small Solar Power Systems (SSPS) project built in the province of Almeria, Spain. This project, performed under the auspices of the IEA by nine countries (Austria, Belgium, Switzerland, Germany, Spain, Greece, Italy, Sweden, and United States of America), consisted of the design, construction, testing, and operation of two dissimilar types of solar thermal power plants: a distributed collector system (DCS) and a central receiver system (CRS). They are constructed adjacent to each other on the Spanish Plataforma Solar in Almeria, southern Spain. Both have the same rated electrical output (500 kW_e design at equinox noon) and have delivered electric energy to the Spanish grid during the three-year period 1981 - 1984.

SSPS PLANT

The SSPS plant operation has produced several unique observations.

* Operational experience has been observed with the functioning of a DCS and CRS power plant.

* Different designs of advanced solar technologies (collectors, heliostats, receivers, storage systems) have been tested comparatively as part of a complete power plant system in different operational modes.

* The grid environment of the Plataforma Solar north of Almeria, with statistically the highest solar irradiation of southern European countries is representative of a wide range of future applications of solar power plants.

* The conventional part of the SSPS power plants, which is the power conversion system, has been tested with respect to its viability for solar applications.

The principal objective of the SSPS project was to examine in detail the feasibility of using solar radiation to generate electrical power. In addition, the project had the following objectives:

* Promote cooperation between IEA members in the field of new technologies.

* Demonstrate the technical feasibility of designing and building solar power plants with available hardware.

* Gather operational performance data on such plants.

* Evaluate the viability of the DCS and CRS concepts.

* Design a plant that was optimized to 500 kW_e, but which had the potential for being scaled up or down.

* Consider different geographical applications and operational modes.

* Minimize the investment costs while achieving reasonable operating expenses, good engineering safety, and a long lifetime.

* Assess the further technical development of solar power plants.

The project consisted of two phases: Phase 1 - the erection of the CRS and DCS system, and Phase 2 - test and operation. The project time schedule shows the main events before and during those two years.

Phase	1977	1978	1979	1980	1981	1982	1983	1984
	I II III IV	I II III IV	I II III IV	I II III IV	I II III IV	I II III IV	I II III IV	I II III IV
SSPS-Specifications trade-offs, feasibility considerations		Stage 1			Stage 2			
Plant (DCS + CRS) final design (Stage 1)								
Stage 2 preparations								
Procurement and installation								
Plant testing, operation and evaluation								
Advanced systems tests				Advanced sodium receiver increased collector field			CAST heliostats	

SSPS Site Location

This particular site was chosen for its geographical characteristics and because this region of Spain promised favorable conditions relative to the annual amount and intensity of solar insolation.

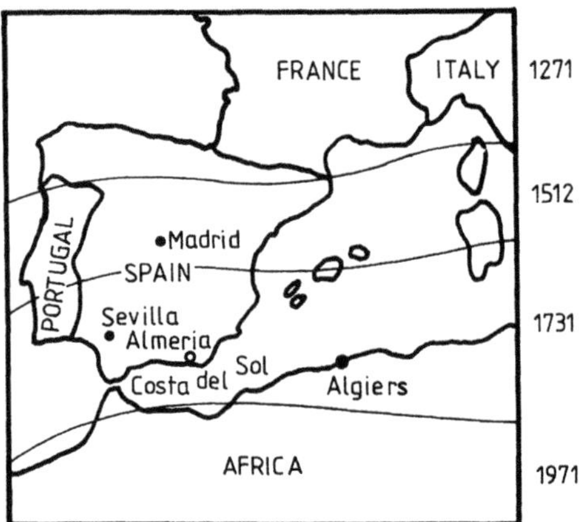

Test and Operation Organization

The testing and operation phase, which was conducted over a period of three years, was organized to collect data on:

- the viability of the selected technical solutions,

- the operational behavior of the systems, and

- the economics of the plants.

This phase of the project was administered by the organizational scheme shown below.

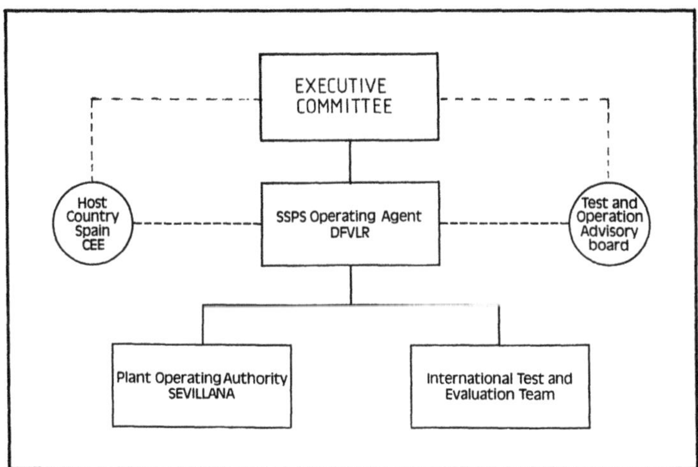

Within this organizational structure, the DFVLR (Deustche Forschungs-und Versuchsanstalt für Luft-und Raumfahrt e.V) served as the Operating Agent and was responsible for carrying out the SSPS project on behalf of the SSPS participating countries. The operational and evaluation activities to be performed were specified in a Basic Test and Operation Program document, as well as yearly updates called the Program of Work. The operation of the SSPS-CRS and -DCS was performed by the regional Spanish utility Compania Sevillana de Electricidad, acting as the Plant Operation Authority.

The scientific testing and evaluation work was entrusted to an international test and evaluation team (ITET) composed of experts from the participating countries that conducted on-site tests and analyses. The ITET was established by the Executive Committee and was headed by Mr. C. S. Selvage. This on-site team has evaluated and reported on test and operation activities and has recommended and advised the plant director on defining, planning, preparing, and conducting tests and operations. The team has performed such functions as:

- recommend tests and modes of operation for the plants

- define criteria to be met for tests and modes of operation and data requirements

- review testing, operation , and maintenance data to assess the validity of the data and potential needs for further data or retesting

- evaluate and report on the results of operation and special tests

- compare the performance of the plants when operating in similar modes of operation

- provide ad-hoc engineering support to the Operating Agent

- at a system level, compare actual performance with design goals

- at a subsystem level, compare the actual performance of the major subsystems with the design goals

- assess the reliability of the various components and subsystems based on an analysis of data.

To summarize, the evaluation consisted of combining and comparing measured, calculated, and reported plant data to determine the plant's performance and behavior over the entire period of the program. The results of these evaluations performed by the ITET have been reported in SSPS technical and internal reports, a listing of which is presented in Appendix A. In addition, four international workshops were conducted on site in order to present the status of the ITET work.

The following are a compilation of new and previously reported studies that represent work done by the ITET performing evaluations of various aspects of both systems that were requested by the TOAB and the Executive Committee. The investigations related to the CRS are described in Volume I of this report; those for the DCS are described in Volume II; and the Site Specific work is in Volume III.

The ITET staff in the years 1981 through 1984 were:

C. Gomes Camacho	Spain	June 1981 - June 1982
A. Baker	USA	summer 1981
R. Stromberg	USA	summer 1981
W. Wilson	USA	summer 1981

M. Loosme	Sweden	September 1981 - June 1983
P. Wattiez	Belgium	September 1981 - December 1984
T. von Steenberghe	Belgium	September 1981 - August 1983
F. Gaus	Germany	December 1981 - December 1983
C. S. Selvage	USA	January 1982 - March 1985
P. Toggweiler	Switzerland	January 1982 - August 1982
H. Jacobs	Germany	February 1982 - March 1985
R. Carmona	Spain	July 1982 - March 1985
M. Pescatore	Switzerland	July 1982 - May 1984
M. Anderson	Sweden	January 1983 - December 1984
J. Martin	USA	May 1983 - December 1984
F. Palumbo	Italy	May 1983 - November 1983
M. Blanco	Spain	January 1984 - March 1985
M. Sanchez	Spain	January 1984 - March 1985
J. Sandgren	Sweden	April 1984 - January 1985
A. De Benedetti	Italy	March 1984 - December 1984
N. Gregory	Switzerland	June 1984 - October 1984
B. W. Swanson	USA	September 1984 - November 1984
W. Schiel	Germany	Part of 1983 and fall 1984
G. Lemperle	Germany	Fall 1984
A. Brinner	Germany	Part of 1983 and 1984

The following specific evaluation reports make up VOLUME I, which is the evaluation of the CENTRAL RECEIVER SYSTEM. This evaluation report VOLUME I, contains reports of thirty (30) specific evaluations of six specific evaluation topic areas referred to as SECTIONS. The evaluation topic areas, by section titles are:

SECTION 3- HISTORICAL ASSESSMENT OF THE SSPS CRS PLANT PERFORMANCE.
SECTION 4- HELIOSTAT FIELD PERFORMANCE.
SECTION 5- RECEIVER BEHAVIOR.
SECTION 6- THERMAL LOSSES/THERMAL INERTIA.
SECTION 7- SYSTEMS ASPECTS/CONTROL.
SECTION 8- POTENTIAL FOR IMPROVEMENTS

A summary of each of the specific evaluation reports and the conclusion of that evaluation is contained in the introduction for each of the sections. The overall conclusions for this project are contained in the foreword.

2. CENTRAL RECEIVER SYSTEM

The 500 kW_e CRS plant has a north field of heliostats directing reflected solar energy to a tower mounted receiver. Thermal energy from the receivers is piped to a hot storage tank and then to a steam generator which produces superheated steam. This superheated steam is fed to a steam motor to produce mechanical energy to drive an electric generator. The CRS plant consists of three major systems: a heliostat field, a sodium heat transfer system, and a power conversion system. A simplified process flow diagram of these systems is shown in Fig.2.-1. The CRS main design features are given in Fig.2.-2 and system data in Figure Fig.2.-3.

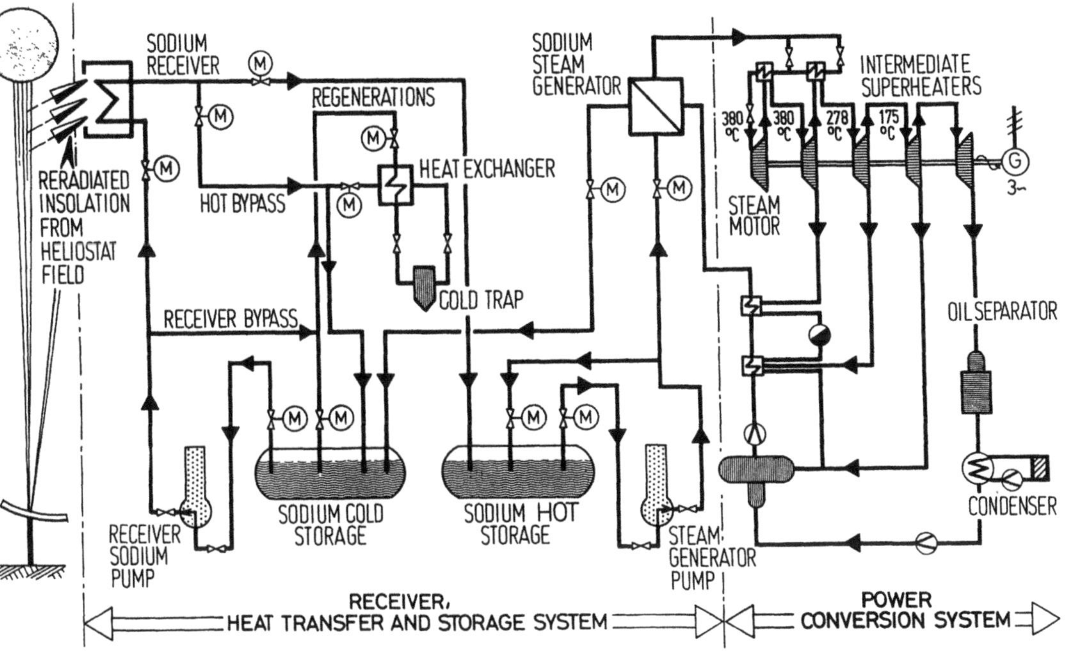

Fig.2.-1: Simplified CRS Process Flow Diagram

Heliostats	Martin Marietta Barstow, 3360 m^2 total reflective area
Receivers	1) Cavity Type with an octagonal shaped aperture of 9.7 m^2; peak heat flux on absorber tubes 62 W/cm^2; inlet / outlet temperature 270/530°C 2) External type with 2.85 x 2.73 m aperture and five panels, each with 39 parallel tubes; peak heat flux 140 W/cm^2
Heat Transfer System	Liquid Sodium
Storage System	Two tank storage equivalent 1 MW_e
Power Conversion System	6-piston steam motor (Spilling), cycle efficiency 27.2% (calculated)
Safety Precautions	Uninterruptable power supply, sodium/waterreaktion and sodium fire protections, lightning protection, design according to possible seismic events
Performance	517 kW_e net output at equinox noon
Design Lifetime	10 years
Guarantee	90% performance guarantee at design point

Fig. 2.-2: CRS Main System Design Features

Design Point	Day 80, 1200 (equinox noon)	
	Solar insolation, kW/m^2	0.92
Heliostat Field	93 heliostats	
	Total reflectice surface area, m^2	3660
Receiver	Aperture size, m^2	9.7
	Active heat transfer surface, m^2	16.9
	Inlet temperature, °C	270
	Outlet temperature, °C	530
	Efficiency (calculated), %	85
2nd Receiver	Aperture size, m^2	7.91
	Active heat transfer surface, m^2	7,91
	Inlet temperature, °C	270
	Outlet temperature, °C	530
	Efficiency (calculated)	94
Thermal Storage	Storage medium	sodium
	Thermal capacity, MWh	5.5
	Hot storage temperature, °C	530
	Cold storage temperature, °C	275
Steam Generator	Sodium inlet temperature, °C	525
	Sodium outlet temperature, °C	275
	Water inlet temperature, °C	190
	Steam outlet temperature, °C	510
	Steam pressure, bar	100
Power (at design point)	Solar input to receiver (insolation), kW	2880
	Thermal input to steam motor (thermal), kW	2200
	Gross electric, kW	600
	Net electric, kW	517
Efficiencies (at design point)	Thermal to gross electric, %	27.3
	Thermal to net electric, %	23.5

Fig.2.-3: Main System Design Data for CRS

The heliostat field subsystem includes the heliostat field and associated controls. The Martin Marietta heliostats used are identical to those at the 10 MW_e Barstow Pilot Plant except that the curvature of the mirror has been increased to shorten the focal length. Each heliostat has a reflective area of 39.3 m^2. The field consists of 93 heliostats with four different focal zones. Fig.2.-4 gives the heliostat field focal zone definition and shows the concentric-circle layout of the field north of the receiver tower. The mirror module is a vented sandwich design of hot-bonded glass mirror, honeycomb core, and steel pan enclosure. The heliostats are controlled by the heliostat array controller (HAC) located in the main control room. The HAC transmits commands to the heliostats via four heliostat field controllers (HFC). Each of these HFC's acts as a heliostat controller (HC) for four heliostats and also transmit data to other HC's which are located on each heliostat. All heliostats had the same aim point on the cavity receiver near the center of the receiver aperture. With the Advanced Sodium Receiver (ASR), three aiming points were used in order to provide a balanced thermal input.

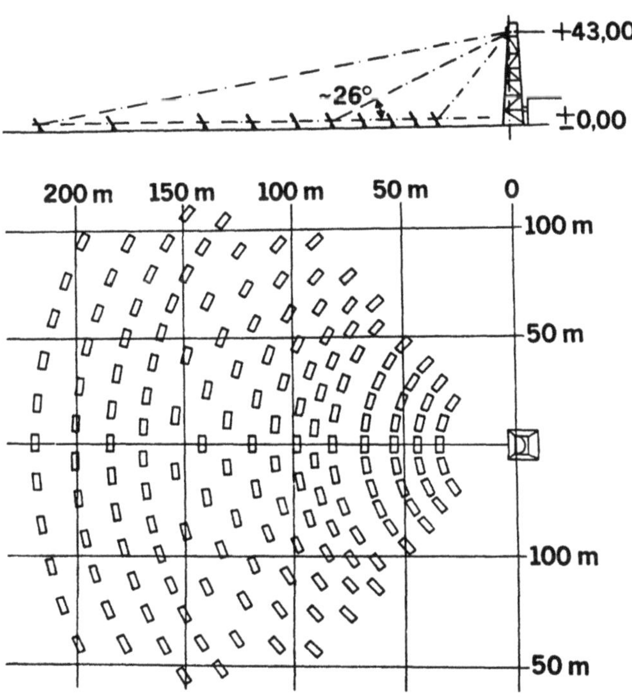

Fig.2.-4: Heliostaat Field Layout

Two receivers were installed at the CRS. The first receiver was a north-facing cavity type with a vertical octagonal shaped aperture of 9.7 m^2. The absorber panel is a 120-degree segment of a right circular cylinder. Sodium flows in six horizontal parallel tubes, which are 38 mm in diameter and -1.5 mm wall thickness, which serpentins from near the bottom of the cavity to the top defining the absorber panel. These tubes are not joined (welded) along their length but are supported by mechanical means. Sodium enters the inlet header at 270°C located at the bottom of the panel and exits the outlet header at 530°C near the panel top. The location of the absorber panel inside the cavity is such that the peak heat flux is about 62 W/cm^2 at equinox noon when 2880 kW_{th} enters the cavity aperture.

Fig. 2.-5: Cavity Receiver (Sulzer)

The second CRS receiver was a 2.7 MW_{th} external type that consists of five panels arranged to form a rectangular absorber 2.85 m high and 2.78 m wide. Each panel consists of a tube bundle with 39 - 14mm diameter vertical tubes, a bottom and top header and a downcomer. The flange of the bottom inlet header and the restraint at the downcomer sodium outlet are attached to the panel. The top header moves vertically to accommodate vertical thermal growth of the panel. The irradiated tubes are assembled together in groups of three and held with four supporting plates to form a 'triplet'. These triplets are connected to the panel framework by means of pins such that the tubes can grow axially with respect to the frame and also rotate, because of the clearance between each pin and its hole in the triplet supporting plate. Gaps are provided between so that each triplet is free to expand independently in the horizontal direction. Liquid sodium is pumped from the cold storage tank at 270°C into the bottom of receiver panel at one edge, through each of the panels in series and out at the top of the central panel at 530°C.

The presence of the gaps between the tubes does allow some concentrated solar flux to pass through the tubes and impinge on the backwall structure. Therefore a double shield of high refractory alumina-based material is located in the back of the tube bundle to protect the back structure from the incident radiation. Behind the second shield, a 175-mm layer of ceramic fiber insulates the hot parts of the receiver from the structure.

Fig. 2.-6: Advanced Sodium Receiver (Franco Tosi)

Storage for the CRS plant is provided by two separate vessels: one cold sodium vessel and one hot sodium vessel. Sodium enters the hot storage vessel at 530°C from the receiver, is drawn from this vessel and pumped to the steam generator, and then is returned to the cold sodium vessel at 275°C. Each of these storage vessels is approximately 3.3 m in diameter, 10 m in length, and has a volume of 70 m^3 and a design pressure of 8.5 bar.

The CRS steam generator is a vertical helical-tube-type with a once-through operation mode. The three heating tubes are coiled around a central displacement chamber filled with nearly stagnant sodium and are housed in a cylindrical shell. Within the tubes, water or steam is flowing from the bottom to the top. Hot sodium (525°C) enters the steam generator at the top, flows downward between the outside shell and displacement chamber around the coiled heating tubes where the heat transfer takes place, and leaves through an outlet at the bottom (275°C). Water enters the three helical tubes at the bottom (190°C, 110 bar) and exits as steam at the top (510°C, 100 bar). The nominal thermal capacity of the steam generator is about 2.2 MW.

Fig. 2.-7: Steam Generator (Sulzer)

The CRS power conversion unit consists of a steam-driven six-piston reciprocating motor and a three-phase alternating current generator. The steam motor uses intermediate heat exchangers to cool the first stage inlet steam from 510°C to 380°C while heating the second stage inlet steam, the exhaust of the first stage which is at 270°C, up to 380°C. Likewise, the second stage outlet steam, which is at 260°C is heated to 278°C before it enters the third stage. Stages four and five are fed directly from the previous stage without reheat. Extraction connections for steam heated feedwater preheaters are located at the exhaust of stages two and three. Degasification steam is extracted from the fourth stage. The steam motor is elastically connected to a brushless self-controlled threephase current generator equipped with an automatic voltage control regulator, for isolated (non-parallel or stand alone) operation and reactive current control for parallel operation. The operating conditions of the power conversion unit (PCU) are:

Thermal input (steam)	$2200 kW_{th}$
Inlet pressure	100-102 bars abs
Inlet temperature	500-520°C
Back pressure	.3 bar abs
Speed	1000 rpm
Motor	845 Hp
Gross electric output	600 kW
Net electricity output	562 kW
Efficiency (gross/thermal)	27.3%
Efficiency (net/thermal)	25.5%

HISTORICAL ASSESSMENT OF CRS PLANT PERFORMANCE AND OPERATIONAL EXPERIENCE

HISTORICAL ASSESSMENT OF SSPS - CRS PLANT PERFORMANCE

INTRODUCTION

The following specific evaluation reports address the actual performance of the CRS in the years 1981 through 1984. The performance is reviewed, analyzed, then summarized, using the operator's daily and monthly logbooks, the daily/monthly meteo reports, and the data tapes that were available and useable. The report, PLANT HISTORY, by N. Gregory, P. Wattiez and M. Blanco, documents this evaluation and lists outage statistics, major causes of outage, weather and operation data, energy collected and delivered. It also presents a discussion of the time taken to heat up the receivers to operating temperature, based on a statistical analysis of the recorded operational data. The consideration of operational procedures, which have a dramatic effect on the system heat-up time was considered late in the evaluation, causing some modification in the conclusions. Consequently, reading this report and comparing this evaluation to the specific evaluations of receiver performance, which is topic Section 5 - RECEIVER BEHAVIOR, brings out the differences between what a subsystem can do and what system mismatch and operating procedures allow it to do.

The conclusions of this report are:

1 - The CRS plant reliability has improved with time. The outage percentage for 1984 was considerably less than for 1981 thru 1983.

2 - The heliostat field system is very susceptible to serious damage from lightning strikes with subsequent long outages.

3 - The data acquisition system seems very difficult. Data processing from the DAS is unreliable.

4 - Statistically, fewer good days have been lost than bad days.

5 - Operation of the plant on weekends in 1984 could have reduced the calculated outage by as much as 20%.

The preceding report stimulated an indepth analysis of performance leading to the next report, DAILY CHARACTERISTICS, wherein M. Andersson and J. Sandgren examined what happens with the CRS on what is defined as a "good day" and how the system handles cloud passage during operation when operating with good day solar conditions. The extensive graphical presentations in this report provide an easy method of learning how the plant operates under these conditions.

The conclusions from this evaluation are:

1 - There is a clear linear relation of calculated energy on the receiver versus energy absorbed by the sodium for the two receivers.

2 - Receiver efficiencies are: 72% for the cavity and 88% for the external,(daily average for a "good"day)

3 - The maximum calculated daily average value of thermal power delivered to the sodium on a per heliostat basis is: 17kw for the cavity and 21kw for the external.

The boundary between sodium and water is always of concern and can be an interesting interface. A detailed description of the sodium/water heat exchanger, the steam generator, is provided by Mr. S. Amacker from Sulzer Brothers, Ltd., the designer and manufacturer of this heat exchanger, with highlights of the experience with this piece of equipment in the report, STEAM GENERATOR EXPERIENCE. Mr. Amacker includes full load and part load performance curves and observes that the steam generator observed performance follows the calculations very closely. This design is "off the shelf" technology and has presented no surprises during these three years of operation.

The conclusions are:

1 - All specified operating conditions are satisfied.

2 - No problems in operation have been observed.

3 - There have been no structural integrity problems.

4 - The design is very flexible regarding system pressure and power needs.

5 - The load change rate could be improved by filling the central cavity with a gas.

6 - Although no problems were encountered, serious attention must be given to the sodium/water interface.

Operational procedures, as developed by the operational team, are described with some discussion of the rationale for their development. Specific problems encountered with subsystems and some of the maintenance problems are discussed.

Conclusions are:

1 - The heliostat field is very vulnerable to lightning.

2 - Heliostat corrosion is a concern requiring special stow.

3 - The external receiver requires venting each operational day.

4 - The Power Conversion Subsystem is very unreliable.

Summarizing all of these specific evaluation reports, it is clear that a Central Receiver Solar System can be built from existing designs of both solar hardware and normal power plant hardware. However, much improvement can result with rather straight forward changes in the solar specific hardware and a considerable change in the power plant hardware. Many of these needed changes become obvious from evaluation reports that follow.

SSPS-CRS PLANT HISTORY AND OPERATION (1981 - 1984)

Neil Gregory, Pierre Wattiez, Manolo Blanco, ITET

1. INTRODUCTION

Since the inauguration of the SSPS project on September, 21, 1981, the CRS plant has been operated by the Plant Operation Authority (POA) and evaluated by the International Test and Evaluation Team (ITET). The plant is under the management of the SSPS/Operating Agent (OA), following the decisions of the Executive Committee of the nine participating countries.

The purpose of this paper is to present an overview of the plant history to date (August 31, 1984) and also to ascertain some of the principal operating characteristics for the major components/systems, heliostats, receivers, steam generator and electrical generator. Details are given of the days of operation, operating hours, and major outages that have occurred at the plant since November 1981. In addition, attempts are made at defining characteristic plant operation under normal operating conditions on days which would be classed as good solar days.

2. PLANT CHARACTERISTICS

The SSPS-CRS plant consists of the three major subsystems: heliostat field, sodium heat transfer system, and the power conversion system. The heliostat field is composed of 94 heliostats, monitored by a central communicating system (HAC). In full operation, the heliostat field tracks the sun into a sodium receiver. A cavity receiver was used until the end of April 1983, at which time it was replaced with a flat plate (billboard) one. The sodium heat transfer system includes, besides the receiver, the two storage tanks (hot and cold) and their respective sodium circulation pumps, the steam generator, and the sodium purification auxiliaries. The power conversion system (PCS) is based on a five piston steam motor and includes the conventional steam circuit, feedwater, and cooling systems. Located at the end of the PCS line is the electric generator designed to produce 500 kWe of electricity.

The main characteristics of the entire plant for the design point (equinox noon) are shown below:

Design	day 80, 12:00 (equinox noon) insolation	 0,92 kW/m²
Heliostat field:	total reflective surface area concentration ratio solar multiple land-use-factor	3655 m² 377 1,11 0,22
Cavity receiver:	heat transfer medium aperture size active heat transfer surface inlet temperature outlet temperature	Sodium 9,69 m² 16,9 m² 270°C 530°C
Thermal storage:	two-tank-system, storage medium capacity equivalent to hot storage temperature cold storage temperature	Sodium 1,0 MWhe 530°C 270°C
Steam generator:	sodium inlet temperature sodium outlet temperature steam outlet temperature steam pressure	525°C 275°C 500°C 100 bar
Power (at design point):	insolation radiation into cavity thermal gross electric net electric	3362 kW 2840 kW 2203 kW 599 kW 500 kW
Efficiencies (at design point):	insolation/aperture plane thermal/gross electric thermal/net electric insolation/net electric (excluding correction for solar multiple) insolation/net electric (including correction for solar multiple)	84,5% 27,2% 22,7% 14,9% 16,5%

3. DATA SOURCES AND SELECTION CRITERIA

The main sources of data used in this paper have been the operator's monthly/daily and meteo reports, and where these were ambiguous, the daily logbooks. Since these were written by many different people over the years, it would be unreasonable to expect a consistent thoroughness and accuracy in the data. However, due to the methods of data evaluation used in this paper, it is hoped that most inconsistencies have been overcome.

These data include comments regarding extremes of solar weather, breakdowns of the various systems, and other remarks which may be found in the operator's monthly/daily reports. In addition, the hours of operation of the heliostat field, receiver, steam generator, and electrical generator have also been collated. For good solar days, data have been included on the number of insolation hours greater than 300 W/m**2. In this way, a complete picture may be drawn of the plant operation or non-operation on any specific day and during specifically defined periods. The data have been set up in a database on the VAX computer in order to simplify searching, collating and summation of operations.

Although the CRS plant has been in operation since September 1981, insufficient data were collected until November. For this reason, it was decided to restrict data evaluation to the period from November 1, 1981 until August 31, 1984. The first complete data sheet found with receiver operation was dated November 16, 1981.

The operation of the different components (heliostat field, receiver, steam generator, electrical generator) and also the insolation level are defined in the database as follows:

3.1 Definitions of Component Operation

The operation of the individual components has been defined in the number of hours as follows:

Heliostat Field	:	Standby operating hours not including wash positioning.
Receiver	:	Heliostat tracking hours, or receiver door-open time taken from the time when at least one heliostat is tracking.
Steam Generator	:	Measured from PCS start-up and including all preheating operation.
Electrical Generator	:	The number of hours of insolation greater than 300/Wm**2 for a clear sunny day (defined as a good solar day).

In order to approximate the receiver and steam generator start-up time, the CRS data tapes have also been analyzed. The definitions pertaining to these data are given below.

3.2 CRS Data Tape and Definitions

The data tapes have been recorded by the Data Acquisition System (DAS). All available on-site data up to August 1984 are processed by the BEXE - code, which creates VAX formatted files. The operational times for the receiver, PCS and alternator are defined as follows:

Receiver	:	Flowrate	5 m3/h and more
		Outlet temperature	500°C and more
Steam Generator	:	Flowrate	5 m3/h and more
		Outlet temperature	460°C and more
Alternator	:	Power	50 kW
		Steam Generator Outlet Temperature	480°C

Due to occasional failure of the DAS these data are somewhat incomplete. In particular, the period from June 1 to July 10, 1984 is affected. Also, in the first few months of 1984, there were days on which correct data were recorded but allocated to incorrect times.

3.3 Search Criteria

The following search criteria have been established in order to select the days which should demonstrate similar and stable characteristics. They have been used for the calculations of the average daily operating hours (section 5), the operational usage factors (see Reference 1 and Section 6) and also the approximation of the start-up times (see Reference 2 and Section 7).

(1) good solar day (on a completely clear day, receiver operation is more likely to be stable)

(2) no HAC problems (tracking computer)

(3) receiver operation with no problems in the sodium circuit

(4) no test operation

(5) no power conversion system (PCS) problems

(6) electrical generator operation

These criteria are based mainly on the operator's comments in the daily/monthly report data sheets. However, it is possible that problems were incurred on other days but were not noted as being significant.

All available data have been recorded, including those for days on which the plant was either not 100% operational or the weather was not particularly suitable for solar plant operation. As a consequence, statistical averages based on this data might be very misleading if the days with similar boundary conditions were not chosen. In an attempt to establish on which days there might have been reasonably stable operation, the following combinations of criteria have been used:

- electrical generator operation (6)
- electrical generator operation on days when neither significant problems, tests nor bad weather occurred (2)+(3)+(4)+(5)+(6)
- only good solar days with electrical generator operation and when neither significant problems, tests, nor bad weather occurred (1)+(2)+(3)+(4)+(5)+(6)

3.4 Errors

There are many sources of error in these calculations, such as those due to logbook data transfer, errors in the DAS system, differences in the operational strategies, and possible errors steming from the assumption that the insolation at time of start-up is always approximately the same. Since only specified day types have been chosen and the average of many values taken, it is hoped that the effects of many of these errors have been cancelled out. On similar day types, the operational strategy is likely to be the same, provided that no tests or problems occur.

4. PLANT OPERATION 1981 - 1984

This section deals mainly with the plant outages and problems, and includes some general plant statistics for the period November 1981 - August 1984. The CRS has suffered from many "teething troubles", many of which have caused quite substantial disruptions in the plant operation. These plant problems may be divided into four categories:

(I) Heliostat Field Problems, Computer (HAC)

(II) Receiver and Tower Problems

(III) Sodium Circuit Problems

(IV) PCS Circuit Problems

If either problems (I), (II) or (III) occur, the plant cannot be operated normally. However, if only problems of category (IV) occur, then, although no electrical power can be produced, the sodium circuit can still be operated for a limited time (approximately 1 day), in the hope that the repair can be completed by the following day.

4.1 Operation Chart

The operation chart shown in Figure 1 in the appendix contains a complete history of the major plant outages affecting sodium circuit operation and operation hours, as defined in 3.1, of the receiver, steam generator, and electrical generator.

4.2 Major Outage Problems Affecting Sodium Circuit Operation

Tab.3.1-1contains a list of the major outage problems which disrupted sodium circuit operation for any period of at least 2 consecutive days. The dates given include those days on which the problem first occurred, although the plant might have been in operation for most of that day.

from	until	duration	kind of event
15-12-81 -->	28-03-82	103	cold tank repair, leaks were patched
29-03-82 -->	17-04-82	19	HAC problems, CRT's blocked
26-04-82 -->	28-04-82	3	HAC problems, computer blocked
29-04-82 -->	10-05-82	12	lightning cause HAC problems
14-05-82 -->	26-05-82	13	sodium pump failure (flooding)
13-07-82 -->	19-07-82	7	sodium pump valve leak (bellows)
24-07-82 -->	25-07-82	2	sodium pump problems (oil fire)
24-08-82 -->	27-08-82	4	work in the tower, thermocouples fitted to receiver
07-09-82 -->	20-09-82	14	sodium tank leak, patches were welded
21-09-82 -->	27-09-82	operation	leakage regeneration vessel
28-09-82 -->	30-09-82	3	PCS leakage regeneration vessel
01-10-82 -->	08-10-82	operation	leakage regeneration vessel
08-10-82 -->	06-03-83	149	cold tank repair, CRT's blocked
09-03-83 -->	13-03-83	4	HAC problems
29-04-83 -->	24-08-83	117	ASR installation
06-09-83 -->	30-10-83	54	ASR repairs (trace heating, tube bundle)
02-11-83 -->	06-11-83	5	ASR repairs (trace heating, tube bundle)
08-11-83 -->	09-11-83	2	ASR repairs (trace heating, tube bundle)
02-01-84 -->	04-01-84	2	HAC problems due to leap year
03-04-84 -->	16-04-84	16	ASR improvements (drained)
03-05-84 -->	07-05-84	5	lightning cause HAC problems
	total	534	days

Table 3.1-1: Major outage occurrences,events longer than one day

There are also many instances when problems overlapped. Tab.3.1-2 has been constructed to show the number of days that any particular problem dominantly affected the plant operation and not the actual days of each type of outage. For example, there were many HAC system disturbances when cold tank repairs were being carried out and the Spilling steam motor was often being repaired at the same time. In this table, only the most dominant level of failure has been counted.

Other outage reasons, such as holidays, weekends, and bad solar conditions, have also been included for those days when there was no other outage caused by component failures, problems, etc. The results are shown as a percentage of the number of days considered in each year.

Plant outage reason	1981 61 days	1982 365 days	1983 365 days	1984 244 days
Outages on days without component failures				
Weekends	12 (20%)	-	1	47 (19%)
Holidays	-	-	1	9 (4%)
Bad solar e.g.cloudy	1 (2%)	4 (1%)	7 (2%)	12 (5%)
High wind	2 (3%)	-	5 (1%)	1
Outages caused by major component failures				
HAC problems	-	25 (7%)	4 (1%)	7 (3%)
ASR installation	-	-	117 (32%)	-
ASR repairs	-	-	61 (17%)	-
ASR improvements	-	-	-	16 (7%)
Cold tank repair	17 (28%)	172 (47%)	65 (18%)	-
Sodium pump failures	-	22 (6%)	-	1
Others incl. PCS	-	33 (9%)	14 (4%)	28 (11%)
No information	10 (16%)	-	-	-
All outages	42 (69%)	256 (70%)	275 (75%)	121 (50%)

Table 3.1-2: Outage statistics 1981-1984

Although most of these outages were basically unavoidable, the outage at weekends, approximately 20% in both 1981 and 1984, could have been reduced with another operational policy. In the years 1982 and 1983 there was a policy of weekend operation. The weekend days included are those days on which there was no mention of either major plant problems or bad weather which might have provided an additional reason for there being no plant operation; i.e., they are days on which the plant could have been operated.

There have also been many outages of the PCS that have occurred at the same time as other major problems, and therefore the following results might be somewhat misleading. In particular, due to other more dominant problems, a PCS outage may sometimes have been considered as irrelevant and therefore not recorded.

1981 61 days	1982 365 days	1983 365 days	1984 244 days
0	159(44%)	24(7%)	76(31%)

Table 3.1-3: PCS Record Outages

4.3 Weather Classification and Considerations

Where it was possible to ascertain extreme weather conditions, either good or bad, this information was included in the data base as follows:

- Good solar days are clear days with no clouds, as defined in the meteo reports.
- Bad solar days are very cloudy, overcast, or very windy days. In general, there can be no plant operation on a bad solar day.

By elimination,

- Medium solar days are any days which do not fall into the above two categories.

The following table gives an idea of how many days were available for use in the restricting criteria.

Day classification	1981 61 days	1982 365 days	1983 365 days	1984 244 days
Good solar days	9(15%)	109(30%)	140(38%)	89(36%)
Medium solar days	46(75%)	184(50%)	164(45%)	106(44%)
Bad solar days	6(15%)	72(20%)	61(17%)	49(20%)

Table 3.1-4: Weather Classification Statistics

If only those days are considered during which the receiver was operational, the following results are obtained:

Day classification	1981 19 days	1982 110 days	1983 91 days	1984 123 days
Good solar days	7(37%)	40(36%)	37(41%)	48(39%)
Medium solar days	12(63%)	62(56%)	47(51%)	59(48%)
Bad solar days	0	8(8%)	7(8%)	16(13%)

Table 3.1-5: Weather Classification Statistics During Operation

A comparison of the bad solar days statistics from these two tables is not justified since many bad solar days have, by definition, excluded plant operation. However, it would appear that the plant has had more than its fair share of operation on good solar days.

As far as the data in Tab.3.1-4/5 are concerned, they may not necessarily coincide with meteo data percentages, but they have been extremely useful in isolating good days and also in providing the reason why there was no plant operation on some bad days.

4.4 CRS Operation Table

From January 1982 until August 1984, the plant and system operation status are recorded on a daily basis in the following way:

- Solar Irradiation Availability: Each day is classified in 5 groups regarding the sun irradiation availability between the sunrise and sunset. These groups are defined in Reference 3 as follows:

⊙	**Shining** **Clear**	◕	**Partially covered**
◔	**Clear with reverberations/hazy** **Clear with sporadic clouds**	●	**Covered**
◒	**Clear with intermittent clouds** **Clouds and clear**	⊕	**High winds**
////	**Rain during sun hours**	—	**No data available**

- Data on Tape: The plant operation data are recorded and saved on tape which can be read with an appropriate software program developed for the VAX computer, and recompiled for any additional investigation. Each day, where operation data are recorded on tape and are readable, is indicated by a black spot.

- HFS Operation: The HFS has been considered in operation when at least one single heliostat has been in any position other than stowed. The days on which this condition occurred have been marked on the table by the black shaded space. Normally the HFS was operated with an irradiation above 300w/m2 except during specific test periods as shown in Fig.3.1-2, HFS Operational Behavior, in the appendix.

- Thermal Energy Collector: Each day that the receiver has been in operation, i.e., heated by heliostats tracking on the tube walls, it has been considered that thermal energy has been collected and indicated accordingly.

- Steam Delivered: The steam generator has been considered to produce steam if the steam pressure and temperature have reached the condition defined in 3.2.

- Electricity Delivered: The gross electrical production has been taken into consideration in order to show when the plant has produced some electricity. In this case, it has been indicated by 100 kwh units.

The operation status, Tab.3.1-6/7/8 , show plant and subsystem behavior without mention of outage problems. A list of these outages has been summarized in Tab.3.1-1 in addition to the daily operation reports. These tables are included in the appendix for the years 1982, 1983 and 1984.

5. PLANT OPERATIONAL CHARACTERISTICS 1981 - 1984

This section is concerned mainly with the plant component operating hours. Specific data are included on the heliostat field (HFS), the two receivers (Sulzer and ASR), steam generator, and electrical generator.

5.1 HFS Availability

The availability of the HFS would be 100% when all 94 heliostats are operational. A graphical representation of this availability is given in Fig.3.1-3/4/5 in the appendix. In the period April 1982 - August 1984, the heliostat field system was severely affected by two lightning strikes (April 29, 1982 and May 3, 1984) which resulted in a shutdown of the entire field. The repair of the damage caused by these natural events took at least 20 days before 90% of the field had been recovered.

As a result of the occurrence of such natural events in the Sierra Nevada, during the period when the sodium cold tank was being repaired and the new receiver installed, the Plant Operation Authority seized the opportunity to provide the site maintenance team with adequate equipment and knowledge to repair the HFS electronics without external help. Since that time, the availability of the HFS has been close to a maximum, and the impact of the second lightning strike was reduced to a minimum.

Between May 1982 and March 1983, 21 heliostats were recorded out of service without including the first lightning strike. From April 1983 until March 1984, 25 heliostats were registered out of service. Nevertheless, during the second year of operation, no more than 5 heliostats have been out of service at the same time.

5.2 Operation Statistical Tables for the Period November 1981 - August 1984

The period November 1981 - August 1984 contains a total of 1,035 days. From this period, the following data are available concerning the days and hours of operation of the three components receiver, steam generator, and electrical generator. These data are also presented on a monthly basis in Tab.3.1-6/7/8 in the appendix.

	Receiver	Steam Generator	Generator
Days of operation	343	244	101
Hours of operation	1885.4	1526.4	238.1
Avg. daily operating hours	5.5	6.26	2.36

Tab.3.1-6: Operation Statistics for the Period November 1981-August 1984

These statistics also include those days on which there may have been either problems, tests or bad weather. If only those days are considered on which electrical generator operation was achieved, then more realistic statistics may be obtained on the average operating hours of the plant. This maybe further improved when restrictions are placed on the data selection, i.e., by exclusion of all days on which significant problems were mentioned or tests performed or those days when weather conditions, i.e., wind, clouds would have had a severe impact on plant operation.

A further criterion of interest is that of a good solar day. In this case, only those days are included which were defined as clear, i.e., a cloudless day. As would be expected the average values for these days are considerably higher.

Restricting criteria	Average daily hours of operation of - Receiver	Steam Generator	Generator
Generator operation	5.54	6.36	2.36
+no problem/bad weather/test	6.23	6.51	2.56
+good solar day	8.40	8.04	3.06

Table 3.1-7: Operating Statistics for a Fully Operational Plant

Since the Sulzer cavity receiver was replaced by the ASR Flat Plate Receiver on April 27, 1983, statistical comparisons of the plant operational characteristics have also been individually made for each of the two receivers.

5.3 PLANT OPERATIONAL CHARACTERISTICS: Sulzer Receiver

The following tables show the total number of operating hours, days and average daily operating time for the three major components: receiver, steam generator and electrical generator. These selections of data also include those days on which there were problems. In addition, the Sulzer receiver was also in operation approximately 300 hours before November 16, 1981 (see Reference 4) but, unfortunately no statistics have been obtainable. The Sulzer receiver was in operation until April 28, 1983.

5.3.1 Days of Operation

During the Sulzer receiver period, the number of days on which the components receiver, steam and electrical generator were in operation were as follows:

Period			Receiver	Steam Generator	Generator
1981	Nov. --	61 days	19(31%)	15(25%)	10(16%)
1982		365 days	110(30%)	88(24%)	25(7%)
1983	Apr.	118 days	43(36%)	35(30%)	20(17%)
Total		544 days	172(32%)	129(24%)	55(10%)

Table 3.1-8: Days of Operation - Sulzer Period

5.3.2 Hours of Operation

During the Sulzer receiver period, the number of hours of component operation, as defined in section 3.1, were as follows:

Period		Receiver	Steam Generator	Generator
1981 Nov. --	61 days	111.0	79.8	15.5
1982	365 days	667.7	518.0	58.2
1983 --Apr.	118 days	226.3	200.5	43.9
Total	544 days	1005.0	798.3	117.6

Table 3.1-9: Hours of Operation - Sulzer Period

A further development of the data contained in Tab.3.1-8/9 . is the average number of hours of operation of the three major components.

5.3.3 Average Hours of Operation

Period		Receiver	Steam Generator	Generator
1981 Nov. --	61 days	5.84	5.32	1.55
1982	365 days	6.07	6.56	2.33
1983 --Apr.	118 days	5.26	5.73	2.19
Total	544 days	5.84	6.19	2.14

Table 3.1-10: Average Daily Operation Hours - Sulzer Period

5.3.4 Average Hours of Operation Under Good Operating Conditions

In the next tables, the days with good operating conditions have been tabulated according to different combinations of criteria.

Restricting Criteria	Receiver	Steam Generator	Generator
Generator operation	5.88	6.17	2.14
no problem/bad weather/test	6.54	5.97	2.12
good solar days	8.57	7.67	3.07

Table 3.1-11: Average Daily Operation Hours - Sulzer Period

From the CRS tape data, it is also possible to obtain similar characteristics of average daily operation hours. In this case, the operation hours do not include any preheating time and therefore should not be compared directly with the data in Tab.3.1-11. A limited comparison is discussed in Section 7.

Restricting Criteria	Receiver	Steam Generator	Generator
CRS tape data	5.04	4.08	2.04
Generator operation	6.48	5.14	2.09
Good solar days	7.13	6.10	2.85

Table 3.1-12: Avg. Daily Operation Hours - Sulzer Period (CRS Tapes)

5.3.5 Component Operation Data Plots

The plots shown in Fig.3.1-6/7 in the Appendix have been constructed with all the data which contained Sulzer receiver, steam generator, and electrical generator operating hours from both the database and from the CRS data tapes respectively. They appear to indicate an operation design characteristic for generator operation. The boundary lines drawn on these plots indicate the minimum requirement of steam generator operation hours for a specified electrical generator operation. The intercept on the "y-axis" indicates the approximate steam generator hours required before the generator has been used.

In particular, the CRS data tapes plot on Fig.3.1-7 tends to suggest that, according to plant operation to date, approximately 1.5 hours of steam production are necessary before electricity generation can take place. This is a function of the time required to heat up the "hot" sodium to its normal operating temperature, a policy conditioned by system design and current operational strategies.

5.4 PLANT OPERATIONAL CHARACTERISTICS; ASR Receiver

The installation of the ASR receiver was completed on August 25, 1983. The following tables show the total number of operating hours, days and average daily operating time for the three major components, receiver, steam generator and electrical generator. These selections of data also include those days on which there were problems.

5.4.1 Days of Operation

During the ASR receiver period, the number of days on which the components receiver, steam and electrical generator were in operation were as follows:

Period		Receiver	Steam Generator	Generator
1983 Apr.--	247 days	48(19%)	24(10%)	7(3%)
1984 -- Aug.	244 days	123(50%)	91(37%)	39(16%)
Total	491 days	171(35%)	115(23%)	46(9%)

Table 3.1-13: Days of Operation - ASR Receiver

5.4.2 Hours of Operation

During the ASR receiver period, the number of hours of component operation, as defined in section 3.1, were as follows:

Period		Receiver	Steam Generator	Generator
1983 Apr.--	247 days	157.5	112.2	9.6
1984 -- Aug.	244 days	722.9	615.8	111.0
Total	544 days	880.4	728.0	120.6

Table 3.1-14: Hours of Operation - ASR Receiver

A further development of the data contained in Tab.3.1-13/14 is the average number of hours of operation of the three major components.

5.4.3 Average Hours of Operation

Period		Receiver	Steam Generator	Generator
1983 Apr.--	247 days	3.28	4.68	1.36
1984 -- Aug.	244 Days	5.88	6.77	2.85
Total	544 days	5.15	6.33	2.62

Table 3.1-15: Average Daily Hours of Operation - ASR Receiver

5.4.4 Average Hours of Operation Under Good Operating Conditions

In the next tables, the days with good operation conditions have been tabulated according to different combinations of criteria.

Restricting criteria	Receiver	Steam Generator	Generator
Generator operation	5.20	6.58	2.62
no problem/bad weather/test	5.95	7.05	2.95
good solar days	8.30	8.28	3.05

Table 3.1-16: Average Daily Hours of Operation - ASR Receiver

From the CRS data tapes, it is also possible to obtain similar characteristics of average daily operation hours. In this case, the operation hours do not include any preheating time and therefore should not be compared directly with the data in Tab.3.1-16.

Restricting criteria	Receiver	Steam Generator	Generator
CRS tape data	4.71	3.96	2.51
Generator operation	6.21	4.74	2.51
Good solar days	6.75	5.31	3.16

Table 3.1-17: Avg. Daily Operation Hours - ASR Period (CRS Tapes)

5.4.5 Component Operation Data Plots

The plots shown in Fig.3.1-8/9 in the Appendix have been constructed with all the data which contained ASR receiver, steam generator and electrical generator operating hours from both the database and from the CRS data tapes respectively. They also appear to indicate an operation design characteristic for generator operation. The boundary lines drawn on these plots indicate the minimum requirement of steam generator operation hours for a specified electrical generator operation. The intercept on the "y-axis" indicates the approximate steam generator hours required before the generator has been used.

In the same way as with the Sulzer data, the CRS data tapes plot for the ASR on Figure 9 also tends to suggest that, according to plant operation to date, approximately 1.5 hours of steam production are necessary before electricity generation can take place. Again this is a function of the time required to heat up the "hot" sodium to its normal operating temperature, and would appear to be independent of receiver performance.

5.5 Discussion

The average operating hours of the ASR receiver are in all cases marginally less (at the worst 12%) than those for the Sulzer receiver. Although the average difference between Tab.3.1-11/16/12 (20 of approximately 25 minutes) is probably below the accuracy which can be expected for this sort of analysis, the ASR results are consistently below the Sulzer values. Several possible explanations are given below:

- The seasonal periods of operation of the Sulzer receiver are not necessarily the same as for the ASR receiver, and therefore no true comparison should be made. However, a method of overcoming this seasonal change is discussed and presented in Section 6.

- The ASR flat plate receiver is more sensitive to tracking errors and greater care must be taken at start-up, particularly if this occurs during the early morning. Further discussion on this point is included in Section 7.2.

- The current operator shift policy may also tend to restrict plant operation hours, particularly on good solar days.

The average generator operation, particularly on good solar days, is approximately the same (about 3 hours) for both receiver periods. This would tend to suggest that the different performance of the two receivers does not significantly affect the generator operation. This was true for both the database and CRS data tapes.

6. OPERATIONAL USAGE FACTOR (OUF)

The previous tables have shown the average operating characteristics of the two receivers. However, these may be misleading since the operation periods do not necessarily coincide and thus seasonal insolation and weather characteristics are not taken into consideration.

The CRS plant is designed for operation at insolation greater than 300W/m**2, and it would be reasonable to assume that on a clear day, tracking and receiver preheating could be carried out before this insolation has been reached. In order to obtain a measure for the actual plant usage, assuming that the plant is fully operational, an operational usage factor for any day (see Reference 1) could be defined as follows:

$$\text{Operational Usage Factor (OUF)} = \frac{\text{hours of receiver operation}}{\text{hours of insolation} \quad \text{300W/m**2}}$$

The operational usage factor has the advantage that it is a measure of the actual operation of the plant on a specific day. Theoretically, if the receiver were to be preheated before the insolation reached 300 W/m**2, full operation should be possible for the total hours of insolation above this limit and would correspond to an OUF of 1. Since the hours of receiver operation from the database also include preheating time, the OUF obtained will always be higher than the true OUF.

6.1 OUF Values

All OUF values have been calculated with the database receiver hours, which correspond to the heliostat tracking time and therefore include the receiver preheating. Data have been included from 1982 until August 1984, since for 1981 no information was available regarding the hours of insolation greater than 300 W/m**2.

6.1.1 Sulzer Receiver

For the Sulzer receiver, 21 days of operation were found which satisfied the criteria for a good solar day on which there were no HAC problems and the receiver was in operation without either sodium circuit problems, PCS problems, or tests being performed. From these days, only 10 were found on which electrical generator operation occurred without any noticeable problems, i.e., full plant operation.

Sulzer Cavity Receiver		Sulzer Full Plant Operation	
Average daily OUF	0.91	Average daily OUF	0.94
Average 1982 OUF	0.89	Average 1982 OUF	0.92
Average 1983 OUF	0.98	Average 1983 OUF	0.98

Table 3.1-18: Sulzer Operational Usage Factors

6.1.2 ASR Receiver

For the ASR receiver, 34 days of operation were found which satisfied the criteria for a good solar day on which there were no HAC problems and the receiver was in operation without either sodium circuit problems, PCS problems, or tests being performed. From these days, 21 were found on which electrical generator operation occurred without any noticeable problems, i.e., full plant operation.

ASR Flat Plate Receiver		ASR Full Plant Operation	
Average Daily OUF	0.92	Average Daily OUF	0.92
Average 1983 OUF	0.93	Average 1983 OUF	0.96
Average 1984 OUF	0.92	Average 1984 OUF	0.92

Table 3.1-19: ASR Operational Usage Factors

6.2 Discussion

All values obtained for the operational usage factor are based on the receiver operation time, which in turn corresponds to the tracking hours. Both receivers show a very similar characteristic for the same search criteria. However, there were some small differences between the results for the receiver operation criteria and full plant operation criteria. The higher OUF, demonstrated by the latter results, is most probably due to incomplete information concerning system problems which might have occurred on these days. Any seasonal differences in site potential should have, for the most part, been taken into consideration by evaluating each day individually.

OUF values greater than 1 are possible, since, on some days, tests have been made to try to establish the minimum possible insolation at which preheating can begin, and also the receiver operating hours used include the receiver preheating time.

From the results it would appear that the operational usage factor does characterize the plant operation, and the fact that the average values are lower than 1 indicates that on many days there could have been a longer plant operation. This may be explained by any one or combination of the following:

- That the design condition for the CRS of insolation of at least 300 W/m**2 is too low and operation at this level can seldom be achieved.

- Failure to find sufficient information for the database on system problems or weather on some days.

- The characteristic could be a function of the operating strategy and as such would indicate that some minor improvement in this area could be made.

It should not be forgotten that, if the true OUF values had been calculated instead of those which include the preheating time, all results of the OUF would have been significantly lower by approximately 15%.

The OUF was also found to be very useful in identifying days of unusually low receiver operation hours, although the insolation values were reasonable and no mention of any serious problem had been made in the daily report. Further investigation of these days in the logbook usually revealed that the low operation hours were due to some component problems which had been omitted from the daily report.

7. AVERAGE COMPONENT START-UP TIMES

In Reference 2, an attempt has been made to approximate the start-up time from the difference between the averaged daily operating hours from the database and the recorded CRS tape values.

For this study, the start-up time for the receiver is defined as the time required to preheat the receiver such that a stable outlet temperature of around 500°C is obtained. For the steam generator, it is the time necessary to achieve such working temperatures and pressures that good quality steam is produced (normally 100 bars, 480°C). These times are very much a function of the operational strategy and vary considerably with recirculation procedures and mass flows.

A higher insolation will normally result in a shorter start-up time, therefore, in order to minimize the possibility of errors caused by start-up at different insolation levels, only clear sunny days have been considered.

Under normal operating conditions, which have been defined by the criteria given in Section 3.3, it would be reasonable to assume that the start-up time of these systems remains more or less constant if similar operational strategies for similar day types have been applied.

The following results have been taken from Reference 2 and were developed from a set of days for which both tape and database data was available. The tape data define the actual operating times and do not include any preheating times. The difference between the two averaged values has been used to approximate the start-up time.

7.1 COMPONENT "START-UP" TIMES

Throughout this section, the connotation "start-up time" for the receiver has been used as a measure of the average time elapsed between the start of heliostat tracking to the time when the receiver operating temperature has been reached. For the steam generator, it is the average time elapsed between the operator start of the steam generator system and the time when the quality steam conditions, as defined in 3.2, have been reached.

Since the Sulzer cavity receiver was replaced by the ASR Flat Plate Receiver on the 27th of April 1983, it has been necessary to calculate different start-up times for each receiver.

It must be emphasized that the "start-up times" of the two receivers presented in this paper, are primarily not related to the characteristics of the receivers but to those of the system as a whole. For a more detailed discussion of this topic - as far as the ASR is concerned - see paper 5.8, "The SSPS ASR: Transient Response: p. 7/8.

7.1.1 Sulzer Cavity Receiver

The Sulzer receiver statistics cover the period November 1981 - April 1983.

Criteria	Days	Avg. Tape Hrs.	Avg. Log Hrs.	Avg. Start-up Time
(1)	48	6.20	7.93	1hr. 43 min.
(1)-(5)	14	7.15	8.77	1hr. 37 min
(1)-(6)	12	7.13	8.64	1hr. 30 min

Table 3.1-20: Sulzer Cavity Receiver

7.1.2 ASR Flat Plate Receiver

The ASR receiver statistics cover the period April 1983 - August 1984.

Criteria	Days	Avg. Tape Hrs.	Avg. Log Hrs.	Avg. Start-up Time
(1)	45	5.82	7.75	1 hr. 55 min.
(1)-(5)	23	6.50	8.55	2hrs. 03 min.
(1)-(6)	18	6.75	8.60	1hr. 51 min

Table 3.1-21: ASR Flat Plate Receiver

7.1.3 Steam Generator

The following two tables contain steam generator statistics for both the Sulzer and ASR receiver periods.

Criteria	Day	Avg. Tape Hrs.	Avg. Log Hrs.	Avg. Start-up Time
(1)	48	5.06	7.30	2 hrs. 14 min.
(1)-(5)	14	5.95	7.99	2 hrs. 04 min.
(1)-(6)	12	6.10	7.97	1 hr. 52 min.

Table 3.1-22: Sulzer Steam Generator

Criteria	Days	Avg. Tape Hrs.	Avg. Log Hrs.	Avg. Start-up Time
(1)	45	4.62	7.90	3 hrs. 16 min.
(1)-(5)	23	4.91	8.40	3 hrs. 29 min.
(1)-(6)	18	5.31	8.51	3 hrs. 12 min.

Table 3.1-23: ASR Sulzer Steam Generator

7.2 Discussion

Tab.3.1-20-23contain the average start-up times for the two receivers and steam generator for the three different combinations of criteria. In most cases the times are comparable. Due to the small difference between the Sulzer and ASR start-up times, it would appear that the Sulzer ceramic wall does not have any significant effect on the start-up time. However, it is important to realize that, for days other than those included in this criteria, the start-up time can vary considerably, and depends very much on the receiver flow rate and the available insolation at that time. A higher insolation, i.e., a later start, can result in a much lower start-up time.

An extreme example of this was demonstrated on April 5, 1983 when the Sulzer receiver was put into operation just after solar noon. In this case, a start-up time of less than 15 minutes was achieved with all heliostats in track. On March 6, 1984 at around solar noon, an ASR receiver start-up was also achieved in less than 15 minutes with all heliostats in operation.

However, the ASR receiver is more sensitive to the number of heliostats in track, and greater care must be taken with start-up to avoid hot spots on the receiver frame. Consequently, the start-up time may be longer. This is necessary because of tracking accuracy errors and in particular the large morning and afternoon sun images, which can cause problems with the smaller available aiming area of the ASR receiver. In addition, the aperture frame of the Sulzer receiver was not as sensitive as that of the ASR receiver.

Although there is no basic difference in the steam generator for the Sulzer and ASR receiver time periods, the ASR operation appears to require a considerably longer start-up time. The reason is - as mentioned before - to be found in the way that the system had to be adjusted in order to guarantee a safe start-up procedure for the ASR (for which it was not designed from the beginning). To be specific, it was probably due to the fact that the minimum level in the hot storage tank has been increased from 300 mm to 450 mm. This change was implemented for tank safety reasons, since it was necessary to reduce sodium tank transients. As a consequence, the actual thermal capacity of the hot storage tank has been increased and more energy must be dispersed before the temperatures necessary for quality steam production have been reached. The thermal inertia of the hot sodium tank has not been directly considered in this analysis.

The lengths of these start-up times are a function of the current operational strategy which requires that sodium is circulated through the receiver until the required cold tank temperature is reached. Any significant change in this strategy could bring about a change in this start-up time.

8. CONCLUSIONS

Despite the high percentage of plant outages that have arisen, it would appear that:

- The outage percentage of 50% for 1984 is a considerable improvement when compared with the outages in 1981 - 1983. It appears that, with maturity, the CRS plant reliability is improving.

- Statistically, fewer good solar days have been lost than bad days, i.e., the plant has had more than its fair share of good solar days.

- The susceptibility of the HFS to serious damage by lightning strikes, and thus long outages, appears to have been reduced with experience.

- A further reduction in outages by approximately 20% should have been possible if the plant were to have been operated on weekends. In addition, the continuity of such operation would definitely prove to be advantageous.

As far as plant operation is concerned, the average number of hours of receiver operation does not differ significantly between the two receivers ($\pm 6\%$). In addition, the average generator operation, particularly on good solar days, is approximately the same (about 3 hours) for both receiver periods. This would tend to suggest that the different performance of the two receivers does not significantly affect the generator operation. This was true for both the database and CRS data tapes. It would appear that the system behavior is restrained elsewhere.

The most probable explanation is an ill-match of the hot storage system operation and PCS. It has been shown in 5.3.5 and 5.4.5 that, during the history of plant operation and regardless of receiver, a minimum of 1.5 hours quality steam production has been required before the Spilling steam motor was operated and electricity generated. This is due both to operation strategies and system design.

The conclusion to be drawn is that performance of individual subsystems, such as receivers, is less important than an overall match of all subsystems.

The definition of an operational usage factor (OUF) has been useful in two ways:

- The OUF has been a useful evaluation tool which has been successfully applied to determine days with unexplainable operation characteristics.

- The OUF values obtained suggest that the CRS design condition of at least 300 W/m**2 insolation is probably too low, and also that improvements in operational strategy might be possible.

As far as the "start-up" times are concerned, the average values presented demonstrate how the system has been operated and not how the system could be operated. Indeed, both receivers could be started up in a relatively short time - this could be governed by the sodium mass flow and the available power into the receiver. However, as stated previously, the good performance of some components cannot be utilized if the performance of other subsystems does not match.

9. RECOMMENDATIONS FOR FUTURE WORK

The simple database used in this paper has proved to be very powerful for data selection purposes. If any similar work is to be made in the future, the current database is, of course, expandable. However, a very efficient and easy-to-use database system (DATATRIEVE), which would be ideal for this purpose, is available for the VAX computer system. Unfortunately, it has not been and maybe cannot be implemented on the SSPS VAX 11/730. The upkeep of such a database on a daily basis would not require much time and input information could be standardized. The current database could also be expanded to contain other data from the daily plant operation reports, e.g.:

- kw electricity produced

- parasitic power required

- more weather statistics e.g., insolation 500, 700, 900 w/m**2.

- number of heliostats in track

In this way, a database could be constructed which would be of interest to all the parties concerned, POA, ITET, OA.

10. ACKNOWLEDGEMENTS

The authors would like to thank Heinz Jacobs, Jose Martin and Clif Selvage for the many consultations which have helped to formulate this paper and also Paco Ruiz and Company for their assistance in obtaining the data.

11. REFERENCES

(1) Neil Gregory, "An Operational Usage Factor for the CRS Plant," SSPS Technical Report R-44/84 NG, October 1984.

(2) Neil Gregory and Heinz Jacobs, "Available Data and Approximation of the Average CRS Start-up Time," SSPS Technical Report R-45/84 NGHJ, October 1984.

(3) Mats Loosme, "Expected and Actual Meteorological Conditions on the Plant Site," SSPS Technical Report 1/83, DCS - Midterm Workshop Proceedings.

(4) Manuel Pescatore, "Comparison of Cavity Receiver and the External Receiver," SSPS Technical Report R-27/84 MP, May 1984.

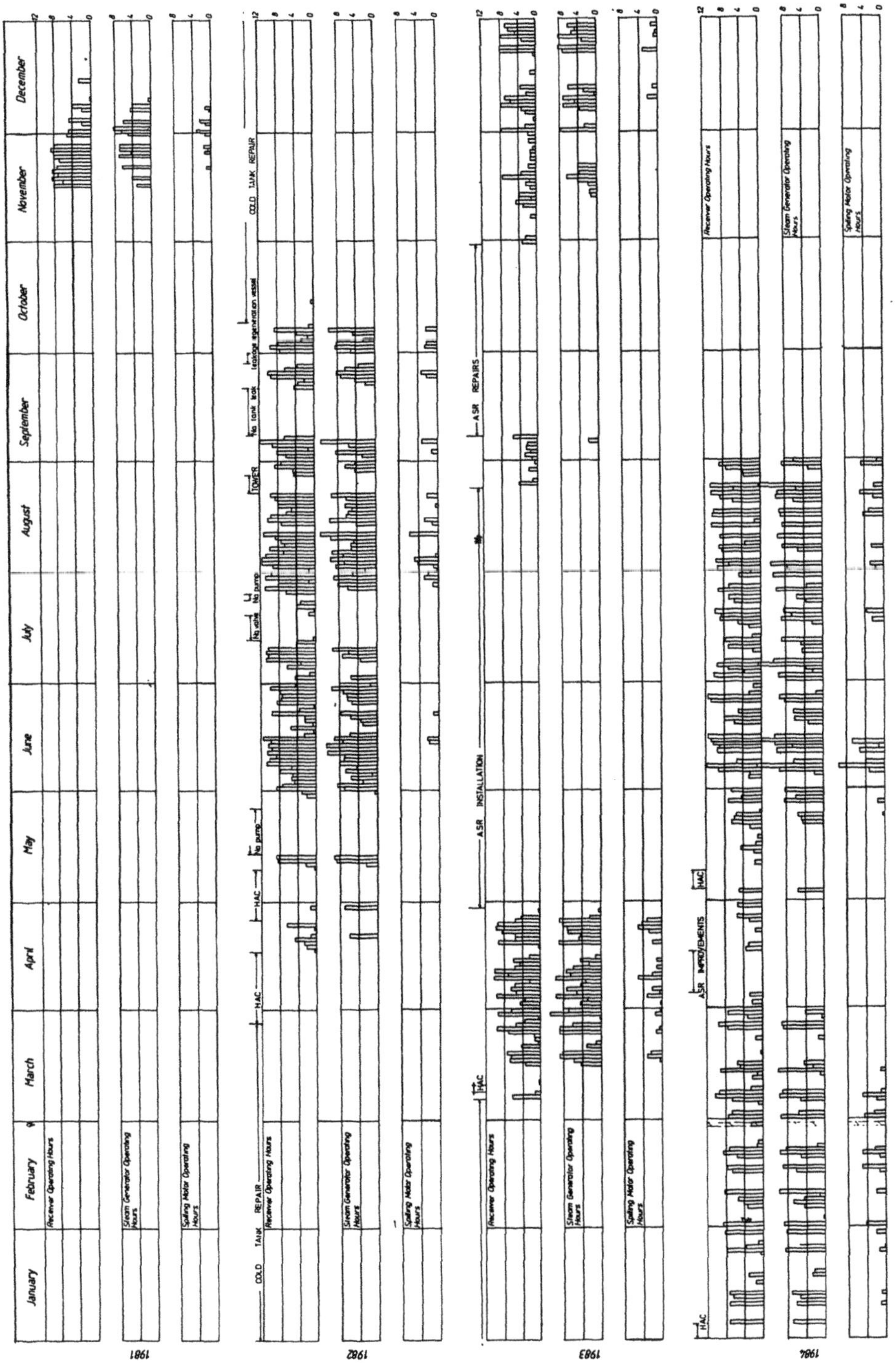

Fig.3.1-1: CRS Operation Chart 1981-1984

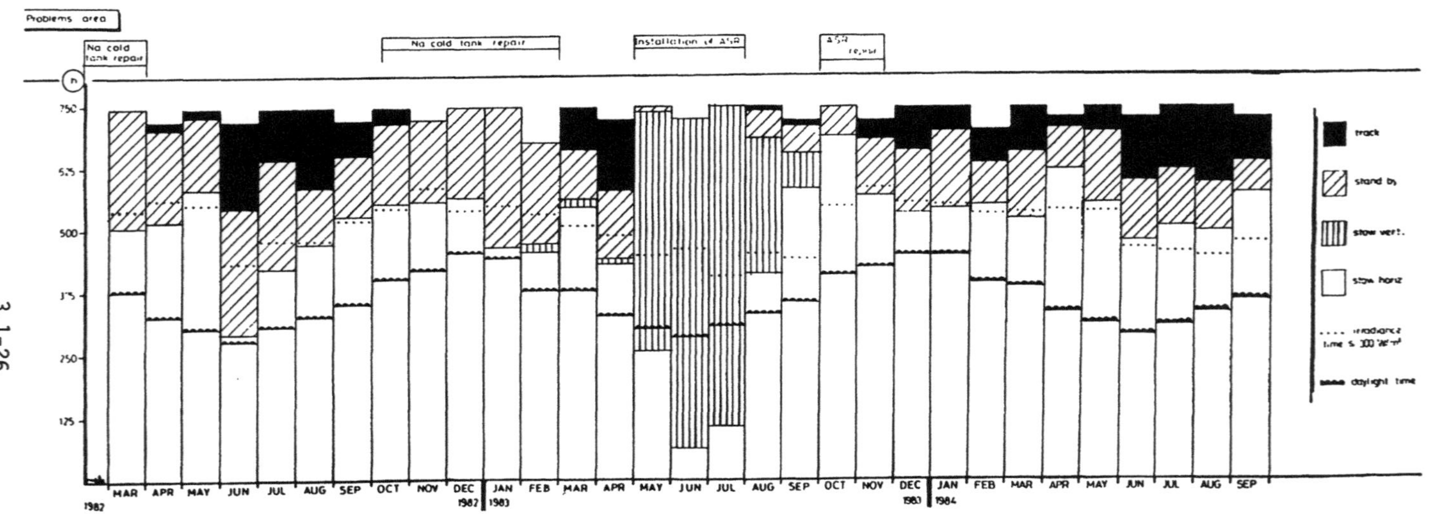

Fig.3.1-2: H.F.S. operational behaviour

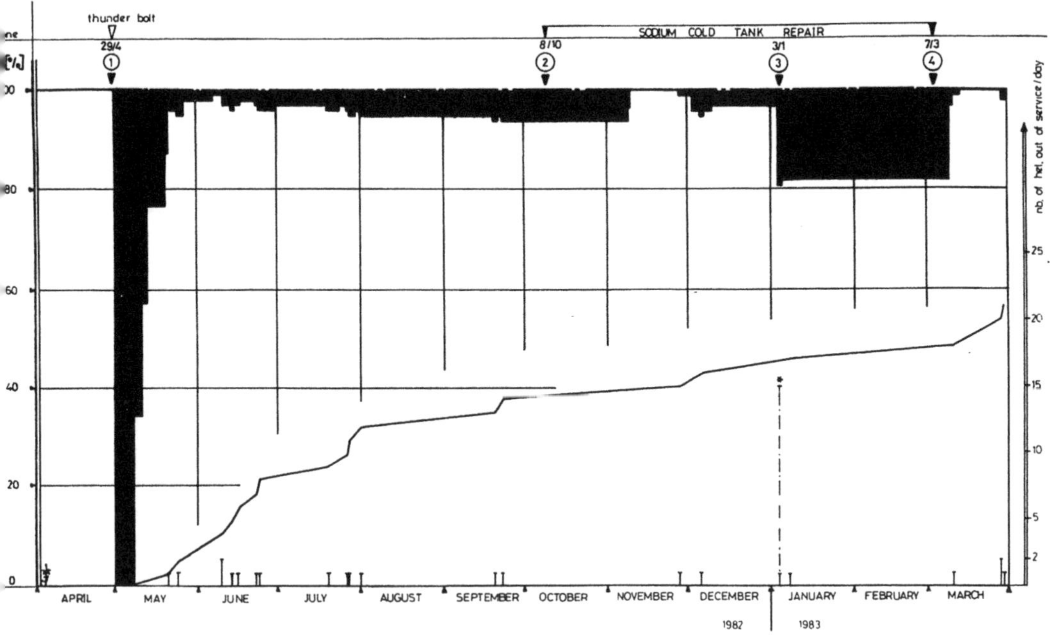

Fig.3.1-3: H.F.S. AVAILABILITY

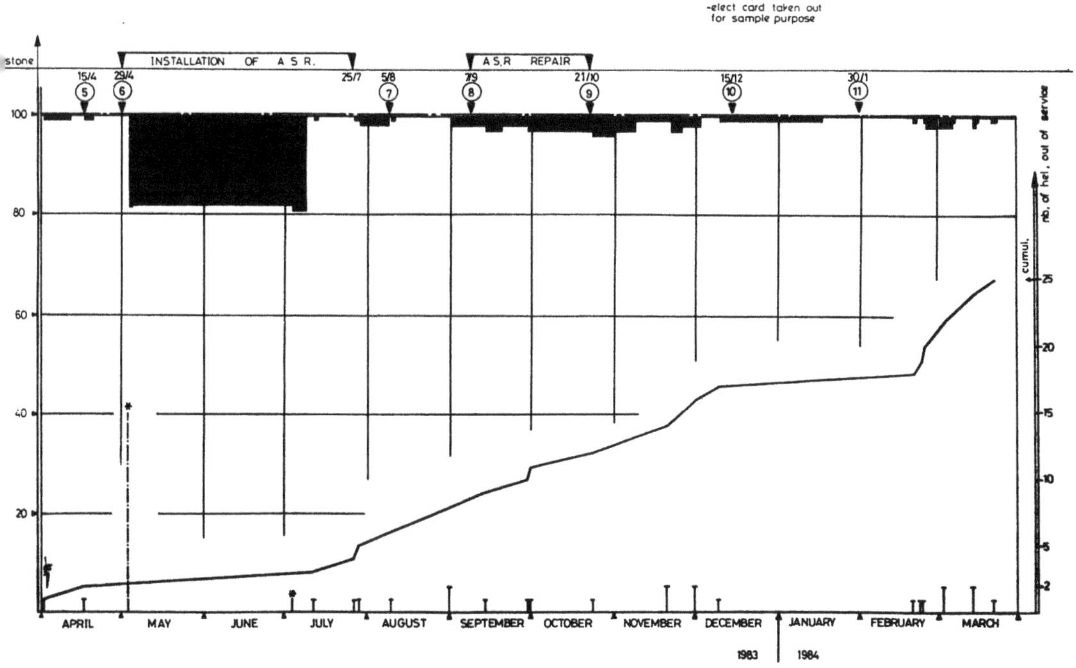

* no failure-
-elect card taken out
for sample purpose

stone
INSTALLATION OF A S R.
A S R REPAIR
15/4
29/4
25/7
5/8
7/9
21/10
15/12
30/1
5
6
7
8
9
10
11
100
80
60
40
20
cumul.
nb. of hel. out of service
25
20
15
10
5
2
APRIL
MAY
JUNE
JULY
AUGUST
SEPTEMBER
OCTOBER
NOVEMBER
DECEMBER
JANUARY
FEBRUARY
MARCH
1983
1984

Fig.3.1-4: H.F.S. AVAILABILITY

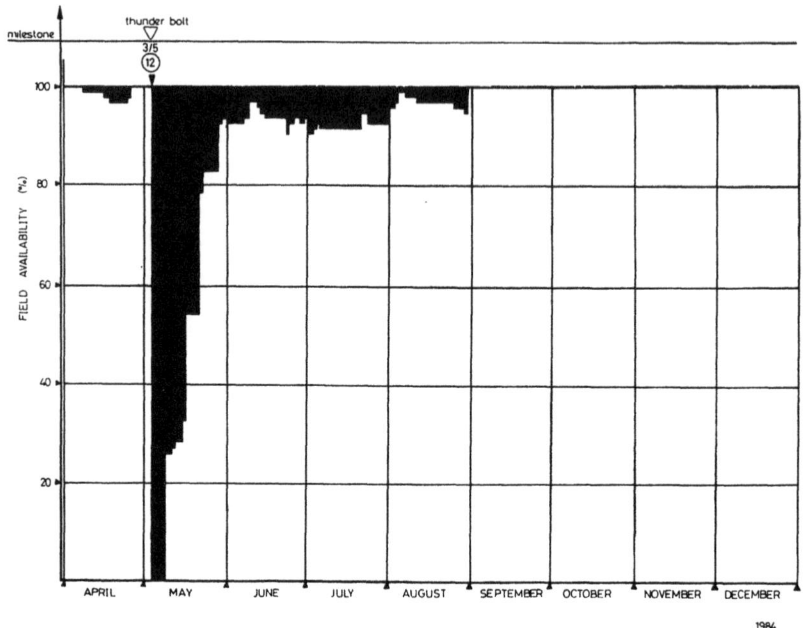

Fig.3.1-5: H.F.S. AVAILABILITY

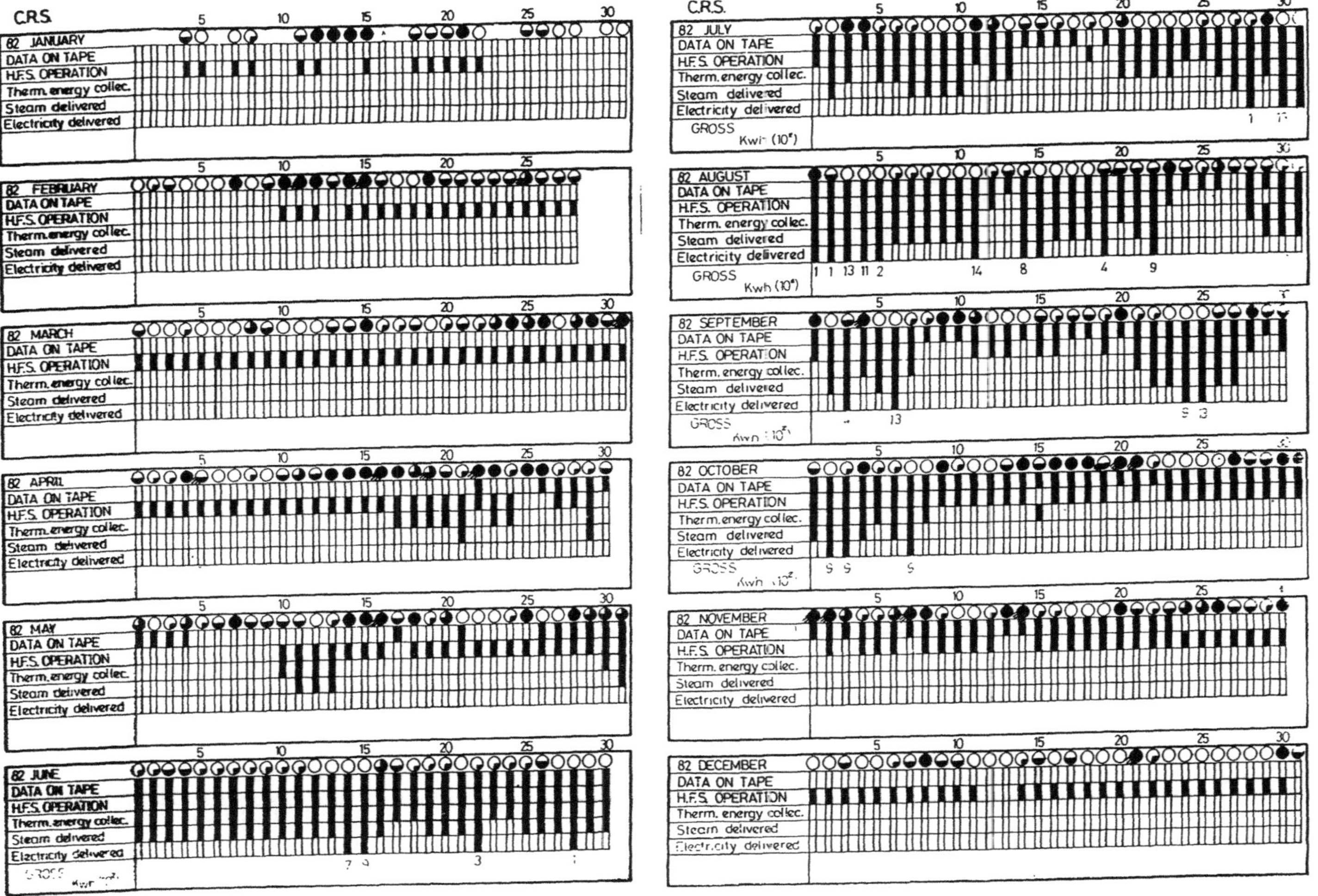

Tab.3.1-6: CRS-1982 operation table

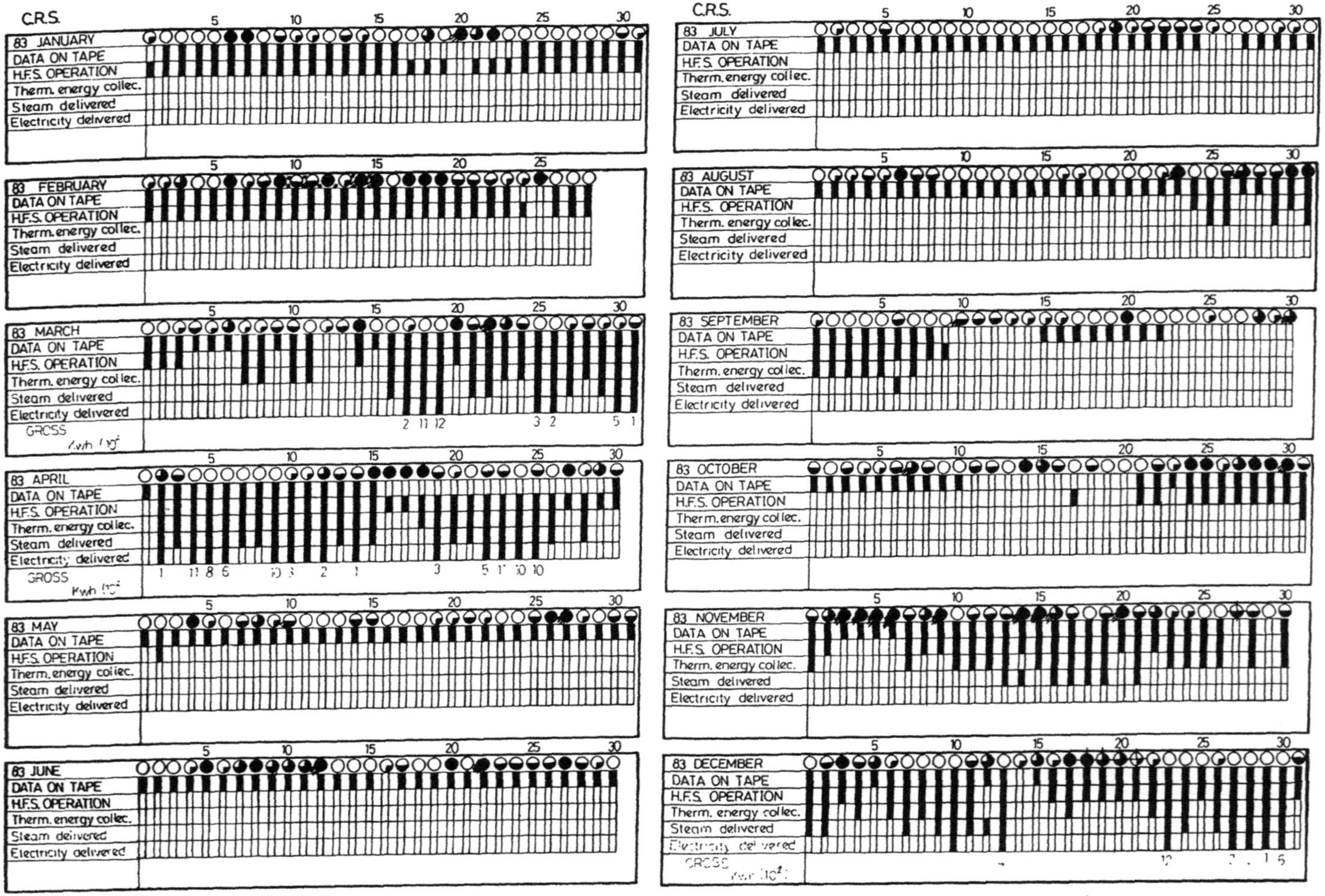

C.R.S.

83 JANUARY
DATA ON TAPE
H.F.S. OPERATION
Therm. energy collec.
Steam delivered
Electricity delivered

83 FEBRUARY
DATA ON TAPE
H.F.S. OPERATION
Therm. energy collec.
Steam delivered
Electricity delivered

83 MARCH
DATA ON TAPE
H.F.S. OPERATION
Therm. energy collec.
Steam delivered
Electricity delivered
GROSS Kwh (10²)
2 11 12 3 2 5 1

83 APRIL
DATA ON TAPE
H.F.S. OPERATION
Therm. energy collec.
Steam delivered
Electricity delivered
GROSS Kwh (10²)
1 11 8 6 10 3 2 1 3 5 1 10 10

83 MAY
DATA ON TAPE
H.F.S. OPERATION
Therm. energy collec.
Steam delivered
Electricity delivered

83 JUNE
DATA ON TAPE
H.F.S. OPERATION
Therm. energy collec.
Steam delivered
Electricity delivered

C.R.S.

83 JULY
DATA ON TAPE
H.F.S. OPERATION
Therm. energy collec.
Steam delivered
Electricity delivered

83 AUGUST
DATA ON TAPE
H.F.S. OPERATION
Therm. energy collec.
Steam delivered
Electricity delivered

83 SEPTEMBER
DATA ON TAPE
H.F.S. OPERATION
Therm. energy collec.
Steam delivered
Electricity delivered

83 OCTOBER
DATA ON TAPE
H.F.S. OPERATION
Therm. energy collec.
Steam delivered
Electricity delivered

83 NOVEMBER
DATA ON TAPE
H.F.S. OPERATION
Therm. energy collec.
Steam delivered
Electricity delivered

83 DECEMBER
DATA ON TAPE
H.F.S. OPERATION
Therm. energy collec.
Steam delivered
Electricity delivered
GROSS Kwh (10²)
12 2 1 6

Tab.3.1-7: CRS-1983 operation table

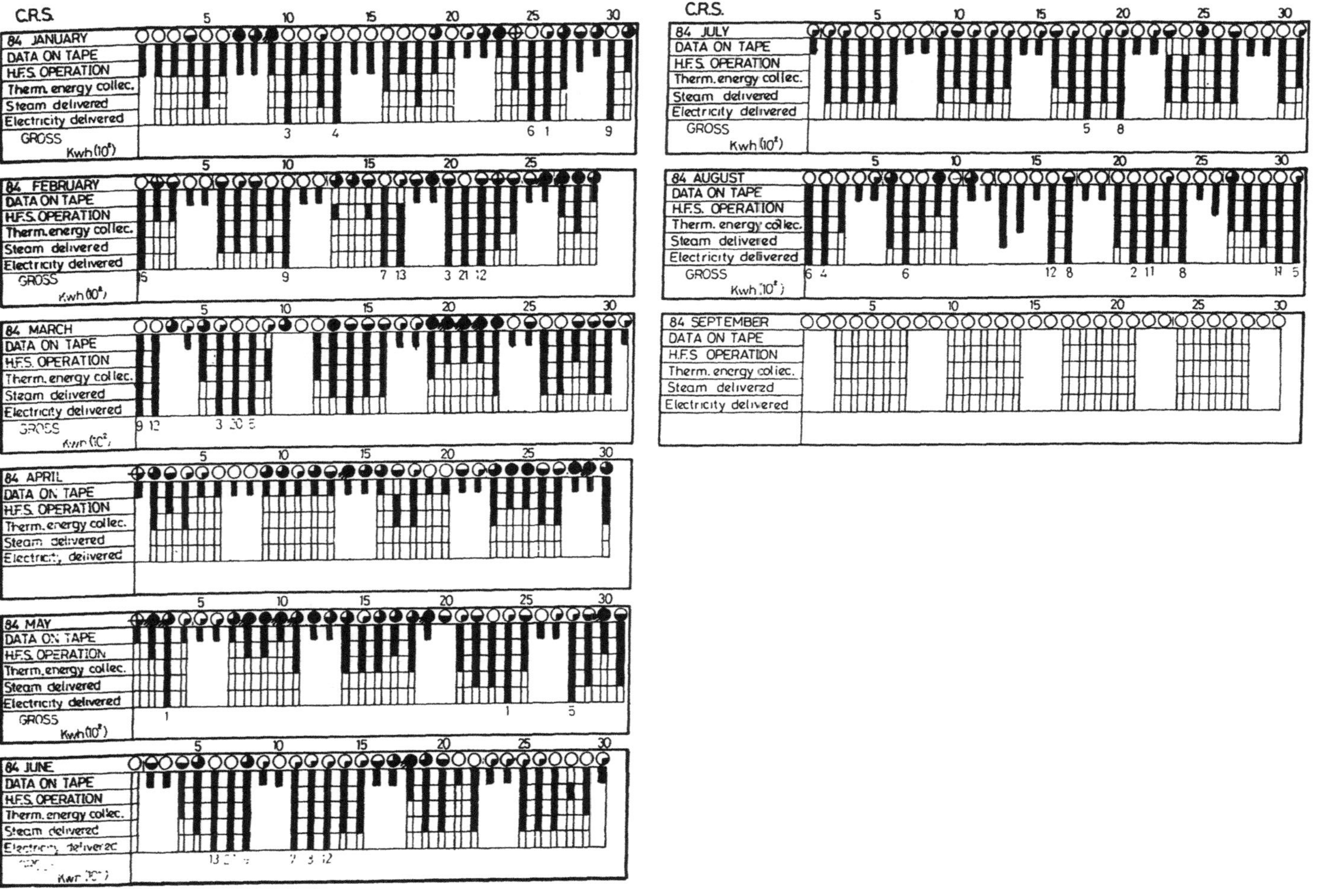

Tab.3.1-8: CRS-1984 operation table

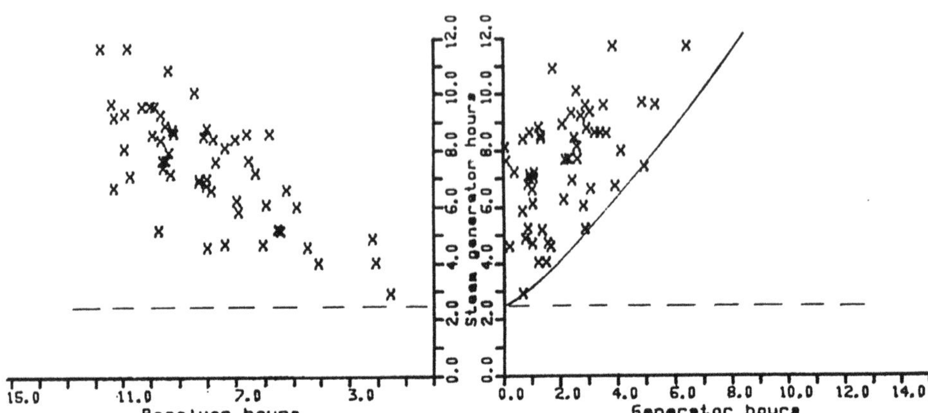

Fig.3.1-6: Sulzer Receiver, Steam Generator, Generator Operating Hours (DATABASE)

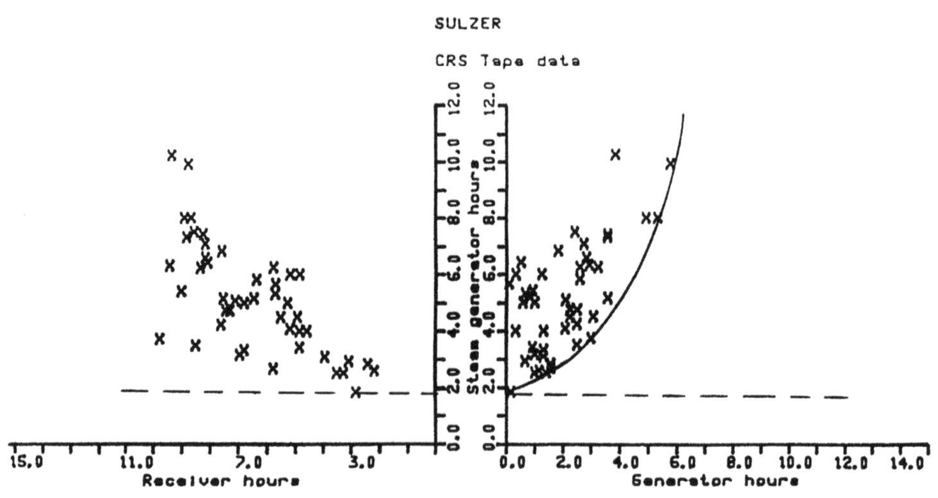

Fig.3.1-7: Sulzer Receiver, Steam Generator, Generator Operating Hours (CRS TAPE DATA)

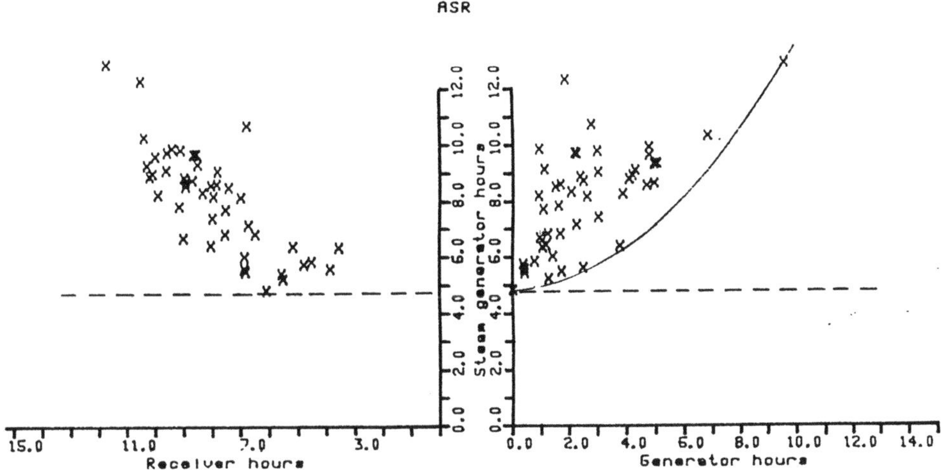

Fig.3.1-8: ASR Receiver, Steam Generator, Generator Operating Hours (DATABASE)

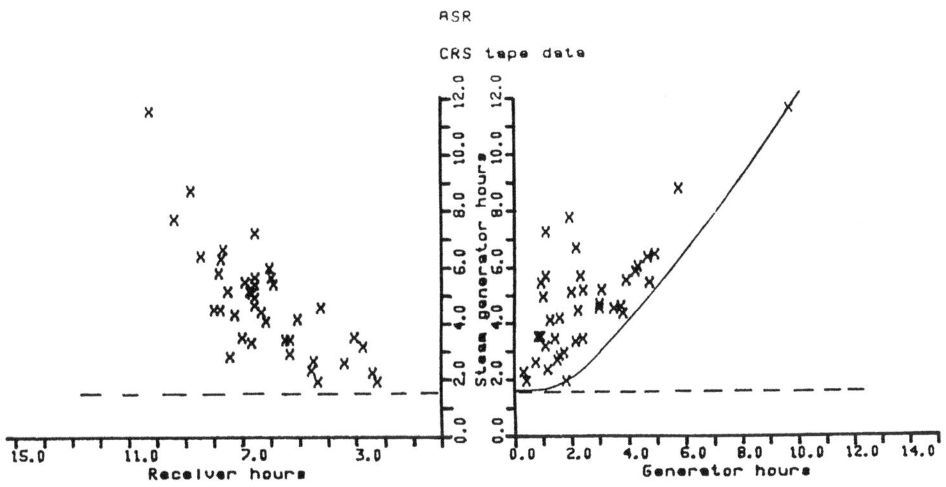

Fig.3.1-9: ASR Receiver, Steam Generator, Generator Operating Hours (CRS TAPE DATA)

DAILY CHARACTERISTICS

Mats Andersson and Jonas Sandgren, ITET

1.0 INTRODUCTION

During the first one and a half years of operation the CRS-plant was equipped with a cavity receiver made by SULZER (November-81 to April-83). At the end of 1983 this receiver was replaced by the Advanced Sodium Receiver (ASR).This change did not lead to any major change in system configuration (see Figure 1), though it did have some operational impacts.

The purpose of this paper is to describe the operation and performance of the CRS-plant on a daily basis and to compare the ASR and the cavity receiver. The evaluation is mainly based on days where electrical energy was produced for at least 50 minutes, though other days were also used for portions of the study. For the SULZER study, 47 days were used during the period November 25 1981 to April 14 1983. For some portions of this study only 22 of these days could be used as there was no information on the number of heliostats in track on the earlier data-sets. The ASR study is based upon 32 days with electrical production from December 10 1983 to August 31 1984.

There were two parts to this evaluation :

- Study of daily values of energies and operation times.
- Detailed study of plots of some selected days.

2.0 OPERATIONAL BEHAVIOUR

2.1 Daily Sums

Daily sums of various energies are calculated from five-minute averages stored on data-tapes by the data acquisition system. The following relationships have been studied:

- Energy into the cavity - Energy absorbed by sodium (Fig.3.2-2) The plots in Fig.3.2-2 show a clear linear relationship, and also show the advantage of the ASR.

- Daily irradiation - Energy absorbed by sodium (Fig.3.2-3) The points form a triangle, one side of which shows the best possible performance. As shown there is a steeper slope for the ASR.

- Heliostat hours (numbers of heliostats times time) - Energy absorbed (Fig.3.2-4) Here the slope is a measure of the maximum available net power per heliostat. The points show a triangular configuration.

- Energy to steam generator - Energy produced by the generator (Fig.3.2-5)

  The energy to the steam generator during electrical production is strongly coupled to the electrical energy produced.

2.2 Conclusions

Fig.3.2-2 shows the receiver efficiences on a daily basis. The lower value for the SULZER receiver can also be seen in Fig.3.2-3 and 4; these figures are based upon relations that do not use the "power into cavity calculations", which have large possible errors. The relations between the numbers presented in the three figures are in the same range.

There was no reason to expect that the change in receivers would have any influence on PCS-operation. As can be seen the power conversion efficiency is calculated to be 20 % in both of the two plots.

3.0 THE DAILY OPERATIONAL ASPECTS

In order to clarify the operational aspects of the plant, temperatures, pressures and energies were plotted for individual days (Fig.3.2-6 through 9 and Tab.3.2-1).

3.1 SULZER Daily Characteristic

July 30 1982 was chosen as beingrepresentativeof a typical clear day for the SULZER receiver. A test was conducted on the PCS but it did not disturb the plant operation. Fig.3.2-6 shows the receiver and PCS performance.

Upper figure: At 8:15, at an irradiance of 340 W/m2 (curve number 1) the receiver doors are opened and the heliostats are put into the tracking mode. After a small drop in the receiver outlet temperature (curve number 2), the temperature starts to increase. At 9:30 500 °C is reached, and the flow is switched over from the cold to the hot tank. This can be seen by looking at curve number 3 which shows the inlet temperature to the hot storage that at this time rapidly starts to increase.

- The hot tank temperature starts at 350°C (curve number 4) and increases constantly throughout the day up to 500 °C.

- Curve number 5 shows the level in the hot tank which starts at 0.25 m and reaches a maximum of 1.7 m at 14:30 when the preheating of the Spilling is started. Half an hour later electricity is being produced.

The PCS operational characteristics for the same day are shown in the lower figure:

At 10:05 the flow to the steam generator (curve number 3) starts changing by mixing the flows from the cold tank (a flow that already exists) and the hot tank. The Inlet temperature (curve number 1) starts to increase, followed by an increase in steam temperature (curve number 2).

- Steam pressure reaches 77 bar (curve number 4) after 30 minutes. (This value is normally 100 bar, but a test was being performed that day).

- At 10:45 the steam generator flow is rapidly increased to its maximum value of 33 m**3/h. This increase is done to remove the "warm sodium" (< 450 °C) that would prevent the content of the hot storage from reaching 480 - 500°C, which is necessary for PCS operation. This operation is of course a very weak point in the performance, and would not be needed if sodium with lower temperature could be used for the electricity production.

- When the hot tank reaches the necessary temperature (see curve 4, upper figure), the flow is reduced to its minimum and maintained at that value while the hot tank is being filled (see curve 5, upper figure).

- When the level is high enough, preheating of the Spilling starts with a high flow, which takes approximately 30 minutes.

- Then the grid is connected and electricity is produced at around 500 kW gross power until 17:45. Hot sodium continues to be produced until 19:40 when the plant is shutdown.

3.2 ASR Daily Characteristic

Fig.3.2-7 presents a similar plot for the ASR. No significant differents can be seen in comparison with the "SULZER DAY", except for the lower steam pressure and the somewhat more careful startup of the receiver.

4.0 CONTROL ASPECTS

Fig.3.2-8 shows two days with changing weather conditions. As can be seen, the outlet temperature is controlled so that there are no significant changes.

5.0 DISCUSSION

During this study, data from approximately 100 days was plotted in order to find days that were representative for the plant. From this data it is possible to study the effect that different operational strategies have had on the plant. For example, one can see that the ASR-receiver has been treated very carefully in that it is started up very slowly in the mornings. As a result, the average startup time is even somewhat higher for the ASR than for the SULZER receiver.

The effect that a larger heliostat field would have on the plant can be seen on the "record day", June 7 1984. Looking again at Fig.3.2-7, one can see that hot sodium was left in the storage tank overnight. Fig.3.2-9 shows how this sodium helped the plant to produce 2.7 MWh gross electric energy on the following day, which is currently the highest value that has been obtained. The electric production started at 10:08 and was maintained throughout the day until 19:58. With a larger solar multiple, this mode of operation could be the normal operational strategy.

It is possible to infer that with a PCS-system that could operate at a lower temperature (400°C), more of the available energy could be used, rather than dumping the "warm sodium", which is the "normal" operational mode with the Spilling motor.

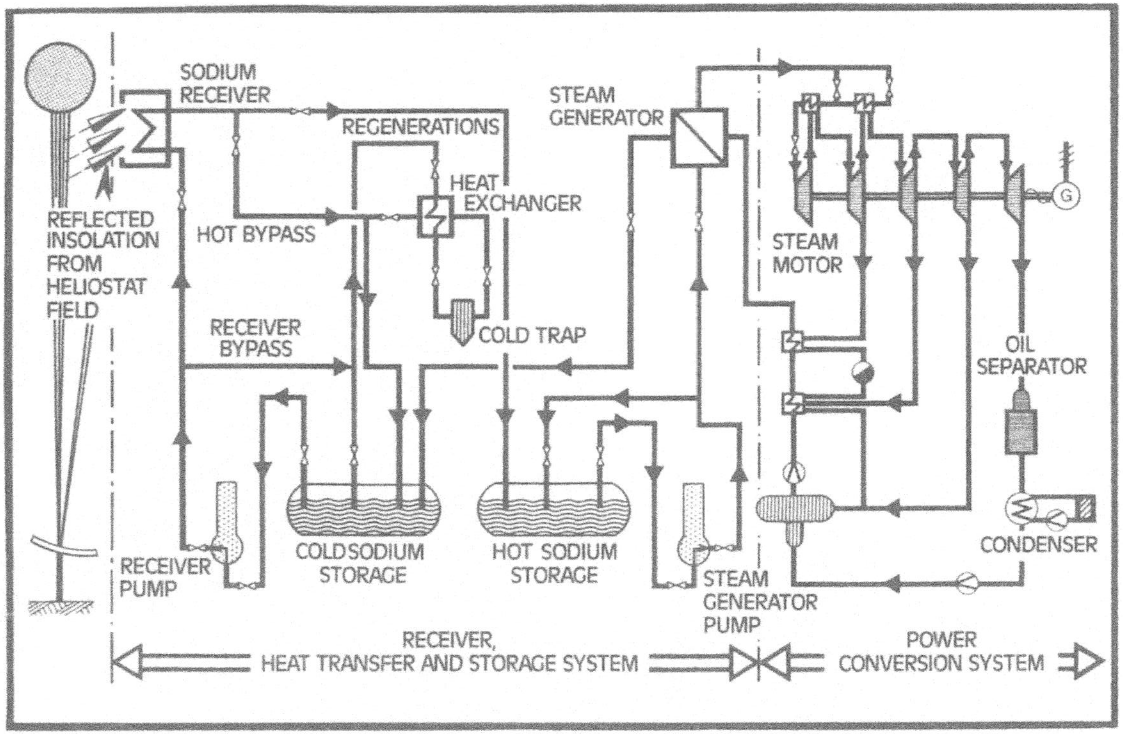

Fig. 3.2.-1: Simplified CRS Process Flow Diagram

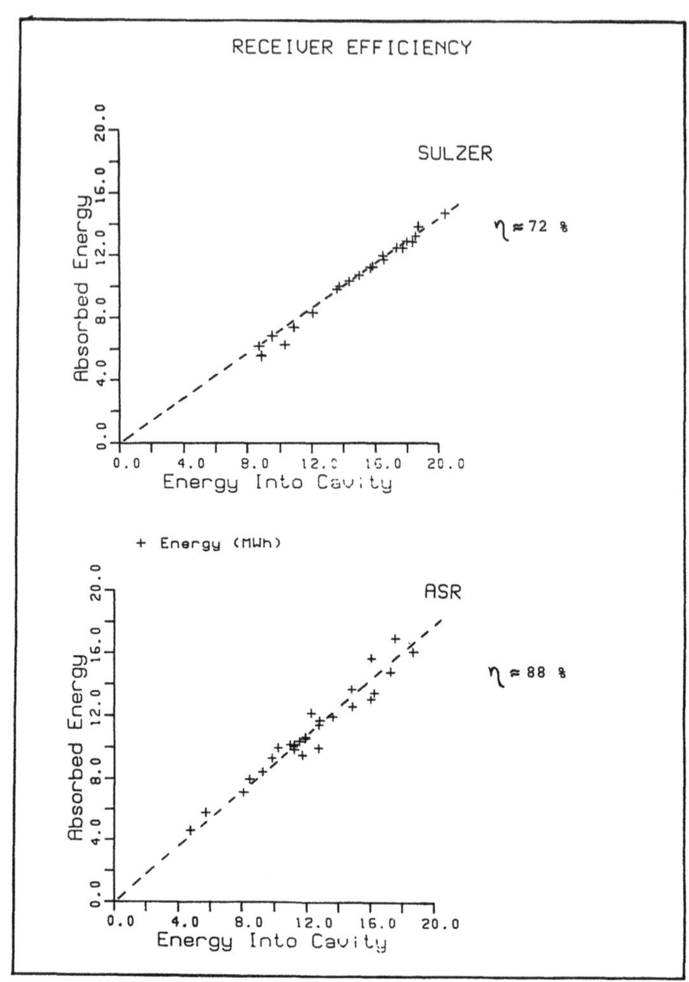

Fig. 3.2.-2: Receiver efficiencies for the both receivers, based on daily energies

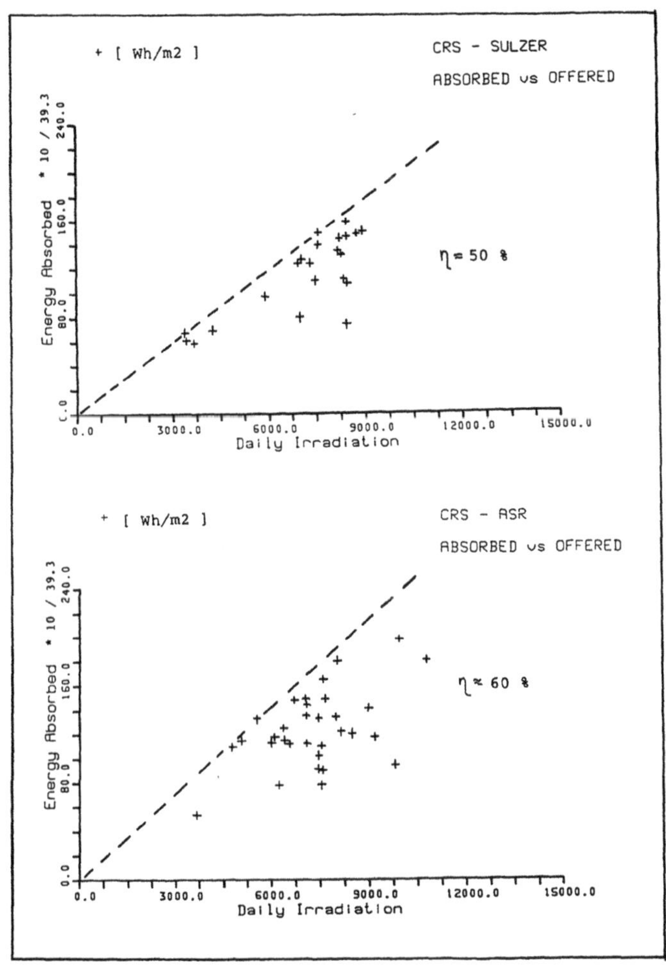

Fig. 3.2.-3: Energy absorbed, calculated per heliostat area versus daily irradiation above 300 W/m2

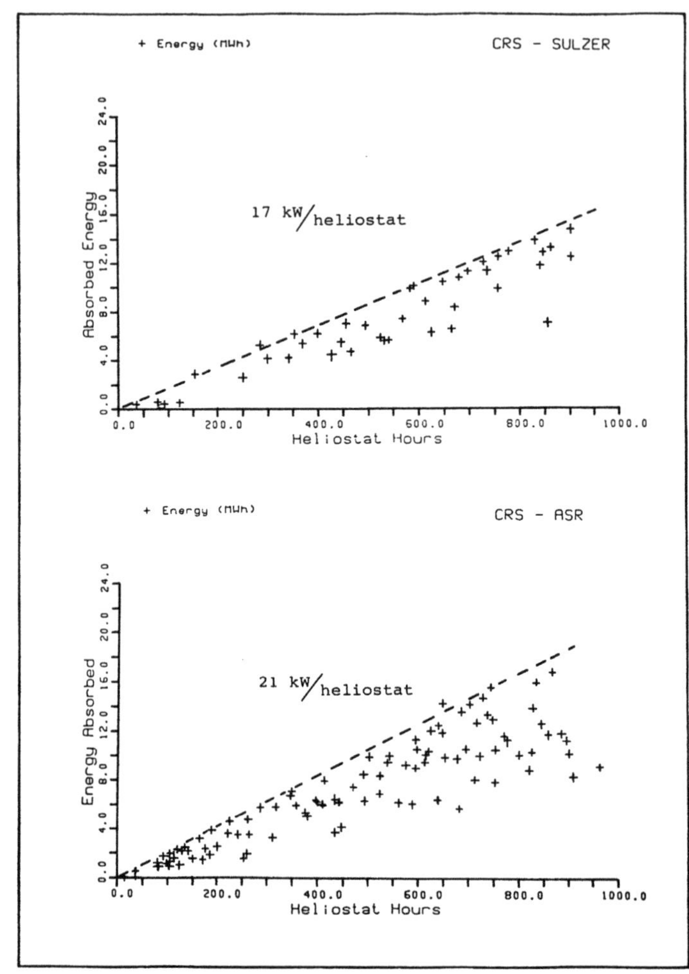

Fig. 3.2.-4: Energy absorbed versus numbers of heliostat hours

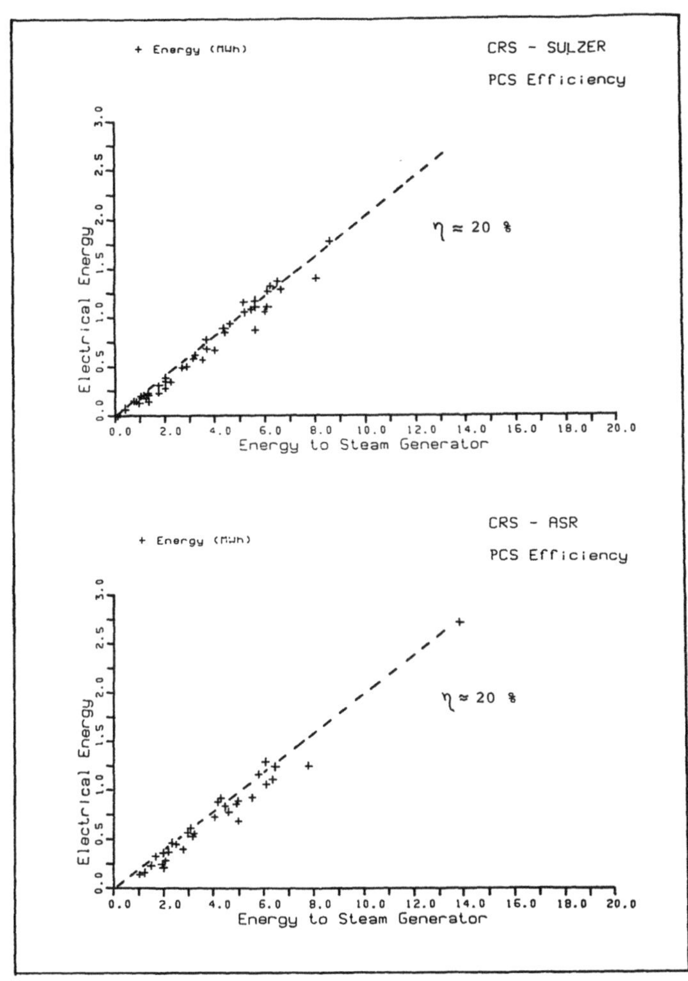

Fig. 3.2.-5: Power conversion efficiency

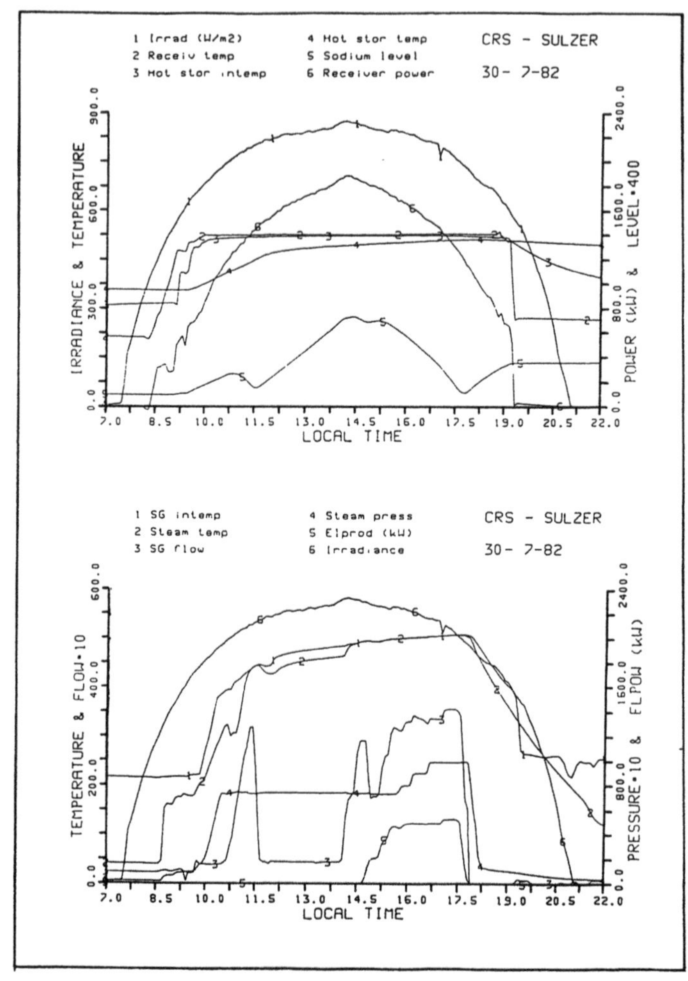

Fig. 3.2.-6: Operation day for the SULZER receiver, July 30. 1982

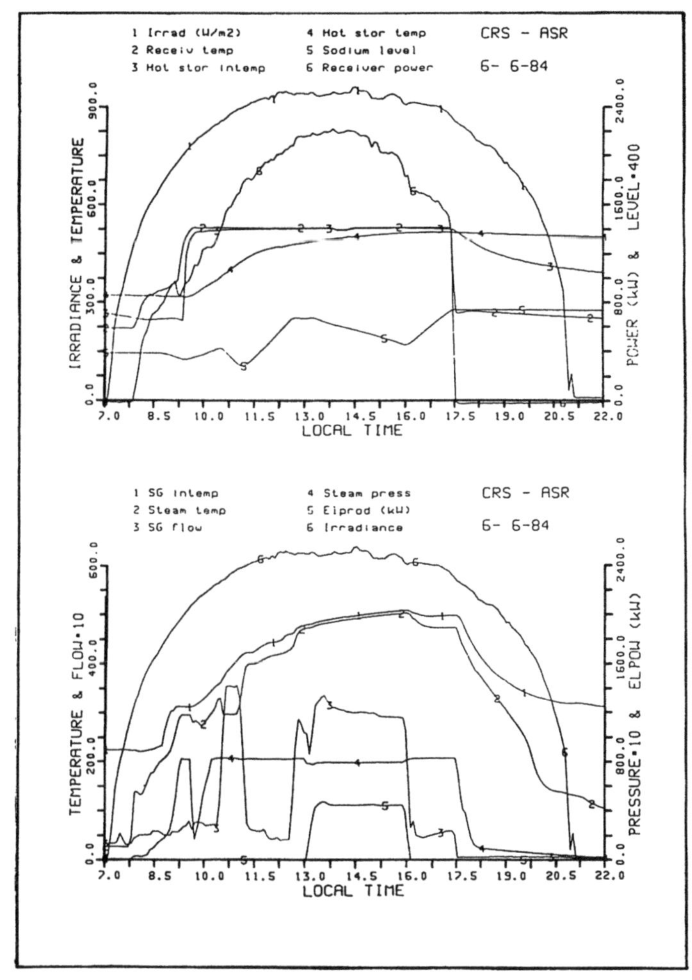

Fig. 3.2.-7: Operation day for the ASR receiver, June 6. 1984

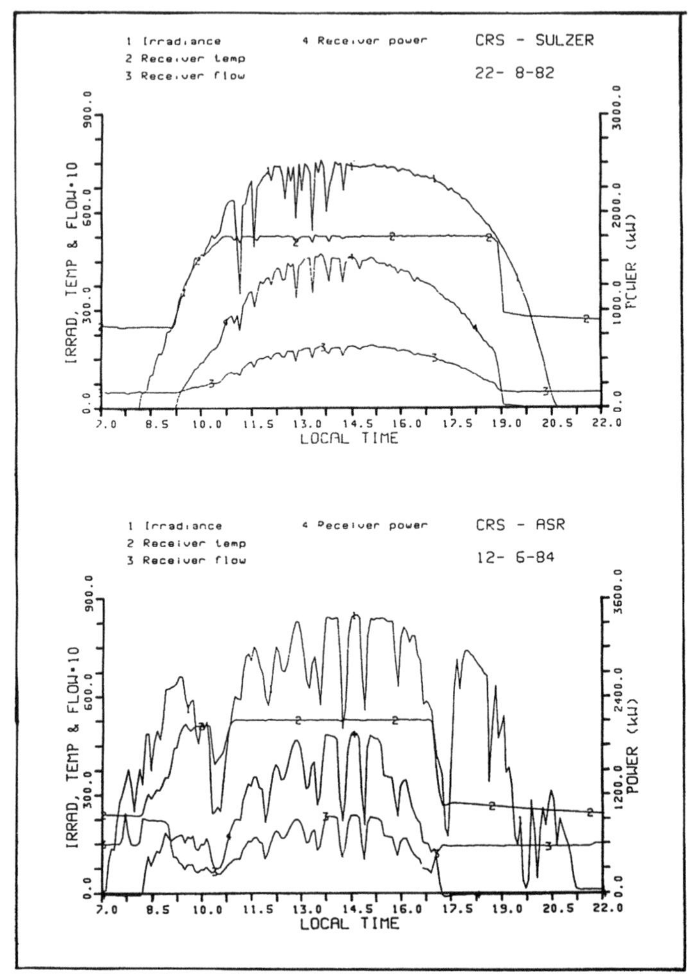

Fig. 3.2.-8: System reaction on cloud passages

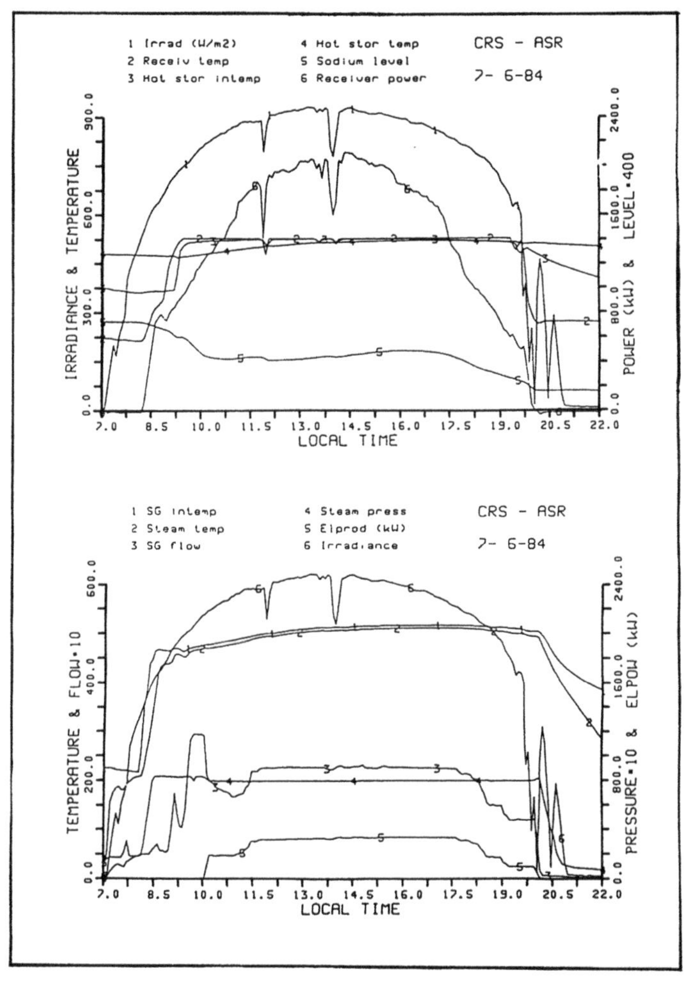

Fig. 3.2.-9: Operation day for the ASR receiver, June 7. 1984

DATE :	30-7-82	6-06-84	7-06-84	
METEO DATA				
Sun Rise (Local Time)	7:16	6:54	6:53	[hh:mm]
Solar Noon (Local Time)	14:16	14:05	14:08	[hh:mm]
Sun Set (Local Time)	21:15	21:21	21:25	[hh:mm]
Irradiation (total)	9.1	10.8	10.2	[kWh/m2]
Irradiation (> 300 W/m2)	8.8	10.7	9.9	[kWh/m2]
Peak Irradiance	873	962	932	[W/m2]
RECEIVER DATA				
Receiver started	8:15	7:53	8:18	[hh:mm]
Time to reach 500°C	75	95	60	[min]
Level in hot tank at start	0.25	0.97	1.79	[m]
Level in hot tank at shut down	0.90	1.84	0.41	[m]
Energy into cavity (start up)	-	1.00	0.60	[MWh]
Energy into cavity (total)	-	16.1	17.6	[MWh]
Energy absorbed	13.4	15.8	17.1	[MWh]
Energy to hot tank	12.7	14.6	16.3	[MWh]
Receiver efficiency	-	97.8	97.0	[%]
PCS DATA				
Flow start to steam generator	10:05	8:58	8:13	[hh:mm]
Max steam pressure during operation	100	84	84	[bar]
Max steam temperature during operation	507	505	512	[degr C]
Start electric production	14:50	13:08	10:08	[hh:mm]
Stop electric production	17:45	16:03	19:58	[hh:mm]
Energy to steam generator (total)	9.2	8.9	15.5	[MWh]
Energy to steam generator before electric production	3.0	2.3	1.6	[MWh]
Energy to steam generator during electric production	6.2	6.1	13.8	[MWh]
Gross electric production	1.3	1.3	2.7	[MWh]
Max electric power	530	485	340	[kW]
PCS efficiency (total)	13.8	14.5	17.6	[%]
PCS efficiency during electric production	20.7	21.4	19.8	[%]

Table 3.2.-1: Operation data for 30/7-82 (SULZER) and 6-7/6-84 (ASR)

STEAM GENERATOR EXPERIENCE

S.W. Amacker, Sulzer Brothers Ltd., Switzerland

1. INTRODUCTION

This paper briefly describes the thermal design of the CRS-steam generator and then concentrates on the data obtained during the operation of the generator.

As will be shown, to date the steam generator has not experienced any problems relative to heat transfer, stability, or structural integrity.

2. SPECIFICATIONS

The specified operating conditions for the steam generator at 100% load are shown in Tab.3.3-1.

	dimensions	normal	upset
Thermal Power	MW	2.22	---
Sodium			
mass flow	kg/s	6.76	---
inlet temperature	°C	525	525
outlet temperature	°C	268.9	---
pressure	bar	8.5	57 max
allowable pressure drop	bar	0.5	
Water/Steam			
mass flow	kg/s	0.87	
inlet temperature	°C	193	
outlet temperature	°C	500	
steam pressure	bar	100	
allowable pressure drop	bar	10	

Table 3.3.-1: Operating Conditions (Ref.1)

These specifications were slightly modified relative to the initial ones (Ref.2). In addition, the following points had to be observed:

- stable load between 25% and 110%
- one load change per day from 100% to 30% to 100% at a rate of 5% per minute for thirty years
- upset condition in case of a sodium water reaction
- insulation of the unit to minimize heat losses
- earthquake loads

3. DESIGN

The design of the steam generator is shown in Appendix (1). A detailed description of the mechanical design can be found in Reference (1).

The heat transfer surface was evaluated using the Sulzer code MIXOU (Ref.3). The following equation was applied on the sodium side

$$N_u = 5.9 + 0.254 \, P_e^{0.635} \quad 30 < P_e < 100$$

Inclusion of the following heat transfer factors in the sizing calculation increased the basic surface by 20%:

- sodium bypass through central tube
- fouling in the evaporator and economizer
- external heat loss
- internal heat recuperation
- geometrical tolerances
- inaccuracies in the heat transfer calculations

In addition, a stability analysis resulted in the introduction of a nozzle at the inlet of each tube that produced a pressure drop at 100% load of

$$\Delta p = 2.1 \text{ bar}$$

In Reference (4) a stability limit at 15% load has been defined. A temperature diagram for 100% and 25% load is shown in Appendix (2).

4. OPERATION

The power conversion system (PCS) was in operation for 230 days (1330 hours). During startup and testing of the plant it is estimated that many load changes took place. However, it is assumed that the number of thermal cycles that occurred did not result in fatigue damage.

Due to a change in the PCS, the part load characteristic of the steam generator was investigated for reduced steam pressure. This study, described in Reference (6), showed that the water side pressure could range from 50 to 100 bar without any power reduction. A lower limit of 50 bars was established due to increasing feed pump power and high velocity of the steam at the outlet. These two parameters are represented in Appendix (3). Today the PCS pressure is held normally at 80 bars.

In the original specifications a load change of 5% per minute was required. Due to the flexibility of the helix design, a higher load change rate of 20% per minute would be possible. This rate would enable the steam generator to be started up within approximately 5 minutes. However, this rate might not be reached due to the large sodium volume in the central cavity. The thermal inertia of this sodium mass dampens the load changes. The content of this cavity amounts to 0.64 m^3 and could be included in a transient calculation. If there is an interest in increasing the startup and shutdown rates, the design of the generator could be changed to eliminate the sodium in the central cavity.

5. EXPERIMENTS

The most important task of the steam generator is to fulfill the specified performance regarding power and pressure loss. A series of measurements were carried out to confirm the thermal design. In addition to the data obtained from the operation control system, thermocouples were mounted on the outer shell of the steam generator and the inlet and outlet tubing as shown in Appendix (4). These thermocouples, which were attached to the outer shroud of the steam generator and insulated, accurately measured the local sodium temperature.

The following data were recorded:

- Inlet water temperature and pressure
- Steam temperature and pressure
- Sodium inlet and outlet temperature
- Sodium pressure
- Sodium and water flow rate
- Sodium temperature distribution

As shown in Appendix (2) and (3), the inlet temperatures of the sodium and steam are very close and temperature measurements create tolerance problems. By lowering the sodium flow, this temperature difference could be increased.

On July 19, 1984, measurements at stable, high load were performed. The identification of the chosen data set is

19. 7.84/15: 48: 23 (Solar time)

The one minute averaged data of this period is shown in Appendix (5). The data from the additional instrumentation is found in Appendix (6). As shown, the data from the data acquisition system (DAS) is in relatively good agreement with the results of the additional instrumentation.

	Data logger	DAS
Sodium inlet (°C)	465.5	462.9
outlet (°C)	263	261.3
Water inlet (°C)	197	195.9
outlet (°C)	414.8	409.1

Using the Sulzer code DISTEMP described in Reference (7), a part load analysis of the steam generator was performed. The results are presented in Appendix (7). The following corrections are included in the analysis:

- temperature differences between data logger and DAS
- heat exchange with inlet and outlet tubing
- influences on heat transfer included in this paper in the Design section

6. CONCLUSIONS

There is very good agreement between measurements and calculation. However, some differences exist within the evaporator part of the steam generator.

Heat transfer was underestimated at the film boiling condition, and overestimated at nucleate boiling. However, the inaccuracies have not affected the transferred power.

An interesting detail is the good agreement between the calculated and measured position of the critical heat flux, which is found within close boundaries.

The operating experiences with the steam generator have led to the following conclusions:

1. The specified operating conditions are met.

2. No problems regarding operation or structural integrity have been encountered.

3. The load change rate of the steam generator might be improved if needed by a design change (central cavity filled with gas).

4. Transient behavior should be investigated.

5. The design is very flexible regarding system pressure and power needs.

6. Although no problems were encountered, the steam generator is an important component of the system. Great attention has to be paid to the boundary between sodium and water.

7. REFERENCES

1. H. Fricker, "Steam Generator for CRS Solar Power Plant Almería", Sulzer report, TA-2/0012, 13.3.80.

2. INTERATOM letter, Zipper/8110, May 3, 1978.

3. MIXOU, Sizing program for steam generators.

4. S. Amacker, "Part Load Characteristic and Stability of the CRS Almería Steam Generator", Sulzer report, TA-3490, 28.1.80.

5. M. Pescatore, private communication.

6. S. Amacker, "Part Load Characteristic with Reduced Steam Pressure, Sulzer report, TA-2/0426, 28.8.82.

7. DISTEMP, Part load program for steam generators.

8. APPENDIX

1. Sulzer Drawing 0-103.073.483, steam generator
2. Temperature Distribution of the Almería steam generator
3. Reduction of steam pressure
4. Additional instrumentation
5. SSPS-CRS-Minute data of July 19, 1984
6. Temperature measurements with data logger Monitor Labs 9300
7. Temperature distribution (analysis and measurements)

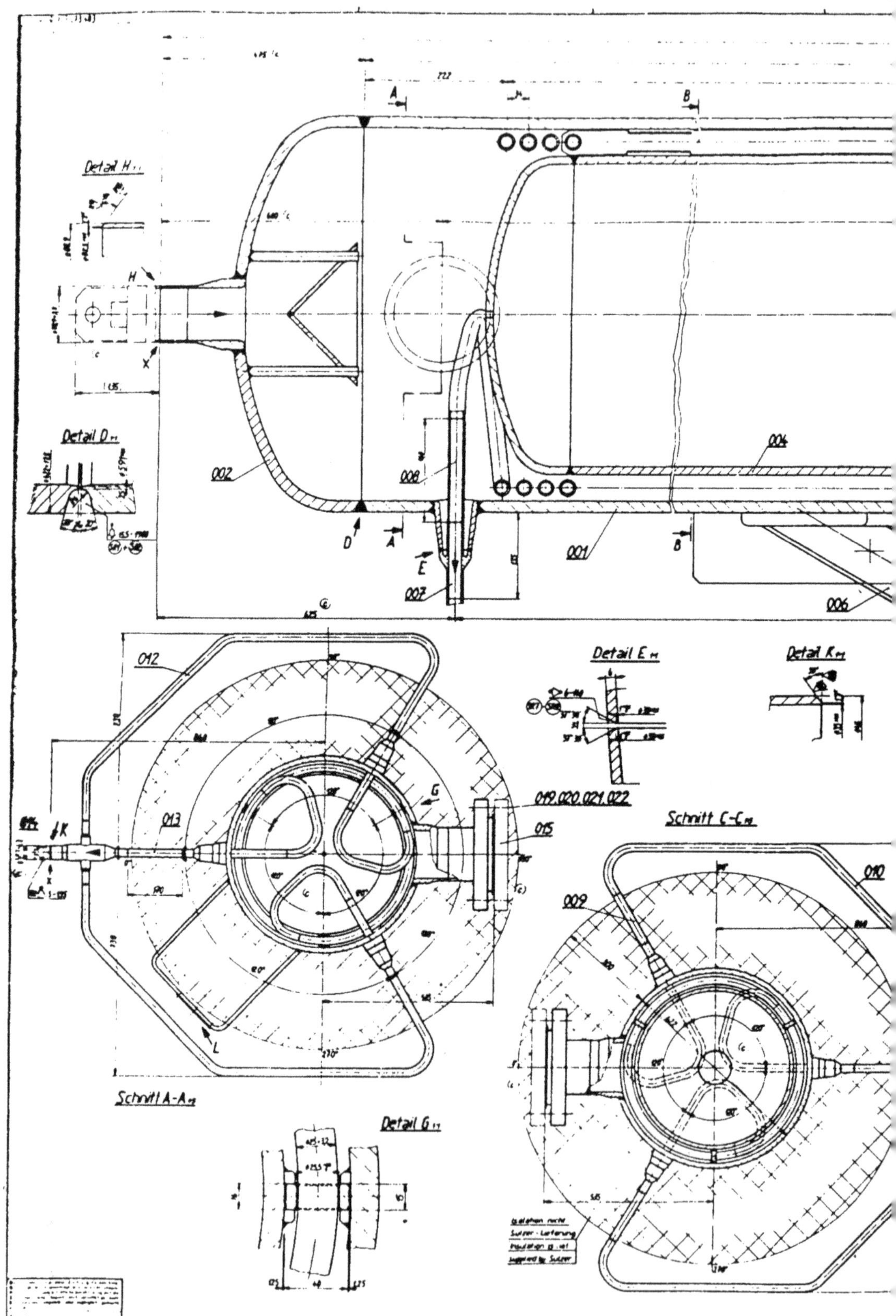

Detail H
Detail D
Detail E
Detail K
Detail G
Schnitt A-A
Schnitt C-C
002
008
007
001
004
006
012
013
015
019.020.021.022
009
010
Isolation nicht
Sulzer - Lieferung
Insulation is not
supplied by Sulzer

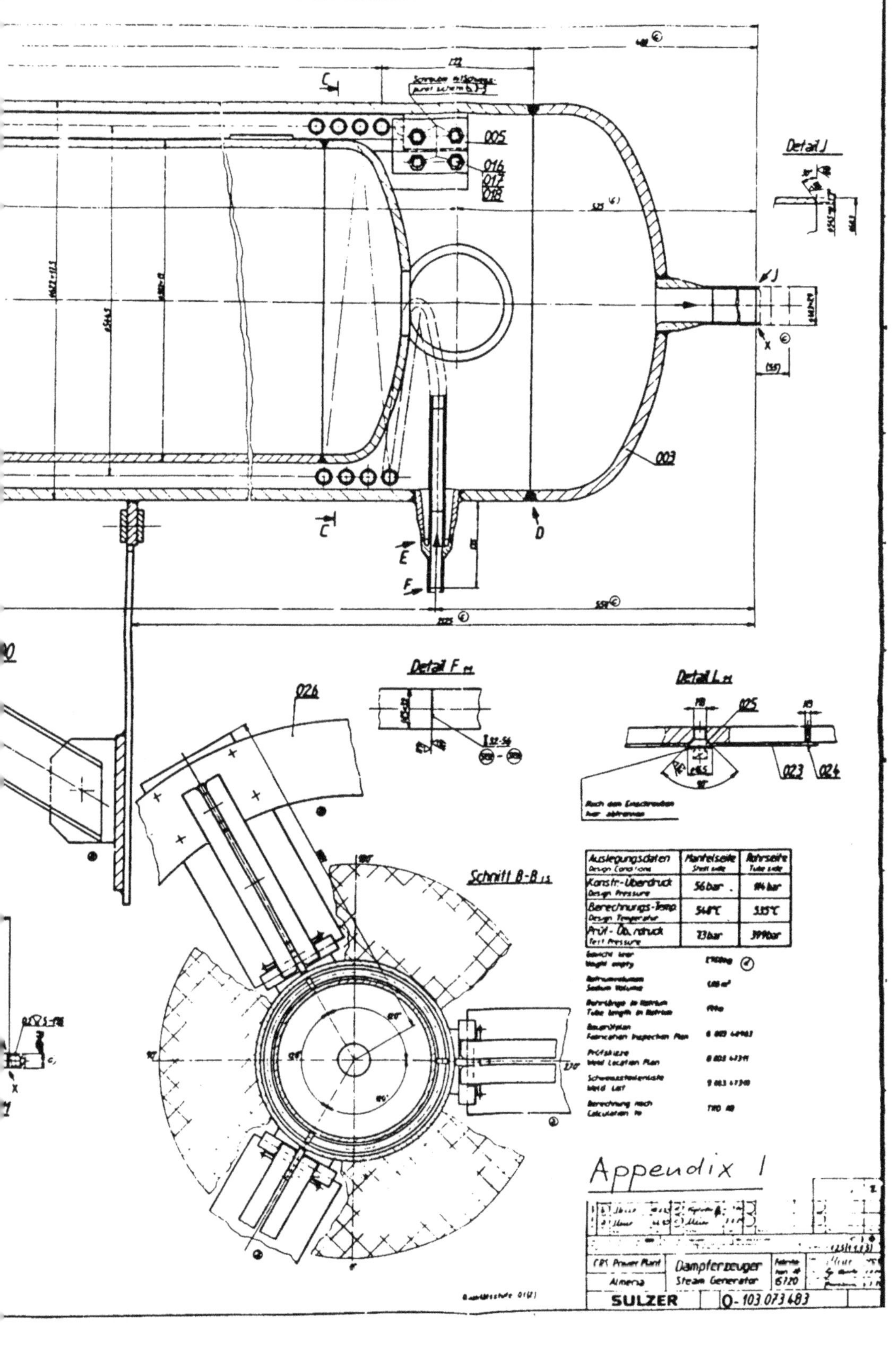

Detail J
005
016
017
018
003
C
D
E
F
Detail F
Detail L
025
023
024
026
Schnitt B-B
Auslegungsdaten Design Conditions
Mantelseite Shell side
Rohrseite Tube side
Konstr.-Überdruck Design Pressure
56 bar
Berechnungs-Temp. Design Temperature
Prüf-Überdruck Test Pressure
73 bar
399 bar
Weight empty
Sodium Volume
Tube length in Sodium
Fabrication Inspection Plan
Weld Location Plan
Weld List
Calculation to
Appendix 1
CRS Power Plant
Almeria
Dampferzeuger
Steam Generator
6720
SULZER
0-103 073 483

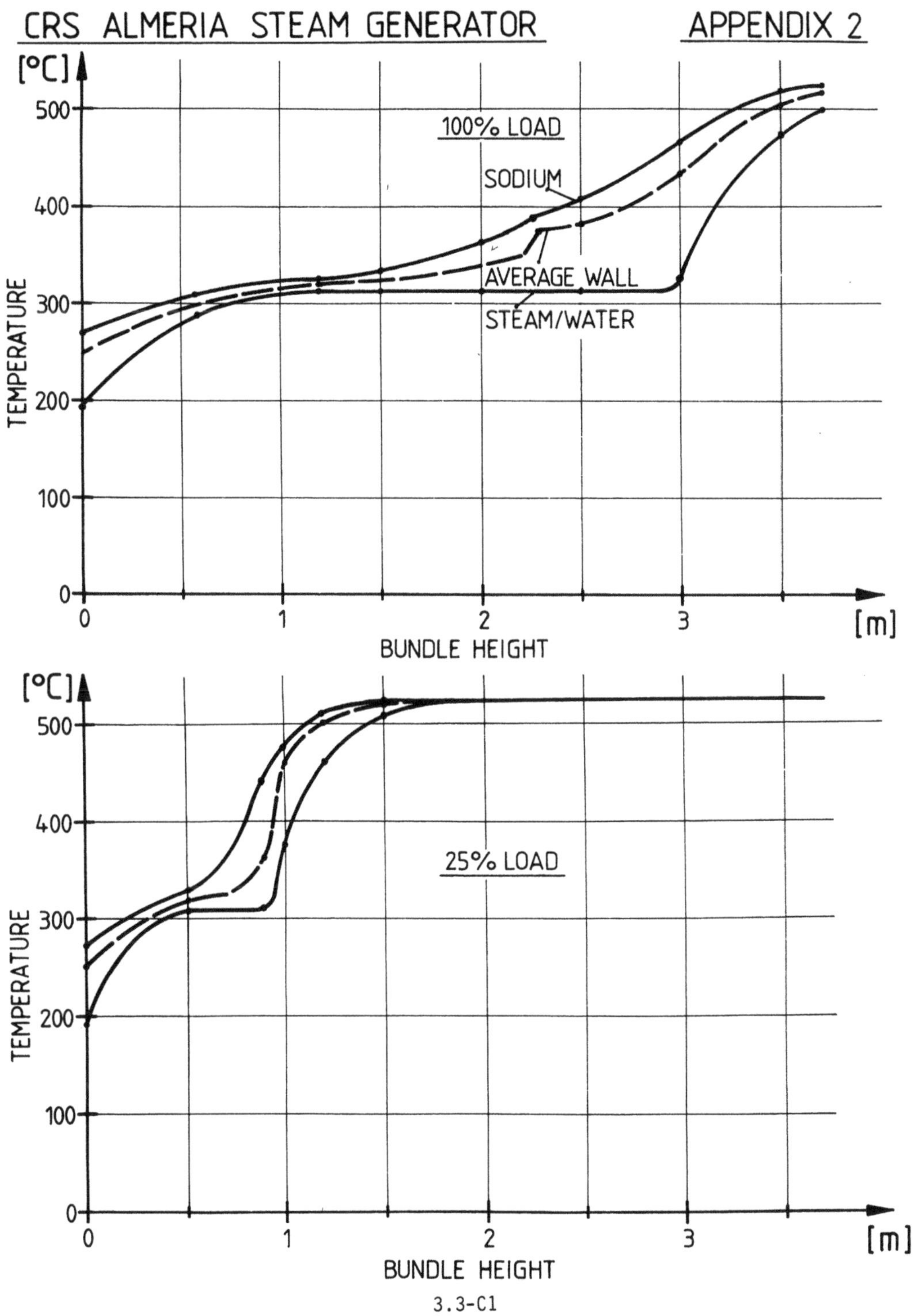

CRS ALMERIA STEAM GENERATOR
APPENDIX 2
[°C]
500
400
300
200
100
0
TEMPERATURE
100% LOAD
SODIUM
AVERAGE WALL
STEAM/WATER
0
1
2
3
[m]
BUNDLE HEIGHT
[°C]
500
400
300
200
100
0
TEMPERATURE
25% LOAD
0
1
2
3
[m]
BUNDLE HEIGHT

REDUCTION OF STEAM PRESSURE APPENDIX 3

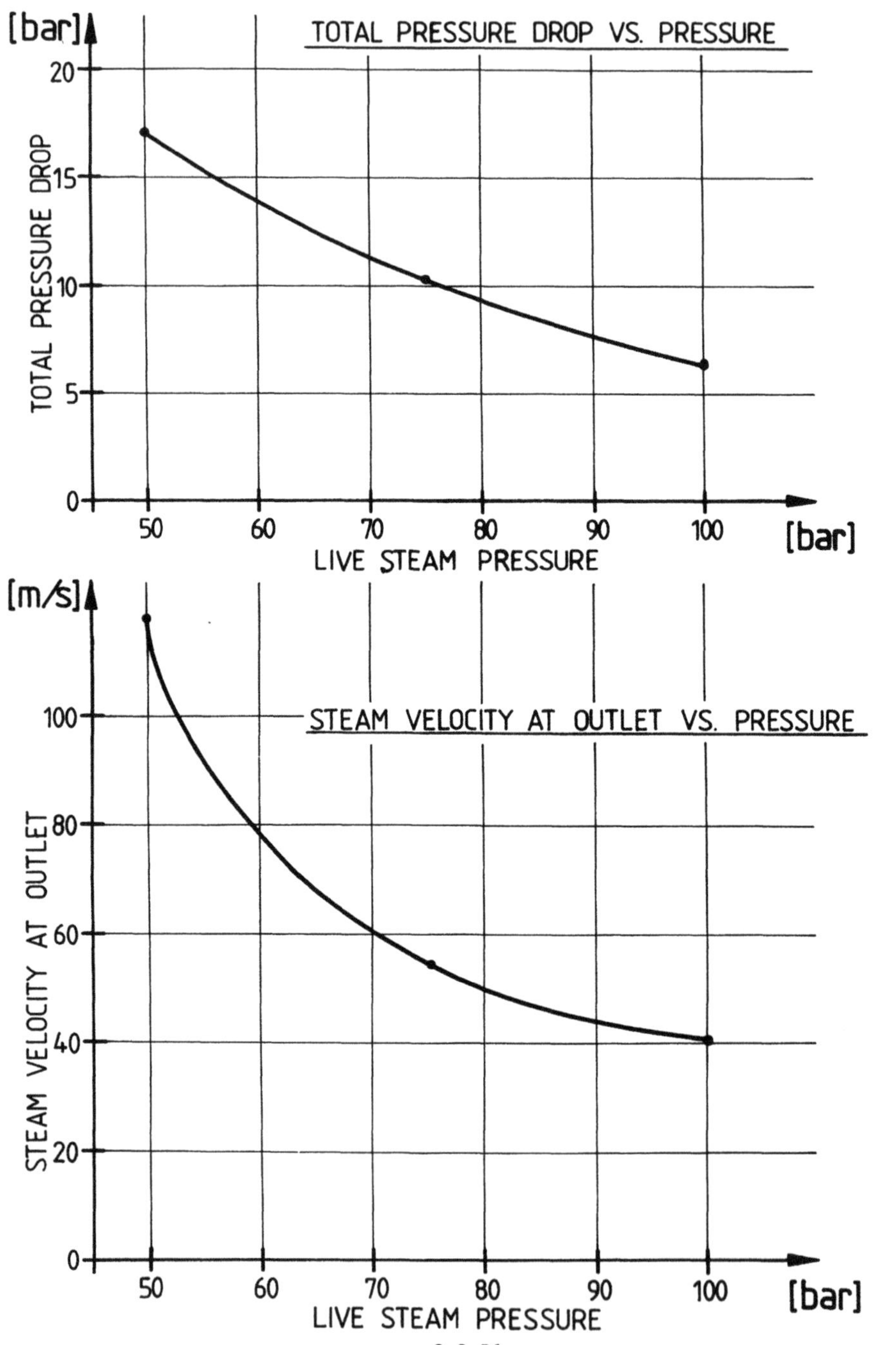

3.3-D1

APPENDIX 4

CRS ALMERIA INSTRUMENTATION (ADDITIONAL)

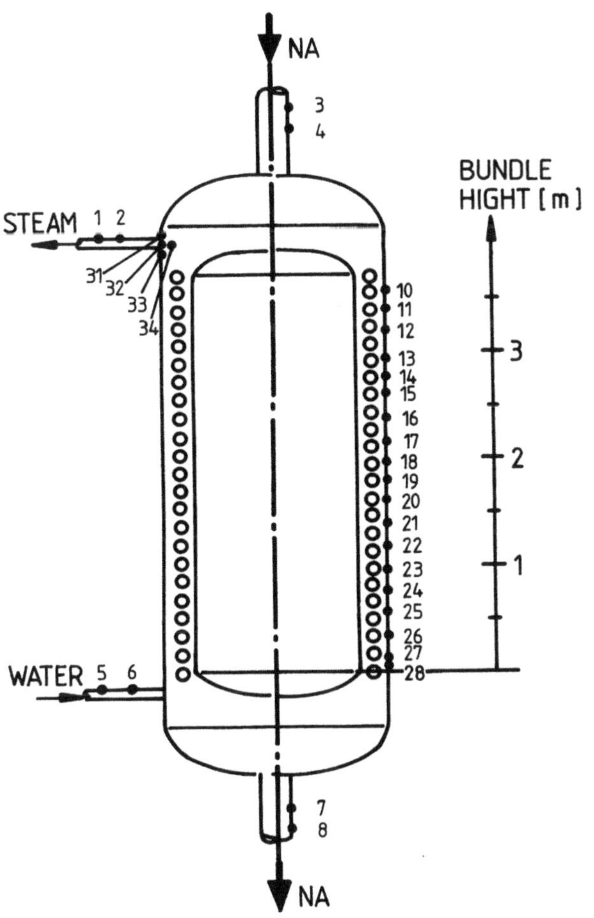

D A T A   L O G G E R   T E M P E R A T U R E S

# TC	Temp
34	404.6 °C
33	412.2 °C
32	407.7 °C
31	411.2 °C
28	275.8 °C
27	276.3 °C
26	284.9 °C
25	291.9 °C
24	296.2 °C
23	300.3 °C
22	303.0 °C
(21	306.8 °C) *
20	305.2 °C
19	308.2 °C
18	310.8 °C
17	316.1 °C
16	319.9 °C
15	326.6 °C
14	336.8 °C
13	362.3 °C
12	384.7 °C
11	423.9 °C
10	448.4 °C
8	262.9 °C
7	263.0 °C
6	197.1 °C
5	196.8 °C
4	465.1 °C
3	465.8 °C
2	414.3 °C
1	415.3 °C

day 201:18:03:57 real time

≙ 19th July ≙ 15:48:00

solar time

* TC not respected

APPENDIX 7

CRS ALMERIA STEAM GENERATOR

• PART LOAD CALCULATION

+ TEMERATURE MEASUREMENTS

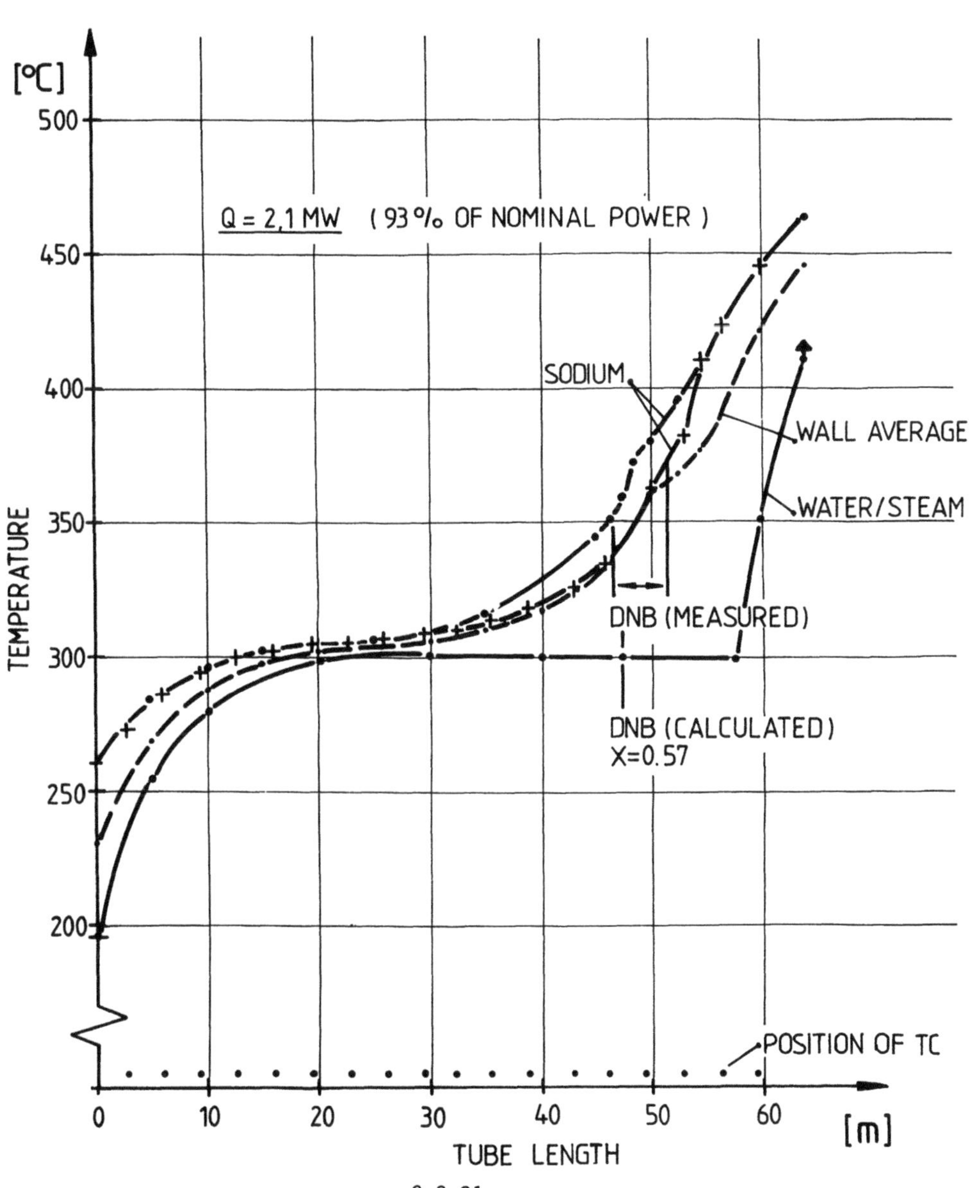

3.3-G1

CRS OPERATIONAL EXPERIENCE

Juan Ramos, Antonio Cuadrado, and Carlos Lopez, Sevillana[1]

SUMMARY

This paper presents the operational experiences collected by the POA[2] in the operation and maintenance of the central receiver system of the SSPS plant. It describes the nonstandard operational procedures that have been developed to allow more efficient operation of the system, the main problems that have appeared, and the maintenance activities it has required.

Special operation procedures are selective defocussing according to the time of day and wind conditions, and modified STOW position to decrease mirror corrosion. In the receiver loop, the ASR requires daily individual venting of the panels, and consequently there has been a need to modify the temperature of the regeneration vessel. A safe procedure to fill the ASR was also developed. The large thermal inertia of the sodium tanks makes necessary the discharge of hot sodium through the steam generator to reach nominal temperatures.

The main problem with the heliostat field has been the damage caused by lightning. The power failure recovery sequence fails occasionally. Installation of the Advanced Sodium Receiver took four months in 1983. The trace-heating system has been the source of many problems due to insufficient power in some sections and lack of redundancy. A manufacturing defect in the cold sodium tank produced several leaks that needed a costly and lengthy repair. In the PCS[3] the steam motor has suffered repeated breakdowns.

1) Sevillana = Cia. Sevillana de Electricidad

2) POA = Plant Operation Authority

3) PCS = Power Conversion System

CRS OPERATIONAL EXPERIENCE

Juan Ramos, Antonio Cuadrado, and Carlos Lopez, Sevillana

1. INTRODUCTION

The CRS plant has been affected by a great number of problems since the start of its operation. This has had the negative consequence that the number of hours of full operation has been much lower than expected. On the other hand, the challenge posed by those problems has forced the operation team to go deeply into the system, with the consequence that it now has a fairly good knowledge of the different subsystems.

2. OPERATION PROCEDURES

The normal operation procedures are sufficiently described in the Plant Operation Manual. Only those procedures which have been developed by the operation team will be commented on.

2.1 Heliostat Field Subsystem (HFS)

All available heliostats are focussed on the receiver. Selective defocussing is applied early in the morning and late in the afternoon, to avoid excessive temperature in the receiver frame, caused by aberration.

For wind speeds of more than 30 km/hr the last rows of heliostats have to be defocussed, since the displacement of their images also produces receiver frame over-heating.

During inactive periods, whenever wind speed allows it, the heliostats are sent to WASH position, in an attempt to decrease the corrosion of the mirrors.

2.2 Receiver Loop

All five panels in the ASR must be vented before the start of normal operation, to eliminate any argon that may have accumulated in the panels and that could impair the flow distribution and the response of the pump. This operation has required the modification of the original venting system, which commanded all five valves simultaneously.

The venting procedure sends sodium to the Regeneration Vessel every operating day. Therefore the tank has to be partially emptied periodically. Since the trace heating system maintained the sodium in the Regeneration Vessel at 220°C, a lot of impurities were dissolved in it, and dirty sodium was sent to the system every few days. The plugging temperature increased very quickly. The procedure now in use is to keep the temperature of the Regeneration Vessel between 115°C and 125°C. In this way the venting of the receiver does not affect the plugging temperature. It requires however, additional attention from the operator.

The filling of the ASR has proved to be a difficult and time consuming procedure. A modified filling method is described in Ref.1. This modified procedure is still unpractical and the receiver has been drained only when it could not be avoided. This fact penalizes the system by increasing parasitics and trace heating consumption, and making impossible a total shut-down of the plant during short inactive periods.

2.3 Sodium Heat Transfer System (SHTS)

The high thermal inertia of the hot tank, together with the low thermal capacity of sodium, make it difficult to reach a temperature high enough to start running the steam motor. If the temperature of the sodium in the hot tank is low (400°C) it is necessary to discharge the sodium through the steam generator at the same rate that hot sodium (530°C) is coming from the receiver (Ref.2,3).

Cleaning the Cold Trap after sodium purification requires the provisional installation of additional trace-heating in the outlet pipe of the Cold Trap.

2.4 Power Conversion System (PCS)

Prior to the start of the steam motor, the flow of water to the Steam Generator has to be raised to 75%, otherwise the steam pressure would drop when the motor started to run. Excess steam is eliminated through the bypass line.

3. DESCRIPTION OF MAIN PROBLEMS

3.1 Heliostat Field Subsystem (HFS)

The CRS heliostat field is particularly sensitive to lightning. On two occasions it has suffered from its effects. The first time in May 1982, 25% of the heliostats were left out of operation. The second time, in May 1984, lightning damaged 75% of the field. Most of the damage happened in the power supplies and in the circuits of the communication interface. The problem seems to be due to insufficient shielding and overvoltage protection of the power supply and communication lines.

The Power Failure Recovery Sequence that should defocus the heliostats in track after a loss of grid power does not work as specified. Even when on most occasions the heliostats leave the receiver according to the programmed sequence, sporadically some of them leave in random directions and on a few occasions a number of heliostats have remained in track, their beams moving towards the east as the sun moved towards the west. An emergency procedure has been developed that takes out of the receiver all heliostats remaining in track when the automatic recovery sequence fails. The reason for the failure is not yet known.

3.2 Sodium Heat Transfer System (SHTS)

The electric trace heating, used to keep the sodium liquid in all the piping of the SHTS, has been the source of a few problems. Insufficient power in critical parts like valves, pipe supports and pipe bends have interrupted on several occasions the flow of sodium through some of the pipes. Being a critical part of the operation of a sodium system, the trace heating should be redundant, at least in the most important sections of the piping and in zones of difficult access like the inner piping of the ASR receiver and the tower pipes.

Due to a manufacturing defect, the cold sodium tank has been the cause of the longest system outages. Sodium leaks that appeared in the bottom of the tank forced two repair actions, the first lasting from December 1981 until May 1982, and the second one, when the bottom of the tank was replaced, from November 1982 until February 1983. This repair is well documented in several SSPS papers.

3.3 Power Conversion System (PCS)

The Steam Motor has been the component causing most of the problems in the PCS. Replacement of the rings in the sliding valve and piston of the first stage are by now considered as routine maintenance. The motor has to be opened for repair after a maximum of 30 working hours. Even this low mean time between failures has been possible only after the pressure at the outlet of the steam generator has been reduced to 80 bar from its design value of 100 bar. Fortunately, the existence of the bypass line has allowed the discharge of the thermal energy collected during receiver operation. Otherwise, the time of receiver loop operation would have been drastically reduced.

The Evaporating Plant needed to purify the exhaust steam of the motor constitutes by itself a very complicated system. It is the source of continuous minor problems.

4. MAINTENANCE

Table 1 gives the number of man-hours spent on the maintenance of the different subsystems in absolute value and percentage for the years 1983 and 1984.

The main maintenance activities are listed below:

4.1 Heliostat Field Subsystem (HFS)

Originally, the facets of the heliostats showed a dangerous tendency to fall from their supports on the heads of people walking under them. After riveting all facets in 1982, the problem disappeared.

Most of the work done until now in the HFS goes into the repair of the electronic controllers. The availability of a test bed has reduced the cost of repair and down time of heliostats caused by electronic controllers.

The high number of man-hours employed for repair of the HFS during 1984 is explained by the incidence of the lightning strike.

During 1983 the field was washed six times. In the first nine months of 1984 the field was washed another six times. Average water consumption for each washing is 4.5 m^3 and manpower requirements are 24 man-hours.

4.2 Sodium Heat Transfer System (SHTS)

Soduim leaks in the Cold Tank required two repair periods, from December 1981 until March 1982, and from November 1982 until February 1983. The repair was done by the manufacturer of the tank (VOEST-ALPINE).

Two sodium leaks in control valves were repaired by the SSPS maintenance team with satisfactory results.

Sodium plugs have been frequent, especially in the Plugging Meter and Cold Trap circuits.

The Hot Sodium Pump was flooded with sodium during routine replacement of the lubricating oil due to a defective valve. Cleaning was done under supervision of the pump manufacturer (BYRON-JACKSON).

From May until September 1983, the Sulzer receiver was dismantled from the CRS tower, and the Advanced Sodium Receiver took its place. The SSPS maintenance team had an active part in all the installation work.

Individual control of the venting valves of the ASR, not originally designed into the system, was implemented in April 1984. At the same time 20 additional thermocouples, corresponding to points of the ASR, were connected to the Data Acquisition System (DAS). The monitoring by the DAS of the temperatures at those points was deemed necessary after the ASR filling incident in November 1983.

Improvements for the Sun Presence Sensors signal, used as feedforward for the control of the Cold Sodium Pump, are currently under way.

4.3 Power Conversion System (PCS)

Periodic replacement of the rings of the sliding valve and piston in the first stage of the Steam Motor.

The DCS Water Treatment Plant had to be connected to the PCS of the CRS, since its own Treatment Plant is unable to meet the demand of the system.

4.4 Electrical System (ES)

A failure in the Uninterruptable Power Supply (UPS) required the presence of a technician from Siemens (Germany). Four days were needed to have the UPS back in operation.

	1983	1984 *	
HFS	390 (11.2%)	983 (42.3%)	Heliostat Field Subsystem
SHTS	1198 (34.4%)	432 (18.6%)	Sodium Heat Transfer System
PCS	951 (27.3%)	466 (20%)	Power Conversion System
ES	509 (14.6%)	338 (14.5%)	Electrical System
DAS	64 (1.8%)	32 (1.3%)	Data Acquisition System
AS	370 (10.6%)	72 (3.1%)	Auxiliary Systems
TOTAL	3482	2323	

* 1984 data are for the first nine months

Table 3.4.-1: Man-Hour Distribution for CRS Maintenance

SECTION 4

HELIOSTAT FIELD PERFORMANCE

HELIOSTAT FIELD PERFORMANCE

INTRODUCTION

The heliostat is the first subsystem in a central receiver solar energy collection/concentration system; consequently, it is the first sub-system evaluation report contained in this report of the SSPS Central Receiver System.The heliostat field was designed and manufactured by the Martin Marietta Corporation.The original field design consisted of 160 heliostats with 39.3m^2 mirrow surface each;however, as a result of a shortage of money during construction, only 93 heliostats were provided and installed. This caused a mayor change in the definition of the threshold for operation of the system and a change in the design point insolation level from 700 w/m2 to 920 w/m2.In addition, this changed the ability of the plant to operate in all of the original modes of operation. This is addressed, in detail, in the evaluation of system performance, Section7, and Potential for Improvements, Section 8.

The first specific evaluation report of this evaluation topic was prepared by J.Ramos and P.Wattiez. With a title, HELIOSTAT FIELD HISTORY AND STATUS, the report describes all the observed events with this field, analyzes the effect these events had on system performance, and reports on the field status. The principle result of this evaluation was to identify areas where significant improvement could be made in design and construction of an advanced heliostat field system.

The concluding observations or recommendations are:

1- The control system must be automated and more flexible.

2- The automated control system must be influenced by the receiver performance such as output temperature.

3- Recovery from power system failure must be rapid.

4- Lightning protection must be provided.

5- Mirrow corrosion is a major problem, at least for the particular mirrow module design used here.

6 - A simple, cost effective, washing system is necessary.

7 - Local maintenance is a must in remote areas.

8 - A beam characterization system would be helpful.

Calculations of the heliostat field performance have been performed by most central receiver system designers using computer codes such as Helios, Mirval, Delsol, etc. The calculations result in values and distribution of energy available at the receiver, but the accuracy of the result is dependent on knowledge of specific heliostats that are in operation, reflectivity of the mirror field, atmospheric conditions, etc. Most central receiver installations have made extensive efforts to measure the energy and its distribution at the receiver interface in order to validate the heliostat field calculations and to obtain realistic input for receiver efficiency determinations. In general, these efforts have experienced extensive equipment difficulties and indications of large errors in the data collected, with the result that most efforts have been abandoned. W. Schiel and G. Lemperle developed a remote flux measuring system that provides frequent calibration in the range of the measurement, and, therefore, energy flux levels and distribution that are accurate within a few percent. This effort is reported in, MEASUREMENTS AND CALCULATIONS ON HELIOSTAT FIELD PROPERTIES OF THE SSPS CENTRAL RECEIVER SYSTEM AT ALMERIA, SPAIN.

The conclusions of this evaluation are:

1 - The measured heliostat field efficiency is in agreement with the calculated values.

2 - Heliostat beam quality measurements disagree with calculated values because of larger than expected sunshapes.

3 - Circumsolar measurements are important in efforts to measure heliostat field performance.

The heliostats installed at the SSPS are essentially the same as the 1818 heliostats that are installed at the central receiver facility SOLAR ONE, in California. Performance of this much larger field is of value to SSPS evaluation and can provide some interesting guidelines in the design of new systems. C. L. Mavis and J. J. Bartel prepared a summation of the Sandia National Laboratories evaluation of the Martin Marietta heliostats installed at Solar One. The report, 10MWe SOLAR THERMAL CENTRAL RECEIVER PILOT PLANT - HELIOSTAT EVALUATION, is included here to support the ITET evaluation.

The conclusions are:

1 - In 1983, heliostat availability (1818 heliostats) was 94 to 99%.

2 - Maintenance hours for 1983 was 160 manhours/month.

3 - Mirror module vents are being installed to control (reduce) corrosion.

4 - The BCS system has been improved and circumsolar (sunshape) measurements are now being made.

HELIOSTAT FIELD HISTORY AND STATUS

Pierre Wattiez, ITET and Juan Ramos, Sevillana

1. INTRODUCTION

This paper presents the evolution of the SSPS heliostat field subsystem during three years of operation. The status of the field is given by a set of performance characteristics which include reflectivity, mirror corrosion, beam alignment, and availability.

Technical problems which have had major influence on the performance of the heliostat field are described, and a survey is given of the experience gained by the Plant Operating Authority.

Finally, conclusions are drawn from these experiences regarding factors that should be taken into account in the design of future heliostat fields.

2. OPERATION

2.1 Field Description

The CRS heliostat field is composed of 94 heliostats. Ninety-three are second generation heliostats manufactured by Martin Marietta Corp., with a reflective area of 39.30 m^2 (12 mirror modules of 1.079 x 3.035 m each). The ninety-fourth uses the same MMC structure, but the reflective assembly is made with Belgian high reflectivity mirrors.

The whole field is controlled by the heliostat array controller (HAC). The HAC computer communicates with all the heliostat field controllers (HFC). Each of the four HFC's distributes messages and commands from the HAC to up to 32 heliostat controllers (HC), and sends status of the field to the HAC. The heliostat field controllers have a limited capability for control of the field in case of failure of the main computer.

The field is designed to be fully operational for wind speeds of 144 km/hr.

2.2 Operational Sequences

The basic operation sequence starts by loading all the heliostat controllers with the initialization data, once the HAC computer is on-line. After finding the marks in the incremental encoders, the field is sent to STANDBY. In this mode the heliostats track a point located 14 m to the east of the receiver. From this position the heliostats can be sent to TRACK mode. In normal operation, the heliostats move from STOW to STANDBY through a safety corridor. In this corridor the heliostats are continuously controlled to make sure that they do not create any hazard for the plant personnel. In case of emergency (DIVE command), all heliostats switch both motors to high speed in the direction required to go to STOW, without taking into consideration any safety corridor. A warning siren is activated during this process.

Heliostats are automatically sent from TRACK to STANDBY by an interlock system whenever there is a serious disturbance in the sodium heat transfer system.

2.3 Operation Strategy

The heliostat field has always been operated with the aim of providing as much energy as possible to the heat transfer system. When the SHTS has not been available for technical reasons or during cloudy days, the field has been operated keeping the heliostats in STANDBY in order to gain a maximum of operational experience and reliability data. With this objective in mind, the electronic controllers of the heliostat field have been powered up to one hundred percent of the time.

Since May 1983, during inactive periods, the field has been sent to the wash position (vertical facing north) instead of to the STOW position whenever meteorological conditions allowed it. This has been done as an attempt to decrease the effects of corrosion of the reflective surface due to the presence of water inside the mirror modules.

Table 1 gives the total number of hours the heliostat field has been in each position since 1982.

	1982	1983	1984
STOW & WASH	5945	7249	4800
STANDBY	2597	1132	962
TRACK	668	379	814
TOTAL	8760	8760	6576*

* Only the first nine months of 1984 are considered

Table 4.1.-1: Number of Hours in Each Operating position

The aiming points of the heliostat field have been adapted to the needs of the two receivers tested in the SSPS plant. The Sulzer (cavity) receiver had only one aiming point in the center of the aperture plane. The ASR (flat) receiver has three aiming points to obtain a more uniform distribution of thermal flux.

Due to the low solar multiple of the heliostat field, all heliostats available are tracking the receiver in normal operation. The only case when heliostats have to be selectively defocussed is early morning and late in the afternoon, to avoid excessive temperature in the receiver frame. This effect is due to the aberration of the mirrors for large angles of incidence. Heliostats to the east of the field have to be defocussed in the morning, and those in the west in the afternoon. The effect is of course more noticeable in the summer.

Since the beginning of 1984 the field is sent to STOW when wind gusts of more than 65 km/hr happen three times in five minutes, or when a peak wind speed of more than 80 km/hr is observed. Before the end of 1983, the criterion for sending the field to STOW was to have a peak wind speed of more than 50 km/hr.

3. PERFORMANCE

3.1 Reflectivity and Washing

Fig.4.1-1 shows the evolution of the heliostat field reflectivity since March 1982. The reflectivity drop is faster in summer than in winter, due to the absence of rain and the larger amount of dust in the air during the summer months. Washing is done using a tank mounted on a truck. Usually the field is washed when the reflectivity is below 80%. On several occasions, due to work overload of the maintenance team, washing has been delayed and the reflectivity has dropped to lower values. On the average, washing the field requires 24 man-hours and 4.5 m^3 of demineralized water.

3.2 Mirror Corrosion

Corrosion of the silvered reflective surface of the mirrors of the heliostat field was first observed in October 1982, after two years of exposure to the SSPS meteorological conditions. A first survey was made in January 1983 to determine the number of mirrors corroded. At this stage 141 modules were affected by varying degrees of corrosion. A second survey was carried out in February 1984. The number of modules presenting some degree of corrosion had increased to 280 (25% of the modules). The last survey was conducted in September 1984. It gives a total of 299 modules showing signs of corrosion. A comparison of the results of the last two surveys is given in Fig.4.1-2 and-3. Fig.4.1-2 compares the number of times corrosion was detected in each of the twelve facets of a heliostat. It can be seen from this figure that corrosion appears more frequently in the left side of the heliostat, and that the frequency of corrosion increases from top to bottom. Fig.4.1-3 shows the position in the facets where corrosion appears. Corrosion frequency decreases as we approach the vent holes. It is smaller in the upper part of the facets.

The effects of corrosion have been evaluated quantatively after each survey. The results and the trends are shown in Fig.4.1-4, and compared to those obtained in Solar One from similar measurements. Even when the current percentage of area corroded is negligible (0.02%), if the exponential trend shown by available data continues, in only four years 10% of the total reflectivity surface will be rendered useless by corrosion.

3.3 Beam Alignment

A survey was made of the alignment of the heliostat field in December 1983. Since no beam characterization is available in the SSPS plant, the results have to be qualitative and subjective. The procedure followed was to send the heliostats one by one to the TRACK position with the receiver doors closed. All heliostats were examined in an interval of two hours around solar noon, to minimize the effects of aberration. At the same time pictures were taken of each image. Only nine heliostats needed realignment.

3.4 Maintenance and Availability

The policy followed for the maintenance of the heliostat field has been to perform most of the maintenance work on site. Since very early in the operation of the heliostats, it became apparent that the main task was the maintenance of the electronic controllers. This became even clearer after the lightning strike of May 1982 that damaged nearly 25% of the heliostat controllers. A simple test bed for both HC and HFC electronic cards was designed and constructed by a local electronic company. The system was completed in February 1983 for the HC cards and in July 1983 for the HFC cards. This led to faster and cheaper maintenance of the controllers. The effects of this test system can be seen in the availability values from March 1983 to May 1984. More impressive is the fact that after the second lightning strike, with 75% of the heliostat controllers damaged, it took only three weeks for the SSPS team to have more than 90% of the field back in operation. Taking into account the remote location of our plant and the scarce resources available, it was a remarkable feat.

Fig.4.1-5,6 and 7 show the field availability in percentage since May 1982. The periods with lower availability between milestones 3 and 4, and 6 and 7 correspond to periods where lower availability of the heliostat field was not critical due to repair works being performed in other related subsystems.

Tab.4.1-2gives the monthly average availability and number of failures from May 1982 to August 1984. Tab.4.1-3 shows the components that were replaced during the same period. It can be seen that most of the failures are due to the electronic controllers and power supplies, and they concentrate at and after the lightning strikes (May 1982 and May 1984).

Excluding those two exceptional events, the reliability of the field can be rated as very good. Mechanical parts have behaved extremely well until now, but of course mechanical degradation has a completely different pattern from electronic component degradation. A high number of mechanical failures at this stage in the life of the heliostats would be indicative of a design error.

Even when the number of failures for the optical encoders seems high, it represents a failure rate of only 0.000003 failures/hour, or one failure every 300,000 hours. The failure rate for the heliostat controllers, excluding the damages of lightning is 0.000011 failures/hour, or one failure every 89,000 hours, which is a fairly good value, taking into account its complexity.

4. CONCLUSIONS

The following conclusions can be drawn after three years of heliostat field operation:

- The control system should include additional automation to ease the task of the field operator and it should be more flexible.

- Increased interaction with the sodium heat transfer system would be desirable to allow for such features as automatic heating up of the tanks and start-up of the receiver, as well as control of the heat flux distribution.

- The time needed to recover from a power failure produces an excessive loss of operation.

- Local maintenance of the electronic controllers has led to better field availability.

- Lightning protection of the controllers is a must and should be considered early in the design of a heliostat field.

- A beam characterization system would be of great help to define and control heliostat tracking errors.

- Mirror corrosion may affect significantly the power sent to the receiver in the near future (3 to 4 years) if the current growth rate continues.

- An improved washing procedure is needed, to reduce the manhours needed for field washing and to improve the reflectivity value after washing.

MONTH	1	2	3	4	5	6	7	8	8	10	11	12	
Average Availability %				96.7	64.1	97.3	96.4	95.2	93.1	92.6	97.0	96.4	1982
Number of Failures				28	7	14	4	2	2	1	1	5	
Average Availability %	1) 82.9	1) 81.7	99.5	99.5	2) 82.9	2) 81.7	2) 95.7	99.5	97.6	96.5	97.9	99.1	1983
Number of Failures	4	0	4	2	0	0	3	3	3	3	2	1	
Average Availability %	99.4	99.5	99.5	99.0	53.9	94.0	93.3	97.2					1984
Number of Failures	0	2	5	3	73	8	4	4					

1) Cold tank repair
2) ASR installation.

Table 4.1.-2: Monthly Average and Number of Failures (April 1982 - August 1984)

YEAR	1982			1983				1984			
QUARTER / COMPONENT	2	3	4	1	2	3	4	1	2	3	TOTAL
HC	27	-	1	5	2	3	4	1	72	6	121
HFC	-	-	-	1	-	1	-	-	4	-	6
AZ ENCODER	2	1	1	2	-	-	-	1	1	1	9
EL.ENCODER	-	-	-	1	-	-	-	2	2	-	7
AZ.LIMIT SWITCH	-	-	-	1	-	-	-	1	-	-	2
EL.LIMIT SWITCH	1	-	-	-	-	1	-	-	-	1	3
AZ.MOTOR	1	-	-	-	-	-	-	-	-	-	1
EL.MOTOR	1	-	-	-	-	-	-	-	-	-	1
5V POWER SUPPLY	8	-	-	-	-	-	-	-	49	-	57
MECH.RELAY	-	1	-	-	-	-	-	-	-	-	1
SOLID STATE RELAY	1	-	-	-	-	-	-	-	-	-	1
WIRING HARNESS	2	1	-	-	-	-	-	-	1	1	5

Table 4.1.-3: Components Replacement (April 1982 - September 1984)

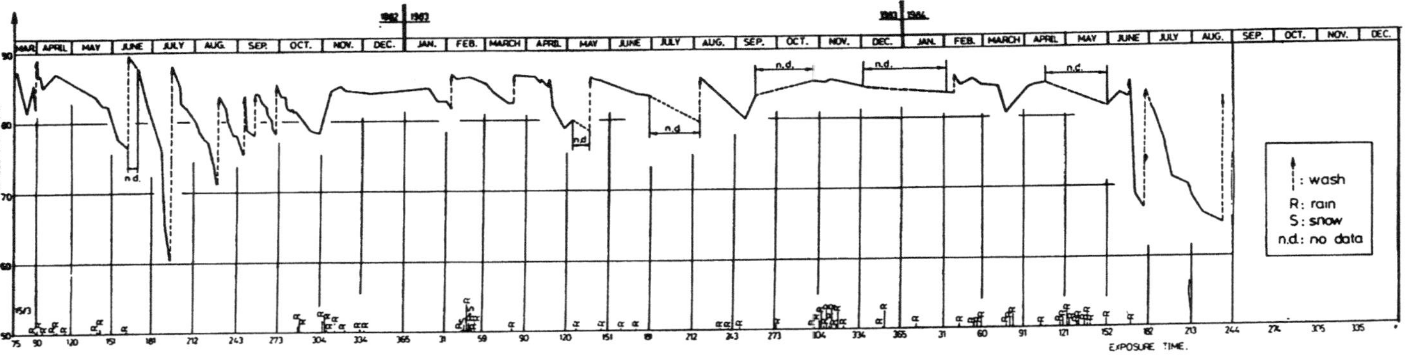

Fig. 4.1.-1: HELIOSTAT FIELD REFLECTIVITY CURVE

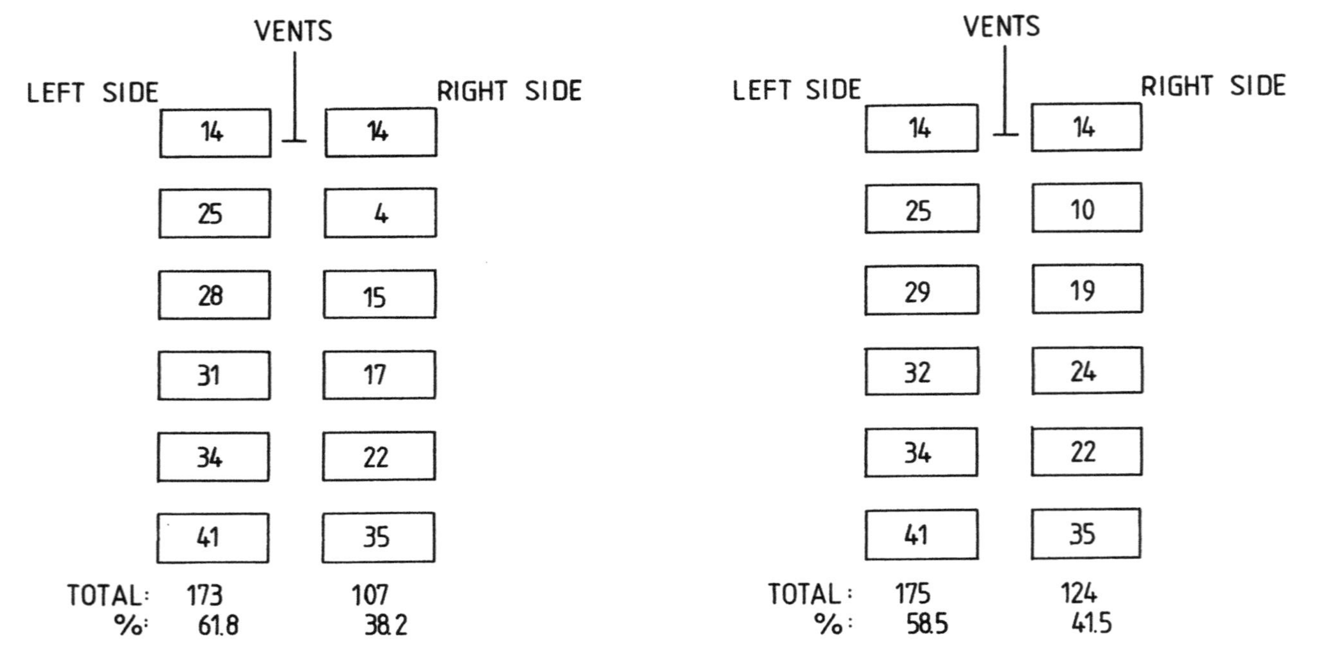

Fig. 4.1.-2: Total of Heliostats Which Show Some Corrosion in that Specific Location

FEBRUARY 1984

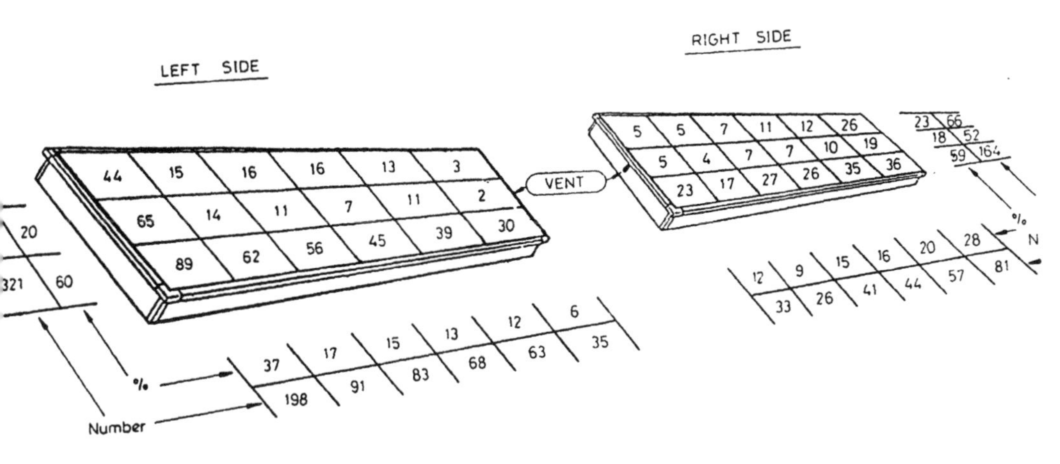

SEPTEMBER 1984

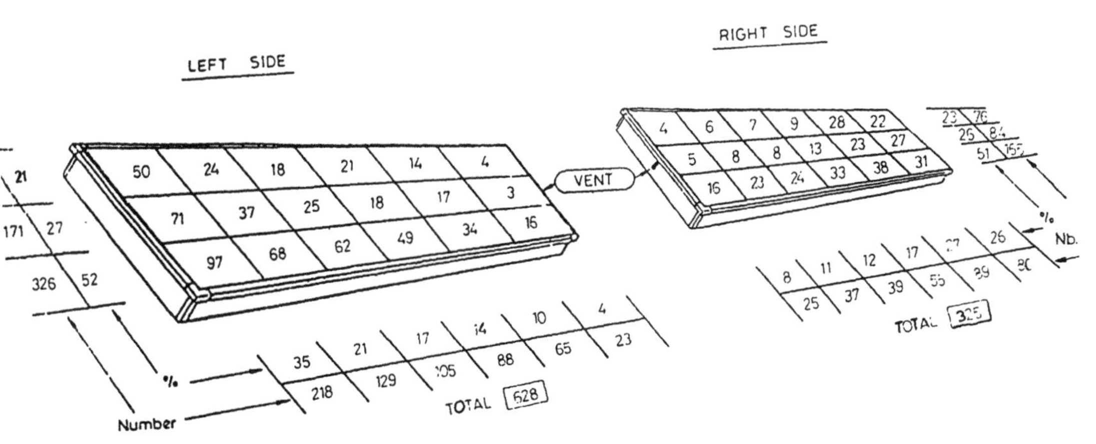

ig.4.1-3: Occurence of corrosion on SSPS mirror moduls

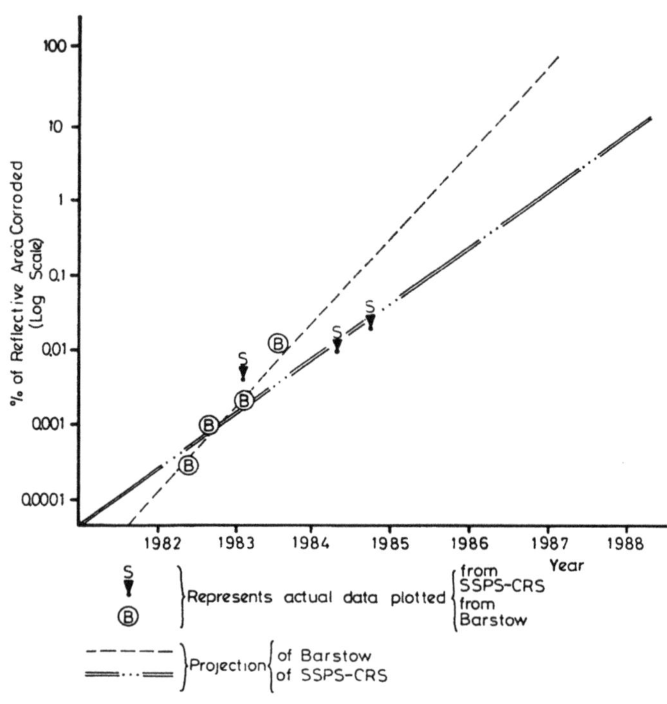

Fig. 4.1.-4: PROJECTED AMOUNT OF CORROSION AT PRESENT GROWTH RATE

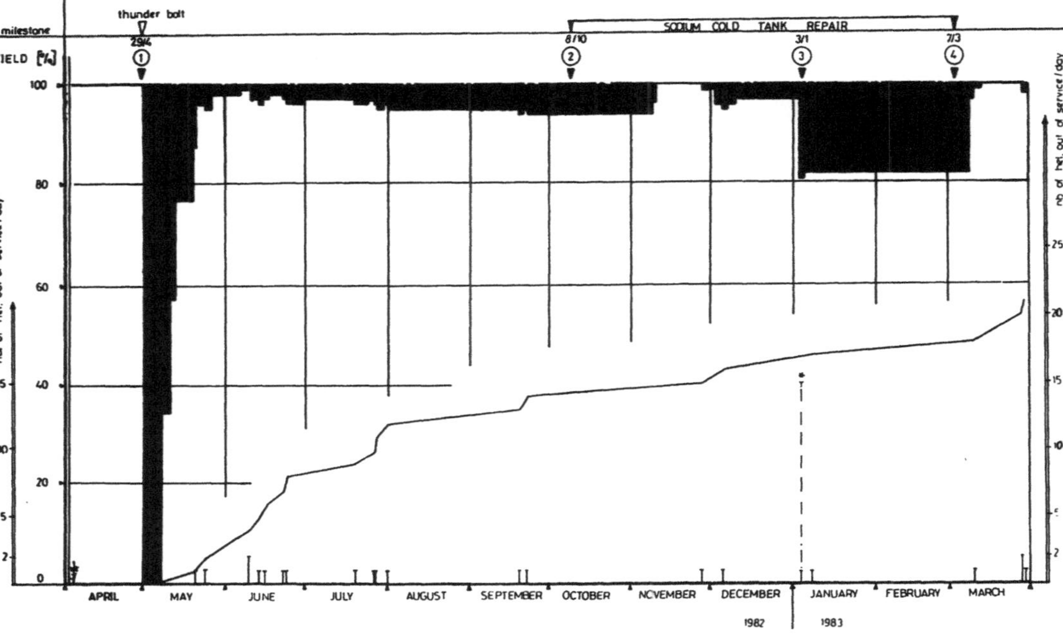

Fig. 4.1.-5: HFS Availability

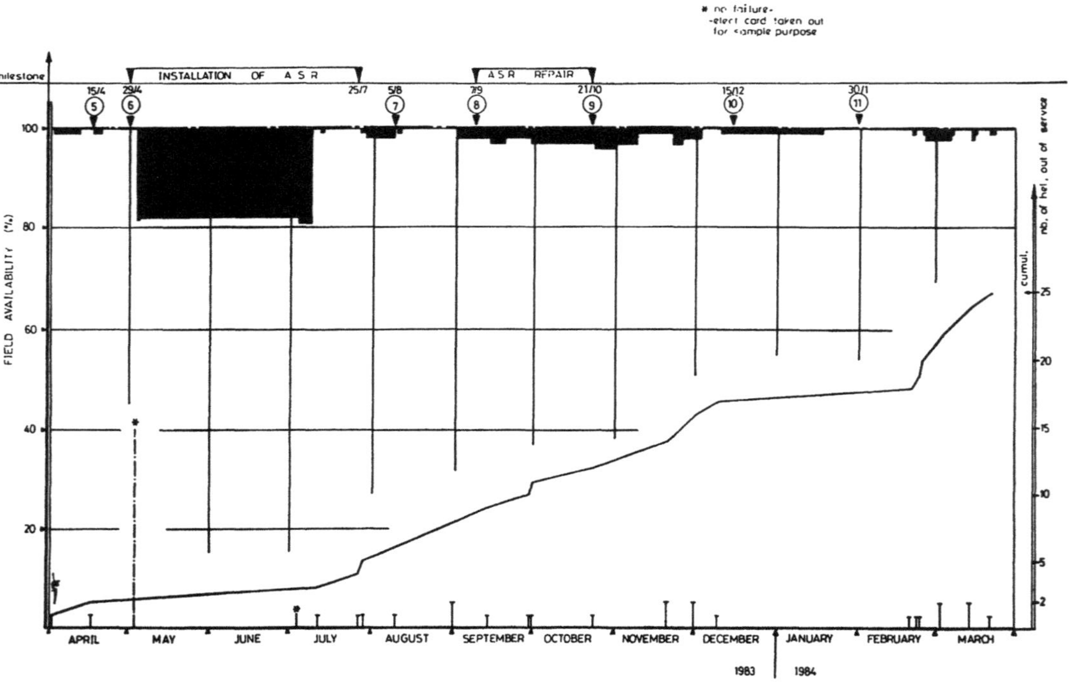

Fig. 4.1.-6: HFS Availability

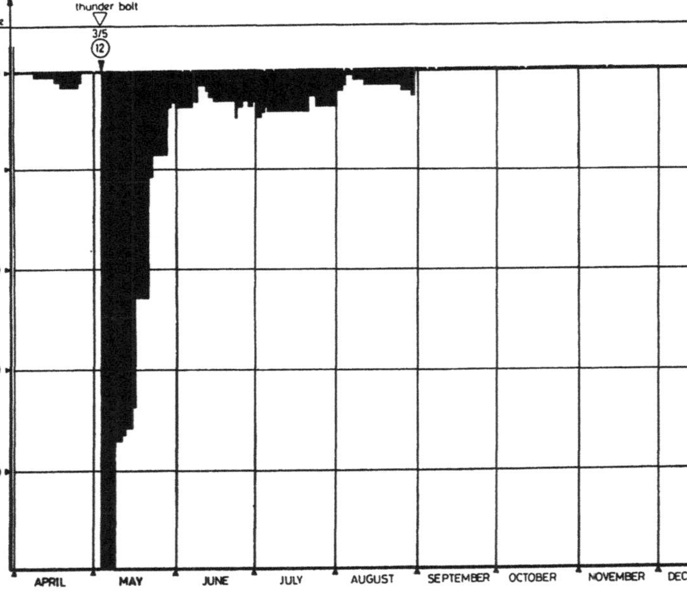

Fig. 4.1.-7: HFS Availability

MEASUREMENTS AND CALCULATIONS ON HELIOSTAT FIELD PROPERTIES OF THE SSPS CENTRAL RECEIVER SYSTEM AT ALMERIA, SPAIN

W. Schiel and G. Lemperle, DFVLR

1. INTRODUCTION

The central receiver system (CRS) at Almería, Spain, operated under the auspices of DFVLR-Cologne is a small solar power system (SSPS) (Ref.1) designed to produce 500 kW_e output at 0.92 kW/m^2 direct insolation. The heliostat field consisting of 93 individually tracking mirrors with a total reflective area of 3655 m^2, redirects up to 2.8 MW of radiative energy into a cavity receiver on top of a tower.

A prerequisite for establishing an energy balance of central receiver type solar tower systems is the availability of exact data on the loss mechanisms. The most important input parameter for the energy balance is the direct insolation measured by means of a Normal Incidence Pyrheliometer with a defined field of view.

The energy leaving the sun and arriving at the earth's surface in the form of radiation must first pass through the earth's atmosphere. There, the direct sunlight is scattered by aerosols (dust, water droplets, or small ice crystals in thin clouds). If the direct insolation is measured in radial direction to the sun's center, the intensity is found to decrease towards the limb of the sun, a characteristic which is known as the so-called limb darkening. As a result of the scattering, under certain circumstances a considerable portion of the total radiation comes from outside the solar disk. This is the so-called circumsolar radiation, the magnitude of which depends very much on the atmospheric conditions prevailing in each case.

Fig.4.2-1 gives a schematic representation of the path the solar radiation takes from the sun to the aperture of a solar power plant. The radiation coming from the angular range of the solar disk and which is not subjected to any atmospheric scattering is reflected by the heliostat onto the receiver. If the aperture has an appropriate geometry, this radiation will pass through the aperture plane within the aperture rim. Some rays coming from the sun's angular range are deviated by the atmospheric scattering such that they do not hit the heliostat at all. On the other hand, some solar rays are only just deviated by the atmospheric scattering onto the heliostat. The heliostat sees this radiation as if it came from outside the angular range of the solar disk. A portion of this circumsolar radiation is also reflected by the heliostats onto the aperture. This portion depends essentially on geometrical dimensions of the aperture.

Besides optical errors of the concentrating system, the circumsolar radiation determines the size and shape of the solar image at the receiver.

The distribution of the radiant flux in the receiver aperture plane and the total power into cavity depends on tracking accuracy of each heliostat, waviness, alignment and reflectance of the mirrors, sunshape, and actual windspeed and direction. In order to verify theoretical calculations with the HELIOS computer code (Ref.16) and the impact of different sunshapes on the performance of the heliostat field receiver it is of interest to measure the incoming power and its distribution in the aperture plane.

The Heliostat and Receiver Measuring System (HERMES) (Ref.3) has been developed to evaluate the heliostat field and receiver and to improve both solar specific subsystems. HERMES measurements were carried out on the SSPS central receiver system during the Fall Measurement Campaign in 1982.

2. EXPERIMENTAL

2.1 HERMES Measuring System

During 1981 and 1982 the HERMES measuring system was developed by DFVLR-Stuttgart. This system is able to determine the optical power characteristic of single heliostats and the radiant energy redirected by a complete heliostat field onto a target in front of the receiveraperture. It is quite similar to the American Beam Characterization System (Ref.4) and the Swiss Flux Analyzing System (Ref.5); however it differs in that it measures the output of a complete heliostat field in real time. The HERMES system, shown in Fig.4.2-12, is subdivided into three parts:

a) video camera system (VCS)
b) computer system (CS)
c) data acquisition system (DAS)

a) The video camera system consists of a special video camera using a 26-mm magnetic focus and deflection video type camera tube with a silicon diode array target structure. It also has a 300-mm lens and a video frame memeory for real-time processing. The video camera signal is first fed into a fast 8 bit A/D converter where the analog video signal is digitized at a rate of 6.34 MHz to give 256 conversion points along the horizontal and vertical scanning line. The digitized video signal is then fed into different multiplexers and Alu's where the real time processing is performed (e.g. averaging, integration, subtraction, non-destructive subtraction, etc.). The video picture is stored in a memory plane which is organized as 256 x 256 x 16 bit. This additional memory depth allows 256 images to be summed up in real-time without losing any intensity resolution. A rapid transfer of the preprocessed results to the host computer is accomplished by direct memory access (DMA).

b) The video camera system is connected via a DMA interface to a PDP 11/44 computer. The computer system is equipped with a 640 KB memory, two 10 MB discs and a 1600 bpi magnetic tape unit, two graphical CRT-terminals for control interactions, and a printer/plotter for graphical data presentation. The computer operating system is a real-time multiprogramming, multitasking executive. Maximum utilization of the video camera hardware features is achieved by software which resides in the host computer. This software provides efficient program control of data acquisition, real-time processing, and input/output between the computer and the video frame memory. Additional software programs were written for data handling and post evaluations as there are two- and three-dimensional flux diagrams, iso-intensity contours, and power versus radius diagrams.

c) The microprocessor controlled data acquisition system is connected via a serial interface (RS 232) to the computer to record all relevant meteorological data needed fo the evaluation of the flux images: actual beam, global and diffuse irradiance, windspeeds and direction, humidity, and ambient temperature. The complete system has been installed in a container with all meteorological instruments on top.

The irradiance E at an aperture point x,y coming from the heliostat field is obtained by measuring the reflectance of E at a Lambertian target screen which has to be placed directly in the beam. This concept was realized by having a 40-cm wide target traversing the receiver aperture plane in a very short time (5 to 8 sec) (Ref.5). The 40-cm width was necessary to reduce the high thermal load changes experienced by the receiver tubes.

Thus, the camera signal V can be related to the target irradiance E at the traversing pixel (x,y) by the expression:

$$V(x,y) = K\, E(x,y)\, p(x,y)\, S(x,y) + N(x,y)$$

N(x,y) can be determined easily by placing the lens cap on the camera and recording the camera output signal. The mean parameter for this dark current is the temperature. The irregular noise distribution is mainly caused by different thermal conductivity at the silicon diode array.

Ideally, S(x,y) would be the normalized camera signal (exclusive of camera noise) produced by a uniformly radiant target screen. This can be done by use of an Ulbricht-sphere (Ref.8,9). The system constant K will then be determined using the known value of E from the screen-mounted flux gauge together with the measured value V from the surrounded screen, and R, S, and N. Taking into account a 3% non-conformity of p(x,y) over the target (Ref.6), a 2% innaccuracy in the Kendall Radiometer, a 2% inaccuracy in thecamera signal V(x,y), a 2% inaccuracy in the dark current N(x,y) and shading S(x,y) measurements, the accuracy of the flux measurements is better than ± 5%. After the measuring campaign through a two month period no change of the reflectance of the traversing bar within the error margin of 3% could be noticed.

2.2 Measurements

During August 1982, the HERMES measuring device was transferred to the Central Receiver System (CRS) of the SSPS project and installed behind the last heliostat row in the north-south direction about 200 m from the tower. The main goals of the measurement campaign were to determine heliostat field tracking accuracy, beam quality, and heliostat field efficiency with respect to the time of the day and different insolation conditions. All of this information can be deduced from the measured flux distribution in the aperture plane. Thus, most of the measurements were concentrated on the flux determination.

In the experimental run, motion of the coated target bar is steered by a radio signal and controlled by the computer system. Just before the target gets its start command, the dark current N(x,y) and the shading S(x,y) are measured to establish the linear region and to compensate for drift and aging of the camera and the digitizer electronics. The dark current picture is obtained by capping the camera lens and adjusting the offset of the camera control unit to obtain the digital value previously determined to be the lower extreme of the system's linear operation region. The white file is obtained by pointing the uncapped camera system into the integrating sphere.

Next, the four corners of the receiver aperture are identified via cursor interaction and stored in a geometric file for off-axis camera angle correction. A radio signal is sent by the computer to the control box on top of the tower, starting the moving target to cross the receiver aperture in about 5 to 8 seconds, scanning an area about 5 x 5 m. Running from west to east the target activates 33 positioning sensors mounted on the rail. As each sensor is passed, a radio signal is transferred to the computer, which in turn activates the camera to take a new picture. When the picture is taken it is first stored in the frame memory in real-time. The computer calculates x and y coordinates of the actual position of the target from a geometric data file. Then these dates are sent to the frame memory and activate the transfer to the target width to a task common in the computer. The system is now waiting for the next radio signal. The resulting 33 single stripes are composed by a selection code into a total picture of the flux distribution in the measurement plane. Before and after each measurement the computer asks for the meteorological data, which is composed of direct, diffuse, and global irradiance; wind speed and direction; ambient temperature and humidity; sun position; and time. These additional data are stored in the head of each picture and used in the final evaluation.

Before each flux measurement, the system constant K is measured by a second, water-cooled bar that slowly traverses through the aperture plane. (This bar was originally to have been equipped with 12 heat flux gauges and used for measurement of the radiant flux) (Ref.10). The absolute, self-calibrating, water-cooled Kendall radiometer is attached to one side of this bar and surrounded by a screen coated with the same Lambertian material as the moving target. This bar is driven by a stepping motor to three different positions in the aperture plane with low, middle, and high radiant fluxes. The absolute irradiance and the response of the camera to the surrounding screen are both measured. This calibration gives the conversion coefficient of the relative grey levels measured by the camera in absolute irradiance values. Finally, the original video data are converted into absolute flux values using the measured constant K, the dark current, the shading factors, and the corners of the aperture.

3.1 Effect of Sun Shape on Flux Distribution

To investigate the effect of sun shape on flux distribution of the SSPS-CRS Sulzer receiver, for typical shapes with circumsolar ratios from

C/(C+S)=.01 (corresponds to Kuiper distribution) up to C/(C+S)=.24 were selected. The investigation was performed by the mean of the HELIOS computer code (Ref.16). As outlined in Ref.11, the sun shape describes the angular distribution of radiant intensity. Although this distribution does not describe an error distribution as, e.g., waviness of the mirrors or sun tracking errors, it represents a stochastic process, namely the origin of the photons on the solar disk. In HELIOS, the sun shape function is convolved with the error distribution of the mirror system to yield the 'effective sun shape'. Its shape only depends on the shapes of the distribution functions. Therefore the algorithm applied in HELIOS can use normalized distributions. Multiplication by the corresponding dimensional quantities is only performed when determining the absolute radiation load on the target.

Out of the large number of HELIOS calculations, two examples will demonstrate the effect of sun shape on flux distribution. In Fig.4.2-2 the flux density distribution for the design conditions of the Sulzer receiver (ideal atmospheric conditions: 920 W/m^2, C/(C+S)=0.01 Kuiper distribution) is presented by a 3-D plot. The plane marked with 'horizontal' and 'vertical' is identical with the aperture plane of the receiver. The plot allows a qualitative assessment of the flux distribution and spillage. For the design conditions the spillage appears to be negligible. On the other hand,Fig.4.2-3 shows the flux distribution for the same conditions only by assuming that we now have unfavorable atmospheric conditions. This condition is simulated by using a sun shape with a large circumsolar ratio and a smaller insolation value (C/(C+S)=.24; Insolation=600W/m^2) for the HELIOS calculation. Since a qualitative comparison would be difficult due to the different insolation, this effect has been suppressed by a corresponding correction in the plot program, i.e., the relative differences in the magnitude of the flux distribution are only caused by the sun shape. The flux distribution produced by the sun shape with the larger circumsolar radiation is obviously broader. For target points outside the aperture, the flux densities were set (equal to) zero. So the aperture contours become visible in Fig.4.2-6. That means the aperture rims out of a certain part of the concentrated radiation.

To allow more qualitative assessments, the flux distributions are also plotted by isolines. The diagrams in Fig.4.2-4 and 5 correspond to a 3.5 x 3.5 m surface located in the aperture plane (it is represented as it would be viewed by an observer looking at the receiver from north to south). The x-axis is parallel to the horizontal and the y-axis parallel to the vertical. Both are scaled in meters. The aperture is characterized by the dashed octagon, and the point x=0, y=0 marks the aperture center. The contour lines correspond to a certain percentage of power which is within the specified curve. For example, 90% of the concentrated power within the curve are marked by the number '5', 70% within the curve '4', and so on. At the right margin of the diagram, the contour lines are designated and additional information is provided. The value named 'POWINT' gives the total power on target; 'APTPOW' specifies the power passing within the aperture rim. The curve diameters characterize the flux distribution. The larger diameters in Fig.4.2-8 demonstrate the broadening of the flux distribution caused by circumsolar radiation. For more results concerning flux distribution, see Ref.11.

3.2 Effect of Sun Shape on Intercept Factor

The intercept factor plays a non-negligible role in the stair step diagram of the SSPS-CRS. Equation (1) defines the field efficiency n_f, which is composed of the efficiencies of various subsystems and determined by Equation (2).

$$P_{apt} = NI \, . \, NHEST \, . \, A \, . \, n_f \tag{1}$$

$$n_f = n_{fico} \cdot n_{refl} \cdot n_{shbl} \cdot n_{atat} \cdot n_{apef} \tag{2}$$

where P_{apt} = power incident into cavity (W)
NI = direct insolation measured by pyrheliometer (W/m^2)
NHEST= number of heliostats
A = reflecting area of one heliostat (m^2)
n_f = field efficiency
n_{fico}= cosine factor of the field
n_{refl}= reflectivity of the heliostats
n_{shbl}= shading and blocking (including shadow cast by tower)
n_{atat}= atmospheric attenuation
n_{apef}= intercept factor and/or aperture efficiency

Each individual efficiency is a multiplicative factor of the overall efficiency, i.e., reduction of any efficiency by 10% results in an overall efficiency decrease by 10%. Hence, according to Equation (1), this reduction would mean 10% less radiant energy gets into the aperture.

The intercept factor depends on several effects described by the following relationship

η_{apef} = f (CRS design, optical behavior of heliostats including errors, — solar — day — and — time, — sun — shape)

Based on flux distribution calculated by HELIOS, the intercept factor has been determined by Equation (3)

$$\eta_{apef} = \frac{P_{apt}}{P_g} \tag{3}$$

where P_{apt} = radiant power into aperture (calculated from flux distribution by integration)

P_g = total solar radiation reflected from the heliostats, from an angular range corresponding to the pyrheliometer, reduced by the power loss due to atmospheric attenuation.

Using the calculated flux distribution, a correlation between circumsolar radiation and intercept factor is represented for the Sulzer receiver in Fig.4.2-6 and for the ASR receiver in Fig.4.2-7. Each curve in Fig.4.2-9 and 10 corresponds to a time, as stated at the diagram's right margin. The set of the three curves has a relatively equidistant characteristic, which suggests that the effect of the circumsolar ratio on the intercept factor is essentially of additive nature.

Fig.4.2-8 and 9 show the dependence of the intercept factor on the solar time for the Sulzer and ASR receiver with the circumsolar ratio as the parameter. Due to the symmetry of the heliostat field, the intercept factors must be also symmetrical to the culmination time (12:00 h). The locations marked with symbols correspond to the calculated values. A circumsolar ratio, indicated at the right margin, can be coordinated to each curve. The parallel characteristic of the four curves refers to the above mentioned additive effect of the circumsolar ratio on the intercept factor. The curve corresponding to circumsolar ratio C/(C+S)=.24 shows that even around noon an additional spillage of 12% has to be expected. Hence the loss is of the same order of magnitude as the cosine efficiency and cannot be neglected for the energy balance. In order to demonstrate how the intercept factor changes in the course of the year, the time-of-day characteristic of the intercept factor for the 355th, 20th, 111th, 141st, and 172nd day is plotted in Fig.4.2-10 and 11. Using three different sun shapes, this results in three sets of curves. Early or late in the day the set of curves broadens, whereas around noon the values are clustering. Coordination to their day is indicated by the dashed arrows. The uppermost curve of a set corresponds to the 355th day, the next one to the 20th day, etc. The lowest corresponds to the time-of-day characteristic at summer solstice (172nd day). This order is produced by the increasing optical aberration towards the middle of the year.

4. RESULTS AND DISCUSSIONS

An evaluation of one complete measurement set is presented in Fig.4.2-13. At the top left a three-dimensional intensity diagram of the corrected video picture is shown. In addition to the meteorological data, the actual sun position and time measured during the test are shown. The second graph shows the iso-contours of the incident power distribution, a theoretical aim point of the heliostat field, the center of gravity of the measured distribution, and the aperture size projected in the measuring plane. The third plot shows the intensity cross section through the centroid in vertical direction. The power versus radii graph was calculated by summing up the power inside concentric cycles around the centroid. The total power was computed by integration over the total measuring plane (5 x 5 m).

The deviations of the measured center of gravity from the aim point in the vertical and horizontal directions yielded the reflected beam tracking error. Fig.4.2-14 shows this error with respect to local time on two different days. It can be seen that the center of gravity moves 10 cm east and about 15 cm down in the morning to 10 cm west and 5 - 10 cm up in the evening. This movement depends on both the position of the heliostats and actual wind loads. The dependence on weather conditions could not be identified because of too little data obtained during the measurement campaign.

For characterizing the beam quality of the heliostat field, the 90 and 95% radii are plotted as a function of solar time in Fig.4.2-15 and 16. For comparison, the theoretical radii calculated with the HELIOS computer code are also plotted. The 0.5 m difference between the experimental calculated results stems from two factors:

- The HELIOS calculation assumes a well adjusted heliostat field. However, the increase of the measured flux distribution indicates a maladjusted heliostat field which obviously does not affect the heliostat efficiency (Fig.4.2-19).

- The change of the circumsolar ratio (Ref.11) results in an enlargement of the reflected beam. The measurements taken on October 4, 1982 with a hazy sky, and therefore a large circumsolar ratio, demonstrates this relation. The total power measurements are not affected by these radiation conditions, but the shape of the flux distribution depends very strongly on this parameter. The 90 and 95% radii deviate much more from the HELIOS calculations when the Kuiper-sun (ideal sun) is applied than the results from a clear day like October 7. These results can be better interpreted only by measuring the sun shape during each experiment.

The measured heliostat field efficiency with respect to solar time is presented in Fig.4.2-19, together with the HELIOS calculated field efficiency for October 5, 1982 based on an ideal 100% heliostat field reflectance and an actual reflectance of 81.5% measured on October 5. Note, however, that this reflectance value is measured punctually on selected heliostats and it is not yet clear whether this value is a representative average value of the total heliostat field. Besides this boundary condition the measurements are in good agreement with the theory.

During these measurements, the software winder was programmed to take one complete picture every ten seconds. Therefore, the relative variation of the flux with respect to time could be evaluated directly on the receiver tubes. The very first results are shown in Fig.4.2-20. This figure shows two different points in the aperture at low and high fluxes. There is a maximum deviation from the average of 15%. Thus, there is even under normal conditions a continuous flux stress or flux pulsation on the receiver tubes. This phenomenon is to be studied in more detail during the next planned measurement campaign in August 1984.

5. CONCLUSIONS

The HERMES video camera measurement system has proved to be well suited for measurements of high solar fluxes from complete heliostat fields. The relevant measured data were heliostat field efficiency and beam and tracking quality of the field. The results have shown that the solar subsystems meet the design criteria during steady-state operation. Measured heliostat field efficiencies were in good agreement with values computed by the HELIOS code; however, the beam quality measurements were in disagreement due to enlarged sun shapes, which are typical for the climatological conditions of the plant's location. For more detailed measurements of the thermal receiver tube stresses the time resolution of HERMES has to be improved.

If unfavorable atmospheric conditions exist, the circumsolar radiation can reduce the intercept factor more than 10%. Not knowing the circumsolar radiation means there is no possibility to establish an exact energy balance by Equations (1) and (2). Therefore, efforts must be made to include the effect of the circumsolar radiation on the energy balance. Several approaches are feasibile:

* Based on sun shape measurements, the curve of the actual sunshape flux distribution in the aperture is calculated explicitly. By integration over the aperture, the incident power is obtained.

* Knowing the sun shape we are able to calculate the circumsolar ratio. By means of diagrams like Fig.4.2-6 and 7, the intercept factor can be determined through interpolation. Equations (1) and (2) then yield the incident power.

* The simplest approach is to measure the direct insolation with a pyrheliometer whose view angle corresponds to the effective acceptance angle of the power plant. This approach was already suggested by Brumleve (Ref.18).

The effective acceptance angle can not only be estimated by rough geometrical considerations neglecting beam quality and optical aberration, but also be determined by the more exact results of HELIOS calculations. The formal conditions for the acceptance half angle δ^* is described by Equation (5).

$$NI \cdot n_{apef} = 2\pi \cdot \int_0^{\delta^*} I \cdot \delta \cdot d\delta \tag{5}$$

where $I(\delta)$ is a measured intensity distribution. By Equation (3) we get the direct insolation NI, while the intercept factor n_{apef} must be calculated by HELIOS using the intensity distribution $I(\delta)$.

Equation (5) is based on the idea that only the direct insolation reduced by the intercept factor can be concentrated within the aperture rims. This energy has its origin in the inner part of the intensity distribution, namely up to the angle δ^*. However this acceptance half angle has to be considered as an averaged value, since heliostats placed near the tower and heliostats in the last rows do have different acceptance angles. To obtain an acceptance angle which is independent of the optical aberration and covers only the effect of the circumsolar radiation, the intercept factor must be correlated by the optical aberration component. On account of the additive impact of circumsolar ratio on the intercept factor, as identified by Fig.4.2-10, the correlation can easily be performed by means of Equation (6):

$$n_{apef,k} = n_{apef} + (1 - n_{apef,kuip}) \tag{6}$$

where $n_{apef,k}$ = corrected intercept factor

n_{apef} = HELIOS calculated intercept factor for a certain time and the intensity distribution $I(\delta)$ with circumsolar ratio greater than that of Kuiper distribution

$n_{apef,kuip}$ = HELIOS calculated intercept factor for same time and Kuiper distribution

Using the corrected intercept factor in Equation (5) yields an acceptance half angle for the SSPS/CRS design of $\delta^* = 15.7$ mrad.

6. ACKNOWLEDGMENTS

The author wishes to thank Mr. S. Merk for assisting in the measurements during the campaign and the system development, Dr. J.M. Kendall for personal help in starting the radiometer's operation, D. R. Köhne for valuable discussion, and Mr. W. Grasse, SSPS project manager, for any support.

7. REFERENCES

1. W. Grasse, M. Becker, and A. Kalt, "500 kW Central Receiver System (CRS) of the IEA Small Solar Power Systems Project (SSPS) - Almería", SAND80-8505 62 C (1980).

2. M. Becker, H. Ellgering, and D. Stahl, "CRS Construction Report", DFVLR (1980).

3. W. Schiel, "HERMES Measurements", SSPS Technical Report 4/83.

4. D.L. King, "Heliostat characterization at the Central Receiver Test Facility", Journal of Solar Energy Engineering, Vol. 103, May 1981.

5. P. Kesselring and N. Real, "Heliostat Test for Solar Thermal Power Plant", Swiss Federal Institute for Reactor Research, Würenlingen, Switzerland, 1980.

6. G.P. Görler, "Messung zum Spektralen Reflexionsgrad Sowie zur Winkelabhängigkeit der Reflexion an Einigen Weißen Oberflächen", DFVLR Interner Bericht IB 353-81/6.

7. J.M. Kendall, "Two Blackbody Radiometers of High Accuracy", Applied Optics, Vol. 9, Nº 5, May 1970.

8. A. Stimson, "Radiometric and Photometric Terms", John Wiley & Sons, New York, N.Y., 1974, p.10.

9. E.B. Rose and A.H. Taylor, "Flux Measurement. Precision Measurement and Calibration", NBS Special Publication 300, Vol. 7, US Government Printing Office, Washington, D.C., November 1971, pp. 409-281 to 454-325.

10. M. Becker, F. Diessner, and J. Bäthe, "Heat Flux Distribution Measurement Device for the SSPS-CRS Project in Almería", paper presented at the STTFUA-Workshop on High Intensity Solar Flux Measuring Techniques, SNLA, October 26-27, 1981.

11. G. Lemperle, "Effect of Sunshape on Flux Distribution and Intercept Factor of the Solar Tower Power Plant in Almería", SSPS Technical Report 3/82.

12. A. Brinner and F. Reich, "Recalibration of Relevant Temperature Sensors and Sodium Flow Meters at SSPS-CRS", to be published as an SSPS Technical

13. J.S. Kraabel, "Convection Testing of the CRS, Fall 1981", Central Receiver System (CRS) Midterm Workshop Proceedings, 1983, pp.537-558, DFVLR.

14. J.S. Kraabel, "An Experimental Investigation of the Natural Convection from a Side-Facing Cubical Cavity", Thermal Engineering Joint Conference, ASME-JSME, Honolulu, Hawaii, 1983.

15. D. Grether, J. Nelson, and M. Wahlig, "Measurement of Circumsolar Radiation, "Proceedings of the Society of Photo-optical Instrumentation Engineers, Vol. 68 (1975), pp. 41-48.

16. F. Biggs and C.N. Vittoe, "The HELIOS Model for the Optical Behavior of Reflecting Solar Concentrators", SAND76-0347.

17. D. Grether, private communication to T. D. Brumleve, June 23, 1981.

18. T.D. Brumleve, private communication to A.C. Skinrood, Subject: Measurements at Barstow, June 26, 1981.

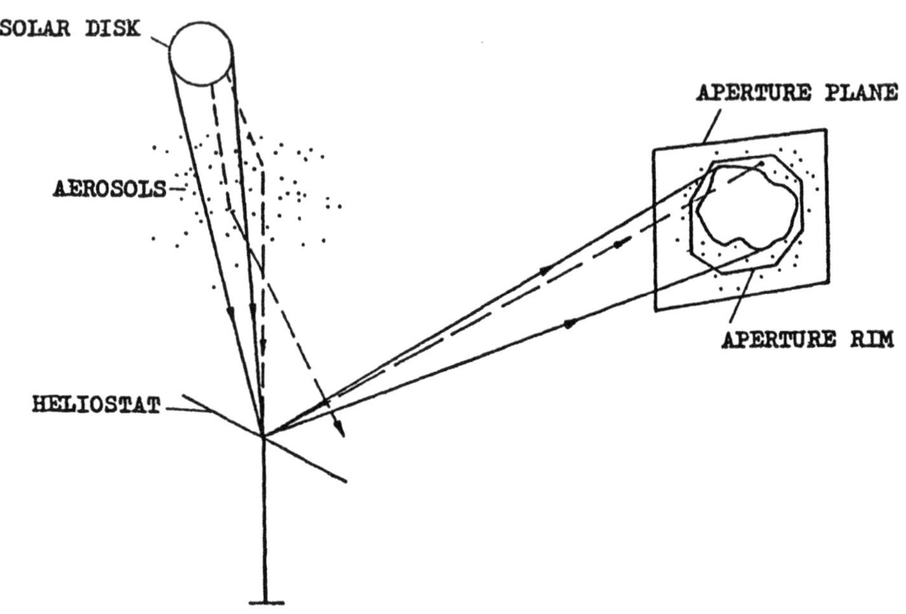

Fig.4.2-1: Schematic presentation of the sun rays (from the sun to the aperture)

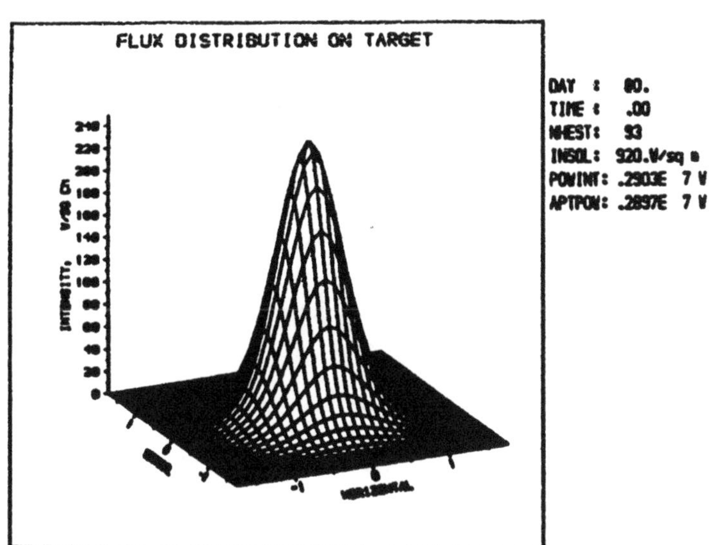

Fig.4.2-2: Flux distribution for design conditions. Generated by 'Kuiper sun shape'; NI = W/m^2 C/(C+S) = 0.01

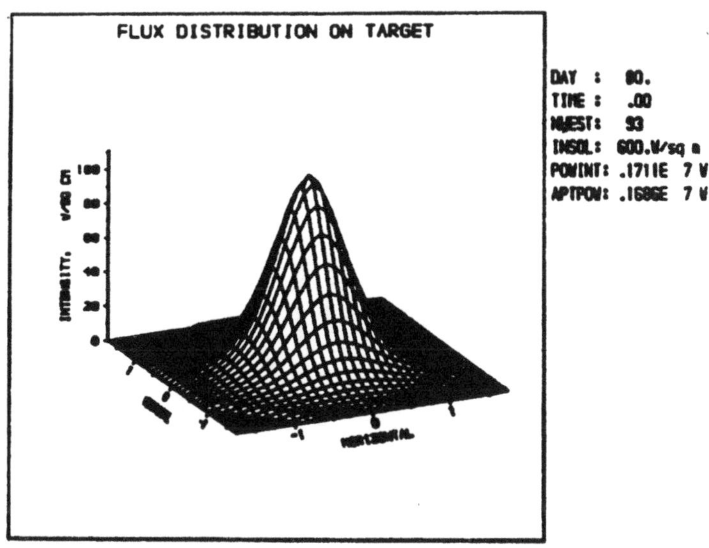

Fig.4.2-3: Flux distribution for design conditions. Generated by 'unfavorable' sun shape; NI = 600 W/m^2 C/(C+S) =0.24

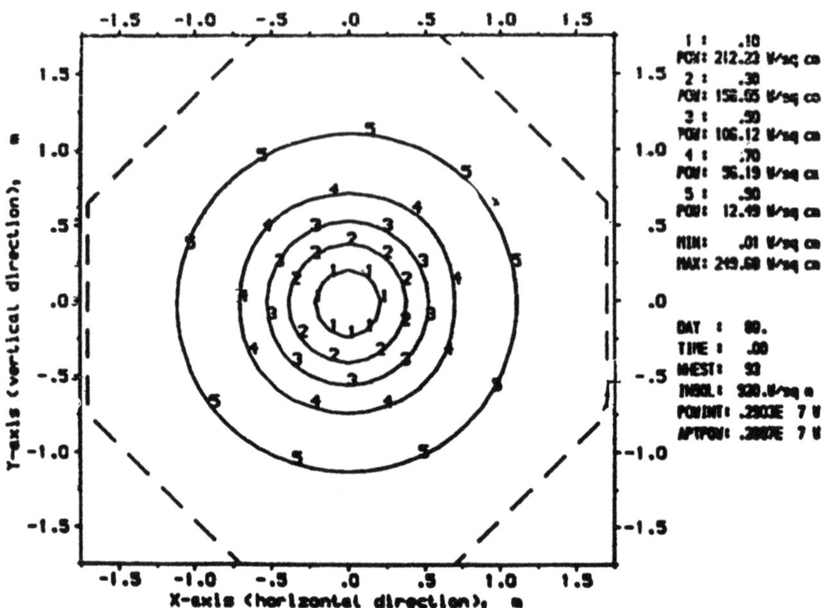

Fig.4.2-4: Isolines of flux distribution for design conditions for the Sulzer receiver. Generated by 'Kuiper sun shape'; NI = 920 W/m^2 C/(C+S) = 0.01

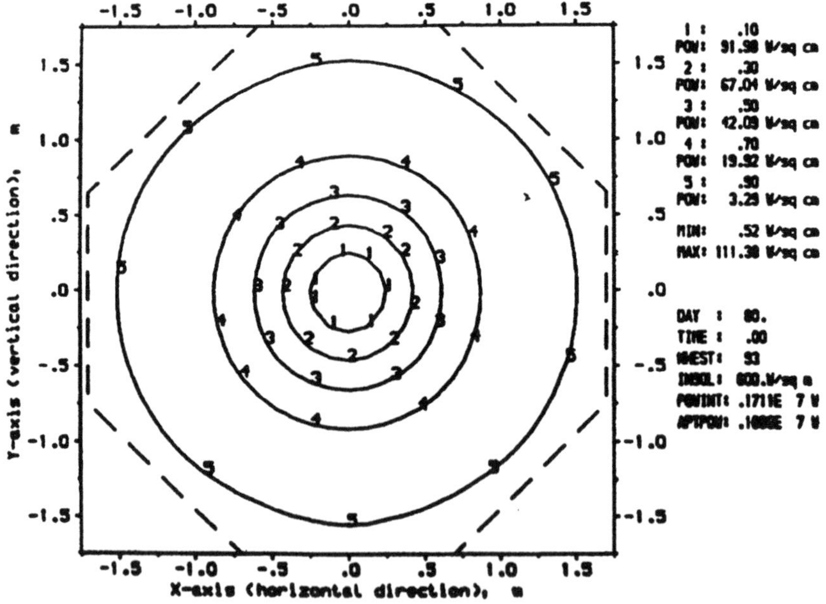

Fig.4.2-5: Isolines of flux distribution for design conditions for the Sulzer receiver. Generated by 'unfavorable' sun shape; NI = 600 W/m^2 C/(C+S)=0.24

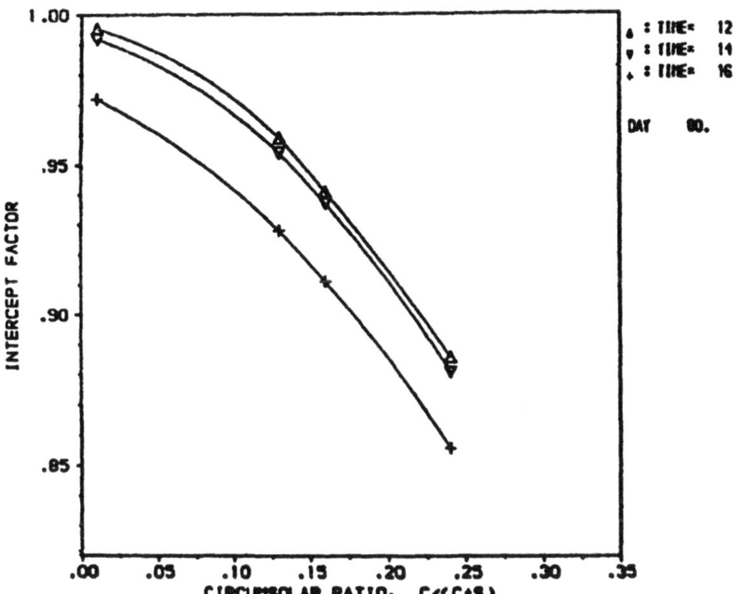

Fig.4.2-6: Intercept factor for the Sulzer receiver as a function of the circumsolar ratio. Day 80 (equinox), for various times.

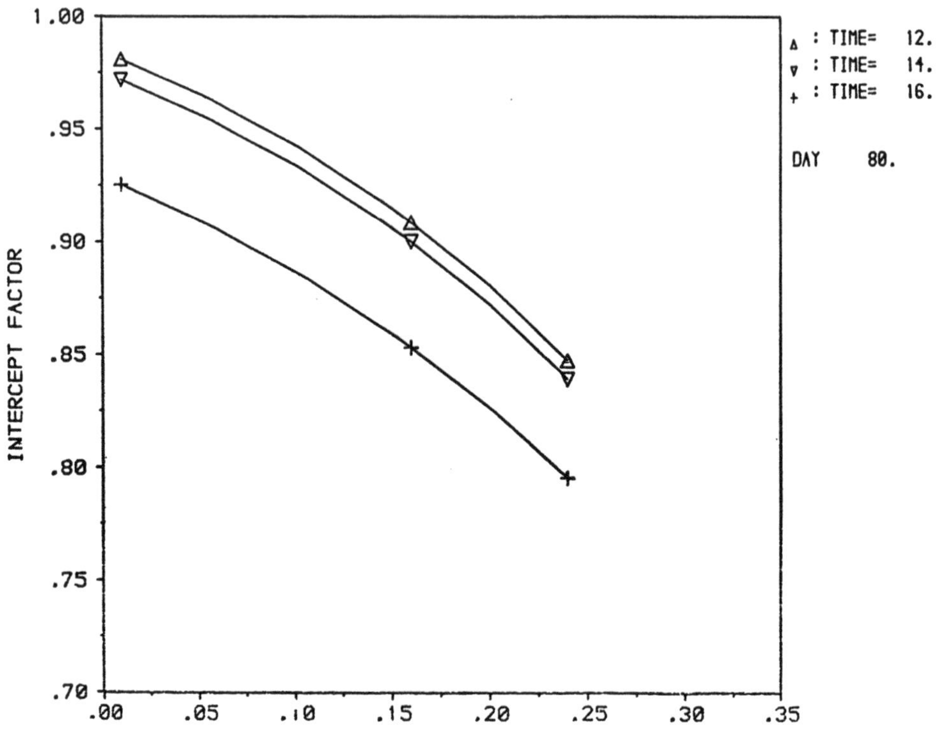

Fig.4.2-7: Intercept factor for the ASR receiver as a function of the circumsolar ratio. Day 80 (equinox), for various times.

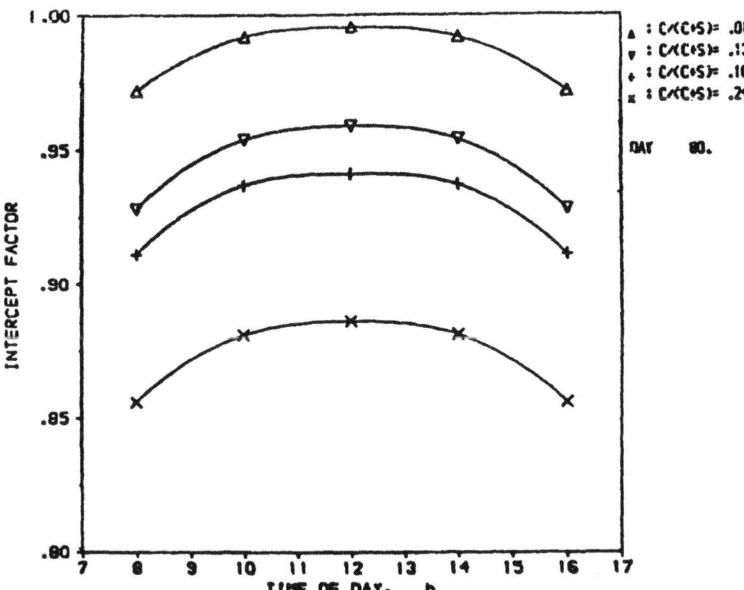

Fig.4.2-8: Intercept factor for the Sulzer receiver as a function of time-of-day. Day 80 (equinox), for various circumsolar ratios.

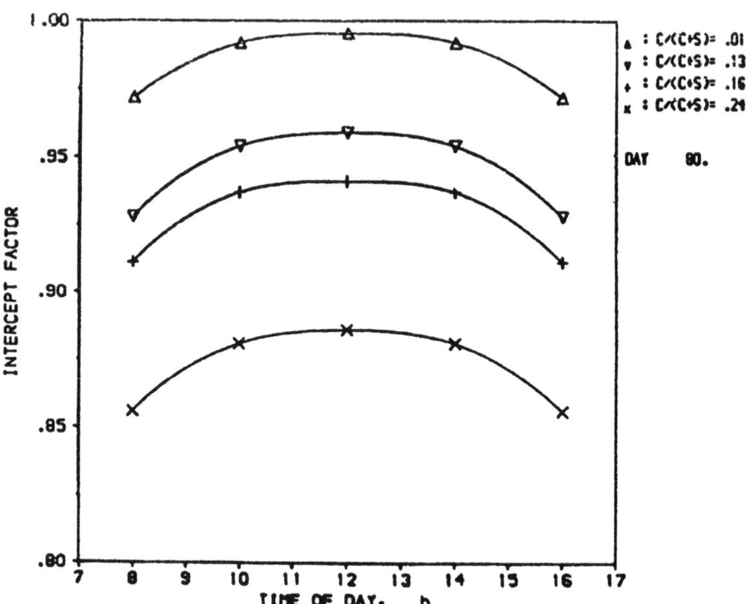

Fig.4.2-8: Intercept factor for the Sulzer receiver as a function of time-of-day. Day 80 (equinox), for various circumsolar ratios.

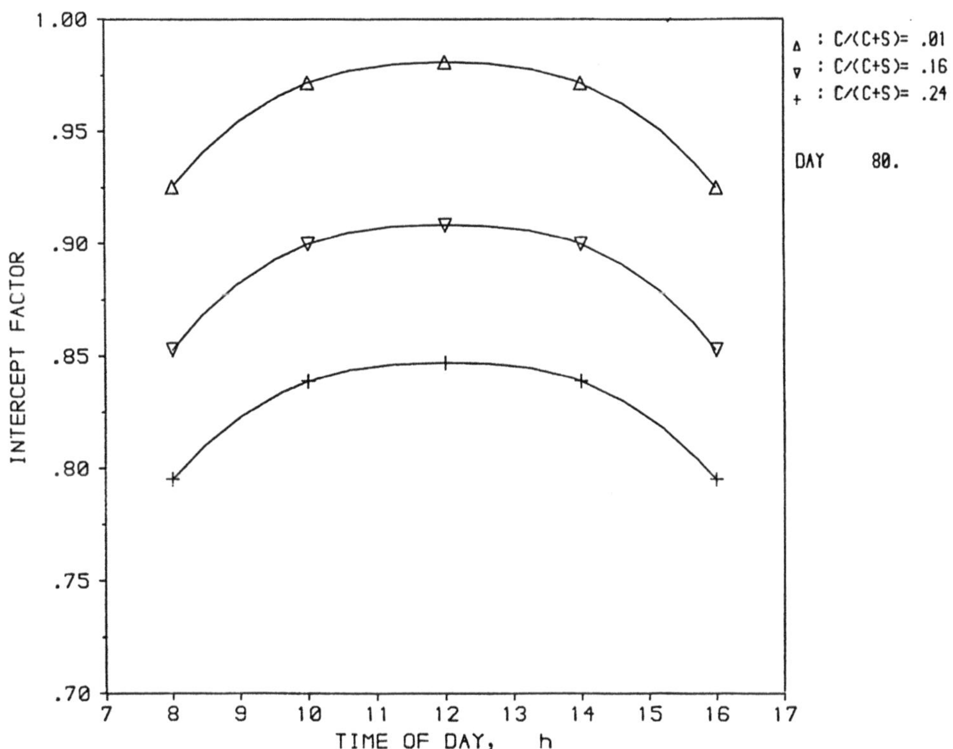

Fig.4.2-9: Intercept factor for the ASR receiver as a function of time-of-day. Day 80 (equinox), for various circumsolar ratios.

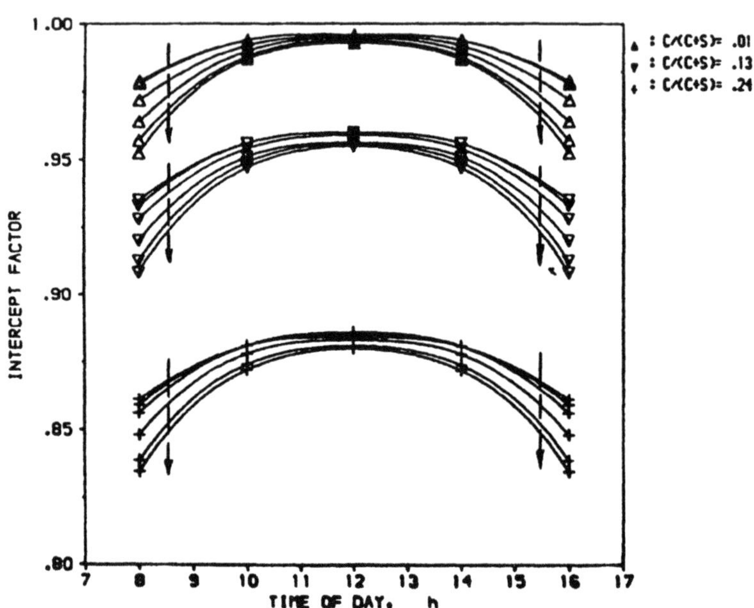

Fig.4.2-10: Characteristics of the intercept factor of the Sulzer receiver for different days and sun shapes.

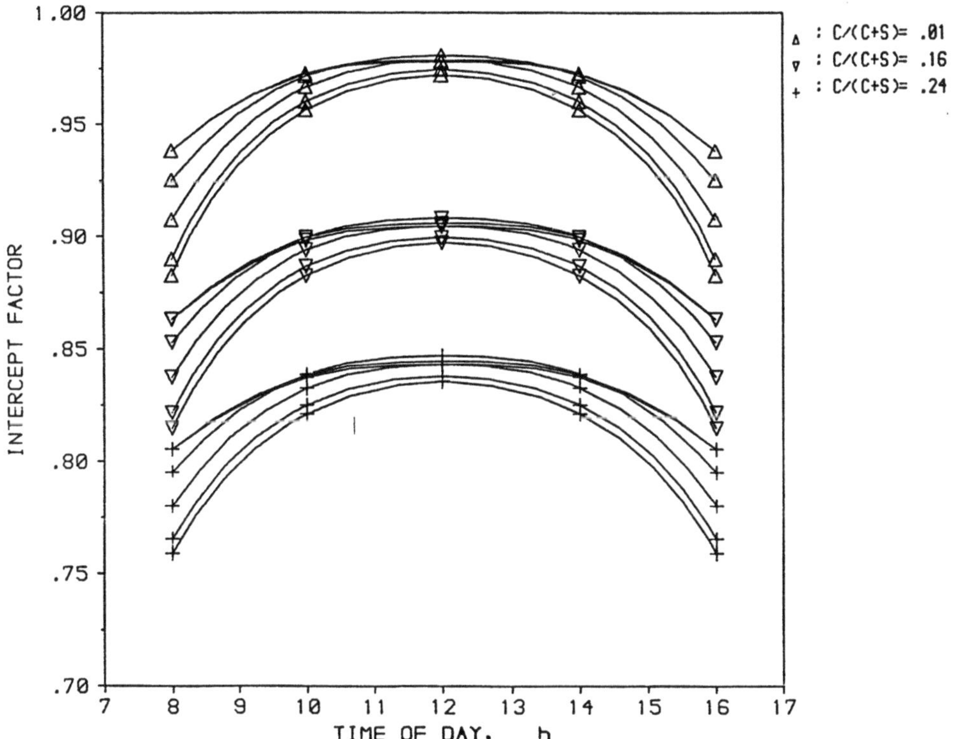

Fig.4.2-11: Characteristics of the intercept factor of the ASR receiver for different days and sun shapes.

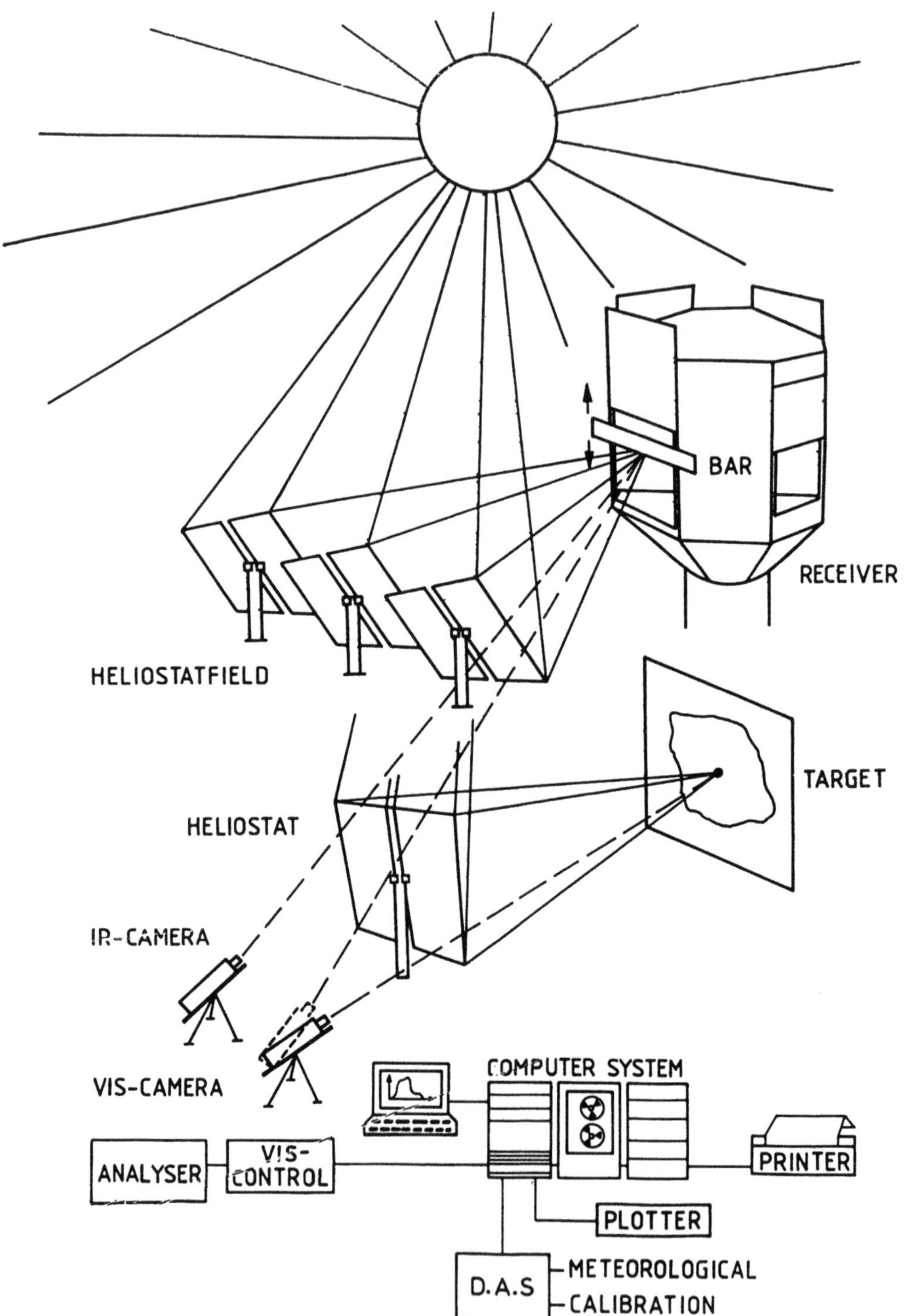

Fig.4.2-12: Block Diagram of the HERMES Equipment

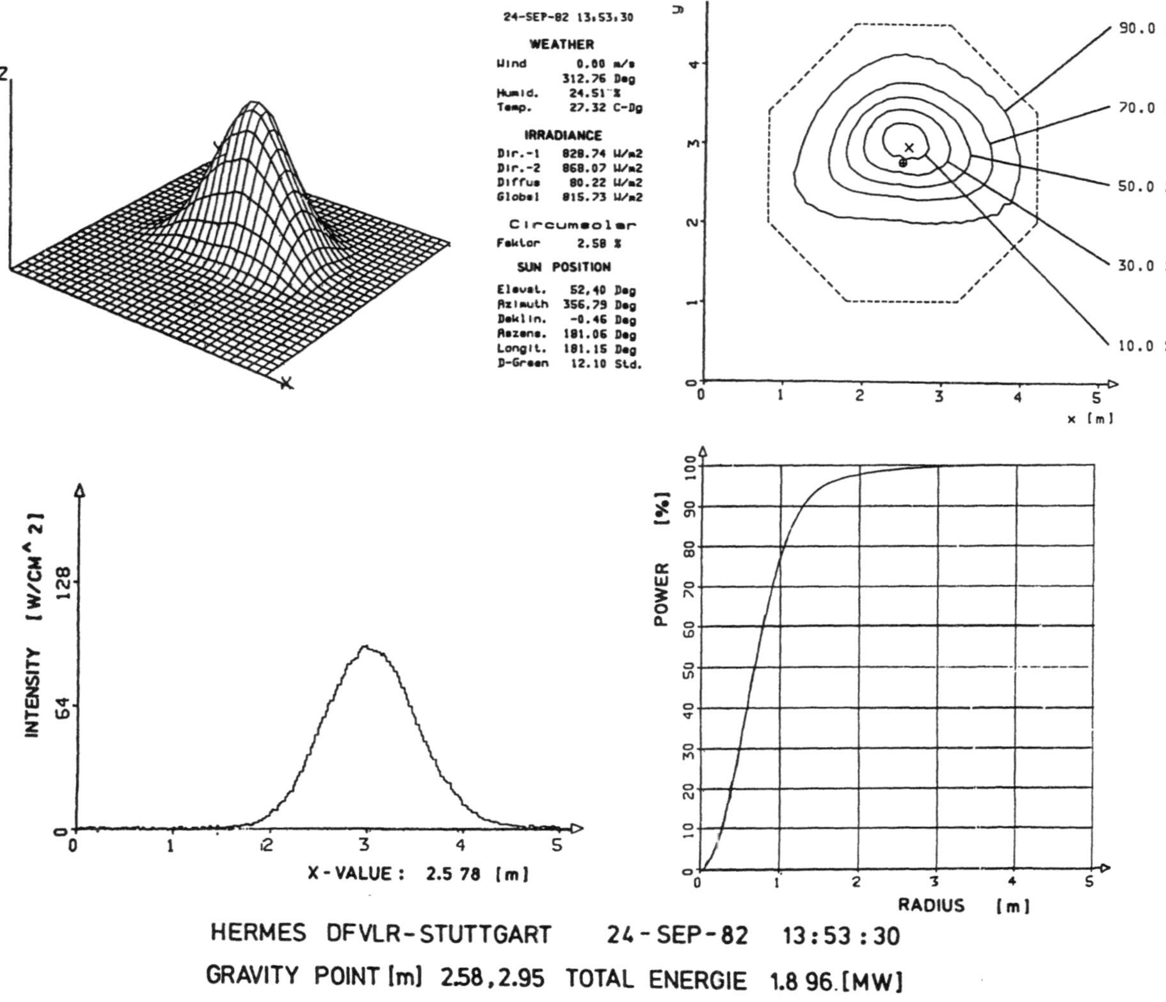

Fig.4.2-13: DAS Complete Set of the HERMES - September

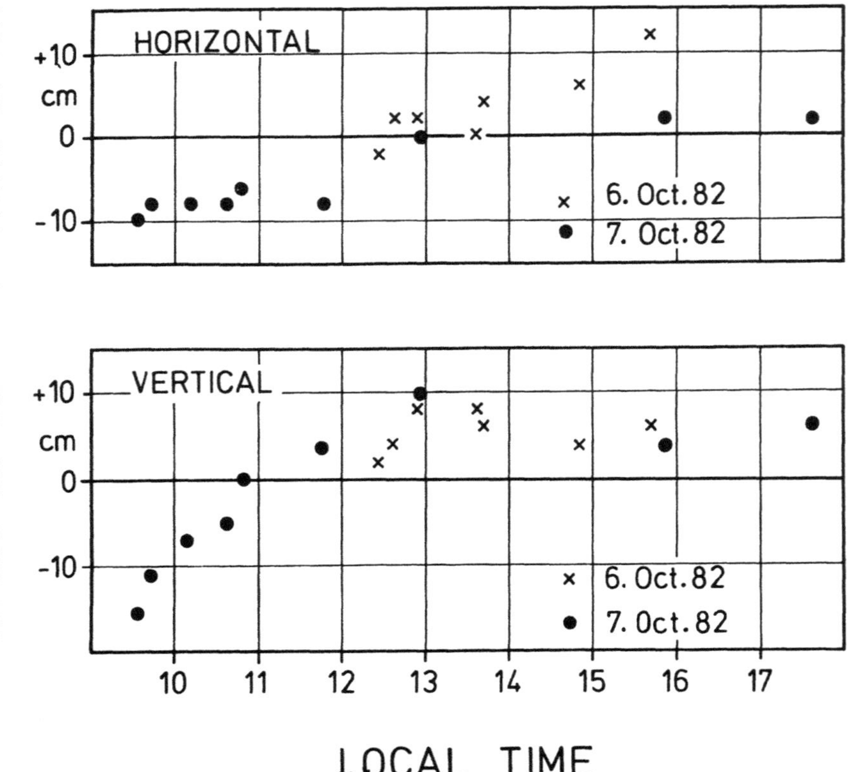

Fig.4.2-14: Tracking Accurary of the Heliostat Field

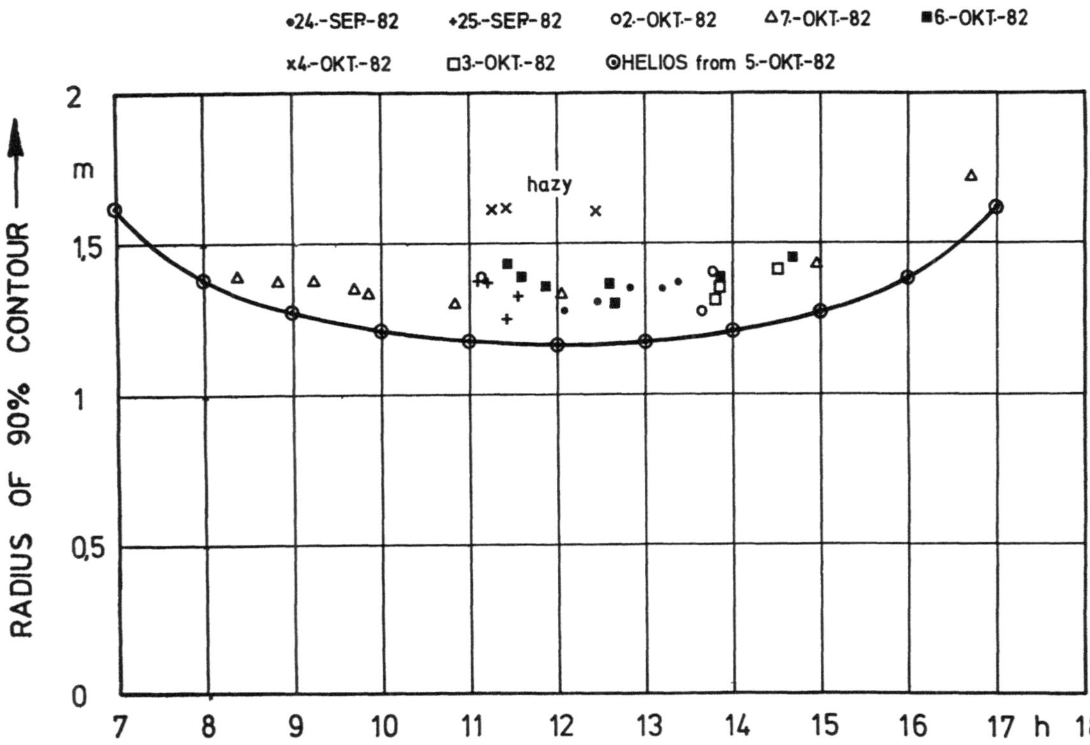

Fig.4.2-15: 90% Contours of the Flux Distribution in the Aperture Plane

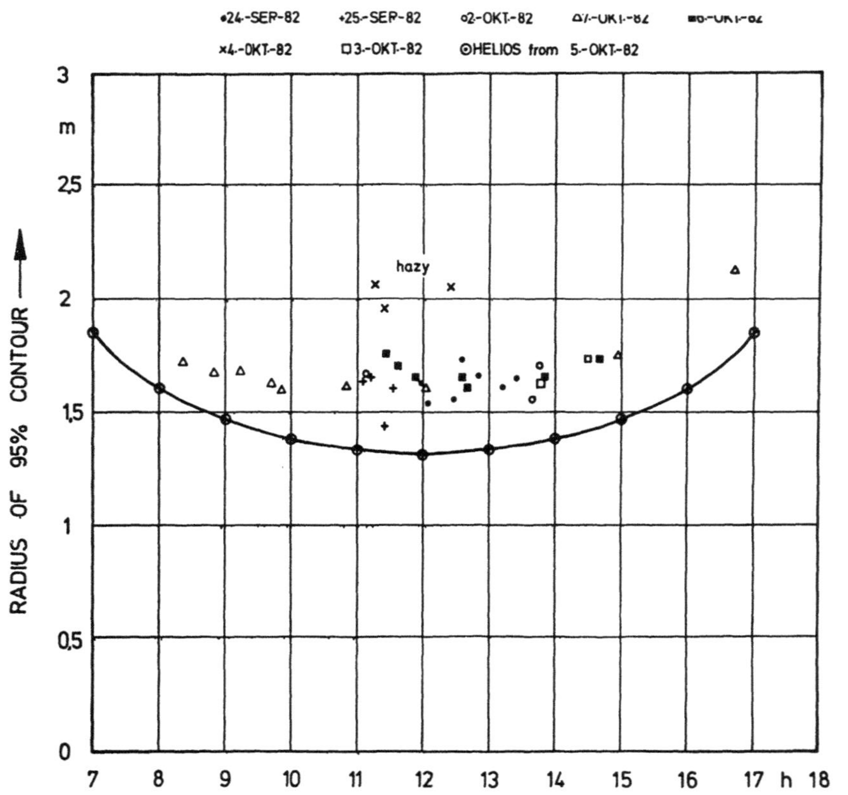

Fig.4.2-16: 95% Contours of the Flux Distribution in the Aperture Plane

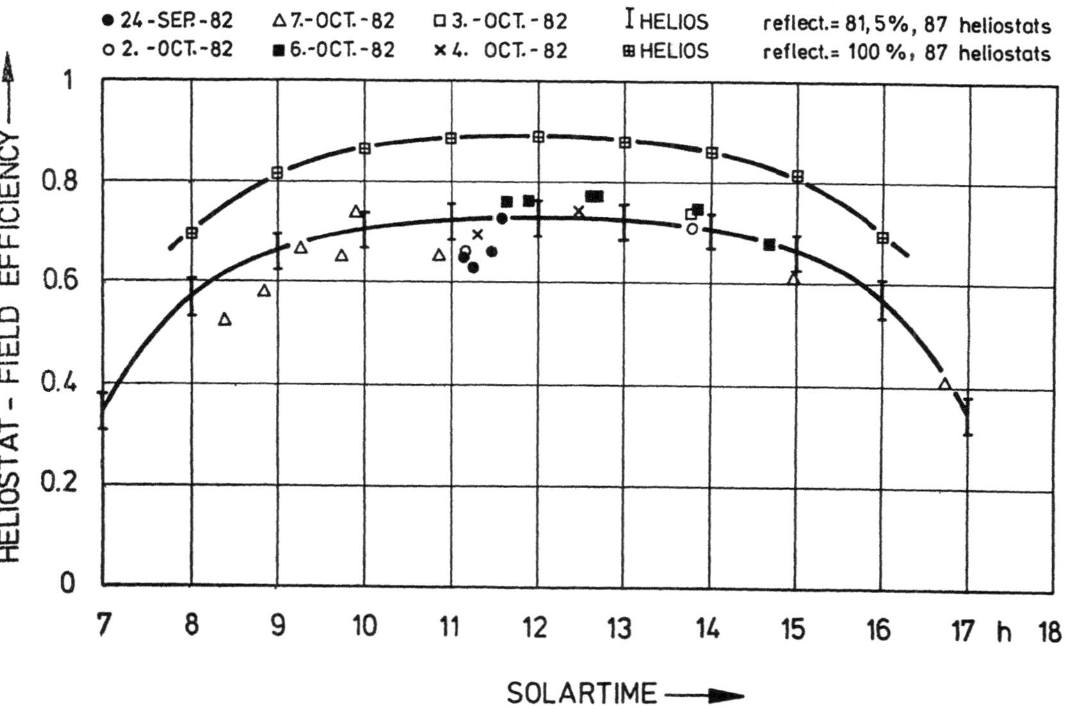

Fig.4.2-17: Heliostat Field Efficiency

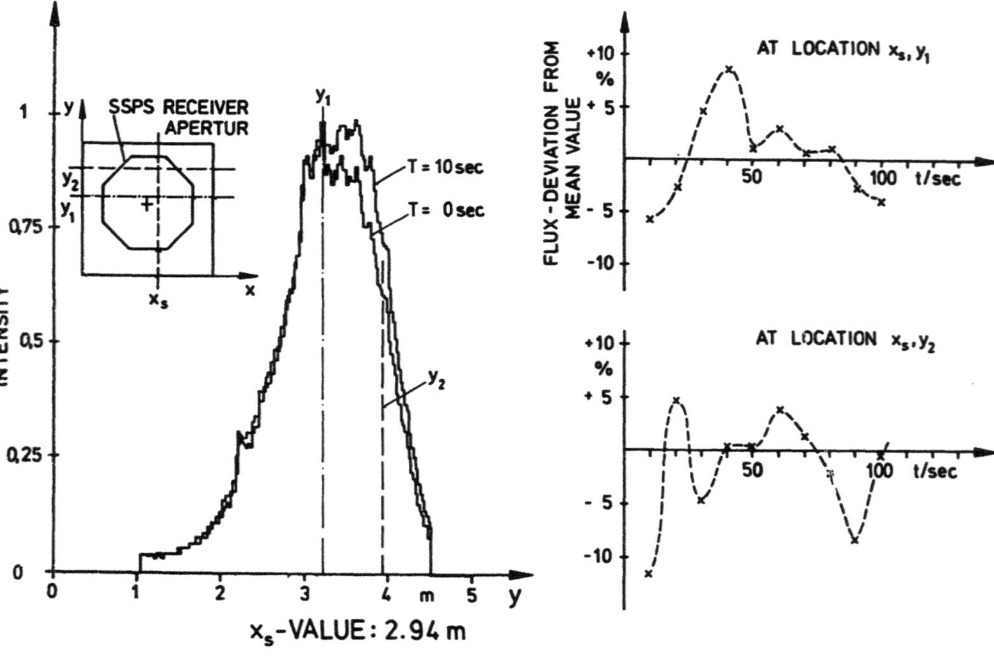

Fig.4.2-18: Variations of the Flux Loads on the Receiver Tubes

10 MWe SOLAR THERMAL CENTRAL RECEIVER PILOT PLANT - HELIOSTAT EVALUATION

C. L. Mavis and J. J. Bartel
Sandia National Laboratories Livermore

Introduction and Summary

Sandia is responsible for evaluating the heliostats at the 10 MWe Solar Thermal Central Receiver Pilot Plant in Barstow, California. The three evaluation objectives are: (1) characterize heliostat performance, (2) identify areas where heliostat research and development may lead to performance improvements and (3) evaluate the need for a heliostat beam characterization system in future plants.

During the past 12 months, reports have been published detailing Barstow heliostat experiences, mirror corrosion survey results, and 1982 meteorological data. A 1983 Meteorological Report was published in May 1984.

The beam characterization system (BCS) has been upgraded and a sunshape measurement system has been added. Heliostat mirror cleanliness has been measured at 2-week intervals, and the effects of rainwashing and spray rinsing of the mirrors have been determined. Mirror module vents are being installed on almost half of the modules to dry out the water that has accumulated inside them and to halt mirror corrosion.

During 1983, 94 to 99 percent of the heliostats were in operation at any one time. Maintenance hours are estimated to be 160 hours per month for 1983. The actual hours have not been determined yet. There were 817 maintenance actions during 1983 as compared with 929 in 1982.

During the next 6 months, evaluation activities will continue; at reduced level of effort after August 1984. An evaluation report for the 2-year plant evaluation period will be published late in 1984.

Accomplishments

Reports

The following reports have been published during the past 12 months. A meteorological report for 1983 will be published in May 1984.

Mavis, C. L., "Monograph Series, No. 1: 10 MWe Solar Thermal Central Receiver Pilot Plant Heliostat Experiences November 1981 - February 1983," SAND 83-8220, Sandia National Laboratories Livermore, May 1983.

Noring, J. E., Mavis, C. L., Decker, E. V., and Skvarna, P. E., "10 MWe Solar Thermal Central Receiver Pilot Plant Mirror Module Corrosion Survey," SAND 84-8214, Sandia National Laboratories Livermore, March 1984.

McDonnell Douglas Astronautics Company, "Solar One Solar Thermal Central Receiver Pilot Plant, 1982 Meteorological Data Report," SAND 83-8216, Sandia National Laboratories Livermore, June 1983.

McDonnell Douglas Astronautics Company, "Solar One Solar Thermal Central Receiver Pilot Plant, 1983 Meteorological Data Report," SAND 84-8180, Sandia National Laboratories Livermore, June 1984.

1983 Meteorological Data Summary

Meteorological data were collected at the pilot plant on 312 days during 1983. The Data Acquisition System (DAS) was inoperative on the remaining 53 days, primarily because plant controls development required use of the DAS computer.

Rainfall was over 2.5 times the normal amount (9.4 inches) during 1983. Temperatures ranged from 27°F to 108°F and the mirror module temperature was at or below the dew point on 32 of the 312 days. Insolation was available and recorded on 289 days. Of this period, 104 days were clear, 52 were light cloudy days and 132 were heavy cloudy days. Daily average insolation for 1983 was lower than in 1976 and lower than the 25-year average values for all months except July, which was a very good insolation month during 1983. The maximum wind speed during 1983 was 60 mph. The wind speed was above 30 mph on 67 of the 312 data days and above 40 mph on 21 days.

Beam Characterization System (BCS)

The BCS uses video cameras in the field to obtain images of heliostat beams on the targets located just below the receiver. These images are analyzed by means of a computer to provide heliostat tracking accuracy and beam quality data. An additional camera was added in 1983 to obtain sunshape data. The BCS was upgraded for better accuracy during 1983; it was converted to a stand-alone system and made operational in April 1984. Further improvements for beam power measurements will be incorporated during 1984. Preliminary tracking accuracy measurements with the BCS indicate that the heliostats perform as expected. Additional data will be taken in the near future to fully characterize the heliostats and evaluate the BCS.

A preliminary assessment shows that a BCS or a similar-type system will be required in future plants. A BCS is probably the best tool available for measuring tracking error and beam quality. However, BCS data cannot uniquely characterize a heliostat or indicate what is wrong, since mirror waviness, mirror canting, contour, and sunshape effects on beam quality are not separable.

Moontracking

A heliostat moontracking capability has been developed at the pilot plant to identify heliostats with large tracking errors. While moontracking, the heliostats are all aimed at one of four aimpoints just below the receiver. Visual observations are made or photographs are taken from the aimpoint to identify the heliostats most in need of a BCS measurement so that tracking errors can be corrected.

Mirror Cleanliness

The mirror soiling rate during 1983 varied between 3 and 5 percent per month. The entire field was spray rinsed once using an electrical substation insulator wash truck. The rinse increased the mirror reflectivity from about 89 percent to 95 percent of clean. Rain washes were used on seven occasions to restore the reflectivity to 95-98 percent of clean. A mechanical wash rig, which uses brushes and a water spray, was delivered to the Pilot Plant in 1983. Modifications and repairs are being made to make the washer usable. The cleaning effectiveness will be evaluated during the next several months.

Mirror Corrosion

Mirror corrosion was first noticed in the collector field in February 1982. Following this observation, a random sample of the mirror modules was surveyed to determine the amount of corrosion, the cause of the corrosion, and the corrosion growth rate. The cause of the corrosion is water from dew and rain which enters the mirror modules through leaks in the edge seals. During the summer of 1983 and 1984, surveys of all 1818 heliostats were completed by Southern California Edison (SCE) to document the extent of the corrosion. The major findings of the survey are:

- 27 percent of the mirror modules have some amount of corrosion
- 0.030 percent of the total reflective surface is corroded
- The area corroded increases by a factor of two each year
- Over 11 times more corrosion was observed on mirrors that were fabricated before July 1, 1981, than on those fabricated after this date. Production process changes on this date caused a dramatic improvement in the quality of the mirror seals.

Mirror corrosion at the current growth rate will not have a significant impact on plant performance for several years; however, attempts are being made to dispel the water in the mirror modules through additional ventilation. Vents that were added to selected mirror modules have demonstrated that the modules can be dried out. Vents have now been added to 10,546 mirror modules. The vents consist of four 1/2-inch diameter aluminum tubes in each module, two on each end.

Heliostat Maintenance

Heliostat maintenance was not a major concern or problem during 1983 and 1984. On any given day, between 94 and 99 percent of the heliostats have been operational. Maintenance hours are estimated to have been 160 hours per month. The actual hours have not been determined yet. The maintenance required from January 1, 1982, through March 31, 1983, was 2586 hours. The number of maintenance orders during 1983 are about the same as those for 1982, as shown in Tab.4.3-1.

Future Plans

Heliostat evaluation will continue through April 1985, however, the level of effort has decreased in August 1984 when the plant entered a power production phase. The items to be completed and ongoing activities throughout this period are listed below:

Items to be Completed:

1. Document BCS
2. Improve BCS power measurement accuracy
3. Evaluate heliostat washing machine effectiveness
4. Analyze wind speed and wind load
5. Publish heliostat evaluation report

Ongoing Activities:

1. Measure tracking accuracy and beam quality
2. Measure mirror cleanliness
3. Monitor mirror corrosion and venting effectiveness

Item	Dec 81 - Dec 82	1983
Azimuth Motor	121	48
Elevation Motor	35	18
Gear Box	2	13*
Heliostat Controller	296	465
Heliostat Field Controller	34	20
Azimuth Encoder	103	138
Elevation Encoder	130	43
Mirror	21	11**
Physical	86	61
Totals	929	817

*Includes 1 Gear Box Failure
**Does not include mirrors identified by Mirror Corrosion Survey

Table 4.3.-1: Heliostat Maintenance Orders

(This page, and the following ones, are copies of view-graphs, and are to support the ITET's evaluation)

BARSTOW PILOT PLANT

HELIOSTAT EVALUATION

OBJECTIVES:

1. CHARACTERIZE HELIOSTAT PERFORMANCE
2. IDENTIFY AREAS WHERE HELIOSTAT R&D MAY LEAD TO PERFORMANCE IMPROVEMENTS
3. EVALUATE THE NEED FOR A HELIOSTAT BEAM CHARACTERIZATION SYSTEM IN FUTURE PLANTS

CLM:4/10/84

ACCOMPLISHMENTS DURING PAST 12 MONTHS

PUBLISHED REPORTS

- HELIOSTAT EXPERIENCES
- MIRROR CORROSION SURVEY
- 1982 METEOROLOGICAL DATA

PREPARED 1983 METEOROLOGICAL DATA REPORT (PUBLISH MAY 1984)

COMPLETED BCS UPGRADE INCLUDING SUNSHAPE CAMERA

OBTAINED MIRROR CLEANLINESS DATA

ADDED MIRROR MODULE VENTS TO DRY OUT MODULES AND HALT MIRROR CORROSION

CLM:4/10/84

REPORTS

MAVIS, C. L., "MONOGRAPH SERIES, NO. 1: 10 MW_E SOLAR THERMAL CENTRAL RECEIVER PILOT PLANT HELIOSTAT EXPERIENCES NOVEMBER 1981 - FEBRUARY 1983," SAND 83-8220 SANDIA NATIONAL LABORATORIES LIVERMORE, MAY 1983

NORING, J. E., MAVIS, C. L., SANDIA NATIONAL LABORATORIES AND DECKER, E. V., SKVARNA, P. E., SOUTHERN CALIFORNIA EDISON, "10 MW_E SOLAR THERMAL CENTRAL RECEIVER PILOT PLANT MIRROR MODULE CORROSION SURVEY," SAND 84-8214, SANDIA NATIONAL LABORATORIES LIVERMORE, MARCH 1984

McDONNELL DOUGLAS ASTRONAUTICS COMPANY, "SOLAR ONE SOLAR THERMAL CENTRAL RECEIVER PILOT PLANT, 1982 METEOROLOGICAL DATA REPORT," SAND 83-8216, SANDIA NATIONAL LABORATORIES LIVERMORE, JUNE 1983

METEOROLOGICAL DATA REPORT FOR 1983 WILL BE PUBLISHED IN MAY 1984.

CLM:4/10/84

1983 METEOROLOGICAL DATA SUMMARY

312 DATA DAYS

RAINFALL: 9.4 INCHES VS. 3.6 INCHES/YR AVERAGE 1956-1970

TEMPERATURE: LOW 27°F
HIGH 108°F
BELOW FREEZING - JAN, NOV & DEC
ABOVE 100°F - MAY, JUNE, JULY, AUG & SEPT
MIRROR TEMP. BELOW DEW POINT ON 32 DAYS

INSOLATION: 289 DAYS RECORDED WITH INSOLATION AVAILABLE
104 CLEAR DAYS
53 LIGHT CLOUDY DAYS
132 HEAVY CLOUDY DAYS

WIND: MAX SPEED: 60 MPH
WIND SPEED: ABOVE 30 MPH ON 67 DAYS
ABOVE 40 MPH ON 21 DAYS

CLM:4/10/84

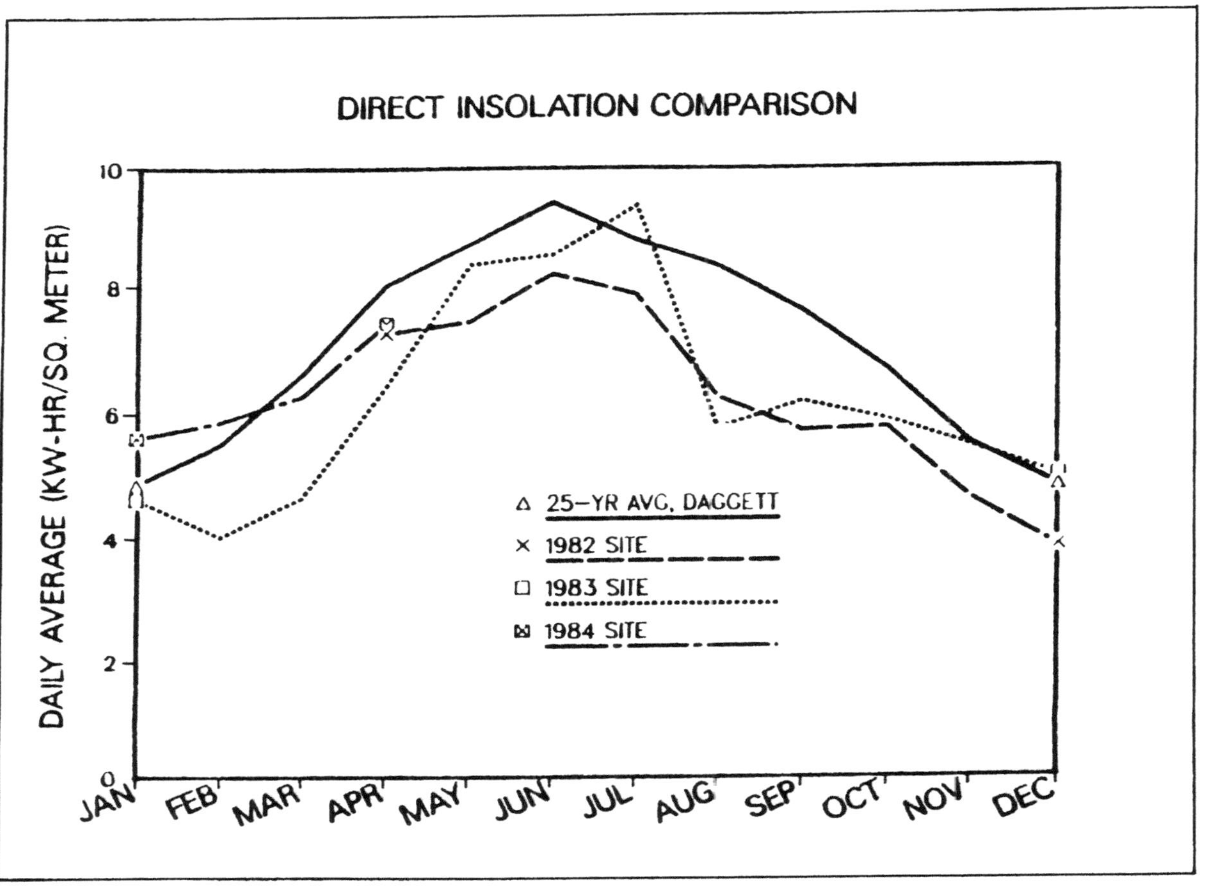
DIRECT INSOLATION COMPARISON
DAILY AVERAGE (KW-HR/SQ. METER)
0
2
4
6
8
10
JAN
FEB
MAR
APR
MAY
JUN
JUL
AUG
SEP
OCT
NOV
DEC
△ 25–YR AVG, DAGGETT
× 1982 SITE
□ 1983 SITE
⊠ 1984 SITE

BEAM CHARACTERIZATION SUBSYSTEM

BEAM CHARACTERIZATION SYSTEM (BCS)

BCS MEASURES HELIOSTAT PERFORMANCE PARAMETERS AND SUNSHAPE

- TRACKING ACCURACY
- BEAM QUALITY
- BEAM POWER

BCS WAS CONVERTED TO A STAND-ALONE SYSTEM AND MADE OPERATIONAL IN APRIL 1984

FURTHER IMPROVEMENT FOR BEAM POWER MEASUREMENTS IS REQUIRED

PRELIMINARY TRACKING ACCURACY MEASUREMENTS INDICATE THAT THE HELIOSTATS PERFORM AS EXPECTED

MOONTRACKING IS BEING USED TO IDENTIFY HELIOSTATS REQUIRING BCS MEASUREMENT AND BIAS UPDATING.

CLM:4/10/84

```
NORMALIZED PLOT

HCNUM.................................2833
DATE..................................10/ 4 83
TIME OF DAY...........................10:29
CENTROID ERROR + - STD.DEV.(HORIZ.)....  -23.79 +/-   9.65 IN
CENTROID ERROR + - STD.DEV.(VERT.).....   14.43 +/-   6.36 IN
THEORETICAL POWER..................... 25408.3 WATTS
POWER EFFECTIVITY + - STD.DEV.........    86.8 +/-  2.8 %
SPILLAGE POWER........................ 0.0 %
INSOLATION LEVEL......................   818.8 W/M**2
WIND VELOCITY.........................  5.64 MPH
LAST HORIZ. CENTROID OFFSET........... -16.8 INCHES
LAST VERT. CENTROID OFFSET............  25.1 INCHES
HAC AIMPOINT.......................... 349.9 FEET
TEMPERATURE........................... 72.11 DEGF
INVALID DATA FLAGS....................0001100000
```

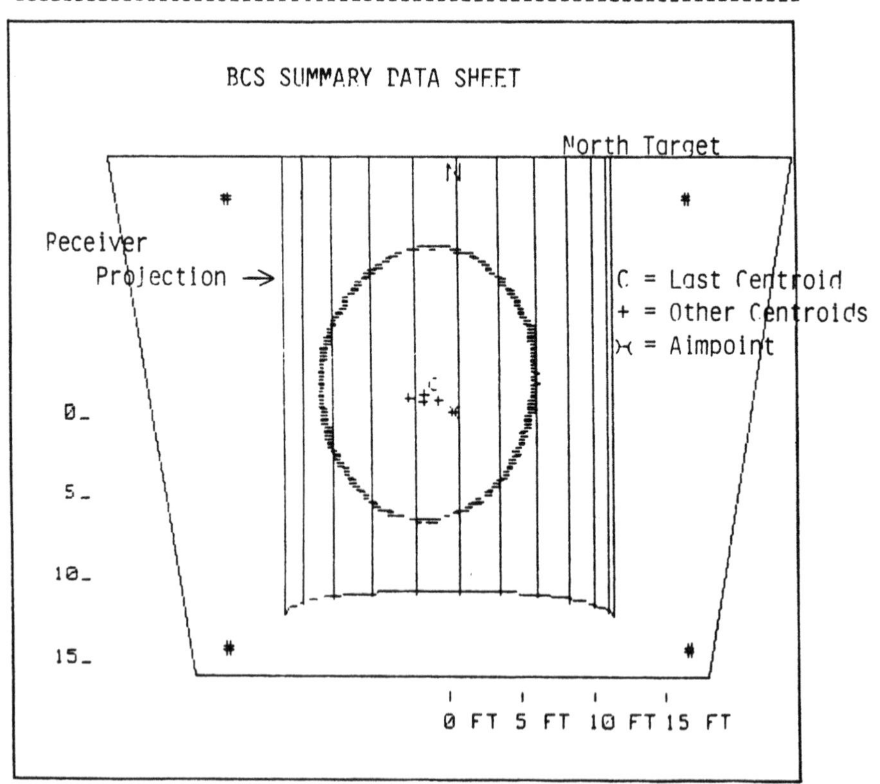

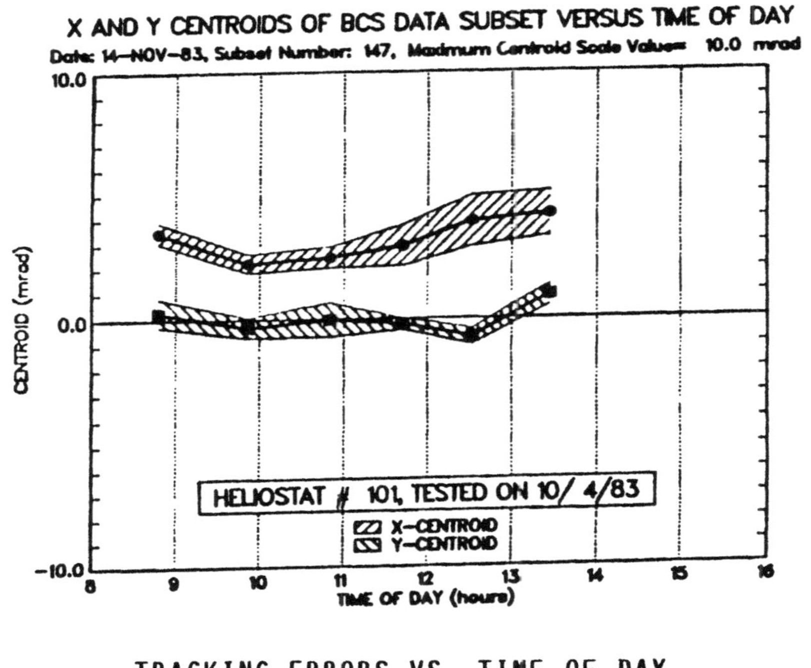

TRACKING ERRORS VS. TIME OF DAY
WITH 1 STANDARD DEVIATION SPREAD IN DATA SHOWN

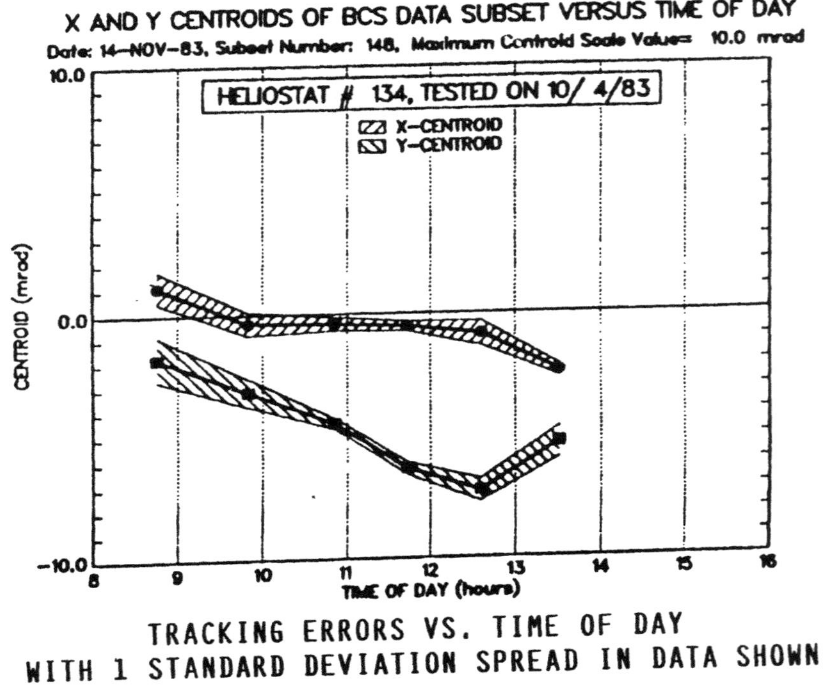

TRACKING ERRORS VS. TIME OF DAY
WITH 1 STANDARD DEVIATION SPREAD IN DATA SHOWN

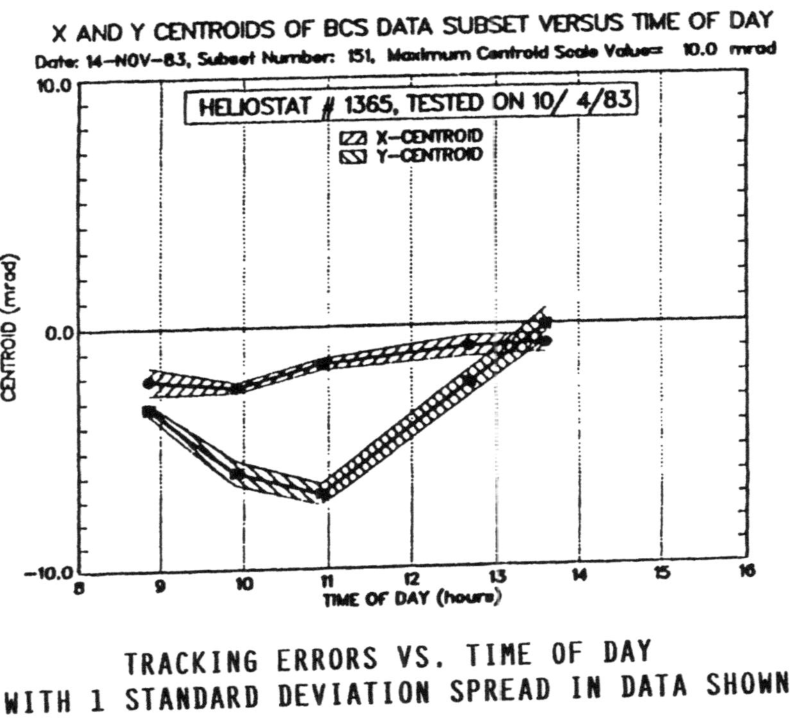

TRACKING ERRORS VS. TIME OF DAY
WITH 1 STANDARD DEVIATION SPREAD IN DATA SHOWN

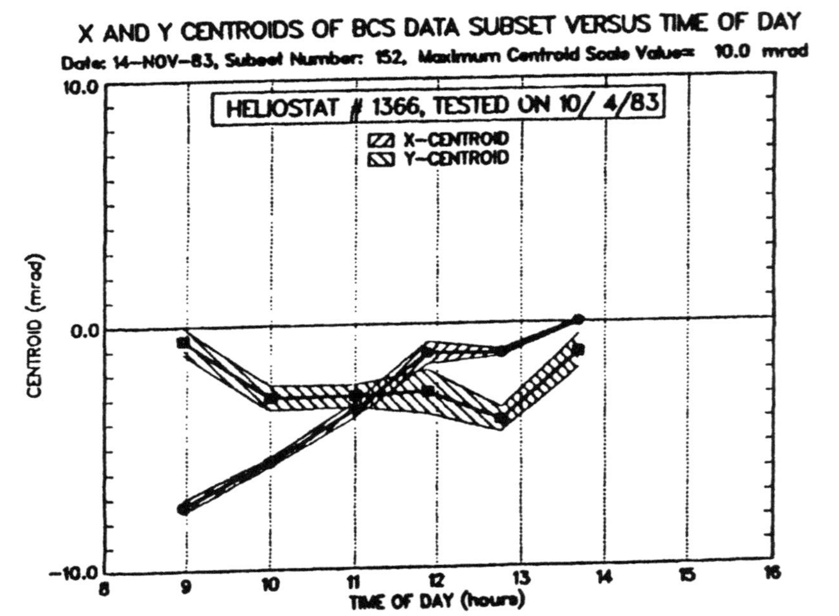

TRACKING ERRORS VS. TIME OF DAY
WITH 1 STANDARD DEVIATION SPREAD IN DATA SHOWN

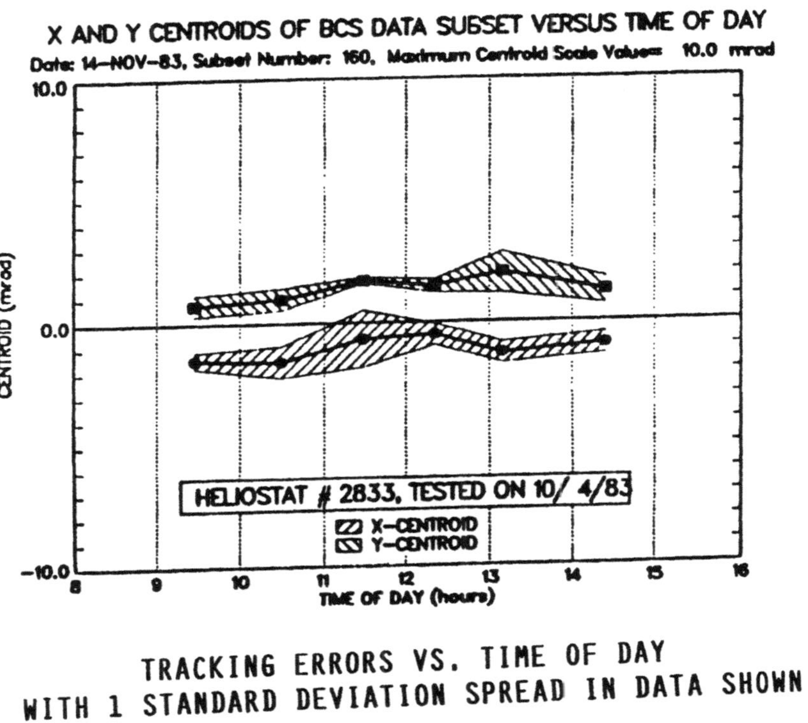

TRACKING ERRORS VS. TIME OF DAY
WITH 1 STANDARD DEVIATION SPREAD IN DATA SHOWN

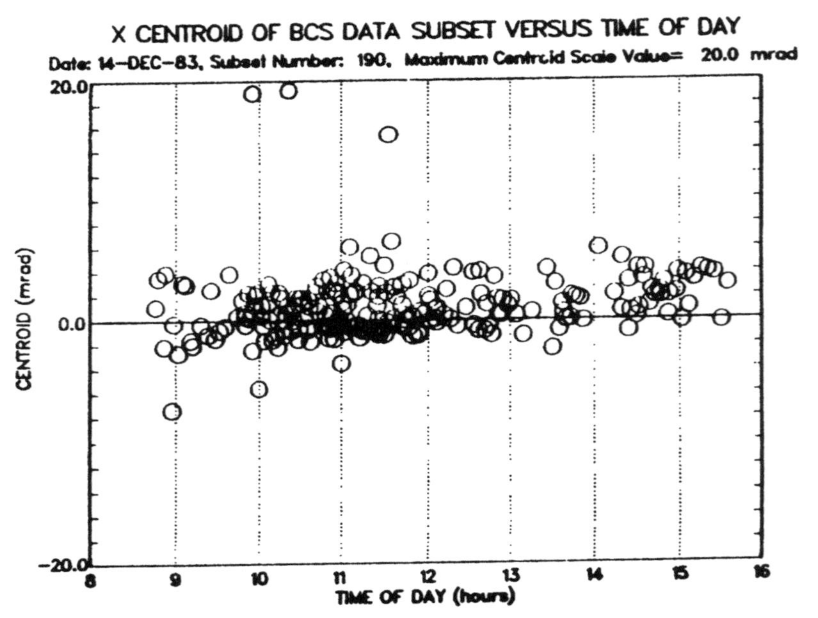

224 BCS DATA POINTS TAKEN ON OCT 4 AND NOV 2, 1983
(CENTROID ERROR IS IN MILLIRADIANS)

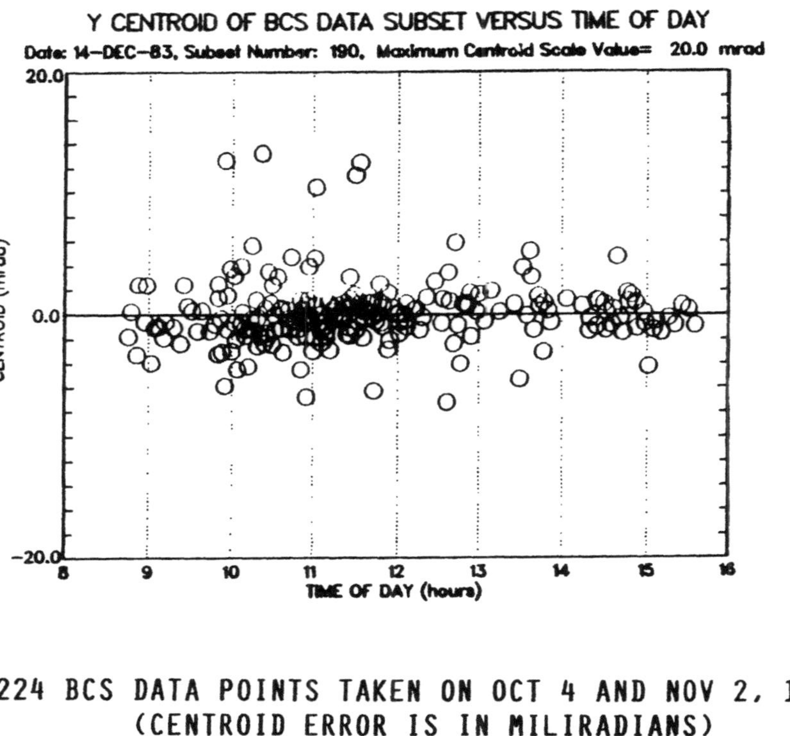

224 BCS DATA POINTS TAKEN ON OCT 4 AND NOV 2, 1983
(CENTROID ERROR IS IN MILIRADIANS)

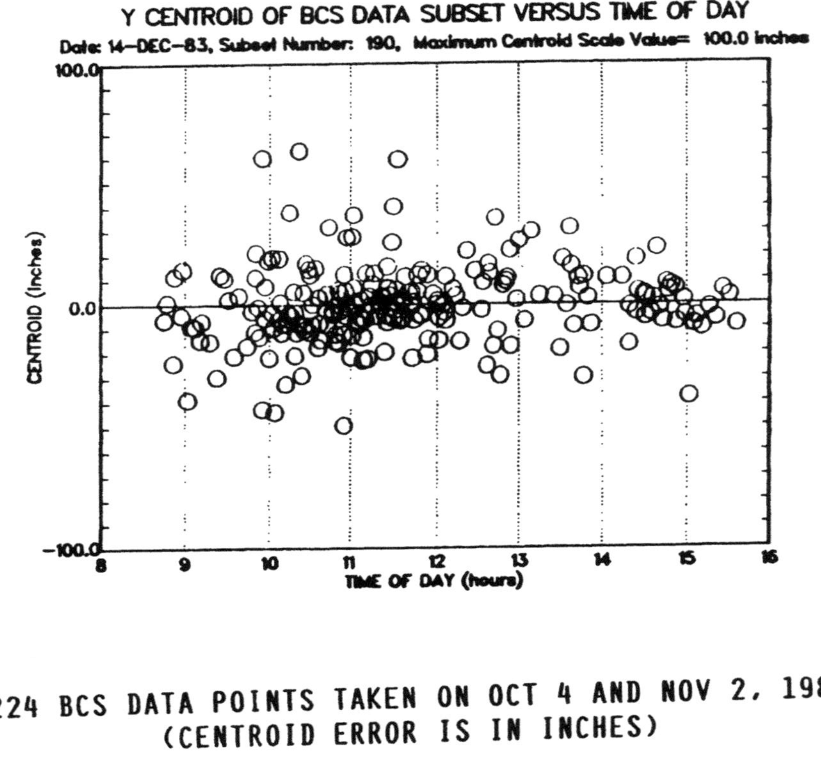

224 BCS DATA POINTS TAKEN ON OCT 4 AND NOV 2, 1983
(CENTROID ERROR IS IN INCHES)

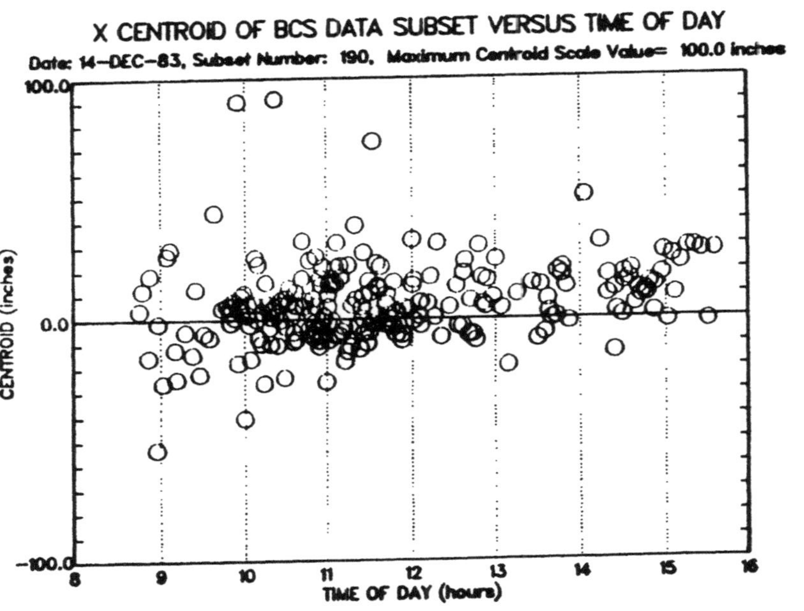

224 BCS DATA POINTS TAKEN ON OCT 4 AND NOV 2, 1983
(CENTROID ERROR IS IN INCHES)

MOONTRACKING

POOR MAN'S HELIOSTAT CHARACTERIZATION SYSTEM

HELIOSTAT FIELD IS PHOTOGRAPHED FROM A COMMON AIMPOINT WHILE TRACKING THE MOON

HELIOSTATS WITH LARGE TRACKING ERRORS CAN BE EASILY IDENTIFIED AND SCHEDULED FOR AN AIMING ACCURACY MEASUREMENT

MISALIGNED MIRROR FACETS CAN BE IDENTIFIED

CLM:4/11/84

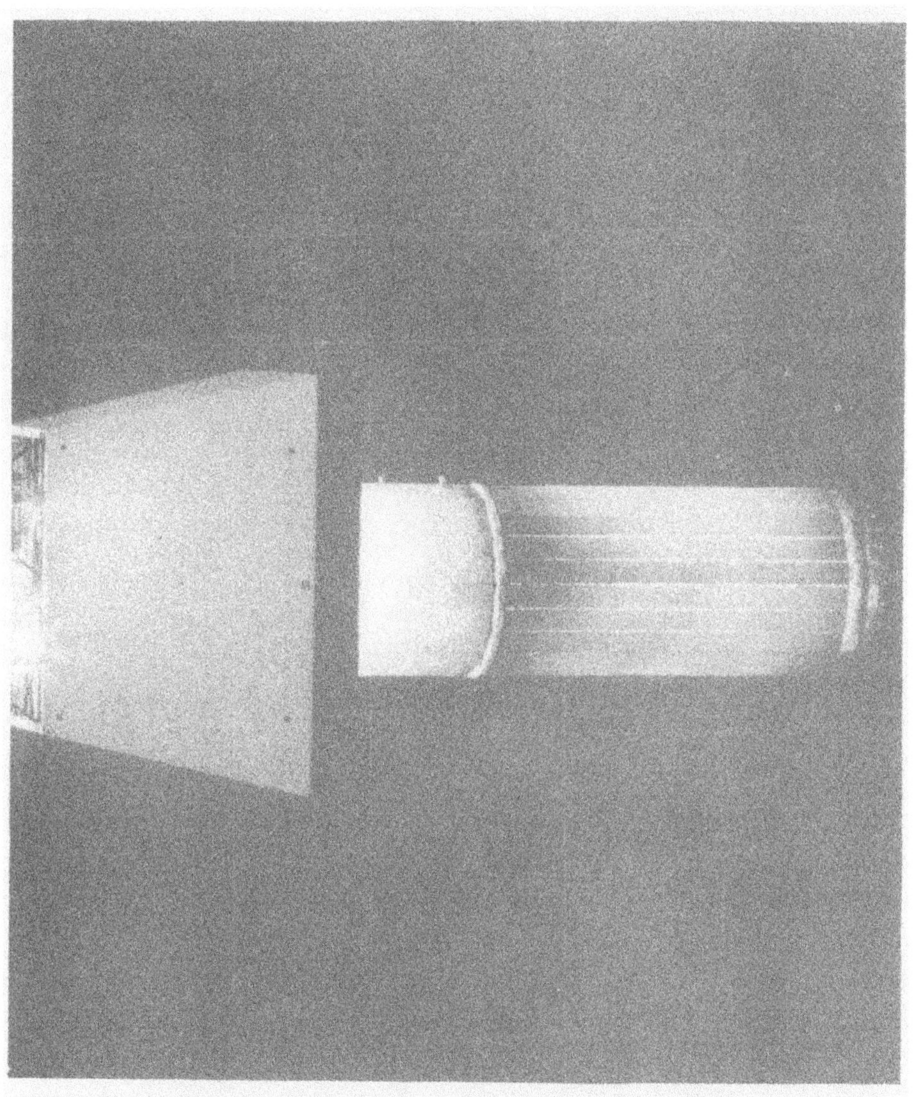

Photo of the receiver at the 10 MWe solar thermal central receiver pilot plant (solar one) under moonlight illumination from the heliostat field.

Photo from the receiver aimpoint of the inner part of the northwest quadrant of solar one heliostats while moontracking. Tracking errors are indicated by the off-center moon image in the mirrors.

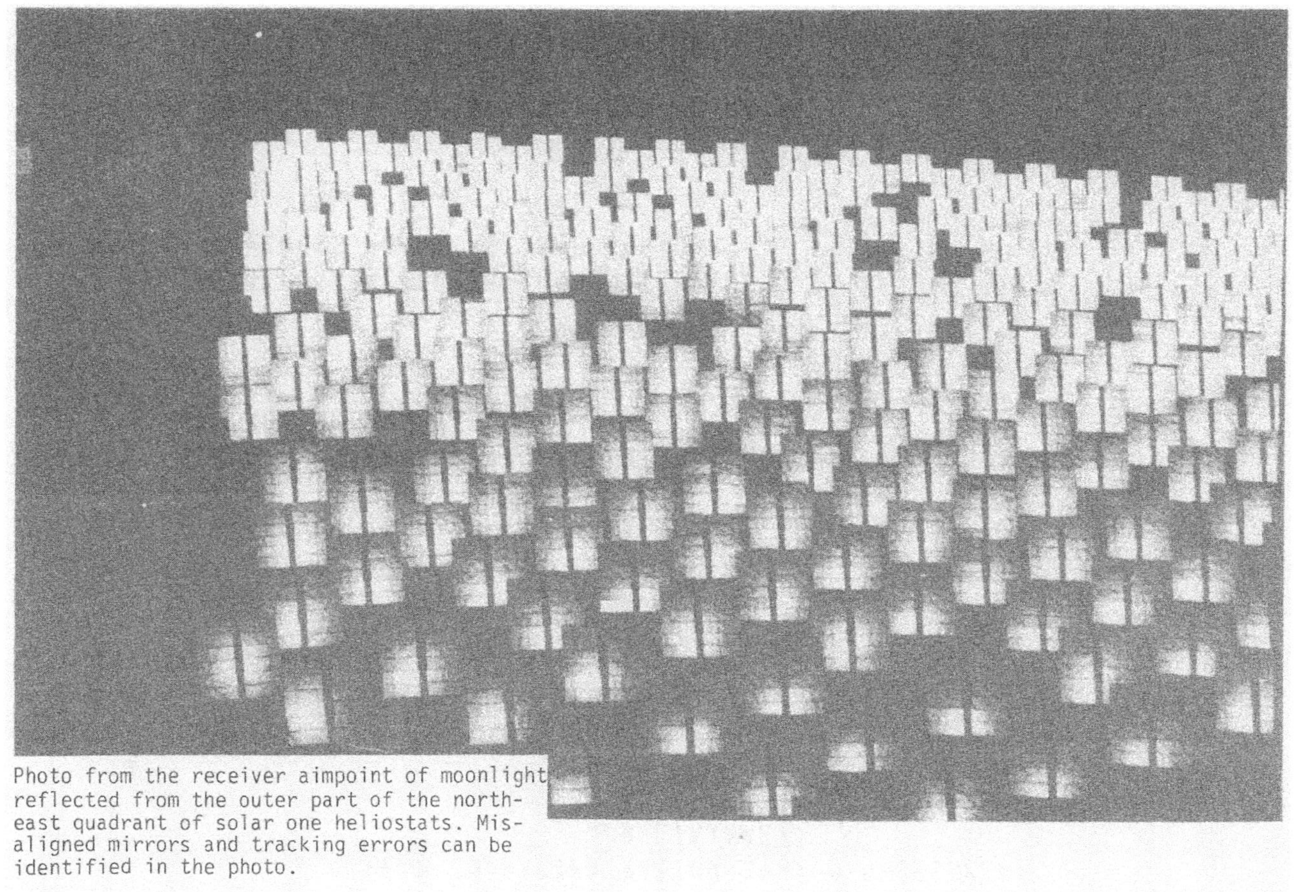

Photo from the receiver aimpoint of moonlight reflected from the outer part of the north-east quadrant of solar one heliostats. Misaligned mirrors and tracking errors can be identified in the photo.

BEAM CHARACTERIZATION SYSTEM ASSESSMENT (PRELIMINARY)

A BCS IS PROBABLY THE BEST TOOL AVAILABLE FOR TRACKING ERROR AND BEAM QUALITY MEASUREMENTS

BCS DATA WILL NOT UNIQUELY CHARACTERIZE A HELIOSTAT

- MIRROR WAVINESS, MIRROR CANTING, CONTOUR AND SUNSHAPE EFFECTS ARE NOT SEPARABLE

A HELIOSTAT CHARACTERIZATION SYSTEM (HCS) IS BETTER SUITED TO ASSESS MIRROR WAVINESS, CANTING, AND CONTOUR

FUTURE PLANTS WILL NEED A BCS AND/OR AN HCS

CLM:4/11/84

MIRROR CLEANLINESS - 1983

SOILING RATE DURING 1983 WAS 3 TO 5 PERCENT PER MONTH

SPRAY RINSING MIRRORS RESTORES 95 PERCENT OF CLEAN REFLECTIVITY

RAINWASHING RESTORES 95 PERCENT OF CLEAN REFLECTIVITY

MECHANICAL WASHING WILL BE COST EFFECTIVE

A MECHANICAL WASH TRUCK WAS DELIVERED TO PILOT PLANT IN 1983, AFTER MODIFICATIONS IT WILL BE USED AND EVALUATED

CLM:4/10/84

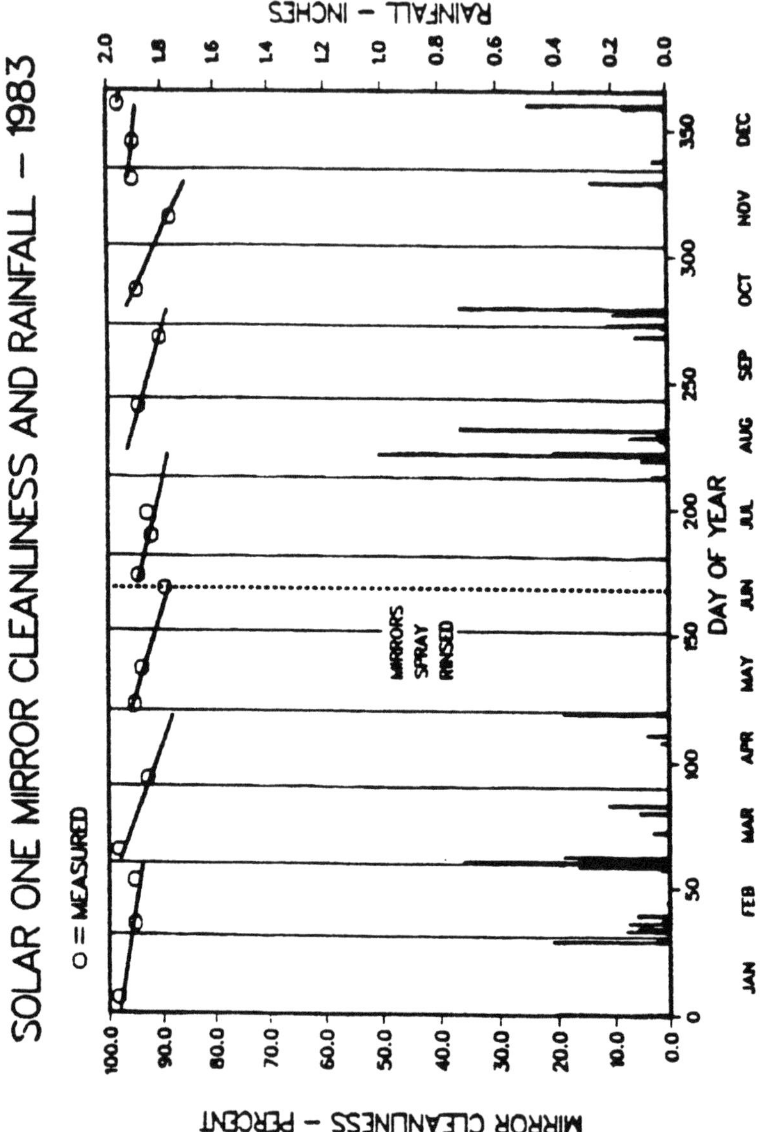
SOLAR ONE MIRROR CLEANLINESS AND RAINFALL – 1983
O = MEASURED
MIRRORS SPRAY RINSED
MIRROR CLEANLINESS – PERCENT
RAINFALL – INCHES
DAY OF YEAR
JAN FEB MAR APR MAY JUN JUL AUG SEP OCT NOV DEC
0 50 100 150 200 250 300 350
100.0 90.0 80.0 70.0 60.0 50.0 40.0 30.0 20.0 10.0 0.0
2.0 1.8 1.6 1.4 1.2 1.0 0.8 0.6 0.4 0.2 0.0

MIRROR CORROSION

SCE COMPLETED SURVEY OF ALL HELIOSTATS IN AUGUST 1983

- .015 PERCENT OF MIRROR AREA IS CORRODED
- 15 PERCENT OF MODULES HAVE SOME CORROSION
- 65 PERCENT OF CORROSION IS IN NE QUADRANT

MIRROR MODULES PRODUCED BEFORE JULY 1, 1981, ARE 10 TIMES MORE LIKELY TO HAVE CORROSION

MANUFACTURING PROCESS CHANGES AFTER JULY 1, 1981, GREATLY IMPROVED MIRROR QUALITY

VENTS ARE BEING ADDED TO 10,546 MIRROR MODULES TO DRY OUT MODULES AND HALT CORROSION

CLM:4/10/84

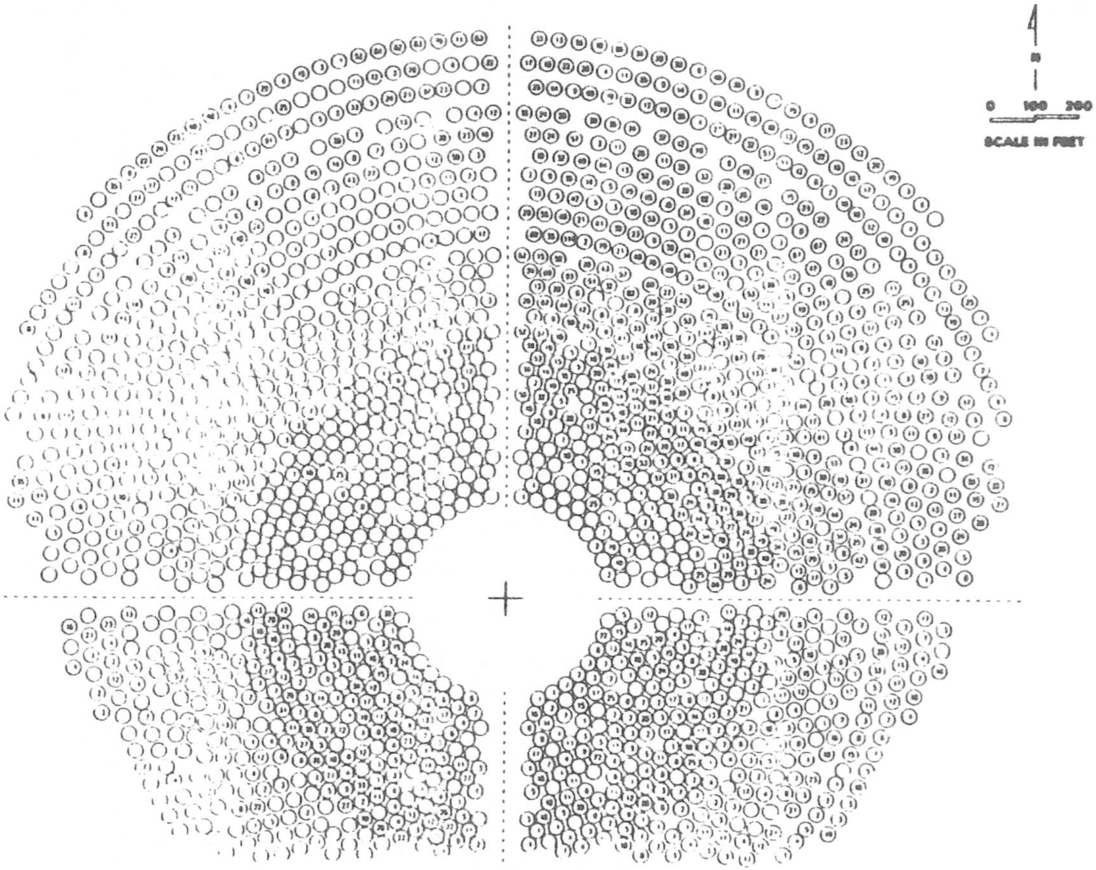

FIGURE 5 - MAP OF HELIOSTAT FIELD CORROSION AREA DISTRIBUTION

Number in Circle Designates the Heliostat Corrosion Area in Square Inches

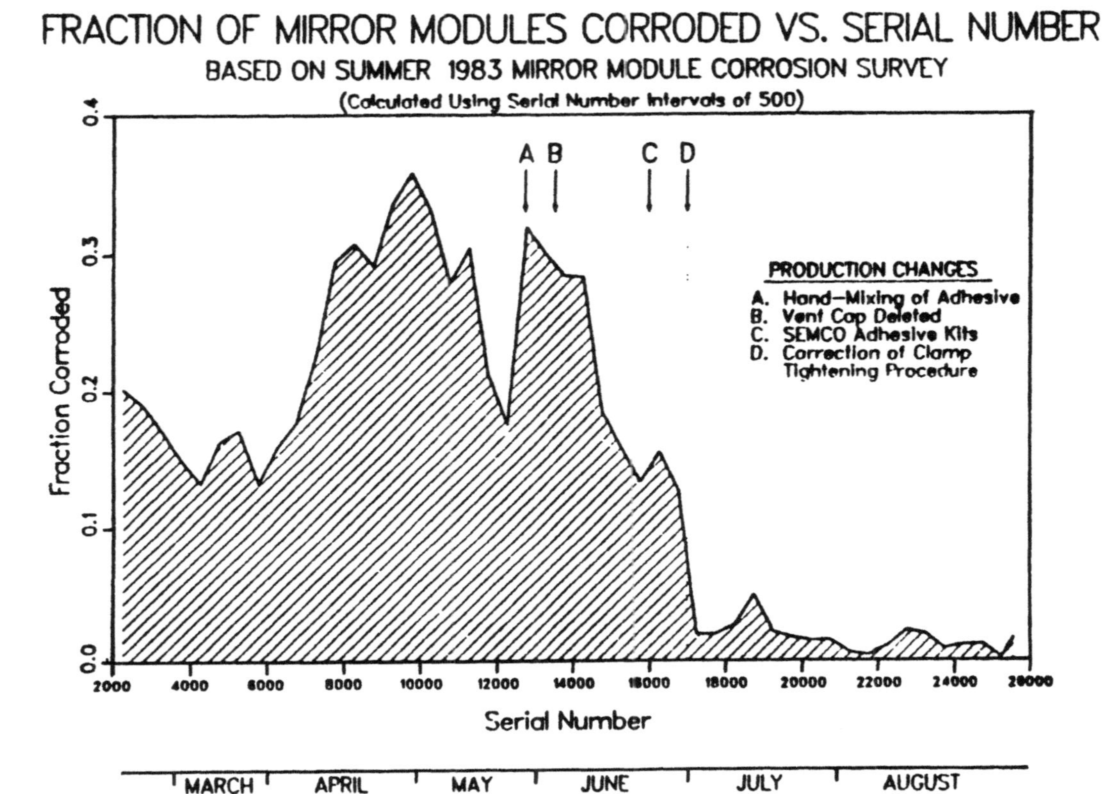

FIGURE 6 - CORRELATION OF CORROSION TO MIRROR MODULE SERIAL NUMBER

HELIOSTAT MAINTENANCE

- 94 TO 99 PERCENT OF THE HELIOSTATS ARE OPERATING AT ANY ONE TIME

- MAINTENANCE HOURS

  JAN 82 THROUGH MAR 83 - 2586 HOURS

  APR 83 THROUGH DEC 83 - TO BE DETERMINED (EST. 160 HR/MO)

- MAINTENANCE ORDERS

	DEC 81 - DEC 82	1983
AZIMUTH MOTOR	121	48
ELEVATION MOTOR	35	18
GEAR BOX	2	13*
HELIOSTAT CONTROLLER	296	465
HELIOSTAT FIELD CONTROLLER	34	20
AZIMUTH ENCODER	103	138
ELEVATION ENCODER	130	43
MIRROR	21	11
PHYSICAL	86	61
TOTAL	929	817

*1 GEAR BOX FAILURE

CLM:4/10/84

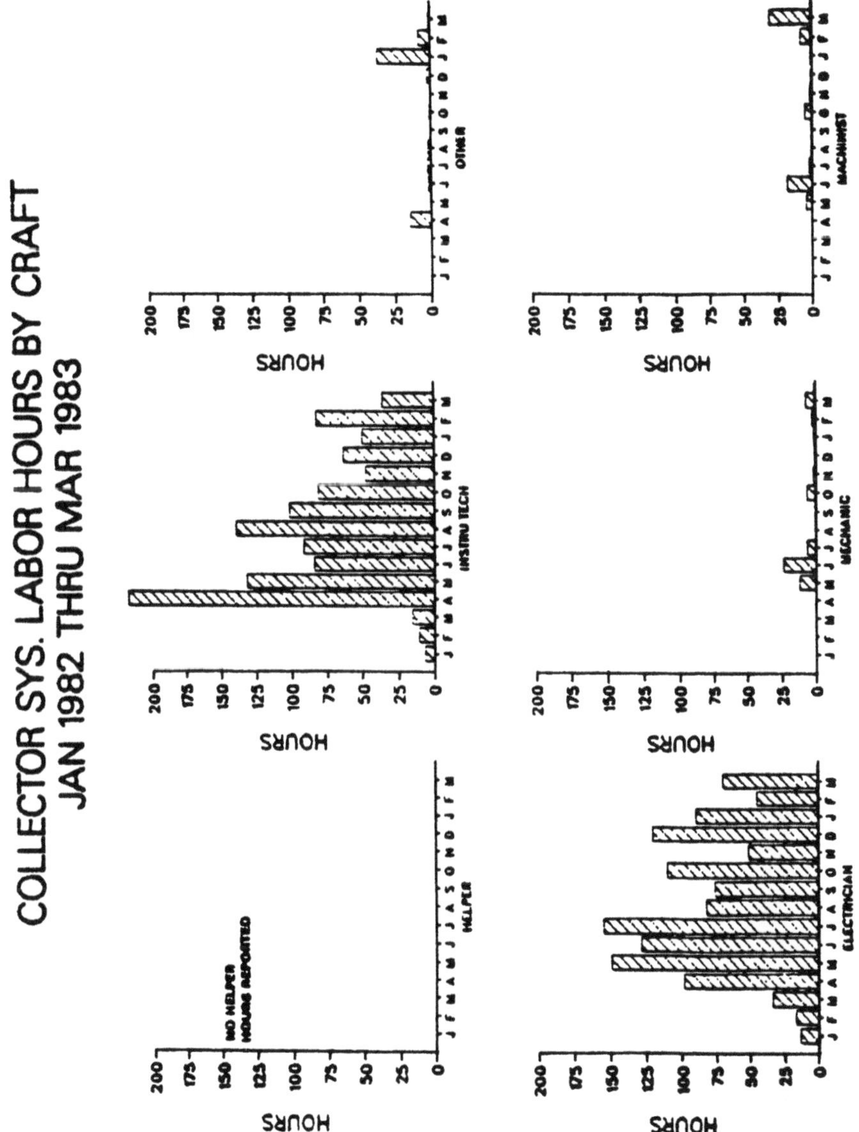

COLLECTOR SYS. LABOR HOURS BY CRAFT
JAN 1982 THRU MAR 1983
HOURS
200
175
150
125
100
75
50
25
0
J F M A M J J A S O N D J F M
NO HELPER HOURS REPORTED
HELPER
INSTRU TECH
OTHER
ELECTRICIAN
MECHANIC
MACHINIST

MAINTENANCE LABOR HOURS BY SYSTEM
JAN 1982 THRU MAR 1983

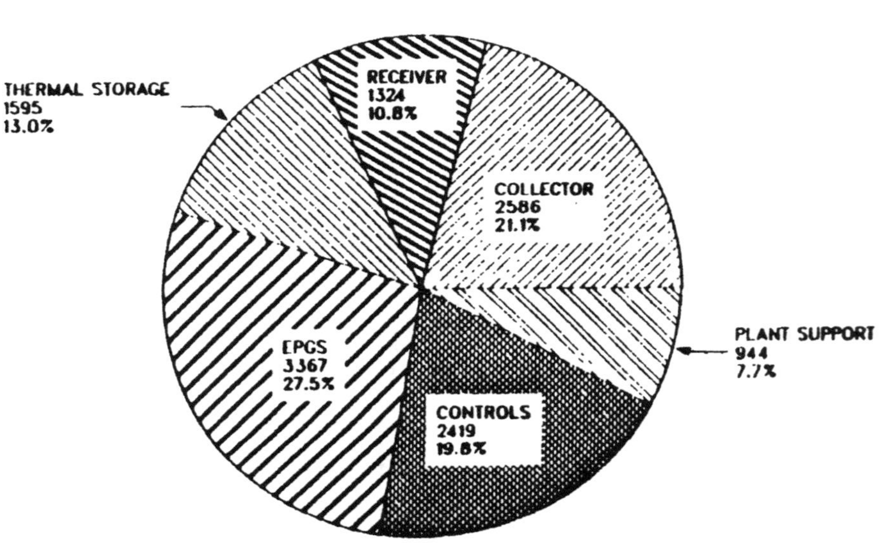

HELIOSTAT EVALUATION
PLANS FOR NEXT 12 MONTHS

TRACKING ACCURACY AND BEAM QUALITY MEASUREMENTS

CONTINUE MIRROR CLEANLINESS MEASUREMENTS

MONITOR MIRROR CORROSION AND VENTING EFFECTIVENESS

WIND SPEED AND WIND LOAD ANALYSIS

EVALUATE HELIOSTAT WASHING MACHINE EFFECTIVENESS

BCS DOCUMENTATION AND IMPROVE POWER MEASUREMENT ACCURACY

PUBLISH HELIOSTAT EVALUATION REPORT

CLM:4/10/84

LESSONS LEARNED

COLLECTOR SYSTEM

1. SEALING MIRRORS TO PREVENT INTRUSION OF WATER IS VERY DIFFICULT WHEN THE MIRROR MODULE CONTAINS A VOLUME OF AIR.

2. PROPER GROUNDING AND SHIELDING AGAINST LIGHTNING IS NECESSARY.

3. RAINWASHING IS AN EFFECTIVE METHOD TO CLEAN MIRROS (97% CLEAN); HOWEVER, MECHANICAL WASHING IS ALSO REQUIRED. SPRAY RINSING (84% CLEAN) MAY BE COST EFFECTIVE. A SPRAY AND BRUSH 99% CLEAN) MAY BE MORE COST EFFECTIVE.

4. DEVELOPMENT OF CONTROL SOFTWARE AND RELIABLE HARDWARE IS TIME CONSUMING AND EXPENSIVE.

5. THE DATA ACQUISITION SYSTEM SHOULD HAVE A HIGH AVAILABILITY AND SHOULD BE INDEPENDENT FROM ALL OTHER SYSTEMS. METEOROLOGICAL AS WELL AS OTHER SENSORS AND INSTRUMENTS REQUIRE FREQUENT MAINTENANCE AND CALIBRATION.

CLM:5/11/84

LESSONS LEARNED

BEAM CHARACTERIZATION SYSTEM (BCS)

1. A BCS IS REQUIRED AND IS WELL SUITED FOR MEASURING HELIOSTAT TRACKING ERRORS AND BEAM QUALITY; HOWEVER, IMPLEMENTATION IS NOT A SIMPLE TASK ($500K+). HELIOSTAT MIRROR WAVINESS, CANTING, CONTOUR AND SUNSHAPE EFFECTS CANNOT BE SEPARATED WITH A BCS.

2. TO ALLOW EARLY TRACKING ERROR CORRECTIONS, THE BCS SHOULD BE AVAILABLE WHEN THE HELIOSTATS ARE INSTALLED. THE BCS SHOULD BE A STAND-ALONE SYSTEM TO MAKE TROUBLESHOOTING EASIER AND MINIMIZE DOWN TIME.

3. A MOONTRACKING CAPABILITY IS A VALUABLE TOOL FOR MAKING A QUALITATIVE EVALUATION OF HELIOSTAT TRACKING ACCURACY AND MIRROR CANTING. HELIOSTATS REQUIRING ATTENTION CAN BE EASILY IDENTIFIED.

4. THERE ARE MANY DETAILS IN A BCS REQUIRING CAREFUL ATTENTION TO MAKE THE SYSTEM WORK.

5. MEASURING ALL THE HELIOSTATS IN A FIELD WITH A BCS IS TIME CONSUMING. AT LEAST 3 DATA POINTS ARE REQUIRED OVER A DAYS TIME AND THE WEATHER OFTEN PREVENTS THIS. A MUCH FASTER SYSTEM IS NEEDED.

CLM:5/11/84

SOME CORROSION STATISTICS FOR THE 10 MW_E SOLAR CENTRAL RECEIVER PILOT PLANT

	1983	1984
FIELD AREA CORRODED PERCENT	0.015	0.029
MODULES WITH SOME CORROSION PERCENT	15	27
SQUARE INCHES OF CORROSION*		
N.W. QUADRANT	2,291	3,880
N.E. QUADRANT	10,905	20,629
S.W. QUADRANT	1.604	3,228
S.E. QUADRANT	2,007	4,471

* THERE ARE APPROXIMATELY 17.8 MILLION SQUARE INCHES OF MIRROR SURFACE IN EACH SOUTH QUADRANT AND 36.8 MILLION SQUARE INCHES IN EACH NORTH QUADRANT.

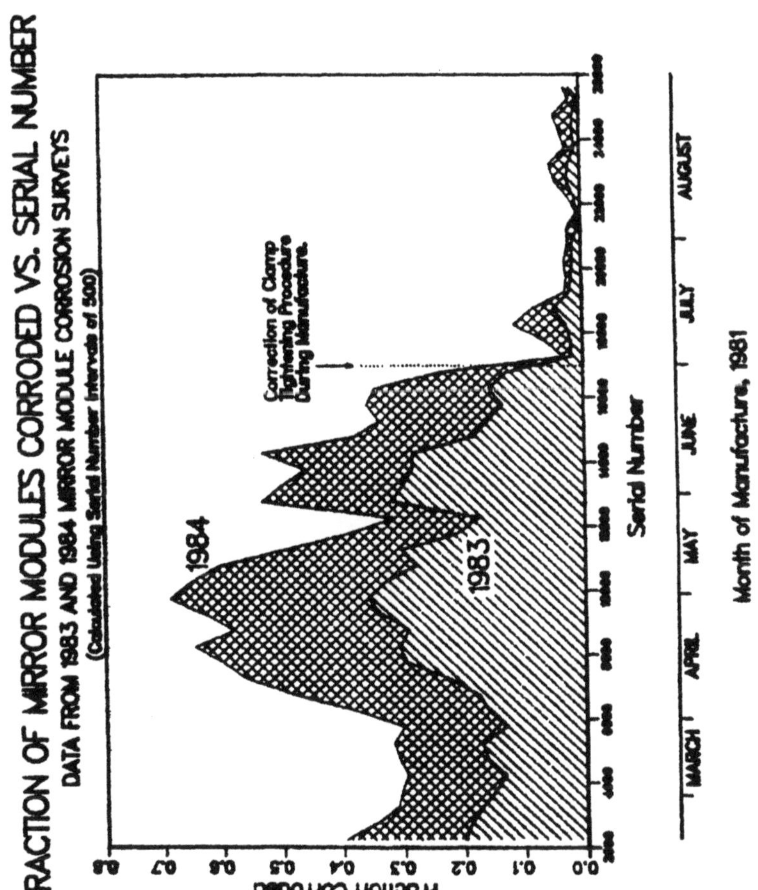
FRACTION OF MIRROR MODULES CORRODED VS. SERIAL NUMBER
DATA FROM 1983 AND 1984 MIRROR MODULE CORROSION SURVEYS
(Calculated Using Serial Number Intervals of 500)
Fraction Corroded
0.0
0.1
0.2
0.3
0.4
0.5
0.6
0.7
0.8
1984
1983
Correction of Clamp
Tightening Procedure
During Manufacture.
Serial Number
MARCH
APRIL
MAY
JUNE
JULY
AUGUST
Month of Manufacture, 1981

SECTION 5

RECEIVER BEHAVIOR COMPARISON

RECEIVER BEHAVIOR

INTRODUCTION

Sodium has very attractive heat transfer characteristics, thus it is an obvious candidate cooling fluid for a high temperature, high performance, central receiver system. The SSPS project has operated with two sodium receivers: a cavity type receiver designed by Interatom, constructed and installed by Sulzer Brothers Ltd., and an external type receiver designed and manufactured by Franco Tosi Company. The cavity was installed first and operated from mid 1981 until April 1983, when it was replaced by the external receiver, the ASR. Even though the design requirements for these two receivers were different, this project provided the first opportunity to make observations of a cavity and an external type receiver under similar operating conditions.

This evaluation area (section) contains detailed descriptions of the two receivers prepared by the designer/manufacturer, several reports of specific evaluation of performance and losses, some details of the methods used in conducting these evaluations, and a few reports which are attempts to make comparisons. The reader is counseled to treat these comparisons with care, simply because the two receivers were designed to very different specifications and for different uses. The first receiver was required to be a conservative design, because this was the first use of sodium as a heat transfer fluid in a solar central receiver system and conservatism was very important, wheras, the external receiver was, at the start, an advanced sodium receiver, the ASR.

The first report in this section, RECEIVER THERMAL PERFORMANCE: THEORY, is an interesting introduction to receiver design considerations and theoretical considerations, written by A. De Beneditti and J. Martin, where they relate the theory to the thermal performance as observed in operation of the SSPS receivers. They expended considerable effort during this evaluation to assure that the theoretical development realistically explained receiver performance, particularly the ASR.

The conclusions are:

1 - Performance analysis of the ASR has confirmed design objectives.
2 - Observed thermal losses are lower than expected.
3 - A factor of 10 scale-up appears reasonable.

The second report is a detailed discussion of the design, analysis and construction of the external receiver as seen by the Franco Tosi/Agip team of P. Cavalleri, V. Bedogni, A. Di Meglio of Franco Tosi, and A. De Benedetti, and C. Sala of Agip. This report, DESIGN AND CONSTRUCTION OF THE ADVANCED SODIUM - COOLED RECEIVER (ASR) FOR THE IEA/SSPS CENTRAL RECEIVER SYSTEM PLANT (ALMERIA, SPAIN), provides a detailed description of the receiver, its instrumentation and operational constraints. The modifications of heliostat pointing and the need for this modification is also discussed. There are no conclusions in the report; however, some recommendations for receiver design are provided.

The third report provides a description of the cavity receiver as constructed and installed by Sulzer Brothers Ltd. The report is prepared by H. W. Fricker who led the team from Sulzer Brothers. Although the intent of this report is to provide a reference description of the receiver, Mr. Fricker has also presented some results of his evaluation of the receiver performance. The report title is: THE CAVITY RECEIVER.

His conclusions are:

1 - The SSPS cavity is a conservative design.
2 - The SSPS cavity has worked well and is easy to use.
3 - Storage capacity of the ceramic wall was not sufficient.
4 - Significant improvement in performance is possible with simple modifications.

Having performed his evaluation of the cavity receiver and then later serving in a consultant role during design of the external receiver, Mr. Fricker proceeded to prepare an evaluation report comparing the two receivers. The reader is again cautioned, in reviewing this report, that the two receivers were designed to very different requirements, and that Mr. Fricker's position, as leader of the design team of one of the receivers, should be considered. The report, RECEIVER COMPARISONS, is clearly very helpful and provides some interesting comparative charts.

Conclusions are:

1 - The SSPS external receiver is more efficient than the SSPS cavity receiver.
2 - Both receivers performed their intended functions.
3 - The cavity is easier to use than the external.
4 - Modifications to either receiver could resolve any identified problem or constraint.

Performance measurements were made on both of the receivers during normal operation and in very specific measurement efforts referred to as measurement campaigns. G. Lemperle and W. Schiel report on their efforts during one of these measurement campaigns where the cavity receiver was undergoing tests and they were developing new interesting techniques for measuring the energy input to the receiver, providing one of the more important inputs to determine receiver efficiency. This report, EFFICIENCY AND TEMPERATURE MEASUREMENTS, effectively describes the techniques used and the result. The conclusion is that: "The measured data compare rather well with calculation of receiver losses for the cavity." Continuing with the evaluation of measurements and tests, R. Carmona, H. Jacobs and M. Sanchez provide a summation and evaluation of testing with the receivers in the report RECEIVER LOSSES; RESULTS OF TESTS. This evaluation takes into account the problems of inaccurate data as produced by the flow meter during early testing, and then provides a realistic efficiency comparison of the two receivers. Again, there are well understood reasons for some of the efficiency differences that can easily be changed.

The conclusions of this evaluation are:

1 - The losses with the cavity receiver are higher than for the external receiver. At 350°C these losses are 450kw vs 150kw.

2 - The difference appears to be the difference in convective losses.

After developing a theoretical analysis of receiver performance, A. Benedetti and M. Blanco developed a thermodynamic model of the external receiver (Theresa), then applied this model in a comparison with acquired performance data. This report, SIMULATION AND COMPARISON shows, in a graphical form, how well the model represents the receiver. The only conclusion is that the model does produce results that compare well with the actual receiver performance data.

In the last days of operation with the cavity receiver, in late March and April 1983, a rigorous test program to obtain comprehensive performance data was implemented. This program was specifically intended to obtain transient response data with this receiver and consequently the data recording was requested to be in increments of 30 seconds, normally referred to as "point and class summaries". Unfortunately, the ability to process this type of data was developed rather late in the evaluation period and, therefore, the data had not been analyzed prior to the summer of 1984. Mr. N. Gregory attempted to unravel this data in late 1984 and, with great difficulties, found sufficient data to prepare the report TRANSIENT RESPONSE OF THE SULZER (cavity) RECEIVER.

The conclusions of this evaluation are:

1 - The data tapes with the point and class summaries are difficuilt to work with.
2 - The cavity operates without problems in transient conditions.

With improved data handling capabilities so that transient data became a more common possibility, the external receiver, the ASR, was subjected to an indepth transient analysis by R. Carmona and J. Martin with the title THE ADVANCED SODIUM RECEIVER (external) TRANSIENT RESPONSE.

The conclusions are:

1 - The ASR can be heated to operating temperature from a warm condition in 2 minutes during mid day.
2 - Early morning start-up requires a longer time because of reflected sun shape and the small receiver.

A comparison evaluation performed by the operating staff from the organization Cia Sevillana De Electricidad, addresses operational constraints resulting from the particular design charateristics of the two receivers. As stated above, the design requirements of the two receivers were very different. Consequently, this comparison is more a critique of the design specifications than of the receivers, as they were constructed. The report DIFFERENCES BETWEEN FILLING STRATEGIES FOR THE SULZER AND ASR RECEIVERS was prepared by F. Ruiz and A. Cuadrado.

The principle conclusion is that the external receiver (ASR) is difficult to fill, so it is not drained each night. This was never stated as a design requirement, because the cavity receiver was not operated that way and was only drained when major shut down was required.

Looking only at the specific operational data, acquired during "normal operation," H. Jacobs and C. Selvage performed an evaluation of the two receivers and reported on this effort in RESULTS ON THE PERFORMANCE OF THE SULZER CAVITY RECEIVER AND THE FRANCO TOSI EXTERNAL RECEIVER. It was necessary to apply corrections to the sodium flow data for the early tests because in the calibration efforts of A. Brinner and F. Reich (DFVLR Stuttgart), the flow meter was shown to be in error. This flow meter was replaced prior to the external receiver testing.

The conclusions are:

1 - The SSPS external receiver is more efficient than the SSPS cavity receiver - 91.9% compared to 77.4%.

2 - The reasons are :

(a) Higher heat flux in the external receiver
(b) Smaller size with the external receiver
(c) Conservative design of the cavity receiver
(d) Poor heat flux distribution with the cavity receiver

RECEIVER THERMAL PERFORMANCE: THEORY

Alessio De Benedetti, AGIP Spa and Jose G. Martin, ITET

1. INTRODUCTION

This paper discusses topics which are relevant to the quantitative evaluation of the thermal performance of a solar receiver. It outlines the general problem of thermal balance by considering the heat transport equation in an element of the receiver wall and describing the difficulties which must be solved or circumvented to arrive at numerical solutions for a real receiver.

Absorption, reradiation, and convection are treated in some detail, borrowing heavily from material which represents the 'state of the art'. The effect of the receiver backwall is also treated.

Analytical solutions for some idealized problems are presented, as well as an overall iterative scheme which can be applied generally. Numerical solutions for cavity and billboard receivers based on this scheme are presented. Finally, the paper draws lessons from the general discussion, and provides a link with the work on simulation and the results of actual experiments.

2. OVERALL THERMAL BALANCE

2.1 Heat Transport Equation

Consider an element of a wall in a solar receiver (with surface dA and thickness s), in contact to the outside with a fluid (cooling fluid or air), and subject in the inside to radiant phenomena and convective heat transfer with the air in the cavity. The equation governing how the temperature T, of the element varies in time t, is the general heat transport equation

$$\nabla^2 T = \frac{\rho . C_p .}{k} \frac{\partial T}{\partial t}$$

where ρ = density of the wall material
C_p = specific heat at constant pressure
k = conductivity

To solve this equation, one must specify the boundary and the initial (i.e., at t = 0) conditions. The boundary conditions are normally specified at the two interfaces of one with the cavity, and the other on the outside.

2.2 Boundary Conditions

On the outside, the boundary condition follows from the requirement that

$$-K \text{ grad } T\big|_e \cdot \hat{n}_e = h_e(T\big|_e - (T_{fe})$$

where $T\big|_e$ = temperature at the wall external surface
h_e = convective heat transfer coefficient
T_{fe} = bulk temperature of the coolant fluid or air

(Hereafter, 'fluid' will refer to the coolant fluid or air.)

The coefficient of convective heat transfer h_e may be calculated from correlations for nondimensional numbers or from analytical/experimental correlations. An example of the latter are the Chen correlations (Reference 2) for heat exchange in boiling fluids. Examples for the nondimensional numbers are the correlations:

$Nu = Re^{a}Pr^{b}$ (single phase exchange) (Ref. 2)
$Nu = Pe^{c} + d$ (exchange with metal liquids) (Ref. 3)

where Nu = Nusselt number
Re = Reynolds number
Pr = Prandlt number, and
Pc = Peclet number.

To determine the bulk fluid temperature it is necessary to solve the equation for the thermal balance in the fluid. If the calculations refer to fast transients, it is also necessary to consider a term for the rate of change of mass in the continuity equation, viz:

$$\frac{\partial \rho}{\partial t} + \frac{1}{Pg}\frac{\partial T_{fe}}{\partial t} = 0$$

$$\rho \cdot C_{pf} \frac{\partial T_{fe}}{\partial t} + \frac{\Gamma_{fe}}{Pg} \, c_{pf} \, \frac{T_{fr}}{l} = he(T\big|_e - T_{fe})$$

where C_{pf} = specific heat of the fluid at constant pressure
Γ_{fe} = fluid mass flow rate
Pg = heated perimeter, and
l = coordinate along which the fluid moves.

These equations give the fluid temperature in terms of the energy entering variation. If there is a change in phase, the relation becomes one at constant temperature. In the stationary state:

$$he(T\big|_e - T_{fe}) = \frac{C_{pf} \cdot \Gamma_{fe}}{Pg} \frac{dT_{fe}}{dl}$$

In the case of phase exchange,

$$he(T\big|_e - T_{fe}) = \frac{\Gamma_{fe}\, dH_{fe}}{Pg \cdot dl}$$

where $H_{fe} = H_{fce} + X \cdot H_{fev}$.

(Here X = liquid fraction, and H_{fel} and H_{fev} are the liquid and vapor enthalpies, respectively.)

When the fluid inlet temperature and flowrate are known, it is possible to calculate T_{fe} along the cooled receiver surfaces.

On the internal surfaces, there is a superposition of radiative and convection phenomena. The external boundary condition results from the fact that, at any location P on the surface, the net emitted flux must be equal to the heat reaching the surface of the wall by conduction. Writing $\hat{n}_i$ as the normal to the surface aimed at the inside of the cavity,

$$-k \text{ grad } T\big|_i \cdot \hat{n}_i = h_i(T\big|_i - T_{fi}) + H(P) + S(P) - B(P)$$

where $-k \text{ grad } T\big|_i \cdot \hat{n}_i$ is the heat conducted to the internal surface,

S(P) = direct solar flux reaching the surface from the heliostat field (this is 'data'),

H(P) = radiant flux reaching the surface from other walls and the medium contained in the cavity (this is an unknown),

B(P) = the total radiant flux (radiosity) leaving the surface. The radiation is the sum of the emitted and reflected fluxes, (it is an unknown).

$hi(T_i - T_{fi})$ is the convective heat flux within the cavity, T_{fi} the temperature of the air in the cavity, and hi is the coefficient for convective heat transfer for natural convection, which should be evaluated from complex correlations based on experiments. Unfortunately, most of the available correlations describe physical situations completely different from those characterizing a solar receiver.

The air temperature in the cavity is not known; in principle it must be calculated. It depends on the radiant energy absorbed by the air and the air circulation driven by density gradients in the cavity.

The treatment of the interaction of radiation with an absorbing and emitting gas is very complex and requires treatment by spectral bands. Further, the velocity at which air circulates in the cavity because of natural convection depends on T_{fi} itself, i.e., on the variable that we wish to evaluate!

Given the complexity of the problem and the many parameters needed to describe it, it may be necessary to neglect the convective heat transfer between the wall and the air to arrive at an approximate solution. Without convection, the external boundary condition for the wall element reduces to

$$-k \text{ grad } T\Big|_i \cdot \hat{n}_i = H(P) + S(P) - B(P)$$

where H(P) and B(P) are not known and must be calculated.

If absorption in the air in the cavity is negligible, the incident flux H(P) will depend on the emission from the internal surfaces and on the geometrical arrangement of those surfaces.

The radiosity B(p) from an opaque surface is the sum of the thermal emission and the reflected fraction of the radiation coming from other surfaces. It is an integral over the solid angle and the whole spectrum of the monochromatic directional intensity $I\,(\rho, \lambda, \Theta_2, \phi_2)$ emanating from a point P in direction $\Theta_2 \phi 2$. For an opaque surface $I\,(\rho, \lambda, \Theta_2, \phi_2)$ has one contribution, from the thermal radiation emitted by the surface and another from radiation which is reflected. The contribution from thermal emission may be written as:

$$\text{thermal emission} = e\,(\rho, \lambda, \Theta_2, \phi_2) \cdot E\,(\rho, \lambda)$$

where

$e\,(\rho, \lambda, \Theta_2, \phi_2)$ = monochromatic directional emissivity and
$E(\rho, \lambda)$ = monochromatic intensity of radiation from a blackbody at temperature which characterizes the surface at P.

The reflected contribution may be written as:

$$\text{reflected contribution} = \int_{\Omega} I_i(\rho, \lambda, \Theta_1, \phi_1) \cdot f(\rho, \lambda, \Theta_2, \phi_2, \Theta_1, \phi_1)$$

where

$I_i(\rho, \lambda, \Theta_1, \phi_1)$ is the monochromatic direction incident intensity,
and

$f(\rho, \lambda, \Theta_2, \phi_2, \Theta_1, \phi_1)$ is the monochromatic bireflectance function.

The term B(P)-H(P), i.e., the absorbed flux, is the difference between the outgoing and the incident radiation, integrated over the wavelengths and angles.

2.3 Summary of the Thermal Balance Considerations

The thermal behavior of the receiver may be formulated in terms of these four equations:

- heat conduction at the wall
- general reflection law
- relation between incident and emitted radiation, and
- fluid thermal balance.

It is possible to assign values to the characteristics of the wall and the fluid (conductivity, density, heat capacity), and to the optical characteristics of the surface (monochromatic directional emittance and bireflectance function). Since the solar flux distribution is also known (from optic computations), there are four unknown functions left:

- wall temperature
- fluid temperature
- emitted intensity, and
- incident intensity.

In other words, the problem is formally solved.

Obviously, there are difficulties in obtaining a numerical solution and a solution must be found at all points P in the cavity.

Iterative Solution

The solution of the general thermal balance equation outlined in Section B is complex, not only for the presence of both integral (for radiation) and differential-equations (for conduction and fluid balance), but because of the complexity of the geometrical and fluid dynamic description of the receiver (this difficulty arises when going from the infinitesimal element to the finite configurations).

In fact, an analytical solution is possible only in a few simplified geometries (conical and cylindrical cavities, cavities with rectangular cross-sections, etc.) and then only if one does not consider the general coupled conduction and convection problem.

Even if all these difficulties were surmounted, one would still have the problem of the time spent in the solution.

This is one motivation for a scheme that solves the radiation problem in a very simple manner and which approaches the receiver evaluation in a 'modular' fashion,

- dealing sequentially with the various problems connected with the analysis and dimensioning of the receiver, and

- achieving a close and comprehensive characterization.

Another motivation is the lack, at the present time, of the detailed information which is necessary for a more exact description. Notably, there is not enough information on the optical properties (emissivity and reflectivity) of materials at high temperatures.

The proposed scheme consists of splitting the problem into two successive blocks, which decouples the problem of the radiating cavity from the thermo-hydrodynamic problem. The solution may be obtained via successive iterations, using the results from a part of the calculations as input for the next. This solution also reflects a definite separation between the physical aspects of the problem. The sequence is as follows:

1. On the basis of the spectral distribution of the radiation incident on the cavity walls in the visible range, one calculates the successive reflections. The aperture is simulated as a black wall at 0^oK, i.e., one which absorbs all the radiation reaching it from the inside. This calculation yields:

   - the energy, Q_k*, in the visible spectrum, absorbed in each surface (Q_k for $k = 2$ to M)

   - the losses, Q_1*, in the visible spectrum (i.e., the radiation leaving the cavity through the aperture)

If the walls are completely black in the visible range, this calculation is not necessary--all the energy from the field is absorbed.

2. From the values of Q obtained in the first step, it is possible to evaluate the thermal balance in the steady state for the walls and the fluid. The result is a temperature distribution in the wall ($T_k*(P)$) and in the fluid ($T_{fe*k}(P)$) due to the contribution from the radiation energy in the visible range.

3. Using the calculated T_{k*}, one calculates the emission in the infrared band. This calculation consists in evaluating the energy exchange in the infrared between the cavity walls to estimate:

   - the energy absorbed by the M walls ($Q_k{}^1, k = 2, M$)

   - the energy leaving through the aperture, Q^1, considered as a black wall. $Q_k{}^1$ is the sum of the energy leaving in the visible and the infrared bands:

$$Q_k{}^1 = Q_{k*} + Q_k{}^1$$

4. From $Q_k{}^1$, one repeats Step (2) above, to obtain new values of T and T_{fe}, which shall be denoted as $T_k{}^2$ and $T_{fek}{}^2$.

5. Using $T_k{}^2$, it is possible to repeat Step (3). Denote the 'new' values for the total radiant energies on each wall as $Q_k{}^2$ and $Q_1{}^2$.

6. From Q_k, one calculates again the wall temperature $T_k{}^3$, and so on.

It is clear that, utilizing the two blocks, i.e., calculating the irradiation, and making the thermo-hydrodynamic calculations sequentially, it is possible to obtain as accurate a solution as desired. The accuracy may be estimated from the difference between $T_k{}^1$ and $T_k{}^{1+1}$. The flow chart for the proposed evaluation scheme is shown in the following figure.

Flow Chart

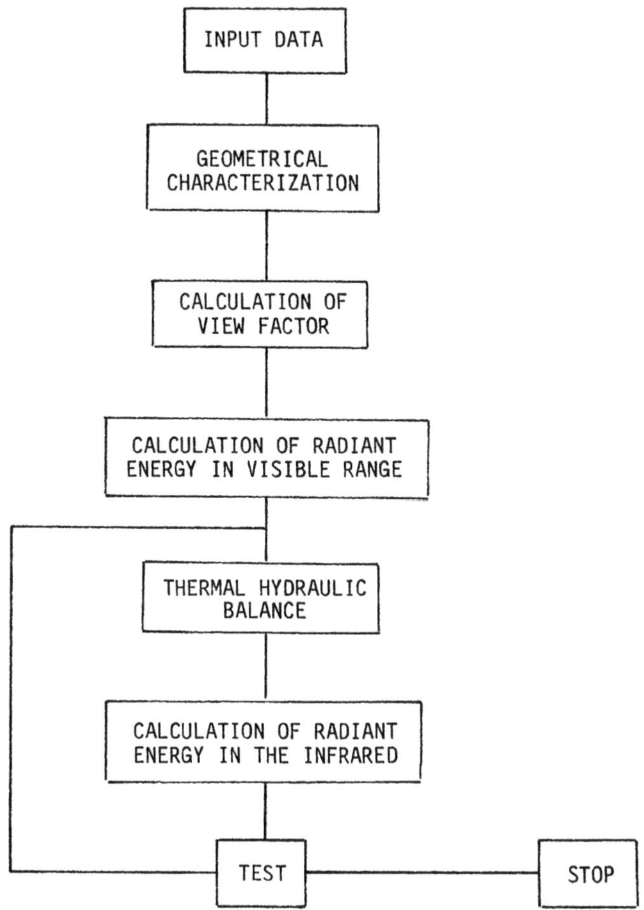

3. ABSORPTANCE

The monochromatic directional absorptance and the directional to hemispherical reflectance are properties which are defined locally, are equal to each other, and depend on the surface temperature and, of course, the surface itself. Detailed data on these two properties are seldom available. Instead, design and evaluations are based on averaged or integrated values of these properties, weighted by some incident flux distribution.

No effort will be made in this paper to discuss these two properties rigorously--the available data does not warrant this effort at the moment. Instead, we shall address some sources of uncertainty which may affect the computations of losses and efficiencies.

'Pyromark Series 2500' is the trademark for a coating formulated from 100% silicone resin by the Tempil Division of Big Three Industries, and which has been used to coat both receivers installed at the site. The monochromatic hemispherical reflectance for Pyromark (i.e., the reflectance integrated over all angles for a certain incident flux) is plotted. At least two observations can be made from these plots: the monochromatic reflectance is relatively constant in the range between 300 and 1000 nanometers, but it also depends on the surface over which it has been applied and on the cycling to which it has been subjected.

The soiling of the surface introduces another uncertainty. At an experiment reported in Reference 14, some surfaces coated by Pyromark were dusted by hand with dust, until the surfaces appeared gray rather than their original dark black. The absorptivity of the dust had been measured at 0.51. The absorptivities of the dusted samples varied from 0.94 down to 0.85 with an average of 0.88 for all samples. After tap water was allowed to run down the surface to simulate rain effects, the absorptivity ranged from 0.9 to 0.95. It is hard to extrapolate from these results to the situation at the SSPS site. Although, in regards to dust, the SSPS environment is not particularly benign. To take absorptance measurements on the receiver, it would have been necessary to empty it, and a decision was made to postpone those measurements to obtain as much actual operating data as possible.

The actual shape of the active absorber surface is another source of uncertainty for the absorptance data. Those data refer to flat surfaces, while the receiver is composed of coated tubes, placed side by side.

Fortunately, it is possible to make a rough estimate of the resulting 'cavity' effect.

Consider a 'cell' composed of two adjacent absorber tubes. If one had the shape viewfactor between the tube quarter-cylinder and the 'aperture surface' with cross-section AB, it would be possible to estimate the effect accurately. A rough estimation is possible if the configuration is approximated by another, where the quarter-cylinders section AC, AB, are submitted by the segments AC, AB.

Neglecting emission, the radiosity of a surface is the product of the incident flux ϕ times the surface reflectance. Assume that the incident radiation (flux ϕ) is in the form of a parallel beam, and that the radiation reflected from each surface is completely isotropic.

Between the surfaces AC, AB, and CB the shape factors, F are equal to:

$$F_{ac,ab} = 1/\ 2,$$
$$F_{ac,cb} = 1 - 1/\ 2, \text{ and}$$
$$F_{ab,ac} = F_{ab,cb} = 1/2$$

The lengths AC and AB are 2 times the tube radius R.

These are going to be two contributions to the flow reaching the 'surface' AC. One is the direct contribution from the incident flux ϕ : because of geometry, this contribution is ϕ/ 2. The other contribution is from B_{cb}, the radiosity from CB. Because of the shape factor, this second contribution is (1 - 1/ 2)B_{cb}. A fraction ρ of this incident flux is reflected to give the 'equivalent' radiosity from surface AC:

$$B_{ac} = \rho \frac{\phi}{\sqrt{2}} + \rho \left\{ \left(1 - \frac{1}{\sqrt{2}}\right) B_{cb} \right\}$$

Because of symmetry, $B_{cb} = B_{ac}$, and therefore

$$B_{ac}\left[1 - \rho \cdot \left(1 - \frac{1}{\sqrt{2}}\right)\right] = \rho \frac{\phi}{\sqrt{2}}$$

and

$$B_{ac} = B_{cb} = \frac{1}{1 - \rho\left(1 - \frac{1}{\sqrt{2}}\right)} \cdot \rho \frac{\phi}{\sqrt{2}}$$

The power 'escaping from AB, W_{loss}, is equal to the 'area' AB, (2R) times the contributions from surfaces AC and AB, i.e.,:

$$W_{loss} = \sqrt{2}R * B_{ac} * F_{ac,cb} + \sqrt{2}R * F_{cb,ab} * B_{cb} = \frac{1}{1 - \rho\left(1 - \frac{1}{\sqrt{2}}\right)} \cdot \rho \frac{\phi}{\sqrt{2}}$$

Since the incident power is 2ϕ R, one may estimate an effective reflectance for the surface, ρ_G , as follows:

$$\rho_G = \frac{W_{loss}}{W_{in}} = \frac{\rho}{\sqrt{2}\left[1 - \rho\left(1 - \frac{1}{\sqrt{2}}\right)\right]} = \frac{\rho}{\sqrt{2} - \rho(\sqrt{2} - 1)}$$

In other words, the effective reflectance is lower (and the absorptance is higher) than the corresponding value for a flat surface.

Using the reflectance for Pyromark, which ranges from 0.05 to 0.08 according to surface treatment, cycles, etc., one has for

$$\rho \simeq 0.05 \qquad \rho_G = 0.036$$

and for

$$\rho \simeq 0.08 \qquad \rho_G = 0.058$$

In other words, the 'grooved' absorber active surface decreases the effective surface reflectance by about 30%.

4. EMITTANCE

The angular monochromatic emittance is a surface property which is equal to the angular hemispherical absorptance and to one minus the monochromatic angular to hemispherical reflectance. When integrated over angle and wavelengths, however, the emittance continues to be a property, while the absorptance and the reflectance do not.

This property is a function of temperature.

Figure shows the total 'normal' emittance as a function of temperature for Pyromark on different surfaces. There is a slight increase in the emittance with temperature in the case of Pyromark on 'as rolled' stainless steel, the increase is from about 0.78 at 315°F to slightly above 0.9 at 1090. The figure, taken from a NASA report, shows that coating emits diffusely and that therefore the measured values of total normal emittance are close approximations of total hemispherical emittance.

When discussing absorptance, it was noted that the actual receiver panel configuration resulted in an increase in the effective panel absorptance over that which would be measured on a flat panel. A related 'cavity effect' can be estimated for emittance.

Referring again to the figure, one notes that there are two contributions to the radiosity from surface AC; one of these is from emission in surface AC, in the infrared band, $\varepsilon \sigma T^4$. The other is from reflection of radiation incident on AC in this band. Radiation incident on AC comes from the infrared radiosity from CB, B_{cb}, multiplied by the viewfactor $F_{ac,cb} = 1 - 1/\sqrt{2}$. There is another incident contribution, 1/2 - T_a*, from radiation from the outside at ambient temperature T_a, which we shall neglect.

Therefore, in the infrared:

$$B_{ac} = \varepsilon \sigma T^4 + (1-\varepsilon)\cdot\left[\left(1 - \frac{1}{\sqrt{2}}\right) B_{cb} + \frac{1}{2}\sigma T_a^4\right]$$

Using

$$B_{ac} = B_{cb}$$

and

$$B_{ac}\left[1 - (1-\varepsilon)\cdot\left(1 - \frac{1}{\sqrt{2}}\right)\right] = (1-\varepsilon)\frac{1}{2}\sigma T_a^4 + \varepsilon \sigma T_k^4$$

one may solve for B_{ac} as follows:

$$B_{ac} = \frac{\varepsilon \sigma T_k^4}{1 - (1-\varepsilon)(1 - 1/\sqrt{2})} = B_{cb}$$

The loss through AB, W_{loss} has contributions from B_{ac} and B_{cb}, multiplied times $1/\sqrt{2}$ (the viewfactor from AC or CB to AB) and the 'area' AB, i.e., $\sqrt{2}R$. Thus:

$$W_{loss} = \sqrt{2}R\cdot\frac{1}{\sqrt{2}}\cdot\frac{\varepsilon \sigma T_k^4}{1-(1-\varepsilon)(1-\frac{1}{\sqrt{2}})} + \sqrt{2}R\cdot\frac{1}{\sqrt{2}}\cdot\frac{\varepsilon \sigma T_k^4}{1-(1-\varepsilon)\cdot(1-\frac{1}{\sqrt{2}})}$$

$$= 2R \frac{\varepsilon \sigma T_k^4}{1-(1-\varepsilon)(1-\frac{1}{\sqrt{2}})}$$

Dividing this by the power emitted from a blackbody of area 2R, 2R T^4, one can estimate an effective 'emittance' for the grooved surface as follows:

$$\varepsilon_o = \frac{1}{2.R\sigma T_k^4} \cdot 2R \frac{\varepsilon . \sigma T_k^4}{1-(1-\varepsilon)(1-\frac{1}{\sqrt{2}})} = \frac{\varepsilon}{\frac{1}{\sqrt{2}} + \varepsilon(1-\frac{1}{\sqrt{2}})}$$

Note that ε = 0 or 1, $\varepsilon_g = \varepsilon$. However, for $\varepsilon \simeq$ 0.08, which may characterize the Pyromark surface,

$$\varepsilon_g = 0.85$$

a 6% increase. This should be taken into account when estimating the radiation losses.

5. RADIATION

5.1 Generalities

Although a receiver should absorb as much of the energy incident on its aperture as possible, some of the incident concentrated flux is reflected from the inner surface and eventually escapes through the aperture. Furthermore, since the receiver surfaces are hot, they emit radiation: some of the emitted radiation escapes also.

Calculation of the Radiant Energy and Its Simplification

It has been noted that the general expression which relates the incident radiation to the radiation leaving the surface in a certain direction is:

$$I(\rho,\lambda,\Theta_2,\phi_2) = e(\rho,\lambda,\Theta_2,\phi_2)E(\rho,\lambda) + \int f(\rho,\lambda,\Theta_2,\phi_2,\Theta_1,\phi_1) I_i(\rho,\lambda,\Theta_1,\phi_1) \cos\Theta_1 \, d\Theta_1$$

The first simplifying assumption consists in assigning 'point' properties to the finite surfaces on which we may assume constant incident and outgoing radiation, temperature, and distribution functions. If k is a superficial element of a cavity with M surfaces, it may be verified that this hypothesis may be written as follows:

$$I_K(\lambda,\Theta_{2K},\phi_{2K}) = e_K(\lambda,\Theta_{2K},\phi_{2K}) E_K(\lambda) + \frac{1}{A_K} \iint_{A_i A_j} f_K(\lambda,\Theta_{2K},\phi_{2K},\Theta_{1K},\phi_{1K}) \cdot I_j(\lambda,\Theta_{2j},\phi_{2j}) \frac{\cos\Theta_i \cos\Theta_j}{R_{Kj}^2} \, dA_j$$

To simplify the problem further, it is necessary to specify the optical characteristics of the surface. Given the complexity of the expressions for the reflectances, and the fact that they are not known experimentally, one must assume limiting forms for the optical behavior, i.e., one may assume that:

- the surfaces reflect in a diffuse manner,
- the surfaces reflect in a specular manner, or that,
- the surfaces reflect in a combined diffuse and specular mode.

The first hypothesis eliminates the angular dependence, and in the second, the material follows rigorously the law of specular reflection.

To consider the walls as 'specular' or 'diffuse' is to presuppose a certain type of receiver model. Specifically, it implies an assumption as to whether the energy is absorbed quickly or is reflected and directed inside the cavity so that the radiation becomes more isotropic.

The following discussion will be based on the first hypothesis not only because it is simpler but because it is a better approximation to real receivers. It also is more suitable to describe surfaces with tube bundles when, because of the curved relief, no reflected energy tends to be isotropic.

6. FORMULATION

The radiant heat transfer in a cavity may be modelled by splitting the electromagnetic spectrum into two ranges: the 'visible' and the 'infrared'. Below, an asterisk (*) identifies a property or magnitude in the visible range; properties or magnitudes in the infrared will have no mark. For example, the rate at which radiant energy leaves a surface A_k per unit area by emission and radiation (i.e., the radiosity) will be represented by B_k (in the infrared) and $B_k{}^*$ (in the visible).

In terms of the viewfactors F_{kj} between surfaces S_j and S_k, the radiosity from a surface A_k can be related to the emissivities ($e_k{}^*$ and e_k) from that surface by the following expressions:

<u>In the infrared range</u>

$$B_k = e_k E_k + (1 - e_k) H_k \ , \tag{1}$$

where E_K is the power emitted from a black body at the temperature T_K of the surface, and

H_k is the total power incident on surface A_k from all other surfaces on the infrared, i.e.,

$$H_k = \sum_{i}^{N} B_j \ F_{kj}$$

<u>In the visible</u>

$$B_k{}^* = (1 - e_k{}^*) S_k{}^* + (1 - e_k{}^*) H_k{}^* \tag{2}$$

where S_k* is the solar flux incident on A_k,
H_k* is the total power incident on surface A_k from all other surfaces A_j in the visible range, i.e.,

$$H_j^* = \sum_i^N F_{ji} B_i^*$$

The notation in the last equation is made shorter by defining $r_k^* = 1 - e_k^*$

By stating that each surface is at some temperature T_k, the problem is considerably simplified: there is no need to solve for the T_k's explicitly.

6.1 Cavity with Three Surfaces

For an arbitrary cavity configuration, the computer program CAVIT 2, which is on line, calculates the infrared and visible radiosities for each surface, and the radiant efficiency of the cavity for any specified configuration. The program also classifies the radiant losses into losses by reflection and losses by re-emission.

In any cavity there will be an aperture surface (A_1), a surface which does not 'see' the heliostats (A_2), and a third area (A_3) which is illuminated directly by the heliostat field.

Let us rewrite Equations (1) and (2) explicitly for the three surfaces.

In the Visible Range (Equation 2)

$$B_1^* = r_1^* S_1^* + r_1^*(B_1^*F_{11} + B_2^*F_{12} + B_3^*F_{1}3_3) \quad (3)$$

$$B_2^* = r_2^* S_2^* + r_2(B_1^*F_{21} + B_2^*F_{22} + B_3^*F_{23} \quad (4)$$

$$B_3^* = r_3^* S_3^* + r_3^*(B_1^*F_{31} + B_2^*F_{32} + B_3^*F_{33}) \quad (5)$$

All the radiation reaching the aperture from the inside will be lost. That is,

$$r_1^* = 0 \quad (6)$$

and

$$B_1^* = 0 \quad (7)$$

Also, by definition, no solar flux reaches surface 2:

$$S_2^* = 0 \quad (8)$$

Substituting these into equations (4) and (5), and solving for the radiosities in the visible spectrum from surfaces 2 and 3, one obtains:

$$B_{3*} = \frac{r_3{}^*S_3{}^*}{1 - r_3{}^*F_{33} - \frac{r_2{}^*r_3{}^*F_{32}F_{23}}{1 - r_2{}^*F_{22}}} \quad (9)$$

and

$$B_2{}^* = \frac{r_2{}^*F_{23}}{1 - r_2{}^*F_{22}} B_3{}^* \quad (10)$$

In the Infrared Range

Writing $E_k = \sigma T_k{}^4$, where σ is the Stephan - Boltzmann constant, Equation (1) reads:

$$B_1 = e_1\sigma T_1{}^4 + (1 - e_1)(B_1F_{11} + B_2F_{12} + B_3F_{13}) \quad (11)$$

$$B_2 = e_2\sigma T_2{}^4 + (1 - e_2)(B_1F_{21} + B_2F_{22} + B_3F_{23}) \quad (12)$$

$$B_3 = e_3\sigma T_3{}^4 + (1 - e_3)(B_1F_{21} + B_2F_{22} + B_3F_{23}) \quad (13)$$

As noted when discussing the visible range, all the radiation reaching the aperture from the inside will be lost. That is,

$$1 - e_1 \pm 0 \quad (14)$$

Also, the outside temperature is very low compared with that in the cavity. It simplifies the equations to set:

$$T_1 = 0^oK \quad (15)$$

In other words, the 'inward' radiosity from the aperture in the infrared range (Equation 11) is:

$$B_1 = 0 \quad (16)$$

Substituting this into Equations (12) and (13), and solving for the radiosities from surfaces (2) and (3) in the infrared, one obtains:

$$B_2 = \frac{e_2\sigma T_2{}^4}{1 - (1-e_2)F_{22}} + \frac{(1-e_2)F_{23}}{1-(1-e_2)F_{22}} B_3 \quad (17)$$

and

$$\frac{e_3\sigma T_3{}^4}{1-(1-e_3)F_{33} - \frac{(1-e_2)(1-e_3)F_{32}F_{23}}{1-(1-e_2)F_{22}}} + \frac{\sigma(1-e_3)F_{32}e_2T_2{}^4}{1-(1-e_2)F_2} \quad (18)$$

Summarizing, for a general cavity with two surfaces plus an aperture, we now have formal expressions for the total flux leaving the cavity surfaces. From this, we may derive a formal expression for the efficiency of the cavity.

Notes on the Radiant Efficiency

The radiant efficiency, n_c, is defined here as:

$$n_c = \frac{\text{power absorbed in the cavity}}{\text{power entering the cavity}} \qquad (19)$$

The power entering the cavity may be written as the product of the solar irradiance on the ground, I, times the area of the cavity aperture, $A_1$2 times some effective concentration factor, C. (This choice means that C is defined as the ratio between the mean radiation flux at the aperture and the irradiance on the ground.)

The power absorbed in the cavity is equal to the power entering the cavity (IA_1C) minus the lost radiant power. The lost radiant power is obviously the power reaching the cavity aperture from the inside in both the visible range (Q_1*) and the infrared (Q_1).

In terms of these symbols, the radiant efficiency (Equation 1) has therefore been defined as:

$$n_c = 1 - \frac{Q_1^* + Q_1}{A_1 IC} \qquad (20)$$

For a '3 surface' cavity,

$$Q_1^* = A_1(B_2^*F_{12} + B_3^*F_{13}) \qquad (21)$$

and

$$Q_2 = A(B_2F_{12} + B_3F_{13}) \qquad (22)$$

which simply restates that energy reaches the aperture from both surfaces A_2 and A_3. Therefore,

$$n_c = 1 - \frac{(B_2^*F_{12} + B_3^*F_{13}) + (B_2F_{12} + B_3F_{13})}{I.C.} \qquad (23)$$

We now have a formal expression for the efficiency of a general cavity in terms of the viewfactors, the radiosities, and the power incident per unit area of the aperture.

6.2 Spherical Cavity

When those surfaces representing the cavity (aperture, nonilluminated, illuminated) can be represented as parts of a spherical surface, there is a straightforward, analytical solution for the cavity efficiency. Consider the viewfactors first.

Viewfactors

The view factor F_{kj} between any two surfaces A_k and A_j is given by the general expression (see Reference 1 or an elementary text on radiant heat transfer):

$$F_{kj} = \frac{1}{A_k} \int_{A_k}\int_{A_j} \frac{\cos\theta_k \cos\theta_j}{\pi \cdot r_{kj}^2} \, dA_k \, dA_j \tag{24}$$

r_{kj} is the length of the vector joining the elements dA_k and dA_j, and θ_k and θ_j are the angles between that vector and the normal to the surface elements.

If the two surfaces are parts of a given spherical surface of radius R, (Figure 3),

$$\theta_1 = \theta_2 = \theta, \tag{25}$$

and

$$r_{kj} = 2R \cos\theta \tag{26}$$

Substituting into (24),

$$F_{kj} = \frac{1}{A_k} \int_{A_k}\int_{A_j} \frac{\cos^2\theta}{4\pi R^2 \cos^2\theta} \, dA_k \, dA_j$$

$$= \frac{1}{A_k} \frac{1}{4\pi R^2} (A_k A_j)$$

$$= A_j / 4\pi R^2$$

This single expression for the viewfactors allows a straightforward representation of the receiver.

Suppose that concentrated light, distributed homogeneously over a solid angle with an aperture 2ϕ, reaches a spherical surface A_1' (= plane aperture A_1), illuminating directly the spherical surface A_3.

Consider the 3 x 3 matrix composed of all the viewfactors between the three surfaces:

$$\begin{bmatrix} F_{11} & F_{12} & F_{13} \\ F_{21} & F_{22} & F_{23} \\ F_{31} & F_{32} & F_{33} \end{bmatrix} = \frac{1}{4\pi R^2} \begin{bmatrix} A_1 & A_2 & A_3 \\ A_1 & A_2 & A_3 \\ A_1 & A_2 & A_3 \end{bmatrix}$$

It is more appropriate to consider the aperture to be the flat surface A_0, and this is accomplished by algebraic matrix manipulation between the surfaces 0, 2, and 3.

The matrix becomes:

$$\frac{1}{4\pi R^2}\begin{bmatrix} 0 & A_2A_1/A_0 & A_3A_1A_0 \\ A_1 & A_2 & A_3 \\ A_1 & A_2 & A_3 \end{bmatrix}$$

Referring to Figure 2, one has

$$A_0 = \pi \ (R\sin\varepsilon)^2 = R^2(1-\cos^2\varepsilon)$$

$$A_3 = 2\pi R^2(1-\cos\varepsilon)$$

$$A_2 = 4\pi R^2 - 2\pi R^2(1-\cos\varepsilon) - 2\pi R^2(1-\cos\alpha)$$

$$A_1 = 2\pi R^2(1-\cos\varepsilon)$$

$$\alpha = \varepsilon + 2\emptyset$$

$$0 \leqslant \emptyset \leqslant \frac{\pi}{2} - \varepsilon$$

or defining $G = \dfrac{A_2 + A_3}{A_0} = \dfrac{2}{1 - \cos\varepsilon}$, or $\cos\varepsilon = \dfrac{G - 2}{B}$

Substituting this in the matrix, one obtains

$$\begin{bmatrix} 0 & \dfrac{\cos\alpha + \cos\varepsilon}{1 + \cos\varepsilon} & \dfrac{1 - \cos\alpha}{1 + \cos\varepsilon} \\ \dfrac{1 - \cos\varepsilon}{2} & \dfrac{\cos\varepsilon + \cos\alpha}{2} & \dfrac{1 - \cos\alpha}{2} \\ \dfrac{1 - \cos\varepsilon}{2} & \dfrac{\cos\varepsilon + \cos\alpha}{2} & \dfrac{1 - \cos\alpha}{2} \end{bmatrix}$$

Note that the viewfactor matrix does not depend on the radius of the cavity.

C2. CONVECTION

Conduction and radiation occur without any mass transfer but convection is always associated with the motion of the fluid. Convection can be visualized in terms of an air stream that flows past the solar receiver surfaces cooling these surfaces.

The local heat flux Φ may be expressed as:

$$\Phi = h * (Tw - T_e)$$

where Tw and T_e are the wall and fluid stream temperatures and h is the local heat transfer coefficient. Because flow conditions on the surfaces vary from point to point, h also varies along the surfaces. Applying the Fourier's law to the fluid, the local heat transfer may also be estimated as:

$$h = - k * \left. \frac{dT}{dy} \right|_{y = 0}$$

because at the surface (y=0) there is no fluid motion and the energy transfer happens only by conduction.

There are two limit forms of convection:

- forced convection, which occurs when the motion of the fluid is due to an external driving force (for example wind) that induces an imposed stream.

- natural convection, which occurs when the flow arises "naturally" because of the effect of a density gradient in a force field (in our case density differences are caused by temperature gradients and the field is the gravitational field).

When both mechanisms are present, the resulting effect is called "mixed convection". This situation is usual in high temperature solar receivers.

Because of the differences between the two forms of convection, different approaches are needed to estimate the heat transfer. In forced convection, the externally imposed flow is generally known, whereas in natural convection the flow results from the interaction of the density difference with the gravitational field and is linked to the temperature distribution. In that case little of the resulting motion is known "a priori" and flow and temperature fields are invariably coupled. This makes the analysis of natural convection, as well as the experimentation, much more complicated.

In recent years we have seen a very rapid increase of the research in the field of natural convection due to concerns about buoyancy-induced motion in the atmosphere, interest in heat rejection in many systems and devices, in environment and pollution analysis, in accidental building fires, etc. Also, in relation to solar receiver design, the studies on natural convection and on the wind effects, which originate mixed convection, have contributed in the last years to improved knowledge on this topic. It is not yet possible to solve the problem in general, but realistic estimates are today possible.

Below, the formal structure of the general convection problem is summarized first, then a review of analyses and experiments specific to high temperature solar receivers (HTSR) is presented.

C2.1 CONVECTION GOVERNING EQUATIONS

The convective process is governed by the basic conservation principles of mass, momentum and energy; applying these principles to a control volume yields the governing equations. The procedure is explained in detail by different authors in books or papers whose subject matter are the mass, momentum and energy transport phenomena (1, 2, 3). Applying these resulting laws of motion to the air flow under the specific boundary conditions of solar receivers (geometry, surface temperatures, wind...) will define the problem completely in principle. We say 'in principle' because the equations of motion governing the air flow in this case are very hard to solve analytically, even in simplified cases.

Also, even with numerical techniques, solutions are not obtainable under general conditions. Assumptions appropriate to the specific problem can reduce the difficulties and simplify the governing equations to a form suitable for numerical solution. Unfortunately some really effective assumptions are not applicable to the solar receiver problem without introducing unacceptable errors.

Conservation of mass (the continuity law) is expressed by the equation:

$$\frac{\partial \rho}{\partial t} + \mathrm{div}(\rho * \tilde{v}) = 0$$

where ρ is the density of the fluid and $\tilde{v}$ is the velocity at x, y, t. The first term gives the time derivative of the net mass of fluid contained in a control volume and the second term gives the net flow of mass out of the boundaries enclosing the elementary volume. The two terms must always add to zero, as no fluid mass is created or destroyed within the volume.

Conservation of momentum is expressed by Newton's law

$$\bar{F} = d(m * \tilde{v})/dt$$

where $\tilde{F}$ is the force acting on the body and $\tilde{V}$ is the instantaneous velocity of the particle of mass m. For an ideal fluid (incompressible and non-viscous) the momentum balance equation can be written as:

$$d(\rho * \tilde{V})/dt = -grad(P) + \rho * \tilde{F}$$

where the substantial (or particle) derivative d /dt gives the change of $\tilde{V}$ for the fluid particle as it moves downstream in the flow and is, hence, associated with the acceleration of the fluid element. ($\tilde{F}$ is the external force and -grad(P) comes from pressure differences). The substantial derivative can be expressed in terms of the local derivative of flow $\delta/\delta t$ as:

$$d\ /dt = \delta/\delta t + \tilde{V} * grad$$

To incorporate the effect of viscosity one must include in the momentum balance equation the surface forces due to the action of the neighboring fluid on the control volume. For each surface a normal stress denoted by σ and two shear stresses denoted by τ must be considered instead of pressure. Viscosity is a macroscopic result of a molecular process, but macroscopic descriptions are generally adequate in order to express surface forces.

The shear stresses are given in terms of the shear rates (or velocity derivatives) through the Stokes' relationships, as follows:

$$\tau xy = \tau yx = \mu(\delta Vx/\delta y + \delta Vy/\delta x)$$
$$\tau yz = \tau zy = \mu(\delta Vy/\delta z + \delta Vz/\delta y)$$
$$\tau xz = \tau zx = \mu(\delta Vx/\delta z + \delta Vz/\delta x)$$

where μ is the coefficient of viscosity of the fluid.

The normal stresses (σxx, σyy, σzz, positive outwards) are expressed by:

$$\sigma xx = -p + 2\mu\delta Vx/\delta x + \lambda\, div(\tilde{V})$$

(σyy, σzz are similar), where λ is the second viscosity coefficient ($\lambda = -2\mu/3$ for monoatomic gas): the resulting force in the x direction is:

$$\delta(\sigma xx)/\delta x + \delta(\tau yx)/\delta y + \delta(\tau zx)/\delta z$$

From this stress relationship and taking into account the continuity equation, the x-component of the momentum equation for constant properties becomes:

$$\rho dVx/dt = Fx - \delta P/\delta x + \mu . \left[V^2x/\delta x^2 + Vx^2/\delta Y^2 + Vx^2/\delta z^2 \right]$$
$$+ (\mu + \lambda) . \delta/\delta x\ (div\ \tilde{V})$$

In vector form,

$$d\tilde{V}/dt = \tilde{F} - \text{grad } P + \mu . \nabla^2 \tilde{V} + (\lambda + \mu).\text{grad}(\text{div } \tilde{V})$$

The energy equation refers to combined convective-conductive heat transfer in the fluid, including the generation of heat due to the viscous dissipation of mechanical energy of the fluid. Energy transfer includes the body force per unit volume, the eventual energy source per unit volume Q', the conductive heat transfer -k.grad(T), and the energy input per unit volume S due to the stress system (product of the surface force times the corresponding velocity). The energy balance per unit volume may be expressed as:

$$d(e+V^2/2)\, dt = \tilde{V}.\tilde{F} - \text{div}(-k \text{ grad } T) + Q' + S$$

where e is the internal energy per unit volume of the fluid at temperature T, V is the velocity at x, y, z and k is the thermal conductivity.

The energy equation may be expanded and simplified with the help of the momentum equation as follows:

$$de/dt = \text{div}(k \text{ grad } T) + Q' - P \text{ div } V + \mu .Iv$$

where Iv, always positive, is called the viscous dissipation and denotes the irreversible part of the energy transfer due to the stress system. Using the thermodynamic expression for the enthalpy

$$h = e + P/\rho$$

the energy equation can be written, as:

$$dh/dt = \frac{1}{\rho}\left[1 - \frac{1}{\rho}\left(\frac{\partial \rho}{\partial t}\right)_p^T\right] dP/dt + Cp\, dt/dt$$

$$= 1/\rho \;*(1-\beta T)dP/dT + CpdT/dt \quad ,$$

$$\rho c_p \frac{dT}{dt} = \text{div}(k \text{ grad } T) + Q' + \beta TdP/dt + \mu \; Iv$$

where Cp is the specific heat at constant pressure and is the coefficient of thermal volumetric expansion.

The continuity, momentum and energy equations give a comprehensive description of the heat transfer in a moving fluid. The full set of these equations is so complex however, that it presents insurmountable mathematical difficulties because the equations have non-linear terms and require a simultaneous solution. Because of non-linearities the superposition principle is not applicable, and complex flow may not be obtained from simple flows. Generally, quite simple solutions are possible when the non-linear terms are assumed to be small compared to other terms (slow motions) or when the effects of viscosity are neglected (potential flows). Unfortunately, these approximations are not rigorously applicable to the case of high temperature solar receivers.

The physical difference between forced and natural convection also affects the equations, influencing their structure. In forced convection (with constant fluid properties) the flow is independent of the temperature field, and hence of the energy transfer. The flow field ($\vec{V}$ and P) can be solved from continuity and momentum equations. If properties are temperature dependent, the energy equation must be considered in addition to the above, and it becomes a coupled problem. In natural convection the problem is always coupled, since it is the energy transfer, and the consequent temperature field that generate the flow. Buoyancies, gravity induced and temperature-dependent, are the coupling term between the momentum and energy equations. Due to the complexity in the analysis of the flow, several simplifying assumptions and approximations are made to allow a more practical procedure for obtaining a solution. The most important of these simplifications in natural convection (the Boussinesque approximation) consists in taking into account temperature effects only on the density and to assume the simplified form $\rho_0 - \rho = \rho\beta(T - T_0)$, disregarding the influence of temperature on other physical quantities as well as the influence of pressure on density. The Boussinesque approximation is clearly valid if temperature differences in the field are small.

A general description of simplifying assumptions can be found in the books by Jaluria (1), which mainly refers to natural convection, and Kakac (2). The book by Schlichting (3) is an excellent reference for the boundary layer simplification theory and practical application. Numerical solution techniques are described in the book from Patankar (4). An outline of the problem of fluid motion inside enclosures is presented in (5), open cavities differ from those in complete enclosures, where boundary conditions are well defined.

C2.2 SEMI-EMPIRICAL CORRELATIONS

The previous section summarized the structure of the basic equations and some approximations that allow analytical approaches in simple geometries. Unfortunately, in several problems of practical interest, the heat transfer and the flow processes are so complicated that even numerical solutions are not possible, and engineers are forced to obtain information from experiments and must turn to dimensional analysis for help.

For this reason heat transfer semi-empirical correlations are developed which allow an estimate of the important heat transfer physical quantities, such as flow velocities, temperature, fluid properties, geometry, etc. in dimensionless groups, such as Reynolds, Prandtl, Nusselt and Grashof numbers. These non-dimentional groups are of considerable importance not only in simplifying the general governing equations, but also in guiding experiments and in permitting an extrapolation of results. They allow comparisons of various flow configurations and processes determining the conditions under which the results are, 'similar.' The adimensional numbers lead also to the characterization of a process in terms of the minimum number of variables and permit the identification of particular states of the flow (laminar or turbulent motion, natural or forced convection, convection/conduction relative importance in heat transfer, etc.).

In convective phenomena velocity and temperature fields can be shown to be dependent on:

the Grashof number (GR = $\beta g L^3 \Delta T / \nu^2$) that is an index of the relative importance of the buoyancy-term ($\beta g L^3 \Delta T$) in comparison with the viscous term

the Prandtl number (PR = $\mu C_p / K$), which is the ratio of the molecular momentum and thermal diffusivities and controls the relative thickness of the velocity and temperature boundary layers

the Reynolds number (RE = $\rho V \cdot L / \mu$), which compares the inertia to the viscous forces. In forced convection when RE is greater than 5*10**5 turbulence is completely developed in the fluid. The ratio GR/RE is an index of the relative magnitude of the natural convection heat transport effects compared to the forced convection ones (natural convection predominates when the ratio GR/(RE)**2 is large, while forced convection predominates when it is small

the Rayleigh number (RA = PR * GR); it is customary to correlate transition in a free convection boundary layer in terms of the value of this number. For vertical plates in an infinite medium turbulence is completely developed when RA 10**9

the Nusselt number (NU = h L / k) which contains the heat transfer coefficient h and represents the ratio of convection heat transfer to conduction in a fluid slab of thickness L.

the Eckert number (EC = $V^2 / C_p (T_w - T_i)$), which compares the kinetic energy of the flow to the boundary layer enthalpy difference

In case of steady state conditions with constant fluid properties (or with the Boussinesque approximation) the velocity and the temperature distributions in the flow depend only on GR, RE, PR, EC and the heat transfer coefficient h (i.e., NU) results in a function of the above mentioned quantities:

NU = h L / k = f(RE, GR, PR, EC)

In pure free convection, as no imposed external velocity is present, the motion in the fluid is buoyancy-caused. It can be demonstrated (1) that GR replaces RE (which contains V) and the group $g\beta$ L/Cp replaces EC. When viscuous dissipation (the irreversible part of the energy transfer due to the stress system) and pressure gradients produce negligible effects, additional simplifications can be introduced.

In case of geometrical complexity, large temperature differences, superposition of natural and forced convection, macroscopic turbulences, etc., the functional dependence is extremely complex: additional correlations are needed and they require 'ad hoc' validation.

C2.3 FIRST ESTIMATE OF A HTSR CONVECTIVE LOSS

It is apparent from a brief literature review that, up to 1978, the estimates of convective loss from central receivers were rough at best. In some cases no account for that loss is made at all. Jarvinen (6), applying the formula:

$$\eta = 1. - \sigma T_w^4 / \Phi_{in} \quad * \quad (Aa/Aw)^2$$

(where Φ_{in} is the flux through the aperture Aa and Aw is the receiver internal area) completely neglects convection effects when comparing the efficiency of windowed and windowless cavity receivers. In the "Status Report on High Temperature Solar Receiver Systems" (7) a convection loss is assigned to external receivers (5 kW/sqm due to buoyancies, 3.15 kW/sqm due to a 18 m/s wind), while any convection from cavity receivers is neglected ("as aperture vertical orientation inhibits the buoyancy induced flow through the aperture and wind effect are neglible"). No information is given in (8) to explain the constant value (2.5% of the total input) assigned "since convection losses are readily computed by hand."

In the analysis presented in (9) the important parameters determining the efficiency are the ratio of the heat transfer area to the aperture area and the heat transfer surface efficiency, but details on this efficiency that should include convection effects, are not given.

Peri (10), analyzing the behavior of anti-radiant and anti-convective devices, estimates the heat convection Qc from a surface at temperature Tw:

$$Qc = (hn + hf) * (Tw - Te) * A$$

where hf is taken from Kreith (11) (V is the wind velocity in m/s and Te is the environment temperature)

$$hf = 9.3 * V\exp 0.6 \quad \text{for} \quad 1. < V < 4.$$

$$hf = 7.3 * V\exp 0.8 \quad \text{for} \quad 4. < V < 40.$$

and hn from McAdams (12)

$$hn = Z * (Tw - Te)^{1/3}$$

(the coefficient Z ranges from 1 to 2 and takes into account air physical properties, temperature and surface angle with respect to the vertical.

Wu and Wen (13), estimating a cavity receiver efficiency, apply heat transfer coefficients given for turbulent flow on vertical plates in ASHRAE

$$hn = 1.3 * (Tw - Te)exp0.333$$

$$hf = 2.0 * V$$

but the interesting point is that the global heat transfer hn + hf is applied to the aperture opening area. In both cases the mixed convection coefficient is obtained by simple addition of free and forced convection coefficients (hn + hf).

In 1977 McDonnell Douglas (MDD) and Martin Marietta (MM), in the general frame of the Central Receiver Thermal Power System contract sponsored by DOE, proposed two different solutions for the receiver (14, 15). Both receivers are water-cooled but the MDD receiver is external with a cylindrical shape while the MM is an octagonal shaped cavity. The design of the two receivers is quite detailed and so is the thermal analysis performed. Experimental data are used to support the evaluation. The two different approaches of MDD and MM, even if affected by a probable lack of precision in measurements and the absence of specific information, are an attempt to improve over simple estimates to obtain quantitative evaluation.

Several reports on the economic evaluation of solar plants (see, for example, Reference 16) show that an accurate prediction of receiver thermal losses is necessary if the cost of solar energy is to be evaluated with enough care. Convection is the aspect of thermal losses about which the least is known and where reports have raised interest, and spurred effort, on the subject.

The following sections describe convective phenomena in cavity and external receivers separately, as the approach is quite different for the two geometries.

Impressive progress has been made, mainly in the experimental area and in the ability to perform convective energy loss estimates. This progress is reviewed by Abrams (17).

C2.4 CAVITY RECEIVER LOSS ESTIMATION

To determine convection loss for the 50 MW water-cooled cavity receiver, MM used data obtained from:

- 1/32-scale electrically heated model (18)
- 1 Mw bench model, tested at Odeillo (19)
- 5 Mw model heated-up by quartz lamps (16)

A linear relation, where the constant coefficients are determined from experiments, describes the convective loss through the cavity aperture as a function of wind velocity. The heat transfer coefficient h (W/sqm C) is expressed as:

$$h = 29.52 + 6.6*V$$

(V in m/s is the wind velocity) and the convection loss as:

$$Qc = h * (Ta - Te) * Aa$$

where Aa is the aperture area. No information is given on what value must be assigned to Ta (the 'aperture' temperature) and the influence of wind direction is not considered. The preliminary estimate of convection losses for the SSPS cavity solution (20) was based on this correlation. In the detailed design activity which followed, convection has been taken into account by a mean heat transfer coefficient of 8 W/ m^2C applied to the receiver internal walls (21).

A convection mechanism is modeled for the analytical computation of convection losses in the 1MW MM bench model (19). A unidirectional flow path is assumed to start in the lower half of the aperture, flow along the cavity floor, rise up the rear wall, and return along the ceiling toward the upper half of the aperture. The fluid velocity is assumed to be 0.3 m/s up at the rear wall, and the heat transfer coefficient between the path and the wall with this hypothesis is 5.7 W/sqm C. The model predicts a 2.3% convection loss in the 1Mw cavity receiver, compared with the 1.5% estimated from the experimental data on the 1/32-scale model (wind and internal conditions are not specified). During the testing at the Odeillo CNRS facility of the 1MW model, a total efficiency of less than 80% was measured at about 750 kW. This was 75% of the maximum input power. The 85% efficiency value that should be reached at full power is in any case less than the 94-95% expected figure.

In (22) the convection loss over a long period is estimated applying on the aperture area:

$$Q1 = ha* (Ta - Te) * Aa$$

where Ta (equivalent aperture temperature) is defined as:

$$Ta = (2 * Tf_{out} + Tf_{in}) /3 + W / (S.hr. \pi / 2)$$

(Tf in/out are fluid temperatures at receiver inlet and outlet, W is the absorbed power, S is the receiver active surface and hf is the tube inside heat transfer coefficient). The value 35 W/sqm C is assigned to ha (equivalent heat transfer coefficient on the aperture). The high probability of underestimating the convective loss is evidenced by the results from the 1MW cavity gas-cooled receiver tested at CRTF (23) in 1978. In the efficiency prediction, convective phenomena have been disregarded because they are estimated to be of secondary importance (from 1% to 2% of receiver input at full load). Heat load discrepancies among the different receiver tube panels became apparent during testing, and a free convection path circulating inside the cavity was assumed. As an outward air velocity just beneath the upper rim of the aperture in the range 1.2-1.5 m/s was measured, a preliminary evaluation demonstrates that this mechanism can account for the increased receiver convection loss (about 7% on the total).

With reference to the results of this experiment, which gave the first evidence that convection loss is not only a problem for external receivers, Clausing presents (24) an analytical model, and Eyler (25) applies a general use computer code.

The model from Clausing assumes an air flow pattern inside the cavity similar to the one hypothesized by Tracy (19) and Zenter (23): air enters the receiver through the bottom part of the aperture, cools down the walls and then escapes from the top of the aperture. The receiver interior is subdivided into two main zones: a stagnant zone at the top, due to the different density between the air entering the cavity and the air in the upper part, and a convective zone below where air is circulating and heated up to Tc.

Energy transfers between:

- top receiver walls and tube (at Tw_2 and Tt_2) and stagnant air zone at bulk temperature $Ts = (Tw_2 + Tt_2)/2$
- the stagnant zone bulk at the top and convective zone just below at Tb
- the heat transfer surfaces at Tt_1 and the convective zone bulk
- the receiver bottom walls at Tw_1 and the convective zone bulk
- the convection zone and outside air at T_e through the aperture of area Aa

are considered (Figure 2).

The energy transfer Qa through the aperture is computed as:

$$Qa(Tc) = Vi * Ai * Cp * (Tc-Te)$$

where Ai is the aperture area through which mass flows into the cavity (Ai=Aa /2) and Vi is the average velocity of the mass influx. Density and specific heat are evaluated at ambient temperature Te. The air velocity Vi is the result of natural convection Vn and forced (wind) convection Vf. For an inviscid flow the velocity due to buoyancies is:

$$Vn = [g * \beta * Li * (Tc-Te)]^{1/2}$$

Vi is then computed as:

$$Vi = 1/2 * [(C1 * Vn)^2 + C2 * Vf)^2]^{1/2}$$

where C1 and C2 are numerical coefficients.

In the stagnant zone the air temperature Tb is assumed to be the mean temperature of the inlet and outlet air stream.

The heat transfer Qc to the air bulk of zone 1 at Tb is:

$$Qc(Tc) = ht*At*(Tt_1 - Tb) + hw*Aw*(Tw_1 - Tb) + hs*As*(Ts-Tb)$$

where At_1 and ht are the area and the heat transfer coefficient of the zone 1 (active area), Aw_1 and hw of the zone 1 (receiver walls), and As and hs of the boundary between zone 1 and zone 2. The heat transfer coefficient in mixed convection is assumed to be given by:

$$h = (hn + hf)$$

where hn and hf are obtained in turbulent flow from:

$$NUn = 0.018 * GR \quad \text{(natural convection)}$$

$$NUf = 0.032 * RE \quad \text{(forced convection)}$$

The problem can be solved by finding the root Tc in the equation (equilibrium condition)

$$Qc(Tc) = Qa(Tc)$$

The analysis of the loss mechanism with the model described above shows that:

- the wind influence (V from 0 to 8 m/s) is small
- the influence of Aw and Tw -Te is much greater than the influence of Aa (convective loss varies almost linearly with Aw and Tw -Te). Hot refractory surfaces in the convection zone must be minimized.
- the orientation of the aperture is critical as La influences the buoyant flow and the height of the convective zone in the cavity.
- an appreciable increase in the heat transfer in the turbulent region is expected when the heat transfer correlations are modified to include the effect of large properties variations (in turbulent flow).

Clausing also estimates the interaction between radiation and air inside the cavity. While the absorption of incident solar radiation is small, the IR radiation emitted by the cavity walls results in an appreciable contribution of between 9 and 82% of the convective exchange, depending on the size of the cavity and the humidity. Also, if some hypothesis in the model produces an overestimation of the interaction, air absorption is demonstrated to be an effective mode of heating air inside the cavity, increasing the convective outflow.

Whereas Clausing describes the influences of the many parameters involved through a simplified model validated by comparison with a single experiment, Eyler approaches the problem by means of a numerical technique. A fully coupled, three dimensional transient finite differences computer code is applied. Unfortunately there are two main shortcomings: the

Boussinesque approximation is used in writing the equations and no turbulence model is incorporated (the calculations are performed at a constant viscosity value). As a consequence of the first approximation, a cavity analysis at high wall temperatures is not allowed, but information on general trends and parameters tendencies are obtained in the case of small differences between air and wall temperature. In a rectangular cavity with a vertical aperture the computed flow follows, which would be generally expected. The buoyancy force induced by the hot back wall drives the recirculation flow and the cold air inflowing in the case of small difference between air and wall temperature. The buoyancy force induced by the hot back wall drives the recirculation flow and the cold air inflowing in the lower portion of the aperture is heated inside and escapes through the upper half of the aperture opening. In front of the aperture, air is still, except for the area near the top. The heat escaping through the aperture is determined by computed velocities Vi and temperatures Ti as:

$$Q1 = \ast Cpa \ast SOMi \ (Vi \ast Ti \ast Ai)$$

where the summation is carried out across the area cells of the opening (surface area Ai).

The effects of cavity size, aperture inclination, and aperture-to-cavity size ratio are investigated and presented in relative form. Scaling up the whole cavity size by a factor 5 results in an increase in convection losses by nearly 5.3 times (a nonlinear geometrical effect). A downward rotation of the cavity by 32 degrees decreases the convective loss by about 36% for the reference case and by about 47% for the scaled up cavity, when compared with vertical aperture. When the opening aperture is halved convection losses go down by 23%.

The presented results above, coupled with the available experimental evidence, indicate that convective losses from a cavity receiver are appreciably greater than previously believed. For certain designs they may have exceeded the losses from similar external receivers. However, in principle, the cavity seems to present advantages (by virtue of protective geometry, radiation losses are reduced and wind seems to have small influence) and possibilities for potential improvements (a better geometrical and thermal design could reduce the flow of cold air into the cavity and the outflow of heated air). To verify these statements, significant scale experiments and computer code simulations, not subject to geometrical and phenomenolagical restrictions, are required.

Le Quere (26) presents a three dimensional scheme to solve natural convection flows at high values of the Grashof number (from 10exp5 to 10exp9). An implicit difference method for the integration of the balance equations is applied without the Boussinesque approximation.

Simplifications are introduced, as follows:

- some terms containing mixed derivatives due to viscous stresses are neglected
- the dissipation function is neglected in the energy equation
- turbulence effects are neglected (viscosity is constant or temperature proportional)

As the emphasis is on the numerical scheme, aspects such as:

- boundary conditions
- time and spatial discreteness
- stability
- computation scheme efficiency

are discussed with reference to a closed cubic cavity and an open two-dimensional groove.

Similar simplifications are suggested by Humphrey (27) who presents a research program supported by experiments (wall temperatures up to 800°C with differences between air and wall temperature of about 500°C and GR up to 10exp7 are foreseen) to provide data to validate the analytical procedure.

An estimation of convection in case of different geometries is presented by Yanagi (28). 1/40-scale (electrically heated), 1/7-scale (heated-up by hot water at 90°C) and 1/3-scale (heated-up by gas to 300°C) models are used, and correlations are obtained from experimental measurements. All the formulas, in dimensionless form, present NU vs RE, where Reynolds is defined in different ways (RE = 8.5*V/ν , RE = 5.7*(2exp2+Vesp2)exp(1/2)/ν). Receiver behavior results are greatly influenced by geometry, which also influences the wind dependence on the loss.

Penot (29) performed significant experiments to measure carefully the overall and internal wall convection losses in cubic cavities (of max. side dimension 1 m) at different aperture inclinations. Wall temperatures (up to 120°C) are held constant and uniform by a controlled electrical tracing. In the range 10exp8 $<$ NU $<$ 5*10exp8 the overall convection loss (at different aperture rotation) is estimated well by correlations as:

NU = 0.034 * (GR)exp0.39 (α = 90 deg)
NU = 0.037 * (GR)exp0.36 (α = 135 deg)

(α is the angle between the upward direction and aperture normal). The results of the overall measurements show a decrease in the loss with an increase in the downward aperture rotation. Loss measurements from the different walls indicate:

- the ceiling is affected by the largest loss, which reaches a maximum at an inclination of about 100/110 deg.

- lateral walls show losses of the same magnitude for an inclination of 90 degrees (about 25% less than the top wall); the back wall presents a faster reduction of the loss when the aperture faces downward

- the smaller losses are from the cubes floor, with a minimum at about 130 degrees.

To perform flow visualization experiments and make flow field measurements in conditions which closely simulate actual cavity receiver, Kraable (30) designed an experimental apparatus consisting of an electrically heated cube with one missing side and inside dimensions of 2.15m. In this experiment the emphasis is on realistic simulation rather than scale modeling.

Temperatures of the walls are in the same range as for a real receiver ($800^{o}C$ max.) in order to investigate the effects of the large temperature differences across the boundary. Also, the size is sufficiently large so that Grashof values are comparable with the ones of actual receivers. From the results of the large cavity experiment, Kraable (31) derives a correlation and outlines a methodology to estimate total convection losses: his suggestions are hereafter summarized. In the range of Reynolds from 10exp5 to 10exp12, the heat transfer from the cubical cavity is described by the correlation:

$$NU = 0.088 * GR^{1/3} * (Tw/Te)^{0.18}$$

The fluid properties are evaluated at ambient temperature Te, and the ratio Tw/Te (where T is temperatures in K) accounts for the effect of the large difference between wall and ambient temperatures.

The exponent 1/3 of the Grashof number shows that the cavity size has no influence on the heat transfer (the characteristic length L appears in NU as L and in Gr as Lexp3). Data from the 0.2 and 0.6 m size cubic cavities tested in different conditions in France confirm this conclusion. The heat transfer area to be considered for convection estimation is the entire inner surface. As the same correlation can be successfully applied to both the cavity vertical internal walls and to vertical plates in free air, it stands to reason that the closed geometry has a small effect in loss reduction. Also, from the Kraable results, the effects of aperture lips on the overall convective loss appear to be small (a 2/3 scale reduction in the aperture area decreases the loss by about 20%).

It would appear ('intuitively') that wind effects should be reduced by the geometrical protection of lateral walls, but, lacking meaningful data, Kraable suggests that this effect be evaluated as though from forced convection from a flat plate with the size of the aperture, and at the average temperature of the inner surfaces. The recommended correlation is:

$$NU = 0.0287 * RE^{0.8} * PR^{1/3}$$

where the fluid properties are evaluated at the film temperature, and the characteristic length L is the aperture width. Effects derived from wind direction are not considered.

The mixed heat transfer coefficient, which includes the effects of both natural and forced convection, is obtained from a general expression suggested by Churchill and applicable to problems with two limiting asympthotic solutions (in this case natural convection is when RE approaches 0 and pure forced convection is when RE assumes high values):

$$h = (hn^{a} + hf^{a})^{1/a}$$

with a = 1.

Kraable suggests that the effect on the heat transfer from tube bundle surfaces should be included as an increase in the effective area of π / 2. This seems reasonable if the back of the tube bundle is insulated; the case in which tubes are fully exposed to the air stream is not discussed. Also, any effect from the tube induced roughness is disregarded (no information is available). These effects could explain the slight underprediction in the estimate of SSPS cavity receiver convection loss presented by Kraable.

Two experimental investigations in which the local temperature and velocity are measured, confirm the hypothesis made in the flow patterns previously presented inside the cavity.

Hess (32) studied the heat transfer characteristics and the buoyancy-driven flow fields in an open cubical cavity (1 m. side). Water is used as the fluid, and so high Rayleigh values (up to 10exp11) are obtained. Only the wall parallel to the aperture is heated. Two geometries were considered: the first corresponds to a cubic cavity with an unrestricted aperture; in the second the opening is restricted by top and bottom lips that reduce the aperture size to 50%. Temperature difference between the wall and the fluid ranges from 2 to 13 C and the corresponding Rayleigh number from 3.*10exp10 to 2.*10exp11.

Relevant data obtained by means of dye visualization, wave shearing interferometer and laser Doppler are:

- the flow ascends for about 90% of the length of the heated plate, is turned by the top wall with considerable mixing, and then follows the top wall and escapes through the aperture

- a large portion of the incoming flow enters right below the escaping flow, forming a shear layer near the top of the aperture

- while in the full aperture cavity, the flow is two dimensional; in the restricted aperture geometry the flow is three dimensional

- the outcoming stream flows only through a small region at the top of the aperture (about 5%). The incoming flow is present in both the geometries in a zone (15% of the aperture) just below the entering stream. The remaining part of the aperture shows negligible fluid velocity or small oscillations

- disturbances in the fluid start at a Rayleigh value of 2.*10exp10, and the flow is fully turbulent at RA = 7.*10exp10, well above the value 10exp9 characterizing the transition in case of vertical flat plates in infinite medium

- local Nusselt vs. Rayleigh number behavior on the lower half of the heated wall has a slope close to the one predicted for vertical plates (RAexp 0.25 and then Raexp0.33). At the top, turbulence and hot mixing produce a completely different slope (Raexp 1.1)

- in the reduced aperture geometry the heat escaping the cavity is reduced about 10% and a thick region of hot stagnant fluid is accumulated from the top of the cavity to the bottom of the upper flap

A similar experiment was performed by Sernas (33), who measured the local heat transfer and velocity profiles along the inner heated vertical wall in an open air cavity at a Grashof number of 10exp7 (laminar flow). The horizontal surfaces are isothermal, with the top at the hotter wall temperature and the bottom at ambient temperature. Cavity dimensions are 12.5 cm high, 12.5 wide, and 18.5 deep. Air flow is visualized by using smoke; heat transfer rates are measured with interferometer and velocities with a laser Doppler anemometer. Heated wall temperatures allowing GR=10exp7 are in the range 70-80 C. Heat transfer data on the inner vertical wall are in good agreement with the heat transfer coefficients obtained by Ostrach (2) for an isothermal vertical plate in free air (NU = 0.505 (GR/4) exp 0.25 when PR-0.75). Some deviations appear on the upper third of the wall (effect of the corner). On the ceiling the heat transfer reaches a maximum in the inner half and then decreases gradually towards the aperture (average values 1.7 W/sgmC). Velocity profiles along the inner hot vertical wall are in remarkably good agreement with velocity profiles predicted by Ostrach (some discrepancy appears again at the top). The authors estimated that the hot air leaves the cavity through the 25% of the aperture at the top, and that cold air enters over the bottom 75% of the aperture.

C2.5 EXTERNAL RECEIVERS

Two different concepts are included in this category: the 360^0 facing surface and the flat plate receiver. For the first concept a circular heliostat field is assumed, and the receiver is shaped as a right-circular cylinder with the surface made of side by side tubes. In flat (billboard) receivers the tubes are oriented on a vertical (or inclined) plane facing a North (or South) heliostat field. The two geometries have a quite different fluid-dynamics, mainly in forced convection, when the cylindrical receiver (for instance 7 m. wide and 10 m. high) is considered to be a vertical tube in a crossflow.

Such an approach is followed by MDD (15) in the convection loss estimation of the water-cooled cylindrical receiver mentioned above. Only wind effects (forced convection) are taken into account by means of empirical correlations normally used for cylindrical bodies in crossflow.

Results from Zukauskas (34) concerning local heat transfer of single tubes in a water-cross flow with Reynolds ranging from 10exp5 to 2*10exp6 are applied. An extrapolation is required because in actual receiver conditions the Reynolds number ranges from 10exp6 to 10exp7. Different heat transfer zones are evident on the cylindrical surface: a stagnant line, a laminar region, a transition to turbulence, and then a separation zone. The heat transfer profile computed in the actual case is then integrated along the circumference and compared with literature data successfully.

The above mentioned approach is valid for cylindrical smooth surfaces, while externally the receiver presents interstices between the side by side vertical tubes. This geometry could be considered as a kind of roughness, characterized by the ratio between the tube radius and the receiver diameter (6.3 mm and 7 m respectively in the MMD case).

Heat transfer rate is usually increased by sand roughness, wherein the protuberances are tightly spaced, while it is decreased by cavities or notches transverse to the flow. In the first case, the laminar sublayer, which caused the highest thermal resistance, is broken by the protuberances; in the other case the stream velocity is reduced inside the notches and the laminar sublayer is widened.

As these effects contradict each other, data from Achenbach (35) for air flow over knurled cylinders are used, as this situation seems to be the closest approximation to the receiver condition. The result is a heat transfer enhancement over the one predicted in case of smooth surfaces. Considering a roughness (relative) of 0.001, properties at film temperature and a wind velocity of 1 m/s, the receiver convection (forced) is estimated at about 2.3%.

General Electric (36) and Rockwell (37) present cylindrical sodium-cooled receivers of similar geometry. In the absorber analyses, a methodology is described to compute heat transfer coefficients accounting for mixed (free and forced) convection effects as well as for surface roughness.

As the ratio GR/(RE)exp2, which gives the relative importance of free and forced convection, is in both cases about 1, free convection effects are not ignored. As results available in the literature on mixed convection, in the required Reynolds and Grashof range, are not applicable to this geometry, the asymptotic formula is again used and the total Nu is computed as:

$$NU^{a} = NUn^{a} + NUf^{a}$$

where the exponent, a, usually lies in the range from 1 to 4 (GE uses in this case a=2).

Pure forced convection Nuf is obtained from Churchill

$$NUf = 0.3 + \frac{0.62 \; RE^{0.5} \; PR^{0.33}}{\left[1.+(0.4/PR)^{2/3}\right]^{0.25}} * \left[1. + \left(\frac{RE}{28200}\right)^{\frac{5}{8}}\right]^{\frac{4}{5}}$$

which fits all the available data and NUn from Churchill and Chu, and then

$$NUn = \left[0.825 + \frac{0.387 \; . \; (GR \; . \; RE)^{1/6}}{\left[1.- \; (0.492/PR)^{9/16}\right]^{8/27}}\right]^{2}$$

that fits available data for smooth surfaces with Grashof ranging up to 4.10exp12.

The effect of the rough cylindrical surface is taken into account with the same method followed by MDD and using Achenbach data. An enhancement factor is extrapolated for the roughness 0.0006 and applied to the force convection heat transfer coefficient computer for the smooth surface.

At design conditions characteristic values for the GE receiver concept are:

Te = 28 C Tw = 620 C

$$Tf = \frac{Te + Tw}{2} = 324\ C$$

r = 1.9 cm. (absorber tube radiius)
D = 16 m. (receiver diameter)
r/D = 0.0006 (roughness)
RE = 1.7*10exp6
GR = 1.5*10exp13
NUn = 2382
NUf = 1920 (smooth surface)
NUf = 3342 (rough surface)
NU = 4100
h = 11.2 W/sqmC

convection loss = 1.2 % of incident power

Rockwell follows the same procedure, only computing the Nusselt natural convection number from a different correlation (11):

$$NUn = 0.13 * (GR * PR)^{1/3}$$

Heat transfer coefficients computed in different wind conditions and at full power are (Vg is the wind velocity at ground level, Vr at receiver level), are given below.

Vg m/s	Vr m/s	hn W/sqmC	hf W/sqmC	h W/sqmC
3.5	5.7	10.	8.	12.
6.7	11.	18.	8.	19.8
10.0	16.3	27.5	8.	28.

The limitations in the analyses by MMD, GE, and Rockwell are:

- high temperatures and large sizes produce conditions well outside the envelope of the available experimental data

- the finite aspect ratio H/D of the receiver is disregarded (data available are mainly limited to $L/D \gg 1$)

- applied correlation does not take into account the large fluid properties variations

- the mixed convection formula in the considered geometry is not supported by experimental data

To investigate these aspects, under conditions approaching reality, a novel technique (the use of cryogenic temperatures) is suggested by Clausing (38). In order to get significant tests with scaled down models, the experimental conditions should keep PR, RE and GR at the same order of magnitude as the actual case. At low temperature, the values of $\beta \cdot \rho^2/\mu^2$ and ρ/μ increase and balance the reduction in T*L**3 and V*D due to the down scaling factor. The final result is that RE and GR are not lowered by the scale reduction if temperatures are adequately reduced.

In his experiment, Clausing used a right circular cylinder with an outside diameter of 0.14 m. and a height of 0.28 m. It is the model of the 10 Mw Barstow receiver, but the obtained results can be also applied to an external flat surface. Ambient temperature in the cryogenic wind tunnel ranges from 80 K (highest Reynolds numbers) to 300 K.

Data obtained during experiments at a small temperature difference between wall and ambient agree well with estimates obtained applying literature correlations derived in constant properties conditions.

For larger temperature differences under laminar conditions (39), experiments results indicate that properties variations have a small influence on the heat transfer rates. In turbulent conditions, on the other hand, large deviations from Bailey correlation predictions are evident. Clausing assumes as an index of temperature difference, the ratio (Tw-Te)/Tw. Below the value ΔT/Tw = 0.3 (ΔT=Tw-Te) no appreciable difference between measured data and Bayley constant properties correlation appears. At ΔT/Tw = 0.75 the correlation underestimates the heat transfer of about 50%.

A factor to correct the Nusselt number resulting from the constant property approximation is proposed in the range 0.3 $< \Delta$T/Tw $<$ 0.8. It results a function of ΔT/Tw and is written as:

$$f(\Delta T/Tw) = 0.8256 + 1.5323\ \Delta T/Tw - 0.8626\ .\ (\Delta T/Tw)^{**}2$$

In forced convection tests the resulting data show an increase of heat transfer rate when compared with usual correlation. This effect (about 40%) is also explained with the influence of the finite L/D ratio, while usual correlation are derived in $L \gg D$ conditions.

Siebers (40, 41, 42) uses a conventional wind tunnel and a large test section to obtain the high Reynolds and Grashof numbers required to reproduce the actual situation of external receivers. The test section is a flat vertical 3 m. side square surface parallel to the horizontal flow; it is electrically heated from 40 to 600 C and cooled down by an air stream with a velocity range from 0 to 6 m/s. Temperatures and boundary layer velocities are measured in the range up to 2*10**6 for Reynolds and 2*10**6 for Grashof. Siebers, after a survey on the previous research on mixed convection and variable physical properties effects, comes to the conclusion that the present knowledge on these subjects is very limited. No information exists on turbulent mixed convection with buoyant and inertia forces directed normal to each other, and conflicting recommendations on how to account for variable properties effects exist. From the data obtained, a set of usual correlations is first checked and then mixed convection and properties variations effects are analyzed.

In natural convection the correlations:

NUy = 0.404 * GRy **1/4　　　GR < 10**9
NUy = 0.096 * GRy **1/3　　　1-**9 < GR < 2*10**12

are used, and the agreement with a set of baseline data (at small temperature differences) is good.

In forced convection:

NUx = 0.453 * REx**1/2 * PR**1/3　　　RE < 2*10**5
NUx = 0.0307* REx**0.8 * PR**0.6　　　REx > 2*10**5

are applied and again the agreement with experimental data is good (all properties are evaluated at ambient temperature).

When compared with results estimated from free convection correlations, data from tests with wall temperatures ranging from 60 to 520 C show an evident influence of the variation in the properties. When properties are evaluated at ambient temperature Te, there is a small decrease in NUy at a given GRy as Tw increases, but, for each temperature, experimental results remain parallel to the correlation. This means that NUy remains dependent on GR**1/3 and only the constant coefficient is changing with the increasing temperatures (it decreases as temperature increases). In the laminar region (GRy 5*10**8) the effect above described is small and is produced by the combination of the effect of variable properties and of a change in the boundary condition (from uniform heat flux to uniform surface temperature). In the turbulent region the correlation:

NUy = 0.098 * GRy**1/3 * (Tw/Te)**(-0.4)

where all the properties are evaluated at T, accounts for the variable properties effects. In case of laminar flow, the correlation:

NUy = a * GRy**1/4 * (Tw/Te)**(-0/04)

based on the data best fit is suggested, where

a = 0.404　　　in case of uniform heat flux
a = 0.354　　　in case of uniform temperature

In forced convection the same discrepancy is evident in case of turbulent flow. In the same way a correction is introduced in the standard correlation obtaining:

NUx = 0.025 * REx**0.8 * (Tw/Te)**(-0.4)

The average heat transfer can be computed by means of the pure forced convection correlation only in a zone extending to GR/ RE**2 = 0.7. Above GR/ RE**2 = 10 the average heat transfer coefficient can be calculated using the pure natural convection correlation. In the mixed convection zone, between GR/ RE**2 = 0.7 and 10, the average heat transfer coefficient is estimated by Siebers as:

hm = (hn**3 + hf**3) **1/3

with an approximation +/- 3% when compared with experimental data. Also, if the heat transfer behavior outside the mixed convection zone can be described as a single phenomena (free or forced convection), Siebers demonstrates that relevant fluid-dynamics effects due to free convections exist well inside the forced convection dominated zone and forced stream effects are evident well inside the natural convection zone.

C3. ASR CONVECTION LOSS ESTIMATE

A convection loss estimation for ASR is performed following the indications and the experimental results by Siebers (41) and Kraable (30).

Receiver nominal condition:

Inlet temperature in the fluid		TFin	: 270 C
Outlet temperature in the fluid		TFout	: 530 C
Average fluid temperature		TF	: 400 C
Aver. temperature delta on the tube wall			: 35 C
Receiver average temperature		TR	: 435 C
Air temperature		Te	: 15 C
Receiver dimensions (m)	high	H	:2.85 C
	wide	W	:2.75 C

The Grashof number is computed at ambient condition:

$$GR = \frac{\beta g \rho^2 H^3 \Delta T}{\mu^2}$$

$= 1.32 * 10**8 * L**3 * \Delta T = 1.27 * 10 ** 12$
(turbulent flow)

By means of the correlation (natural convection)

NU = 0.098 * (GR)**(1/3) * (TR/Te)**(-0.14)

as TR / Te = (435+273)/(20*273) = 2.14

we obtain

NU = 954

and with the air conductivity value 0.00246

h = NU * (k/H) = 954 * 0.00246 / 2.85 = 8.23 w/sqmC

In case of circulation test for evaluation of losses, we have

TW = 200 C, TR/T = 1.61, NU=991, h = 8.55 W/sqmC
TW = 300 C, TR/T = 1.95, NU = 965, h=8.33 W/sqmC

To compute the loss induced by wind (forced convection), Siebers suggests:

NU = 0.037 * RE**0.8 * PR**0.6 * (TR/Te)**(-0.4)

in which adimensional number must be evaluated at T.

As

RE = 1.71*10**5 * Vexp0.8

where V is wind velocity in m/sec.

we obtain

NU = 490 * V**0.8 * (TR/T)**(-0.4)

Computing for different wind and temperature conditions the heat transfer coefficient (W/sqmC), we obtain:

		Receiver Average Temperature		
		435 C	200 C	300 C
Wind	1 m/s	3.1	3.5	3.25
	5 m/s	11.3	12.6	11.7
	8 m/s	16.4	18.4	17.1

If the formula

$$h = (hn^{3.2} + hf^{3.2})\exp(1/3.2)$$

is applied, the mixed convection heat transfer coefficients are:

		Receiver Average Temperature		
		435 C	200 C	300 C
Wind	1 m/s	8.3	8.7	8.4
	5 m/s	12.4	14.4	12.8
	8 m/s	16.9	19.2	17.6

Convection loss (kW) can be estimated by means of

Q1 = h * (TR - Te) * A

in which A = 7.83 sqm (ASR surface)
Te = 20. C (air temperature)

at different wind velocities and the results are:

		Wind Velocity		
		1 m/sec	5 m/sec	8 m/sec
TR (C)	435	26.9	40.3	54.2
	300	18.5	28.1	38.6
	200	12.2	20.2	27.1

To take into account the actual tube bundle area this value should be multiplied by $\pi/2$.

Performing analogous computations applying Clausing correlation (40) we have:

$$TF_{lm} = \frac{20 + 400}{2} = 210^{\circ}C$$

RA = PR * GR = 0.68 * 10**7 * L**3 * ΔT = $9.27 \cdot 10^{10}$

f (TR/Te) = 1.95

NU = 0.082 $(RA)^{1/3} \cdot f$ = 720

h = k*NU/H = $\frac{3.86 \cdot 10^{-2}}{2.85}$ W/sqm C = 9.7 W/sqm.C°

REFERENCES

1. Y. Jaluria, "Natural Convection and Mass Transfer HMT Series," Vol. 5; Pergamon Press (1980)

2. S. Kakac, "Convection Heat Transfer," Middle East Technical University (1980)

3. H. Schlichting, "Boundary Layer Theory," 4th Edition, McGraw Hill (1960)

4. S. V. Patankar, "Numerical Heat Transfer and Heat Flow," McGraw Hill (1980)

5. S. Ostrach, "Natural Convection in Enclosures - Advances in Heat Transfer" (1978)

6. P. O. Jarvinen, "Window vs. Windowless Solar Energy Cavity Rec." 11th I.E.C.E.c. (September 12 - 17, 1976), Stateline, Nevada (USA)

7. A. Skinrood, T. Brumleve, "Status Report on High Temperature Solar Energy System," SAND74-8017 (1974)

8. Boeing, "Closed-Cycle High Temperature Central Receiver System," EPRI-ER-629

9. Black and Veatch Consulting Engineers, "Solar Thermal Conversion to Electricity Utilizing a Central Receiver, Open Cycle Gas Turbine Design," EPRI-ER652 (March 1978)

10. G. Peri, J. Desantel, "Capteurs a Concentration du Rayonnement Solare Electricite' Solaire Int. Conf.," Toulouse, (March 1-5, 1976)

11. F. Kreith, "Principles of Heat Transfer," Dun-Donnelly Publishing Corp. (1973)

12. McAdams, "Heat Transmission," McGraw Hill

13. Y. C. Wu, "Solar Receiver Performances in the Temperature Range of 300 to 1300 C," J.P.L. (1978)

14. McDonnell Douglas, "Central Receiver Thermal Power System," SAND1108-8, (November 1977)

15. Martin Marietta Corp., "Central Receiver Solar Power System;" Vol. 4, SAND1110-77-2 (April 1977)

16. D. Siebers, "Natural Convection Heat Transfer from Ext. Rec.," SAND78-8276

17. M. Abrams, "The Status of Research on Convective Losses from Solar Central Receivers," SAND83-8224 (June 1983)

18. T. Tracy, "The Design of a Solar Cavity Steam Generation for Electrical Power Generation," I.E.C.E.C. (1975)

19. T. Tracy,"1 MW Solar Cavity Steam Generator Solar Test Program 12," I.E.C.E.C. (1977)

20. SSPS Project - CRS Final Documentation Phase 1 (1980)

21. D. Stahl, H. Fricker, "Detailed Design of a 2.7 Sodium Cooled Receiver Int. Symposium on Solar Thermal Power," Marseille (June 15-20, 1982)

22. B. Rivoire, "Bilan Annuel de Centrales Heliotermoelectriques Entropie," n.103 (1982)

23. R. C. Zentner, "Evidence of Free Convection Effects in EPRI/Boeing Solar Receiver Test," DOE/Sandia Convective Loss Workshop, Dublin, CA (USA) (April 17-18, 1979)

24. A. M. Clausing, "An Analysis of Convective Loss from Cavity Solar Receivers," Solar Energy, Vol. 27, (1981)

25. L. L. Eyler, "Predictions of Convective Losses from Solar Cavity Receiver," PNL-SA-8070 (December 1980)

26. P. LeQuere, "Three Dimensional Numerical Evaluation of Heat Loss through Natural Convection in Solar Boiler GAMM Conference Proc.," Cologne (1979)

27. J. Humphrey, "Investigation of Free-Forced Convection in Cavity Type Receivers," Semiannual Report on CRS (1981)

28. K. Yanagi, "Estimation of Thermal Efficiency of Solar Tower Receiver," Int. Symposium on Solar Thermal Power, Marseille (June 15-20, 1980)

29. F. Penot, "Etude Experimentale des Echanges de Chaleur par Convection Naturelle dans une cavite ouverte," Int. Symposium on Solar Thermal Power, Marseille (June 15-20, 1980)

30. J. S. Kraable, "An Experimental Investigation of the Natural Convection from Cubical Cavity," ASME Thermal Eng. Joint Conference Honolulu, USA (March 20-24, 1983)

31. J. S. Kraable, "Estimating Convective Energy Losses from Solar Central Receivers," SAND84-8717 (April 1984)

33. V. Sernas, "Natural Convection in an Open Cavity," 7th Int. Heat Transfer Conference, Munich (1982)

34. A. Zukauskas, "Local Heat Transfer of Tubes in the Critical Flow Region for PR 1," 5th Int. Heat Transfer Conference, Tokio (1974)

35. E. Achenbach, "Heat and Mass Transfer Source Book," John Wiley

36. General Electric, "Conceptual Design of Advanced C.R.S.," SAND20500-1 (June 1979)

37. Rockwell International, "Conceptual Design of Advanced Solar C.R.S.," SAND/1483-1 (March 1979)

38. A. M. Clausing, "An Experimental Investigation of Convective Losses from Central Receivers," DOE/Sandia Semiannual Review Meeting (1981)

39. A. M. Clausing, "The Influence of Property Variations on Natural Convection from Vertical Surfaces," Journal of Heat Transfer, Vol. 103 (1981)

40. D. Siebers, "Experimental Mixed Convection from a Large, Vertical Plate in Horizontal Flow," SAND83-8225

41. D. Siebers, "Experimental Mixed Convection from a Large, Vertical Plate in Horizontal Flow," 7th Int. Heat Transfer Conference, Munich (1982)

42. D. Siebers, "Experimental, Variable Properties Natural Convection from a Large, Vertical, Flat Surface," ASME Thermal Eng. Joint Conference, Honolulu, Hawaii (USA) (March 20-24, 1984)

DESIGN AND CONSTRUCTION OF THE ADVANCED SODIUM-COOLED RECEIVER (ASR) FOR THE IEA/SSPS CENTRAL RECEIVER SYSTEM PLANT (ALMERIA, SPAIN)

P. Cavalleri, V. Bedogni, and A. DiMeglio,
Franco-Tosi Industriale, Legnano (Italy)

and

A. De Benedetti and C. Sala,
AGIP SpA, Milan (Italy)

1. INTRODUCTION

The ASR receiver was designed and constructed by Franco-Tosi Industriale and AGIP SpA with fundamental contributions from ENEL (Italian National Board of Electricity) for the receiver dynamic analysis and control system design and from CNR (National Research Council), the Italian official representative. The receiver installation at Almería Central Receiver System power plant was completed in August 1983.

The aforementioned plant was built in the framework of a project named SSPS (Small Solar Power Systems) promoted by the International Energy Agency whose goal was the demonstration of technology and maturity for electricity generation from solar energy (1).

The decision to participate in the SSPS Project, taken in Italy in June 1979 and formalized in the Implementing Agreement, became official with the contract signing by the Italian partners CNR, Snamprogetti, and Franco-Tosi (afterwards changed into AGIP and Franco-Tosi Industriale). Since the Italian participation was decided after the assignments of all the plant units and components, the design and supply of a second receiver with more demanding performances (Advanced Sodium Receiver) was commissioned. The conceptual design was completed in October 1979 while preliminary specifications for the two receiver sizes (2.7 and 4.2 MW_t) were presented in May 1980.

The project contract guidelines specified that only proven hardware would be installed; therefore, a full-scale qualification test was required by the EC. Qualification procedures were prepared for testing an ASR panel first at CRTF of Albuquerque (July to October 1981) and then, as CRTF was unavailable, at the CRS Almería plant. However, plant modifications, time schedules, costs, and the long period of plant shut-down resulted in the second solution being unfeasible. The alternative solution, approved by the EC, was to perform the ASR qualification by design instead of by test as originally established.

The group of international experts following the ASR activity was charged to make a step-by-step examination of the design and manufacturing of the receiver. The final solution for the proposed and selected design constituted the initial phase of the activity: an external type receiver of 2.7 MWt, consisting of 5 panels arranged to form a rectangular shape absorber of 2.85 m height and 2.78 m width. Liquid sodium from the cold storage tank at 270°C is pumped by a feed pump through the receiver, where the sodium is heated to 530°C. Basic design data are presented in Tab.5.2-1.

2. BASIC DESIGN OBJECTIVES

The primary objectives of ASR design are:

- incident heat flux peak density in the range of 100 - 150 W/cm^2
- average incident heat flux in the range of 30 - 50 W/cm^2
- upscaling aspects: the ASR should include basic technological aspects of large future receivers.

3. INCIDENT FLUX, TARGET POSITION, AND HELIOSTAT AIMING

Both cavity and external receivers were considered in order to investigate how an average flux in the range of 30 - 50 W/cm^2 and a peak flux of 100 - 150 W/cm^2, together with acceptable spillage and distribution, could be obtained on the absorbing surface. Computations performed with HELIOS code (Ref.2) point out the impossibility of meeting the abovementioned figures with a cavity-shaped receiver; for instance, on a cylindrical vertical axis surface, 1.5 m radius and 18.6 W/.cm^2 average, 74 W/cm^2 are obtained. To fulfill these flux figures, a flat vertical target was considered with a single aiming technique. Computations at different distances from the aiming plane indicated the possibility of obtaining the reference peak flux, but the resulting average value is low and the flux profile too peaked. Little amplitude targets tilted toward the field produce only limited improvements.

In order to obtain a flux distribution which is as uniform as possible, so as to avoid large temperature differences among the tubes, a multiple point aiming strategy has to be considered. Three aiming points are a satisfactory compromise between the abovementioned requirements and simplicity. Back row heliostats aim to the target center, first row heliostats aim to two lateral points (Fig.5.2-1) whose position depends on heliostat field error. With a reference total error (beam quality and tracking) of 3.1 mrad, the final optimized aiming technique allows for the attainment of:

- peak heat flux density 138 W/cm^2 (d.p.)
- average heat flux density 35 W/cm^2 (d.p.)

An additional sensitivity analysis verifies the impact of heliostat errors on the computed flux distributions and shows that aiming points can be arranged in the interval 2.5 - 3.6 mrad, leading always to acceptable distributions. Considering both the need to reduce both spillage with thermal emission and the thermal conditions of the receiver edges, the ASR active surface was optimized.

Figs. 5.2-2-7 present the incident peak flux, the incident power, and the spillage behavior vs. time (3 typical days) and day, and settle the basis input data for both the receiver design and the performance estimation (2).

4. RECEIVER CONFIGURATION SELECTION

During the conceptual design and optimization process of the receiver absorber unit, several factors were taken into account. These mainly include:

- high flux levels are acceptable on receiver tubes, provided that a suitable selection of the receiver tube geometry is made.

- flow instabilities advise against low sodium velocities and downward flow in the irradiated tubes.

- pressure drop should be at most 1.5 bar, according to plant requirements.

- upscaling aspects are a basic design objective.

- conventional, well-established materials and techniques should be used.

Based upon preliminary analysis, the following absorber tube size and material were selected:

-absorber material : AISI 316L
-tube OD/thick : 14/1 mm

With these dimensions and with a sodium velocity of 1.9 m/sec, a maximum wall temperature of 596°C is reached in the most irradiated section of the absorber tubes, which is cooled by sodium at 495°C. To assure the required cooling (sodium velocity) of the tubes, 39 of these are connected in parallel.

The possibility of modeling the absorbing surface (in total or only the lateral parts) with parallel tubes directed in a serpentine fashion from bottom to top, with a suitable number of turns was investigated. However, upscaling conditions made modular solutions more interesting, with vertical panels laid side by side, each consisting of 39 tangent tubes. In fact, in the use of a commercial receiver, the power incident on a single panel is sufficient to bring the sodium from the inlet to the outlet temperature in one pass only and the panel solution is particularly simple and attractive. In the case of the ASR, the final configuration of the active area consists of five equal panels connected in series. Sodium flows upward in all of the panels, first in the two lateral ones, then in the two intermediate panels, and finally through the central one. Sodium circulation is shown in Fig.5.2-8, and allows for the attainment of three main objectives:

-receiver efficiency optimization

-better dynamic behavior

-the central panel will work in the typical operating conditions of a large power plant with regard to sodium temperature, heat flux density peak, and maximum metal temperature.

5. RECEIVER THERMAL ANALYSIS

Temperature distribution in the absorber tubes and performance of the receiver, in steady-state and transient conditions, were investigated at different partial loads by employing ad hoc developed computer codes based on the finite difference method. The active length of each tube was subdivided into 19 finite elements. Each panel was characterized by 741 modal points. For each element, the heat transfer processes taken into account for heat balance are:

- radiation absorption and emission (the surfaces of the tubes exposed to solar flux are coated by Pyromark paint 2500.)

- conduction through tube thickness

- convection tube-to-sodium inside the receiver tubing

- convection tube-to-air outside the receiver tubing

In the tube axial direction, the heat transfer is assumed to be due to sodium mass flow only. No axial heat conduction has been considered. In Fig.5.2-9, the temperature profile in the central tube of the central panel is shown as a typical plot of the computer code evaluation.

The plot of the receiver efficiency versus absorbed power is shown in Fig.5.2-10. Sodium local velocity distribution in the receiver panel has been investigated too, to determine the influence of geometric arrangement and hydrostatic pressure on the maximum sodium temperature nonuniformity at outlet headers. The thermohydraulic analysis has been limited to lateral panels which are subjected to the maximum transverse power gradient. The results showed that the thermohydraulic conditions do not cause any mismatch problem, and that outlet header temperature profiles are sufficiently stable at every load condition.

Denoting:

TA = sodium outlet temperature in the first tube of a lateral panel,
TB = sodium outlet temperature in the 39th tube of a lateral panel,
TM = sodium average outlet temperature of a lateral panel,

the ratio (TA -TB)/TM is maximum at full load and reaches the value of 14%.

For this panel, the tube-to-tube outlet temperature variation does not exceed 1°C, while the maximum outlet temperature difference between adjacent tubes of different panels is 137°C (3rd - 5th panels). Both of these values are of primary importance in the panel detailed design.

6. RECEIVER DESIGN

Three main parts of the construction are worth describing:

- sodium wetted parts
- main frame structure
- backwall insulation system

The sodium wetted parts of a panel are the tube bundle, the bottom and top headers, and the downcomer. The fixed points of a panel are the flange of the bottom inlet header and the restraint at the downcomer sodium outlet. The top header moves vertically to accomodate vertical thermal growth of the panel. The irradiated tubes are 3 x 3, assembled together with a supporting plate at four levels so as to form a 'triplet' (see detail 1 of Fig.5.2-11).

The triplets, in correspondence to the four plates, are connected to the panel framework (see detail 2 of Fig.5.2-11) by means of pins. This kind of tube restraint allows the tubes to grow axially with respect to the frame and also to rotate, because of the clearance between the pin diameter and the hole in the triplet supporting holder.

In order to allow lateral thermal expansion of tubes and panels, gaps are provided between tubes in addition to manufacture tolerances. With this concept for the supporting system of the irradiated tubes, each triplet is free to expand independently of the other ones so no problem arises from the heat flux gradient in the horizontal direction.

An alternative solution consisting of joining the tubes by full length longitudal welds (tube wall concept) was examined but not adopted because of the lower capability to withstand horizontal flux density gradients and manufacturing difficulties. In the selected solution, however, the presence of gaps allows some concentrated solar flux to pass through the tubes and impinge on the backwall structure. In the back of the tube bundle a double shield of high refractory alumina-based material protects the back structure from the entering radiation (see detail 3 of Fig.5.2-11).

Behind the second shield, a layer of 175 mm ceramic fiber insulates the hot parts of the receiver from the cold back structure. A wide range of values of gap width was investigated in the backwall thermal analysis. The selected setback distance of 45 mm, together with the high material refractoriness, assure that no problem of dangerous overheating of the back structure can arise.

The triplets, in groups of two or three are anchored with beams to the single panel frame, which can withstand, besides its dead weight, wind and/or earthquake loads and forces and movements due to panel tubes and connecting pipes. A structural analysis was performed to ensure that placements and distortion would not occur.

For the same reason, particular attention was paid to limit conduction heat transfer from hot sodium wetted parts to cold panel frame through supporting elements, mainly triplet beams, header supports, and piping restraints.

Each panel is then assembled in the main frame structure of the receiver, for which the same structure analysis performed for the panel was carried out to check the stability and system operability in all load conditions.

7. OPERATING CONDITIONS

In the structural analysis of the receiver, it is important to consider the operating conditions.

From the analysis of the site meteorological data, the number of sunny hours per year has been assumed to be approximately 3000. The incident peak heat flux on the receiver absorbing panels changes during the day and reaches a maximum at noon; the noon peak value changes during the year too, ranging from 100 to 138 W/cm^2. In order to have a realistic estimate of the damage done during the lifetime of the receiver, the probability of having the receiver operating at peak flux value between F1 and F2 has been evaluated using the equation.

$$p(F1,F2) = \sum_{1}^{12} (t_m(N_m-N_{mc})r)/H$$

where H = number of operating hours per year

t_m = daily time at peak flux between F1 and F2

N_m = number of days in month m

N_{mc} = number of covered days in month m

r = factor taking into account the reduction due to cloudiness of "cloudy days" (partially covered)

The probability distribution of the incident specific heat flux has been applied in the fatigue damage evaluation.

The estimation of the component life has been carried out considering 10 years of operation with 327 morning hot start-ups per year (no start-up during covered days) and 50 cold start-ups per year. Three kinds of cloud passages at three different cloud velocities have been taken into account with a number of events reported in Fig.5.2-20.

Besides the abovementioned normal operating transients, two other kinds of possible events during ASR lifetime have been considered:

- Upset Operating Conditions: this category includes the loss of control in the receiver control system and the loss of sodium supply due to electrical failure of sodium pump drive.

- Faulted Conditions: blocking of sodium pump.

8. RECEIVER STRUCTURAL ANALYSIS

Important locations at which a careful stress analysis is requested, corresponding to the most irradiated tube section, to the tube-supporting plate connections, and to the tube-header connections.

For all these parts, a detailed thermal and structural analysis has been performed with an extensive use of the finite-element technique.

The evaluation of the results has been performed utilizing criteria in accordance with ASME Pressure Vessel code Section VIII, Div. 1 and 2.

For component service lifetime evaluation, consisting of evaluation of the creep-fatigue interaction effect, Code Case N47-17 has been assumed.

Particular emphasis in this analysis was given to the most irradiated tube section, the analysis of which has been performed first on an elastic basis, and then, because of high thermal stresses beyond the yield limit, on an inelastic basis to check the validity of the previous analysis and in order to evaluate its safety margin. Taking into account the real material behavior, a generalized plain strain analysis has been performed and a reference strain range of $1.42\ 10^{-2}$ for the daily cycle has been found.

In Fig.5.2-13, the temperature and equivalent plastic strain distribution is represented; in Fig.5.2-14, the hysteresis stress-strain loop is reported.

The triplets, in groups of two or three are anchored with beams to the single panel frame, which can withstand, besides its dead weight, wind and/or earthquake loads and forces and movements due to panel tubes and connecting pipes. A structural analysis was performed to ensure that placements and distortion would not occur.

For the non-irradiated parts of the receiver, a detailed analysis has been carried out in order to have the right picture of the thermal stresses because of the very strong transients and the high level of thermal shocks.

Tube supporting plate connections have been carefully analyzed by the finite element method in transient conditions. Particularly the evaluation of the shrink effect of the supporting plate on the triplet tube has required a detailed 3-D analysis. In Fig.5.2-15, the 3-D mesh plot is shown in 20 modes, brick type, finite elements: the mesh was used for both the thermal transient calculation and the mechanical evaluation. Another critical point is the tube-header connection represented in Fig.5.2-16.

For all these locations and others of minor importance a successful analysis has been carried out on an elastic basis using the fatigue curves T1420 of the ASME Code Case N47-17, with a safety factor of 5.

9. SUPPORTING TECHNOLOGICAL TESTS

Before starting the receiver construction, an extensive development program was completed to ensure the feasibility of the proposed design. This work included mainly:

- Header prototype,
  - °to investigate the manufacturing problems and to set up improved technological procedures with the objective to attain thickness uniformity as required by fast sodium thermal transients.

- "Triplet" prototype,
  - °to investigate the influence of the supporting plate thickness and of the welding bead length on the straightness of the triplet;
  - °to investigate the gaps between tubes;
  - °to investigate the tearing strength of the tube-plate connection.

- Oxidation test,
  - °to investigate the oxidation possibility in the pin material in order to avoid blocking in the supporting system of the triplets (stirrup system).

10. MANUFACTURING

The major steps of the manufacturing process are:

- tube inspection, cutting, and bending
- setting up of "triplets" in fixture
- inlet/outlet headers fabrication
- panel support structure fabrication
- welding of triplets to headers
- gap width adjustment (by regulating the position of the stirrup beams)
- painting, curing, and verification
- insulation installation
- assembling of panels with interconnection piping
- preparation for shipment

Several manufacturing problems were encountered with this new project. The most interesting and the most important was tube-tube supporting plate welding. Owing to the very small thicknesses to join, plasma jet manual welding procedure was adopted. It was found that the weld profile was very sensitive to parameter variations and as this is an important feature for the acceptance of the weld, considerable work was required to stabilize variables to obtain the desired profile on production.

A full program of destructive testing on welding specimens was completed. Similar precautions were taken for the welding of the triplet tubes to the header nozzles. In this case, an automatic orbital inert gas welding procedure was adopted (Fig.5.2-17).

Another interesting aspect was the machinery of the headers and tube nozzles. Particular care was required for the execution of the inner and outer fillet radii, so as to minimize stresses generated by thermal shocks due to fast sodium temperature transients (Fig.5.2-18).

Considerable inspection effort has also been applied by an extensive quality control program exercised on the quality of materials from the point of view of mechanical and chemical properties, on the manufacturing, on the welding, and generally on all the most relevant activities related to the attainment of a certain confidence in the standard of unit.

To ensure that the required quality is achieved, an extensive documentation of the performed examinations was provided too.

The results showed the large conservatism of the Code Case elastic creep-fatigue analysis for the particular "strain controlled" situation, and underlined the necessity for the development of special design standards for solar components (Ref.4).

The method of inspection adopted for each weld started with a visual examination followed by a dye penetrant and radiographic inspection extended to 100% of the welds. For the tube-tube supporting plate welding, endoscopic visualization from the tube inner side was employed.

On the completion of each panel tube bundle, pressure tests and helium leak test were carried out.

After assembling the five panels with the connecting pipes a final pressure test and helium leak test were performed of the whole unit before conservation and packing for shipment.

11. RECEIVER INSTRUMENTATION AND CONTROL

For controlling and monitoring purposes 53 K-type thermocouples were installed on the ASR. Thermocouple junctions are ground connected to reduce time response delay as much as possible, and covered with A15136L to withstand high temperatures and thermal fluxes.

The thermocouples were positioned in such a way to:
- satisfy control system requirements and optimization.
- allow controls and checks of the reference conditions assumed in the design phase during operation.
- provide a full monitoring for evaluation with special care of the central receiver panel.
- make unusual and non-routine operation easy.

Thirty thermocouples were installed on the backside of the active tubes (the top level for regulation and check of temperature differences across the panels; the bottom level mainly to monitor preheating), eleven inside the walls corresponding to the ten panel headers (for safety reasons, two thermocouples are installed at the outlet), four on the receiver edges, two on the door frame, and six on the receiver panel frame. All thermocouples were tested and some of them, installed later and corresponding to the most important positions, were calibrated.

The receiver controller, directly using 18 temperature signals (13 from active tubes and five from headers), regulates sodium temperature at the receiver outlet (controlled variable with a set point value of 530°C) by the action on the pump speed to obtain the required sodium flow rate (Fig.5.2-20).

An original control design (Ref.5), set up by means of accurate process modelling (Ref.6), was adapted because ASR operations, in addition to the normal aspects of solar receivers (uncontrollability and large variations of the thermal power entering the coolant and automatic operation in a wide load range) presented peculiar aspects such as the low thermal capacity of the active parts, high impinging flux, and the unavailability of a direct estimation of the power impinging the receiver.

The control action is performed by a microprocessor with an ad hoc implemented software, which also manages the monitoring and alarm protection tasks.

12. DESIGN AND CONSTRUCTION PHASE FINDINGS

The findings drawn from the design and construction phase activities, by light of the results of the first year of operation, are as follows:

- Transient absorber performance is a determining factor in selection of the best solution: the dynamics of the external type receiver is very fast. However, suitable stress analysis (creep-fatigue interaction) and manufacturing processes have shown that these transients are acceptable. The thermal stress analysis may be very complex, however when a clear strain controlled situation can be assumed, an elastic stress analysis seems good enough to evaluate the strains in the structure even if the yield limits are exceeded.

- The tube supporting system in groups of three is a better solution than joining the tubes by full-length longitudal welds.

  By the light of the actual operation conditions of the absorber (heat flux and sodium flow distribution) as registered in this year of operation, even the triplet system as designed for Almería should be made somewhat less rigid so as to accept greater thermal differences between adjacent tubes than those expected.

- Due to gap presence, thermal analysis of the back structure should be very accurate. Assumed evalution models, design details, and materials are satisfactory.

- nonroutine operation and eventual anomalies during these phases must be considered with great care. Preheating and filling as demonstrated by sodium-cooled receivers are critical operations and potentially dangerous. As the light reflected by only one heliostat is able to produce in a limited zone temperatures up to 300°C, more thermocouples on the active surface are advisable in order to avoid cold zones or local overheating. The possibility of on-line adjustment for heliostat aiming coordinates could reduce considerably the preheating difficulties.

13. REFERENCES

1. W. Grasse, IEA Small Solar Power Systems Project, International Symposium, Cologne, 1975.

2. F. Biggs and C. Vittitoe, "The HELIOS Model for the Optical Behavior of Reflecting Solar Concentrators", Sandia National Laboratories Report SAND76-0347 (1979).

3. A. De Benedetti and C. Sala, "The Advanced Sodium Receiver (ASR) for the IEA/SSPS Central Tower Plant: Operative Conditions, Control System Design, and Performances", 2nd International Workshop on the Design, Construction, and Operation of Solar Central Receiver Projects, Varese, Italy (1984).

4. V. Bedogni and A. De Benedeti, "Almería Advanced Sodium Receiver: Stress Analysis Considerations", ISES Solar World Congress, Perth, Australia (1983).

5. G.A. Magnani, S. Quatela, and A. De Benedetti, "Dynamic analysis and Control of the Almería Advanced Sodium Receiver (ASR)", ISES Solar World Congress, Perth, Australia (1983).

6. C. Maffezoni, G.A. Magnani, and S. Quatela, "Process and Control Design of High Temperature Solar Receivers: Integrated Approach", IEEE Transaction on Automatic Control (in course of publication) (1982).

Design Point:	equinox noon
Maximum Power Input:	2.7 MWt (d.p.)
Sodium Inlet Temperature:	270°C
Sodium Outlet Temperature:	530°C
Sodium Flowrate:	7.3 Kg/sec
Design Pressure:	6 bar
Maximum Pressure Drop:	1.5 bar
Receiver Lifetime:	10 years
Number of Operating Transients:	6343 per year

Table 5.2.-1: Basic Design Data

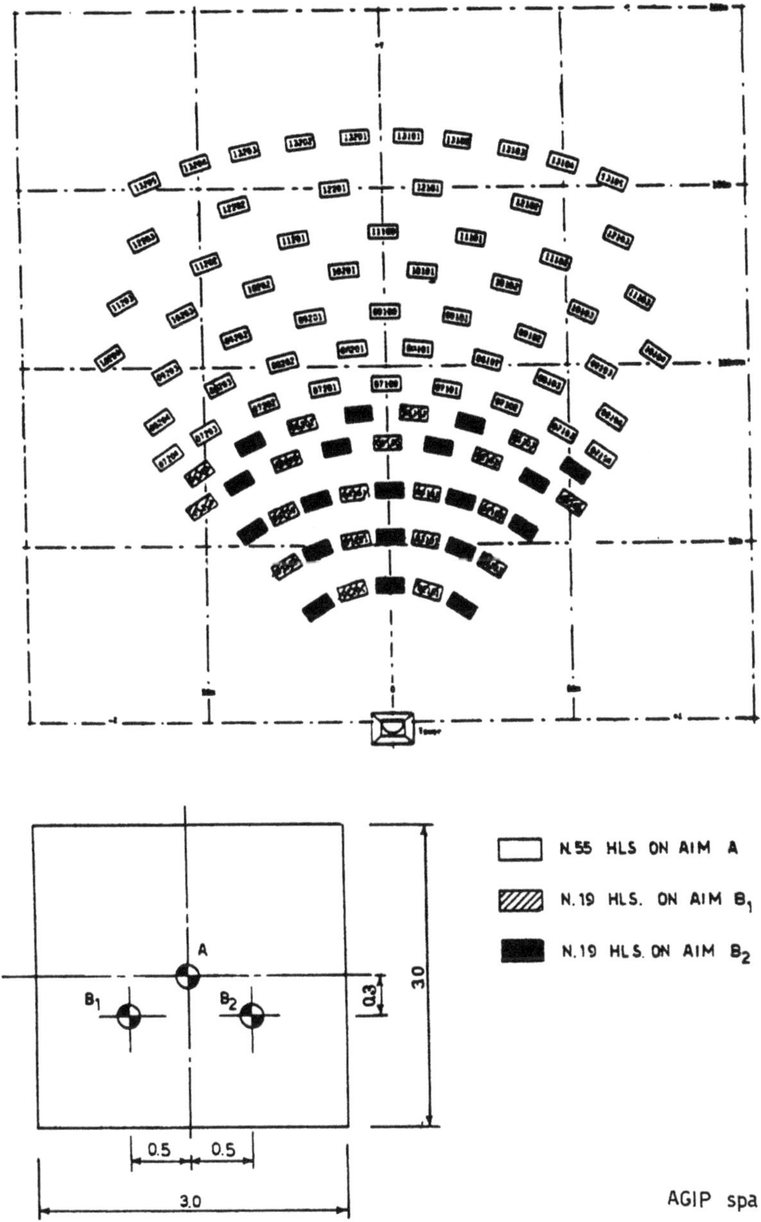

Fig. 5.2.-1: Heliostat Field And Aiming Strategy

ALMERIA - ASR PERFORMANCE

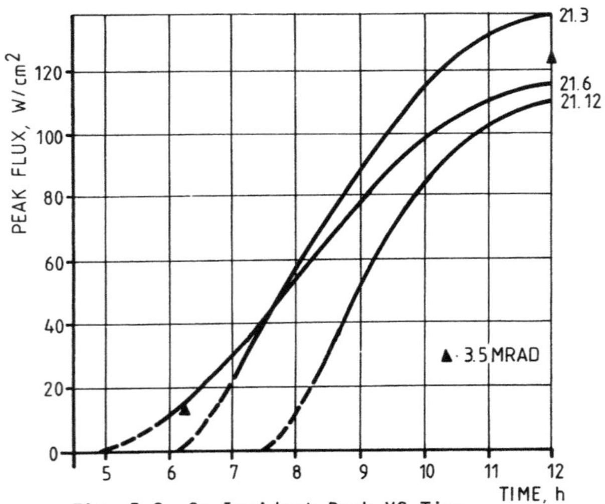

Fig. 5.2.-2: Incident Peak VS Time

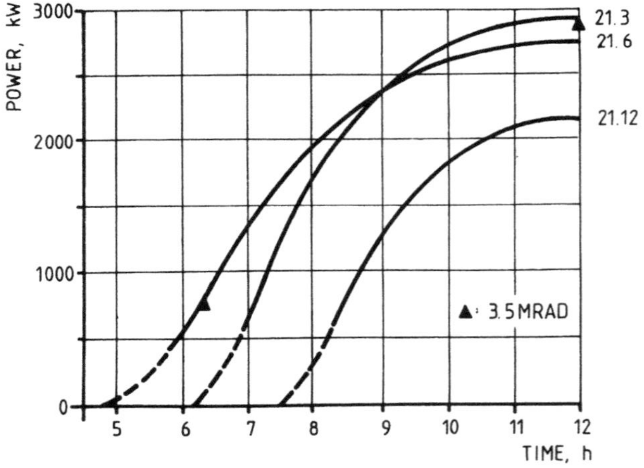

Fig. 5.2.-3: Incident Power VS Time

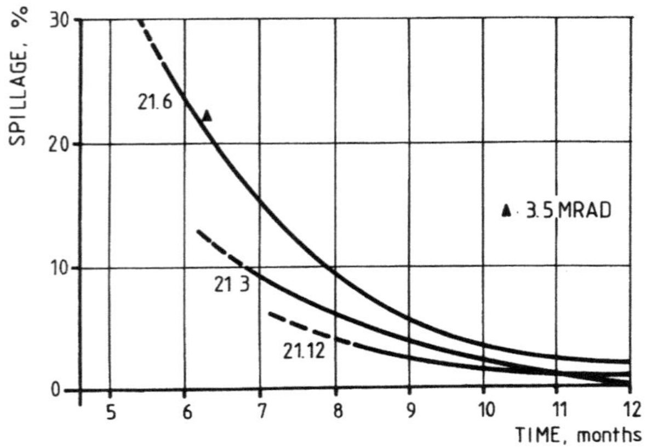

Fig. 5.2.-4: Spillage VS Time ON 3x3 m Target

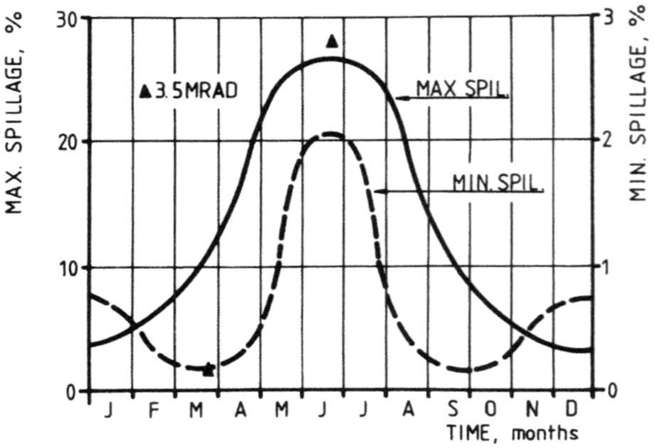

Fig. 5.2.-5: Max. And Min. Spillage On

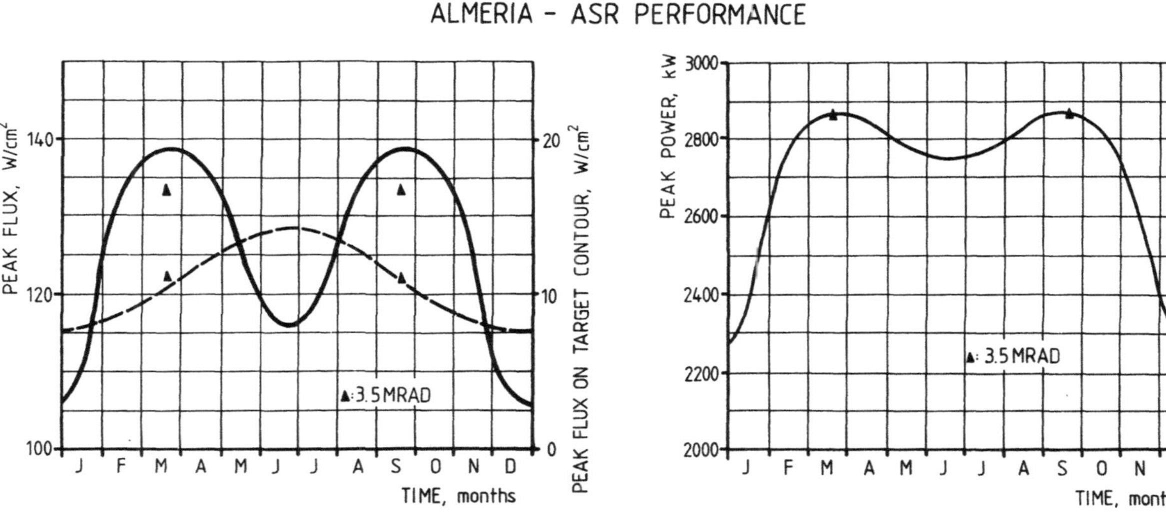

Fig. 5.2.-6: Max. Incident Flux On 3x3 m Target

Fig. 5.2.-7: Max. Incident Power On 3x3 m Target

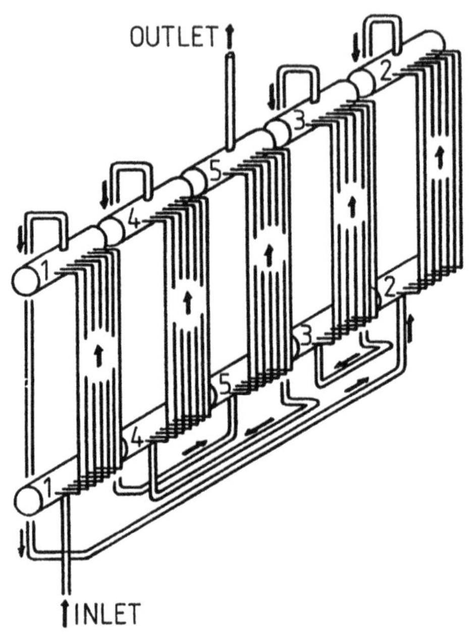

Fig. 5.2.-8: Sodium Circulation in the Panels

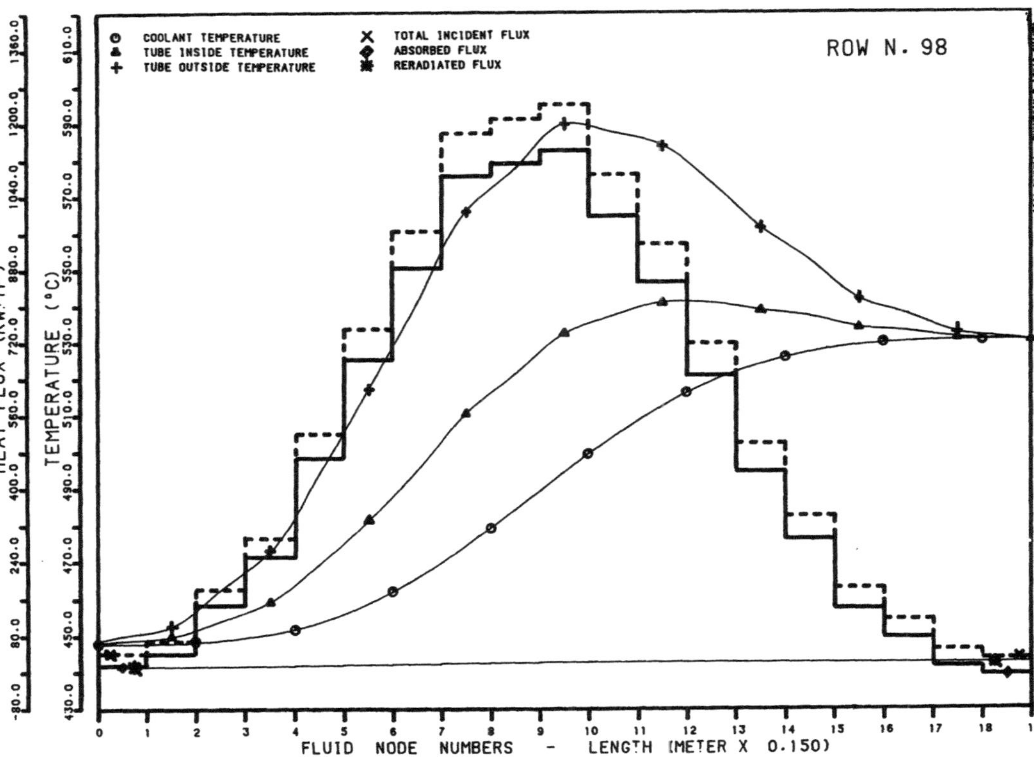

Fig. 5.2.-9: Heat Flux and Temperature Profiles on the Central Tube of the Central Panel

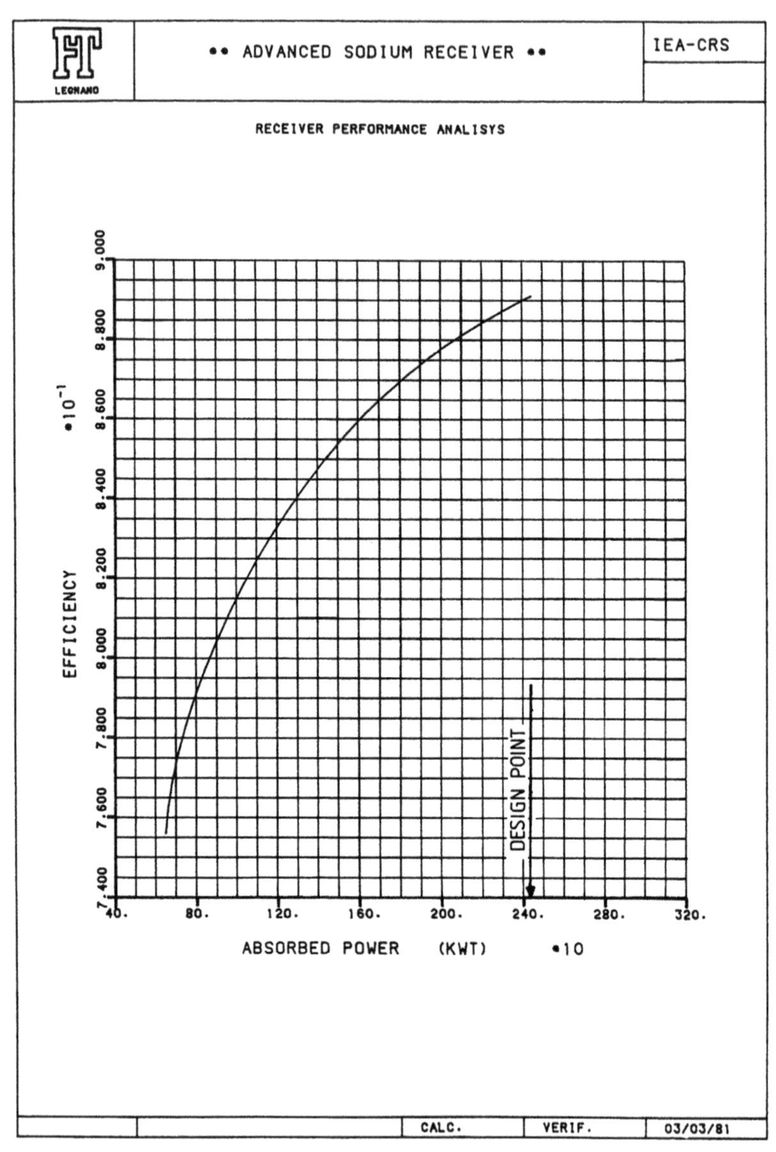

Fig. 5.2.-10: ASR Calculated Efficiency versus Absorbed Power

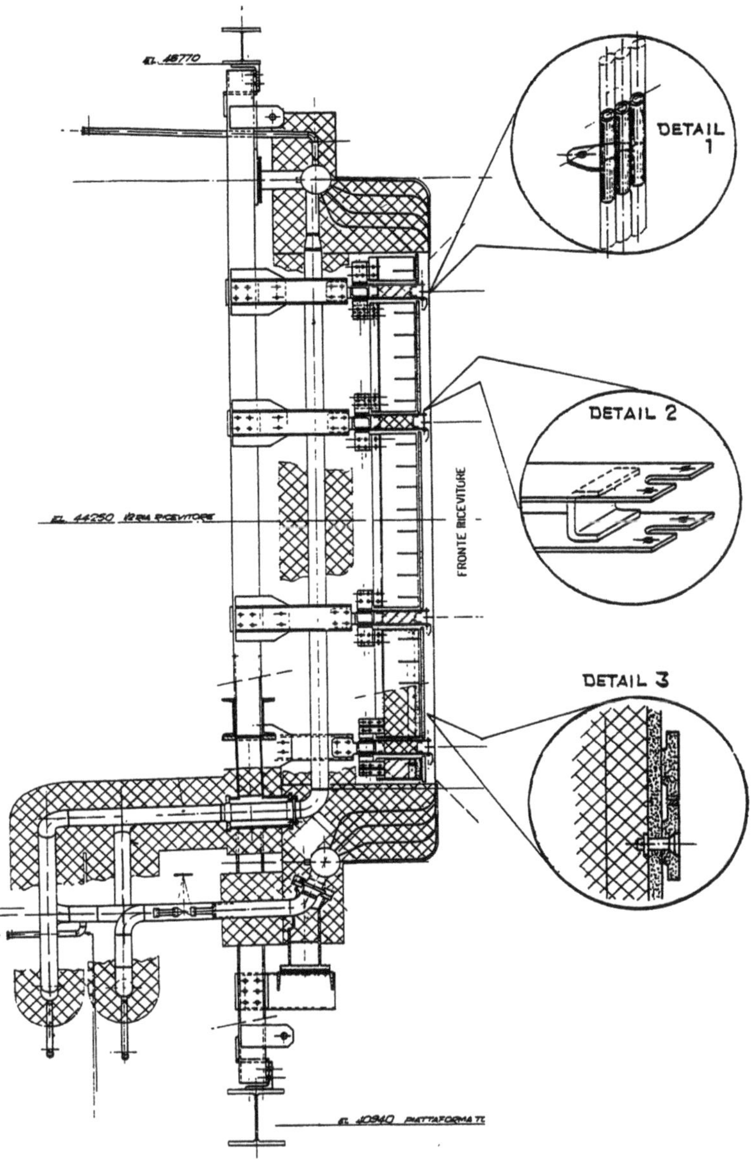

Fig. 5.2.-11: Absorber Back Structure Details

TYPE		CLOUD VELOCITY (Km/h)	10	23	50
		t_s (sec)	54	24	11
a	1	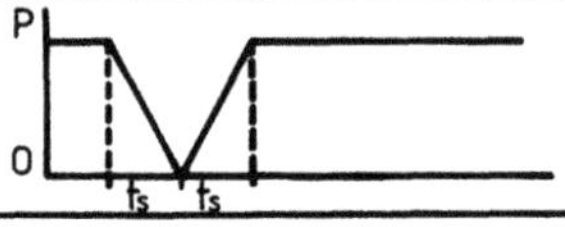 P, 0, t_s, t_s	1000	1000	1000
	2	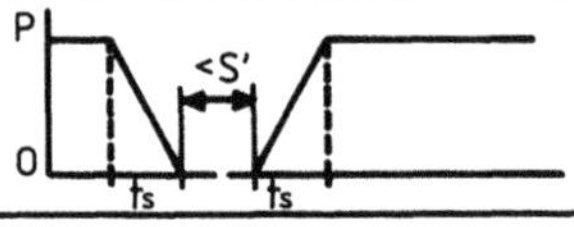 P, <S', 0, t_s, t_s	1000	1000	1000
	3	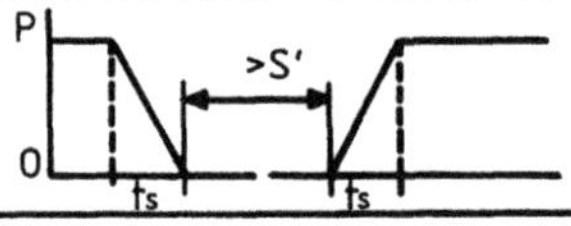 P, >S', 0, t_s, t_s	150	150	25
			2150	2150	2025

Fig. 5.2.-12: Cloud Passage Transients (per year)
(P=peak indicent flux; events=6325)

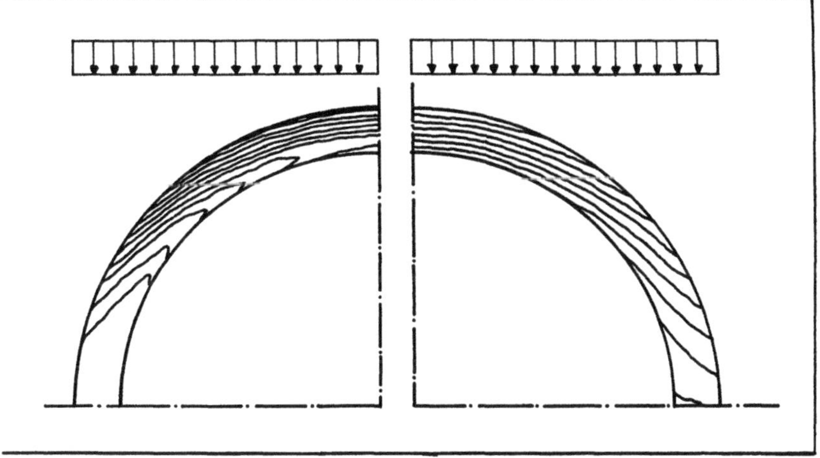

Fig. 5.2.-13: Equivalent Plastic Strain (left) and Temperature Distribution (right) in the Tube Section

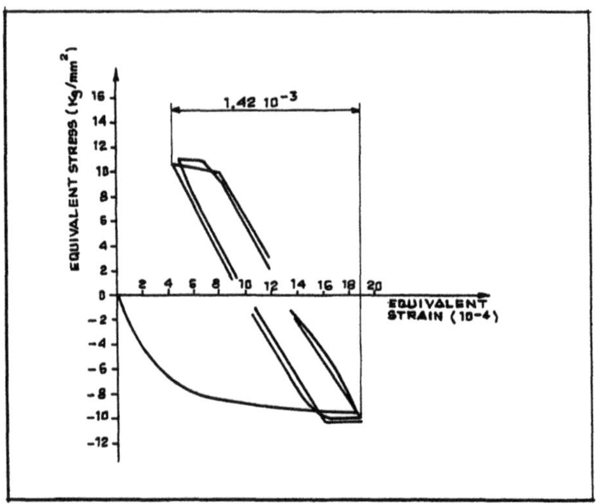

Fig. 5.2.-14: Stress-Strain Loop

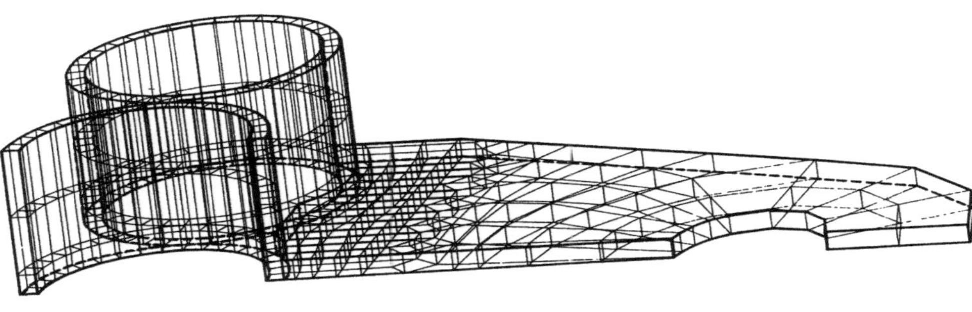

Fig. 5.2.-15: Tube-Supporting Plate Three-Dimension Mesh Plot

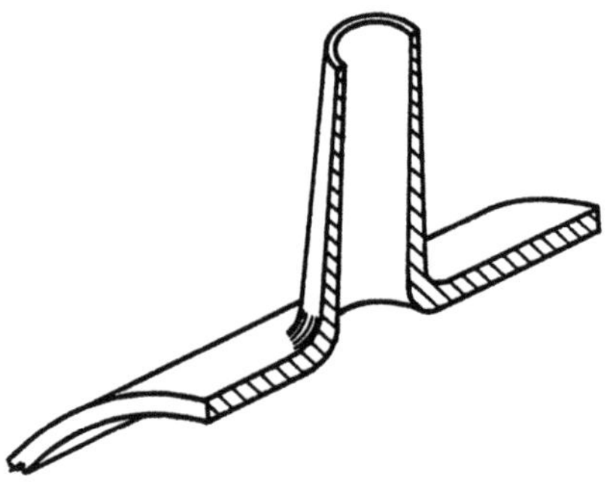

Fig. 5.2.-16: Tube Header Connection

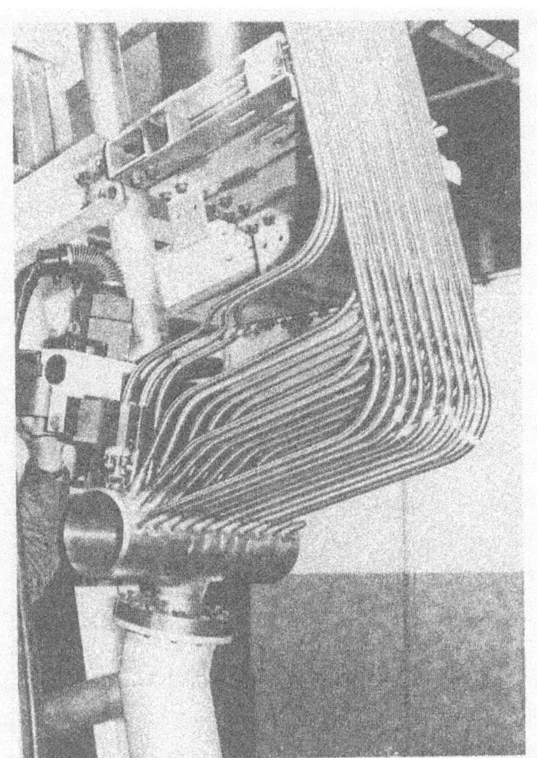

Fig. 5.2.-17: Automatic Orbital Welding of Absorber Tubes to Bottom Header

Fig. 5.2.-18: Top Header Details

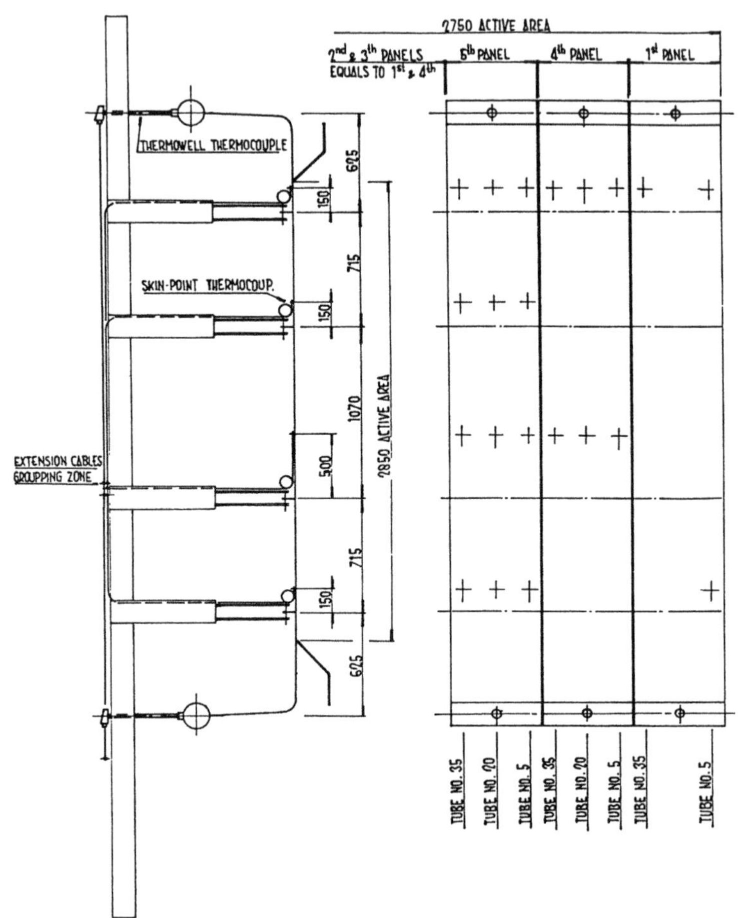

Fig. 5.2.-19: Thermocouple Arrangement

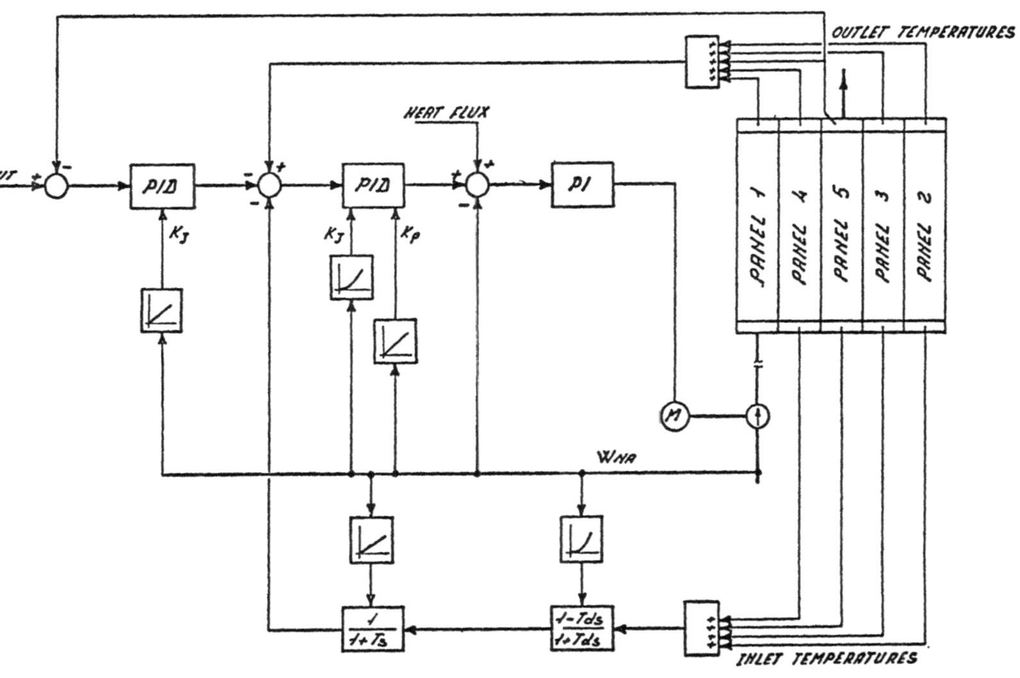

Fig. 5.2.-20: Control Circuit Diagram

S S P S

CENTRAL RECEIVER SYSTEM (CRS)

CAVITY RECEIVER

By

H. W. Fricker
Sulzer Brothers, Ltd.
CH-8401 Winterthur
Switzerland

Abstract

A cavity receiver was chosen to be the first receiver tested in the 500 kWe CRS solar power plant of the SSPS Project. Its basic layout and the thermal analysis were performed by Interatom.

The detailed design with stress analysis, the manufacture , and the erection on site, was performed by Sulzer within the time period July 1979 to December 1980.

1. OPERATING CONDITIONS

The nominal operating conditions for full load are:

Thermal power:	input	2.8	MW
	output	2.44	MW
Sodium:	mass flow	7.34	kg/s
	inlet temperature	270	°C
	outlet temperature	530	°C
	pressure (max. 8 bar)	2.6	bar
	pressure drop	0.45	bar
Maximum heat flux on tube o.d.		63	W/m^2

2. DESIGN

2.1 Tube wall

The heating surface (Fig.5.3-1) consists of a curved tube wall with a mean radius of 2250 mm, a height of 3607 mm and an active angle of 120°.

Six parallel tubes of 38 mm outside diameter and 1.5 mm wall thickness of a total length of approximately 87 m each are arranged as parallel meanders. Their spacing is 43 mm.

The tubes are held by four vertical supports:

- a fixed support beam on the east side, which also carries the inlet and outlet header.
- a support beam which is movable in a perpendicular direction on the west side of the tube wall.
- two movable support stacks spaced 40°.

The return bends and the connecting tubes to the headers are butt welded with an automated orbital welding tool. This ensures a regular and high quality weld.

2.2 Tube Supports

All four support beams are free to expand in the vertical direction and are carried on pedestals which penetrate through the floor insulation onto the base platform.

2.2.1 The fixed support on the east side consists of a square, hollow column flanged to the pedestal at its lower end and connected to the casing by two perpendicular links at its upper end. The receiver tubes are clamped onto this column, see Fig.5.3-2, by means of welded and bolted saddles.

2.2.2 The movable end support on the west side is similar to the fixed support; however, instead of being flanged to the pedestal it is connected to it via a moving link which permits movements parallel to the aperture plane. The top end of the column is fixed to the casing by one link only, permitting approximately the same parallel movement. In this way the gross expansion of the tubes can be permitted without the tubes being stressed.

2.2.3 The two intermediate movable supports are to carry some of the weight of the tubes. They consist of a stack of tube support lugs which are held together by a tie rod with spring washers, Fig.5.3-3. The tie rod is supported on the pedestal by a double ball joint link. This permits it to move in any direction the tubes dictate with very low restraint due to the favorable frictional geometry.

The tie rod is sufficiently flexible to allow some individual tube movements by bending. Vertical expansion of the tube support stack in relation to the tie rod is permitted by the spring washers at the top, while at the same time maintaining the necessary clamping force to avoid tilting and jamming of the lugs on the guide tubes. These latter are segmented, one per six receiver heating tubes. Thus, differential expansions between supports and guide tubes remain small and the deleterious effects of a possible jamming are minimized.

2.3 Headers

The inlet and outlet header are of identical design. Great care has been taken to achieve as closely as possible: thin walls, constant walls, and sodium wetted walls. The intention was to produce a design which would be unaffected by fast temperature transients.

The result of this is shown in Fig.5.3-4. The basic wall thickness is 4 mm, reducing to the 1.5 mm of the heating tube wall thickness by a long tapered nozzle. The flat bottom is 5 mm thick. The header is supported on the two blind nozzles near its bottom. The center point between the two blind nozzles constitutes the fixpoint. The support structure is welded on the fixed tube support beam; therefore, only minor expansion differences between header nozzle and first tube support must be taken up by the connecting tubes.

For complete venting and draining the headers are mounted at an angle of 5° to the horizontal.

Tubes and headers are made of X 6 Cr Ni 18 11 (DIN 1.4849, approximately equivalent to AISI 304 H) austenitic heat resisting steel. The support columns are made of X 2 Cr Ni Mo 18 12 (DIN 1.4435 approximately equivalent to AISI 316 L), whereas the support stack is made of X 10 Ni Cr Al Ti 32 20 (DIN 1.4876, Incoloy 800).

All welds of sodium wetted parts are fully penetrating and 100% radiographed.

2.4 Ceramic Wall

A ceramic brick wall behind the tube wall absorbs the 5% radiation through the gaps between the tubes and provides a certain heat reservoir for extended zero power conditions. It is held in a steel frame and attached by tie strips to the corrugated sheet metal wall behind it.

The bricks are made of Mullite with 70% Al_2O_3 content. This composition exhibits excellent thermal shock resistance. The individual bricks are mechanically interlocking and curved according to the general curvature of the ceramic wall. In addition, they are joined by a special heat resistant cement.

2.5 Casing

The casing around the tube wall has a semi-circular ground projection, is 6003 mm high, 6000 mm wide and 3100 mm deep.

The front opening (aperture) is octagonal and measures 9.69 m^2, see Fig.5.3-5. It is surrounded by freely supported austenitic steel sheets of 2 mm thickness, which have a reflective white coating of inorganic paint. A spillage irradiation of at least 3 W/cm^2 can be accepted.

The steel frame consists of the floor grid and the vertical part, and the roof. Attached to the surrounding U-profile of the floor grid are vertical T-beams which are themselves attached to two more surrounding U-profiles: one at the top and one above the door opening. The removable roof is again a grid consisting of beams. All structural parts are outside the insulation in order to avoid uncontrolled temperatures.

The thermal insulation is generally rock wool and is contained between the inner hot and the outer cold sheet metal casing. The inner casing consists of several austenitic steel sheets attached to the steel frame in such a way as to permit thermal expansion. A welded trough on top of the floor insulation is able to contain the sodium in case of a leakage. At locations where high temperatures are to be expected, ceramic fibres (behind the ceramic wall) or bonded ceramic fibres (lower layer in the roof) are used.

The outer casing consists of zinc plated metal sheets. They are bolted onto the steel frame in a relatively leak-tight manner and in this way form the outer casing. This design permits the insulation to be fitted from the outside on site.

2.6 Doors

The sliding door is made up of two halves; each half door has a separate driving mechanism with an electric motor, a gearbox and a chain drive. The doors roll on a rail at the bottom and are guided in a rail at the top. The door frames carry asbestos seals which make contact with the stationary seal frame. This frame is fixed to the steel structure of the door opening. With the door closed the seal is outside the irradiation from the hot receiver tubes, which ensures a relatively cold seal with small thermal distortions. In order to obtain a defined sealing pressure the doors are slightly inclined to their running track, thus permitting a door movement without seal rubbing for the largest part of its trajectory. The time needed to fully open or close the door is about 15 seconds.

2.7 Painted Surfaces

The receiver tubes are coated with black Pyromark 2500 paint. Great care has been taken in the application of this paint, resulting essentially in the following steps:

- degrease tube
- sand blast with new sand, grain size 0.3 to 0.6 mm
- clean with alcohol
- spray first coat 0.03 mm thick
- air dry overnight
- spray second coat 0.03 mm thick
- air dry overnight
- oven dry
  slowly heat to 260 ^{o}C
  cure for 1 hour
  slowly heat to 540 ^{o}C
  cure for 1/2 hour
  slowly cool to room temperature

Absorption measurements on flat steel plates, treated in this way and exposed to 650°C for 10 hours, show 92% at 0.5 μm wavelength decreasing linearly to 88% at 2 μm wavelength.

The reflective surfaces of the door frame and the doors consists of white Pyromark 2500 paint applied in a similar way.

3. STRESS ANALYSIS

A stress analysis to confirm the operational and abnormal conditions has been performed. The structure of the casing and the ceramic wall have been designed and analyzed to withstand a wind speed of 40 m/s and an earthquake of 1/3 g in the horizontal direction.

A simple stress analysis was sufficient for the hot header under transient temperature conditions. Because of its design characteristics it was found that the header can sustain more than the required 15,000 cycles with a temperature transient of 6 $^{o}C/s$ for the full range 530 ^{o}C to 270 ^{o}C.

The tube bundle stress analysis was performed for the 100% load case. In spite of the flexible support, design forces occur between tubes and supports because of the non-linear temperature distribution.

As shown in Fig.5.3-6. a mean temperature for each of the 42 tube elements has been defined based on the temperature calculations by Interatom. Six parallel tubes of one pass are assumed to have the same mean temperature, which is a certain optimistic simplification.

The finite element model used to calculate the stresses contains these 42 tube elements as well as 52 support elements.

The loads are:

Internal pressure	Bottom	3.4 bar
	Top	2.6 bar
Dead weight		
Earthquake acceleration:		
Horizontal	Tubes	3.4 m/s^2
	Supports	4.2 m/s^2
Temperatures		Figure 6

The sodium side heat transfer coefficient has been calculated according to the formula:

$$\alpha_i = \frac{\lambda}{d} (4.48 + 0.0238 \quad P_e^{0.8})$$

For 7.34 kg/s sodium this amounts to:

$$\alpha_i \approx 20'000 \ W/m^2K$$

Considering the location of the highest insolation of 63 W/cm^2, a loss of 10% for reradiation, 5% for convection and the tube thermal conductivity of 22 W/m^oC the temperature differences are:

Tube outside to inside:	$\Delta \vartheta = 36\ ^oC$
Tube inside to sodium:	$\Delta \vartheta_i = 27\ ^oC$

These temperature differences are approximately proportional to the temperature increase in each tube element.

For the tubes pipe bend elements STIF 60 of the ANSYS program have been used, because they allow the uneven temperature distribution around the tubes. The pillars are modeled by beam elements STIF 4; their temperatures are assumed to be in between the ones of the adjacent tubes.

The results of the stress analysis can be summarized as follows:

3.1 Primary Stress Analysis

3.1.1 Tubes

Internal pressure	2.9 N/mm^2
Weight	3.9 N/mm^2
Earthquake (very conservative)	55 N/mm^2
Total stress	62 N/mm^2

The highest tube temperature is 530°C at the top of the tube bundle. The allowable stresses are:

$S_m = 2/3\ \sigma_{1\%} = 76.5\ N/mm^2$

$S_t = \sigma_{1\%/262'000\ h} = 79.5\ N/mm^2$

The primary stress stays below these limits.

3.1.2 Fixed support

Weight	0.8 N/mm^2
Earthquake	19.6 N/mm^2

Both stresses are well below the limit of $\sigma_{0.2\%}$= 98 N/mm^2 for X 10 Ni Cr Al Ti 3220 at 530 °C.

3.2 Creep Fatigue Analysis

The peak value of thermal stress from membrane and bending action across the whole tube is:

$$\sigma_b = 124\ N/mm^2$$

at mid-height at the fixed support.

The highest value due to the temperature difference of 36 °C across the tube wall is 84 N/mm^2.

The maximum stress in the supports occurs at mid-height of a movable support and is 157 N/mm^2.

The stresses have been evaluated according to case N-47 of the ASME B + PV Code.

3.2.1 Tubes

The primary and secondary stresses are superimposed very conservatively:

$$\sigma = 2.9 + 3.9 + 55 + 124 + 84 = \underline{270 \text{ MPa}}$$

Assuming n = 15'000 cycles, the allowable cycle number and the fatigue damage are:

$$\varepsilon_t = \frac{270}{170'000} = 1.59 \quad 10^{-3}$$

$$N_D\ (1000^oF) = 70'000 \text{ (Fig. T-1420-1A)}$$

The damage fraction then is:

$$D_f = \frac{n}{N_D} = 0.21$$

This is a very conservative value due to the superimposition of stresses.

In the center of the bundle, where the thermal stresses are significant, the temperature is still too low to produce significant creep damage. At the upper end of the bundle, the stresses again are too low to produce significant creep damage.

3.2.2 Pillars

$$\sigma = 3.2 + 19.6 + 157 = 180 \text{ MPa}$$

$$\varepsilon_t = \frac{180}{170'000} = 1.06 \quad 10^{-3}$$

$$N_D\ (1100^oF) \quad 10^6 \quad \text{(Fig. T-120-1C)}$$

$$\longrightarrow D_f = \frac{15'000}{10^6} = 0.015$$

The creep damage is negligible, as after first relaxation of the stress, the stress is not shifted much at shutdown and damage due to a reversed relaxation cycle is small with this range.

3.2.3 Displacements

In Fig.5.3-7 the calculated displacements at 100% load against the cold condition are plotted for the four support locations:

- lower end of tube wall
- mid-height of tube wall
- upper end of tube wall.

3.2.4 Conclusion

It can be concluded that the stresses at 100% load are well within allowable limits. A change from the cold state to the full load can be performed 15'000 times with no more than 21% fatigue damage, provided no stresses greater than those in the full load condition are set up during load changes. In order to fulfill this last requirement the sodium flow should at all times match the heat input into the receiver. If this is not the case, different temperature distributions to the one calculated could produce different stresses. In particular, underfeeding the bundle with sodium should be avoided because this results in local hot spots on top of possibly higher stresses.

Whenever the correct sodium flow cannot be maintained for some reason, overfeeding should be chosen or the heat load should be removed by diverting the heliostats.

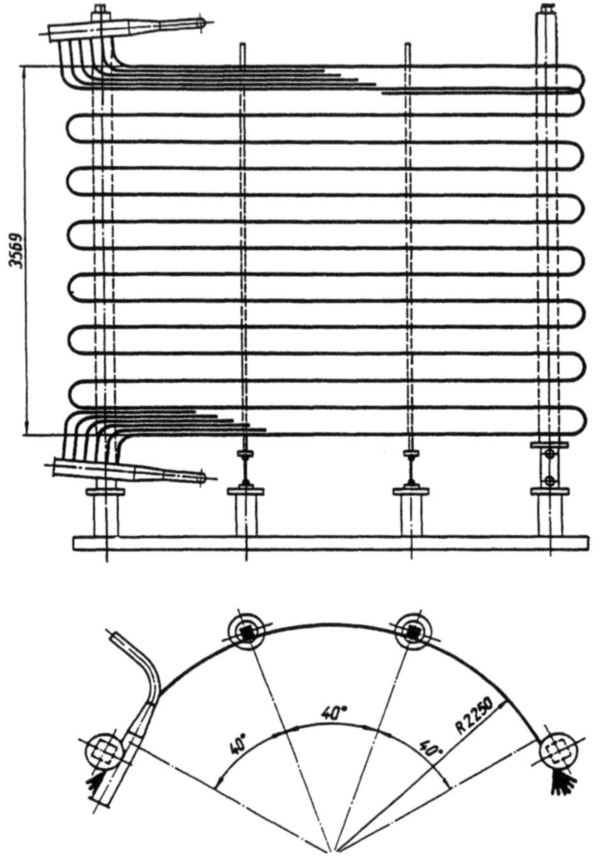

Fig. 5.3.-1: Tube bundle with supports

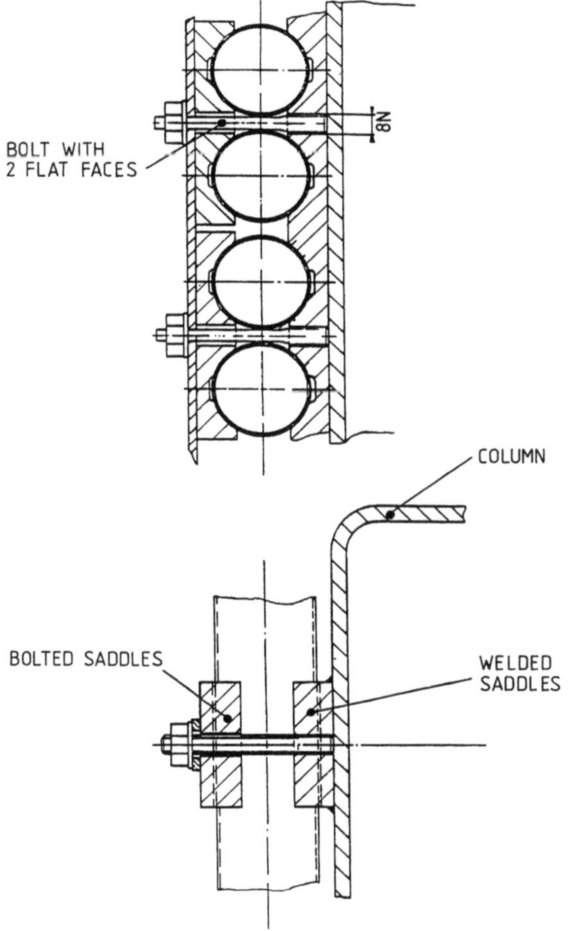

Fig. 5.3.-2: Fixed tube support

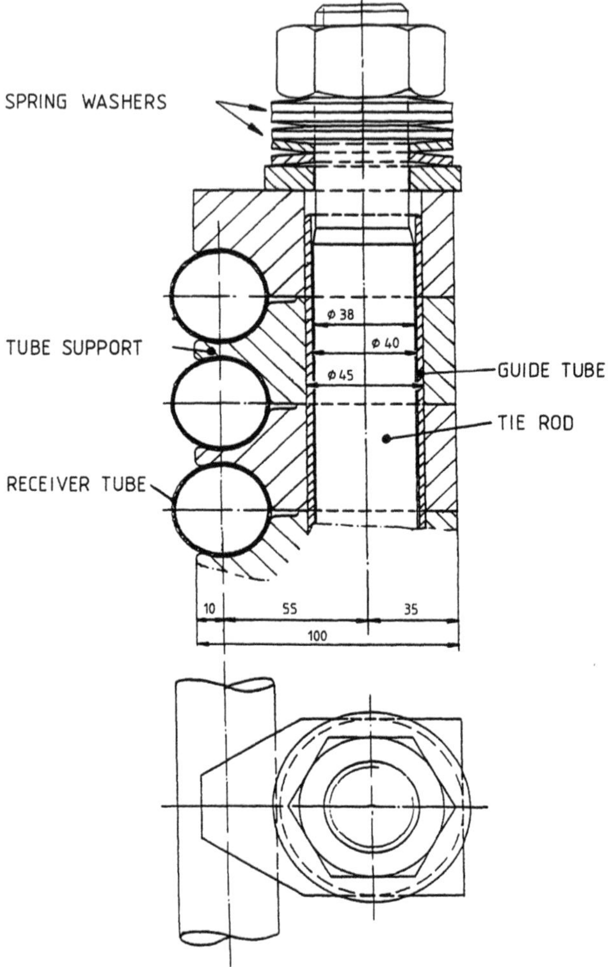

Fig. 5.3.-3: Intermediate tube supports

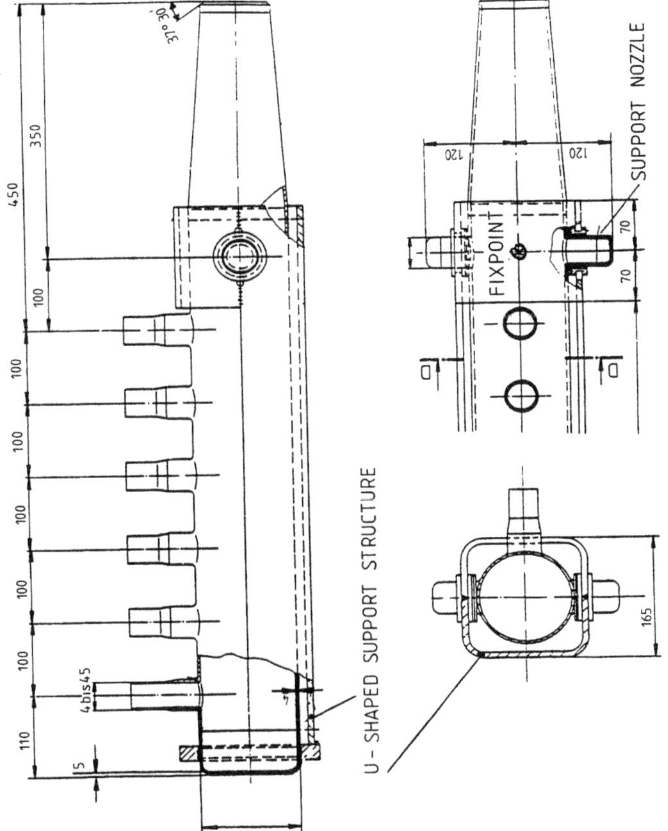

Fig. 5.3.-4: Sodium header with support

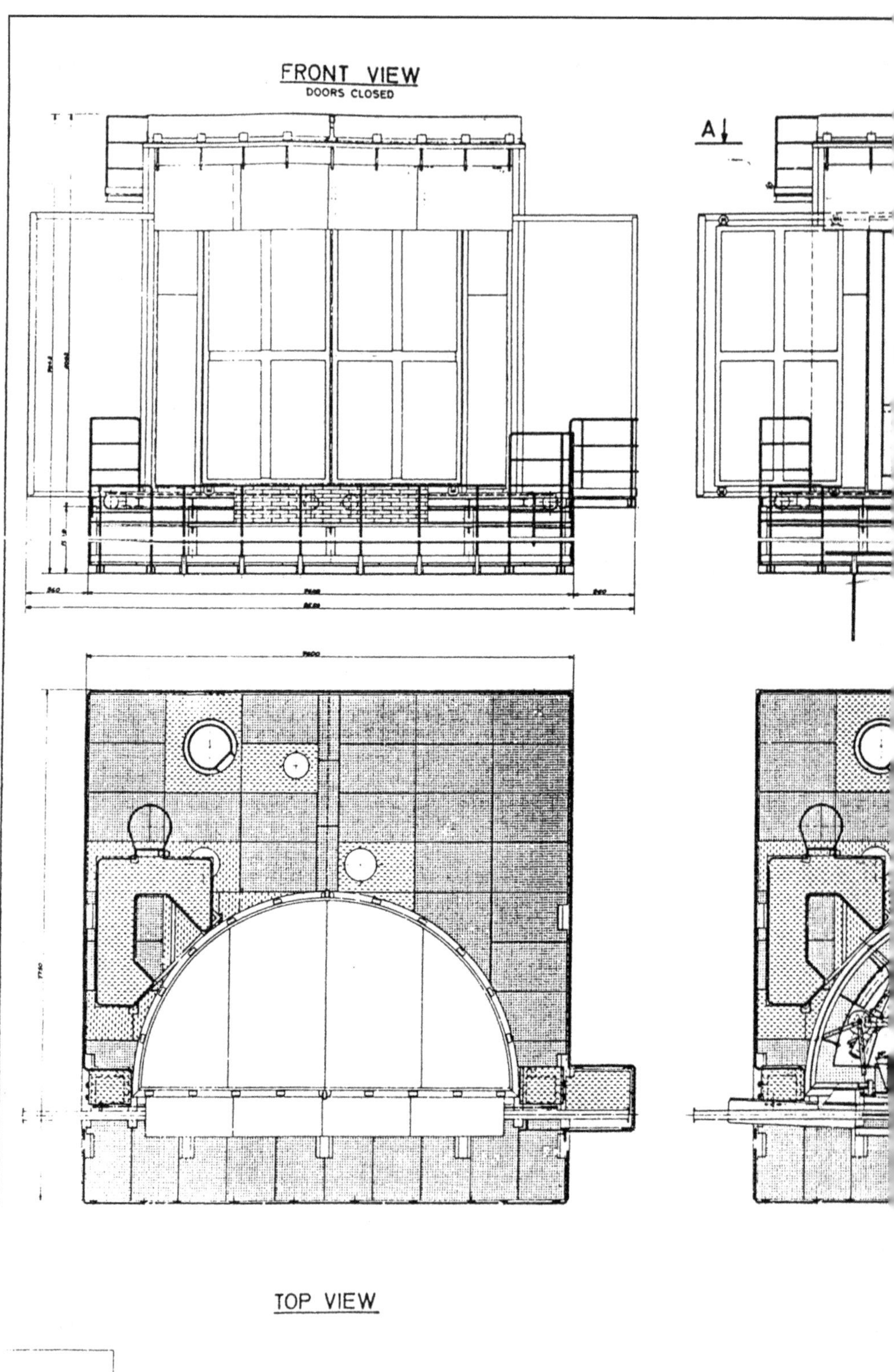
FRONT VIEW
DOORS CLOSED
A
TOP VIEW

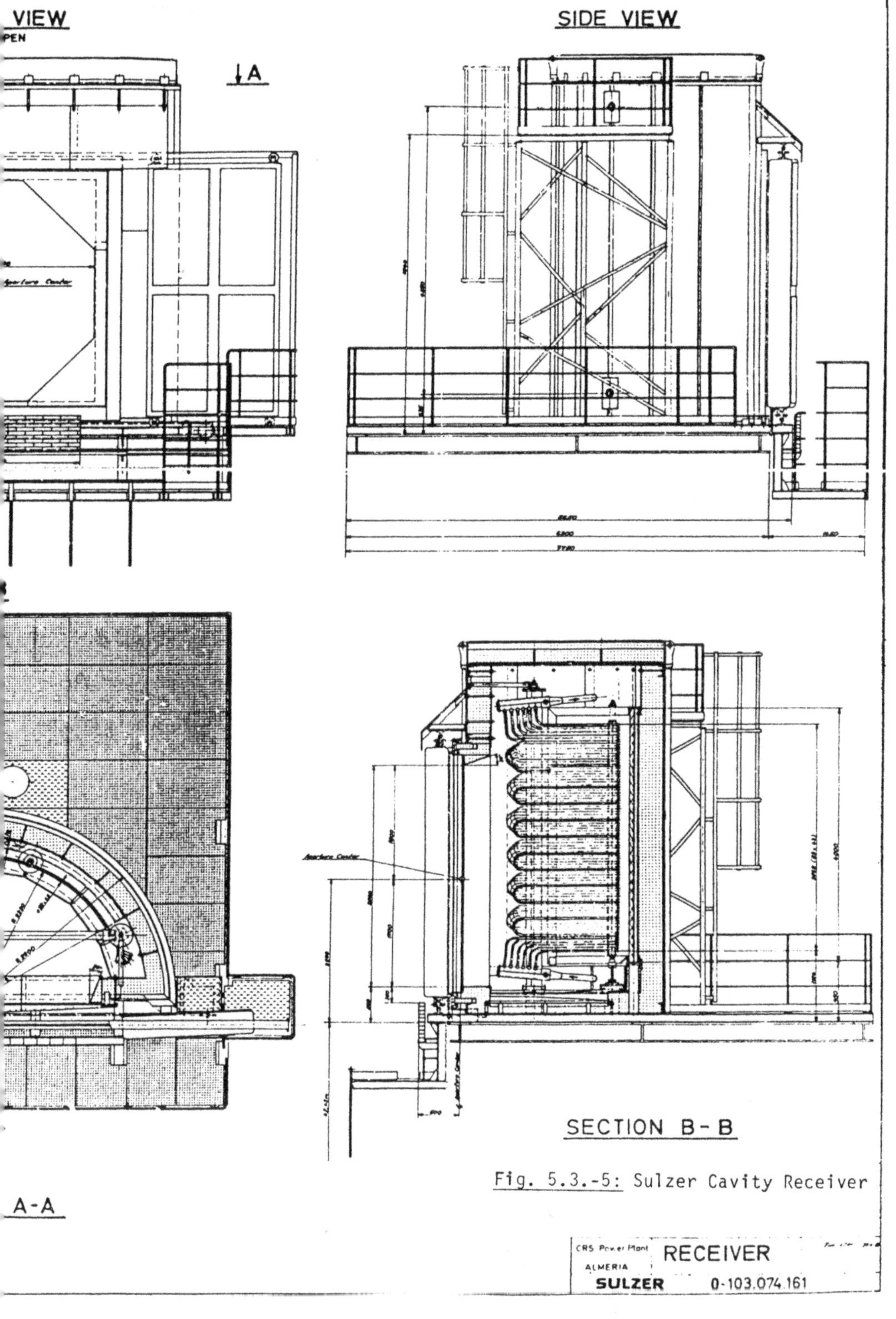

Fig. 5.3.-5: Sulzer Cavity Receiver

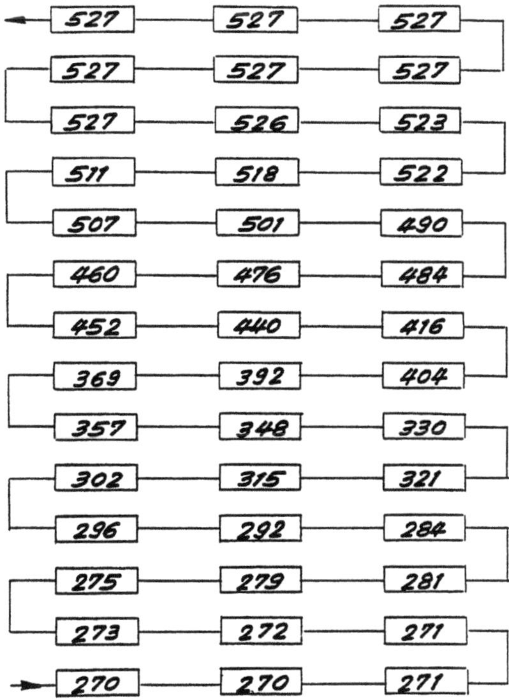

Fig. 5.3.-6: Sodium temperatures in tubes at 100% load

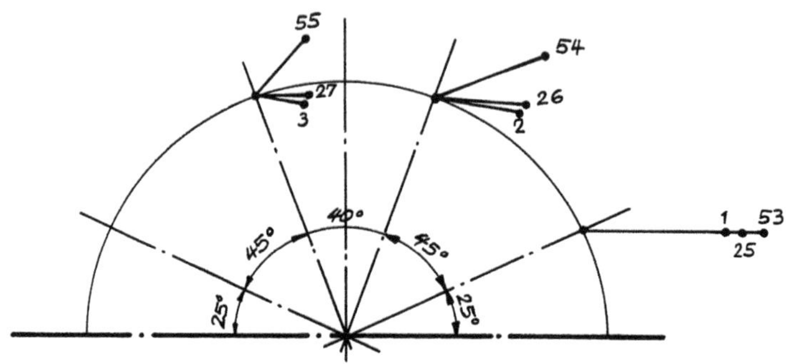

tube wall	node №	deflection [mm]				
bottom	3	10.9	2	18.3	1	31.1
mid-height	27	11.5	26	20.1	25	34.8
top	55	16.0	54	26.1	53	39.4

Fig. 5.3.-7: Calculated support deflections for 100% load

S S P S

CENTRAL RECEIVER SYSTEM (CRS)

COMPARISON OF THE TWO RECEIVERS

By

H. W. Fricker
Sulzer Brothers Ltd.
CH - 8401 Winterthur
Switzerland

1. COMPARISON OF THE TWO RECEIVERS

A description of the external receiver, the Advanced Sodium Receiver (ASR) is given by DiMeglio and Bedogni in 5.2 of this section, and a description of the cavity receiver (SULZER) is in 5.3 of this section.

In the following, a qualitative comparison of the two receivers is attempted. This comparison is by no means complete and only covers the headings: design philosophy, some operating results, efficiency and complexity, and from the point of view of the designer of the cavity receiver.

1.1 Design Philosophy and Operating Results

The cavity receiver was the first sodium receiver of this size in the world. One of the most important design requirements was the need for safe, reliable operation. Therefore, the maximum heat flux was set at a high, but still somewhat conservative, 63 W/cm^2. The result was a design using relatively thick tube walls, which in turn permitted the use of a small number of relatively large and long tubes.

Another criteria was flexible operation. This second criteria was fulfilled by the tube support structure which was chosen. Uneven temperature distributions are tolerated since small stiffnesses create only relatively low stresses.

Operation showed that generally, the flexibility was sufficient; however, some distortions at mid-height of the tube wall indicated that for the operating conditions, as applied, the flexibility could have been better. It is, however, strongly suspected that the usual preheating cycles, with 20 or even more heliostats in operation, resulted in gross overheating of the empty tubes and very severe temperature transients when filled with cold sodium. Almost certainly a more careful preheating, as practiced with the ASR, would have avoided the deformations.

The storage capacity of the ceramic wall was not sufficient to maintain tube wall temperatures for any length of time, which was a design goal. The expected temperature profile of the ceramic wall was not achieved, probably due to the heat distributing effect of convective air currents. However, this did not have any negative consequences; the ceramic wall simply acted as a heat shield behind the tubes.

A third criterion was low standby losses. This was achieved by providing an air tight casing with good door seals, and by applying generous insulation.

The main goals of the ASR were the achievement of high heat fluxes and a design which could be easily scaled-up to larger units. The first criterion led to a design with small wall thickness and small tubes. The second criterion demands an external receiver element, which could be repeated when scaling-up the receiver (upward).

Straight vertical tubes were chosen for this design and, with the constraint of a given sodium pumping capability, resulted in a multi-panel unit with a relatively large number of small parallel tubes in each panel with the five panels connected in series. One consequence of this design is the need for continuing knowledge of the tube temperatures under all operating conditions, since the inherent capacity to absorb abnormal temperatures is relatively small. This was achieved by a thorough stress analysis under carefully determined operating conditions, and installation of a large number of thermocouples on the tubes to monitor tube temperatures and aid in sodium flow control.

The cavity receiver gave excellent performance during the operation period from spring 1981 to April 1983. It operated for over 1300 hours, on approximately 250 days. During this time it caused no operating restraints.

Excellent performance has also been observed with the ASR. It has, up to the end of August 1984, operated for 174 days. However, in contrast to the cavity receiver, some operational restrictions exist. When the insolation is diffuse, power reductions become necessary because of overheating of the aperture frame due to increased spillage. This type of behavior is to be expected, because the ASR has been designed much closer to the limits than the cavity receiver.

1.2 Efficiency

Although an exact measurement of receiver efficiency is very difficult and has not been possible for both receivers, it is clear that the ASR has a considerably better efficiency than the cavity receiver. Since a main reason to build a cavity receiver in the first instance was the desire to achieve a good efficiency, this seems to be a case where expectations have not been fulfilled. The situation may be more favorable for the cavity receiver if the comparison is on overall efficiency: in contrast to the ASR it does not impose operational limits such as reducing power because of excessive spillage at high winds or high diffusivity. Furthermore, it is easily drainable and needs less trace heating.

In the ASR only the front half of the panel tubes is exposed to the convecting air currents. The surface affected is approximately 12 m^2. In the cavity receiver practically all tube surfaces, approximately 60 m^2, are exposed to convection. The fact that the estimated convection losses are only about two times higher than for the ASR, indicates some effectiveness of the cavity as a convection suppressor.

1.3 Complexity

Tab.5.4-1lists some of the main design characteristics of the two receivers.

The more sophisticated ASR requires a 23% higher total tube weight and twelve times more welds on pressure carrying tubes. On tube welds alone, the number is more than eight times higher than for the cavity receiver. Since this is reflected strongly in manufacturing costs, a careful total cost analysis would be required to decide which way to take in the future.

DESIGN IMPROVEMENTS

Design improvements are possible with both receivers. Some of the most cost or efficiency effective improvement possibilities are indicated below.

In the ASR a modified connection of the panels could reduce the length of the connecting pipes substantially. The two outer panels 1 and 2 could be purged with parallel sodium flows; panels 3 and 4 could be purged in parallel in downward direction, both feeding panel 5. The schematic of this is shown in Fig.5.4-1 As a side benefit, the number of headers could be reduced from 10 to 6. This modification would be very effective on first cost, but increases the failure risk in operation because of the downward flow.

In the cavity receiver the ceramic wall could be replaced by a ceramic felt insulation contacting the rear of the tubes. In addition, the header, tube bends, and tube wall corners could be insulated front and back. The two bottom tube rows could be eleminated or insulated, since they do not contribute to the energy gain according to the operating results.

These measures would be cost effective and reduce thermal losses considerably, since the exposed surface drops from 62 m^2 to approximately 24 m^2. With the latter measures it should be possible to increase the cavity receiver efficiency considerably.

1. Tube Panel			
Tube dia. x wall	mm	38 x 1.5	14 x 1.0
Number of parallel tubes	-	6	39
Length of each tube	m	87	4.7
Total tube length	m	520	916
Total tube surface	m^2	62	40.4
Total tube weight	kg	710	300
Number of tube welds	-	180	780
Number of support welds	-	0	780
Total number of welds	-	180	1560
2. Headers			
Dia. x wall	mm	127 x 4	162 x 7
Total header weight	kg	28	220
Number of nipples	-	12	400
Total number of welds	-	24	430
3. Connecting Pipes			
Dia. x wall	mm	88.9 x 3.2	88.9 x 5.5
Total length	m	1.78	36
Total weight	kg	12	400
Number of welds	-	2	67
4. Sodium containing parts			
Total weight	kg	750	920
Number of welds	-	206	2057

Tab.5.4-1: Comparison og characteristical data of cavity receiver and ASR

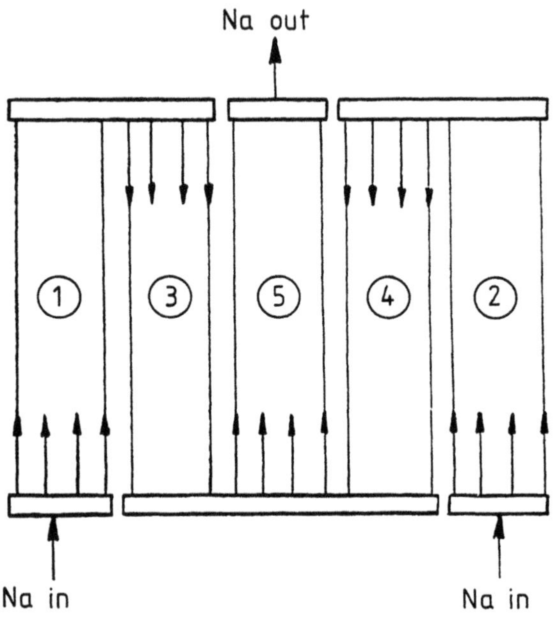

Fig. 5.4.-1: Alternative ASR flow

solation of 900 W/m^2, a receiver efficiency of about 87% and a thermal efficiency of about 60% were measured. The thermal efficiency averaged over the time of operation on a clear day (October 7, 1982) is 57%. Converting this value to the total day from sunrise to sunset (11'25") by 8'50" of operation, one gets an average efficiency of 50%. The receiver efficiency as a function of power into the cavity is presented in Fig.5. 5-3. The solid line corresponds to the design curve, calculated by INTERATOM (Ref.2). The measured data points show that the receiver essentially met its specifications.

To describe the losses, Kraabel (Ref.3) used some simple relations. The convection losses could be expressed by

$$\dot{q}_{conv.} = 1.14\ A_c/A_R\ (T_W - T_\infty)^{1.426} \tag{1}$$

and the IR-radiation losses may be written as

$$\dot{q}_{IR} = \varepsilon_{eff} \cdot 5.67 \cdot 10^{-8}\ (T_W^{\ 4} - T_\infty^{\ 4}) \tag{2}$$

Since the conduction losses are negligible, this yields to a simple expression for a first estimation of the thermal efficiency.

$$\eta_{th} = \alpha \bullet \eta_F - \frac{1}{C \bullet I} \{ \dot{q}_{IR} + \dot{q}_{conv.} \}. \tag{3}$$

Comparing the measured data with the calculated efficiencies (Table 1), a good agreement is obtained. However, it is necessary to prove this simple relation by more measured data.

I (W/m²)	η_F	η_{TH} meas.	η_{TH} calc.
900	0.72	0.60	0.62
800	0.68	0.55	0.57
700	0.60	0.48	0.49
600	0.54	0.41	0.42
500	0.43	0.30	0.30

Table 5.5.-1: Comparison of calculated and measured efficiencies of the cavity receiver

3. NUMERICAL SIMULATION OF THE SULZER RECEIVER

Receiver efficiency strongly depends on the design. The designer has to consider efficiency as well as material constraints. Solar central receiver tubes are characterized by a non-uniform temperature distribution, especially along the circumference. The distribution results in thermal stresses. Therefore, knowledge of the temperature distribution is a prerequisite for calculating thermal stresses. In the following, the results of a numerical simulation are provided.

3.1 Basic Equations

The simulation is based on the following equations:

Heat Conduction Tube Wall

$$\frac{\partial T}{\partial t} = a\left(\frac{\partial^2 T}{\partial r^2} + \frac{1}{r}\frac{\partial T}{\partial r} + \frac{1}{r^2}\frac{\partial^2 T}{\partial \varphi^2}\right) \quad (4)$$

Energy Balance Sodium

$$\frac{\partial T'}{\partial t} + w'\frac{\partial T'}{\partial z} = \frac{\alpha'}{\rho' c_p' R_i \pi} \int_0^{2\pi} (T_{R_i}(\varphi, z) - T'(z))\, d\varphi \quad (5)$$

Energy Balance Ceramic Back Wall

$$\frac{dT_{ce}}{dt} = \frac{1}{\Delta V \cdot \rho_{ce} c_{P,ce}} \Delta A \left\{K(T_{am} - T_{ce}) + \dot{q}_{ce}\right\}. \quad (6)$$

Boundary Conditions

$$r = R_a: \quad \frac{\partial T}{\partial r} = \frac{1}{\lambda} \cdot \Sigma \dot{q} \quad (7)$$

$$r = R_i: \quad \frac{\partial T}{\partial r} = \frac{\alpha}{\lambda} (T_{R_i}(\varphi) - T') \quad (8)$$

$$z = 0: \quad T = T(t, r, \varphi, 0) = T_e \quad (9a)$$

$$T' = T'(t, 0) = T_e \quad (9b)$$

All three basic equations are connected by the laws of heat transfer, especially radiation and forced convection. By the boundary condition for the outer tube side, the solar and thermal radiation affects the tube temperature. On the other hand, the sodium temperature is affected by the tube temperature. The spatial angle dependent solar radiation is

state within six hours. Although there is 5% space between each tube, the effective space for solar radiation is only about 1% due to the finite diameter of the tubes and the angle between the surface normal of the tube bundle and the direction of the solar radiation. The temperature distribution of the back wall is similar to the flux distribution of the concentrated solar radiation.

Thermal deformation of receiver tubes are caused by a non-uniform temperature distribution, especially along the circumference since the solar heat flux hits the tubes only on the front side. In Fig.5.5-11, the solar heat flux along the circumference of one tube which is situated in the upper central region of the receiver (at $\sim$47 m tube length) is plotted versus azimuthal angle. The angle zero corresponds to the front side, at $\pi/2$ the tube surface facing the ground, and angle π corresponds to the back side of the tube. A maximum heat flux of about 50 W/cm^2 is reached for this tube section. On the back side of the tube only a small heat flux which was reflected by the ceramic wall hits the tube.

This heat flux profile results in a temperature distribution shown in Fig.5.5-12. The uppermost curve corresponds to the outside surface temperature; the lowest to the inside surface temperature. The other curves are inside wall temperatures. On the back side, the tube temperatures are close to sodium temperature. The radiation from the back wall is not high enough to heat the tube. The difference between front and back side is about 80 K. Thermal radiation between backwall to tube or tube to tube does not effect the temperature distribution. The heat flux caused by thermal radiation from back wall to tube versus the circumference of the tube is plotted in Fig.5.5-13 (at the tube length 47 m, upper central region of the receiver). Since the ceramic wall temperature in this area is higher than the tube temperature, there is a heat flux from the ceramic wall to the back side of the tube. But the dimension is much smaller than the solar radiation on the front side.

The heat flux from the adjacent tubes versus the circumference is presented in Fig.5.5-14. From 0 - 2 rad the tube radiates energy to the tube below. On the back side there is nearly no temperature difference between the tubes. No radiation will be exchanged. From 4 - 2π rad, the tube absorbs from the tube above.

3.3 Conclusions

The conclusions of the numerical simulation are very simple:

- The ceramic back wall cannot fulfill the task to produce a more uniform temperature distribution along the circumference of the tube as it was understood to do.

- Having space between the tubes does not make sense with only one tube bundle. The convection losses may increase and the dimension of the receiver is larger than necessary.

4. TEMPERATURE MEASUREMENTS WITH INFRARED-IMAGING SYSTEMS

To verify such simulation as described above, it is absolutely necessary to compare it with temperature measurements. Infrared-imaging-systems combine the advantage of not disturbing the temperature distribution and getting a comparable description of the front surface temperature. The measurement is physically based on the fact that any material with a temperature higher than 0 K radiates electromagnetic energy. The intensity depends on temperature and wave length.

To avoid large errors by reflected solar radiation from the tubes, the detector of the camera operates in the favorable atmospheric window at a wave length of 8 - 12 µm. At the moment the camera has an optical resolution of 0.5 mrad; using an optical system made of germanium, this will be improved by up to 0.25 mrad. The radiated intensity measured by the detector is digitized by an A/D convertor to give 256 conversion points. The infrared-image is stored in the memory frame organized as 256 x 256 x 8 bits. The infrared camera is now a part of HERMES, and the format of its image corresponds to the video camera system of HERMES (Ref.1). Therefore all image processing abilities of HERMES could be used.

There are three main sources which may cause errors if one measures the temperature distribution of the front side of the receiver tubes:

1. Reflected solar radiation from the tubes. At 100 W/m^2 flux, an absorptivity of the tubes of 0.9 and a reflectivity of the mirrors in the wave length of 16%, one obtains an error of + 4 K. The reflectivity of the mirrors was determined by measuring the sun's temperature looking at it through the mirrors of the heliostat with the camera.

2. Calibration of IR-camera. The calibration was done using a black body ($\varepsilon \geq 0.99$). The estimate for the calibration yields ± 5 K.

3. Non-uniform emissivity on panel. Assuming a variation of the emissivity of ± 1.5%, this yields an error of ± 8K. It has to be proven by detailed emissivity measurements on the panel, whether this assumption can be justified.

Taking into account all three errors, the total estimate for this measurement yields ± 10 K.

In the last week of September 1984, the IR-camera was installed on site and some very first results were available. Since August 1983, the ASR has replaced the Sulzer receiver of the SSPS-CRS plant. The ASR is described in detail in Ref. 7 and 8. The main characteristic of the ASR is its billboard design consisting of five identical tube panels. The panels are connected in series so that sodium flows upwards through each one, first through the outer panels in the low flux areas and finally through the hottest center panel.

The temperature measurements were taken 50 m in front of the tower. The temperature distribution can be displayed on a color monitor. Using the image processing abilities of HERMES, one can plot isothermal lines, temperature cross sections, and three-dimensional temperature diagrams. Fig.5.5-15 represents a horizontal temperature cross section of the lower part of the ASR. Coming from the left side, the temperature increases at the aperture rim up to the sodium inlet temperature of about 275°C, corresponding to Panel 1. The temperature of Panel 2 on the outer right side is only a little bit higher. In Panels 3 and 4 the temperature is increasing and in the center panel there is a maximum. Since this is a cross section of the lower part of the receiver, the temperatures of each panel correspond more or less to the outlet temperature of the proceeding one. The vertical surface temperature profile along the tubes of the center panel is plotted in Fig.5.5-16. The maximum temperature of 630°C is reached at ~ 2 m tube length. For comparison, the solar heat flux is plotted in the same diagram.

Assuming an emissivity of ε = 0.9 in the infrared range, the total emission can be calculated by applying the Stephan-Boltzmann law on each data point. The integration over the whole receiver area yields the total emission of 78 kW for this special case.

Since these are the results of the very first 'on site' measurements with the IR-camera, here may still be a lack of experience, hence the data has to be evaluated more carefully. Nevertheless, IR-imaging systems seem to be an important tool for temperature measurements of solar systems.

5. REFERENCES

1. W. Schiel, "HERMES Measurements", SSPS Technical Report 4/83.

2. H. Weitzenkamp, "Receiver Thermodynamics", op cit.

3. J.S. Kraabel, "Convection Testing of the Central Receiver System", op cit.

4. F. Biggs and C.N. Vittitoe, "The HELIOS-Model for the Optical Behavior of Reflecting Solar Concentration", Sandia National Laboratories, USA, SAN76-0347.

5. G. Kasparck, "Der Energieaustausch durch Wärmestrahlung zwischen Feststoffoberflächen" BWK 24 (1972) № 6, June 1972.

6. H. W. Fricker, "Cavity Receiver Design" SSPS Technical Report 4/83.

7. Agip Nucleare and Franco-Tosi Industriale, "The Advanced Sodium Receiver (ASR)", op cit.

8. J. Hansen, "ASR Construction Report", SSPS Semiannual Report, September 1984.

NOMENCLATURE

A_c	interior surface of the cavity	m^2
A_f	mirror area	m^2
A_r	receiver area	m^2
a	diffusivity	m^2/sec
C	geometric concentration ratio A_f/A_r	
C_p'	specific heat capacity of sodium	J/Kg/K
C_{pce}	specific heat capacity of ceramic back wall	J/Kg/K
I	beam irradiance	W/m^2
$\dot{q}_{ce}$	sum of heat fluxes of ceramic back wall	W/m^2
$\dot{q}_{conv}$	convection losses	W/m^2
$\dot{q}_{ir}$	infrared reradiation losses	W/m^2
R_a	outer radius of the tube	m
R_i	inner radius of the tube	m
r	coordinate in the radial direction	m
T	tube temperature	K
T'	sodium temperature	K
T_{ce}	temperature of ceramic back wall	K
T_w	average receiver wall temperature	K
T_∞	ambient temperature	K
ΔV	finite volume element	m^3
w'	sodium velocity	m/sec
z	coordinate parallel to tube axis	m
K	U-value	W/m^2K

GREEK SYMBOLS

α	absorptivity of tube surface	
α'	sodium film coefficient	$W/m^2/K$
ε_{eff}	effective cavity emissivity	
η_F	heliostat field efficiency	
η_{th}	thermal efficiency	
λ	thermal conductivity of tube material	W/m/K
ρ'	density of sodium	Kg/m^3
φ	polar angle	rad

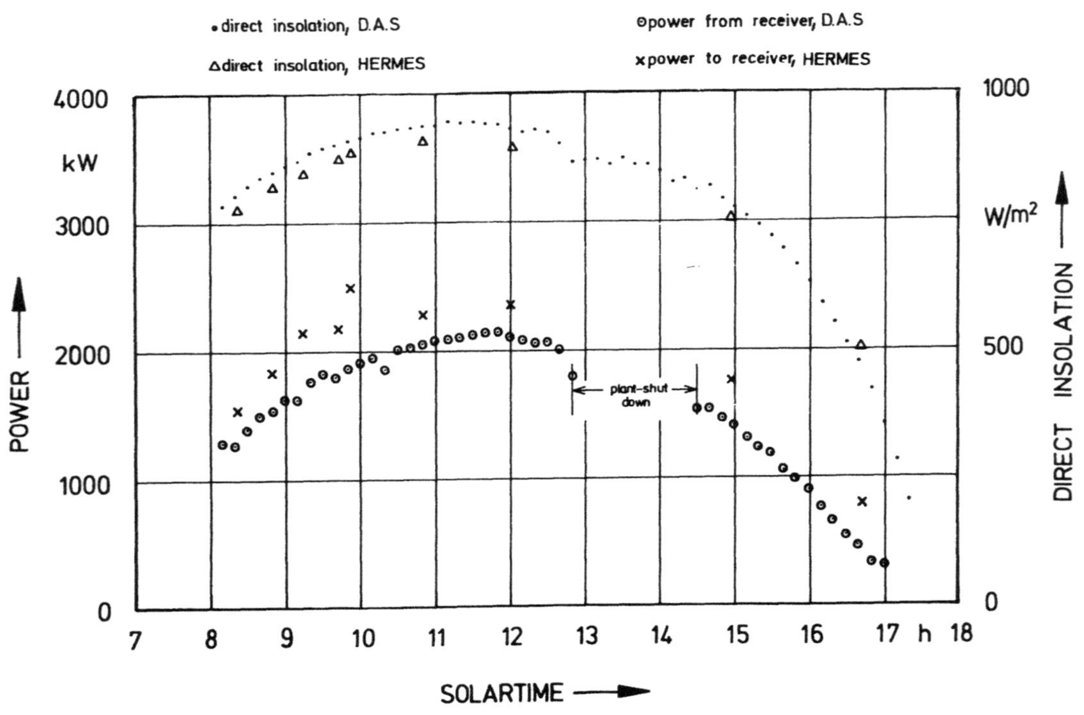

Fig. 5.5.-1: Measurements from 7.- Okt.- 82

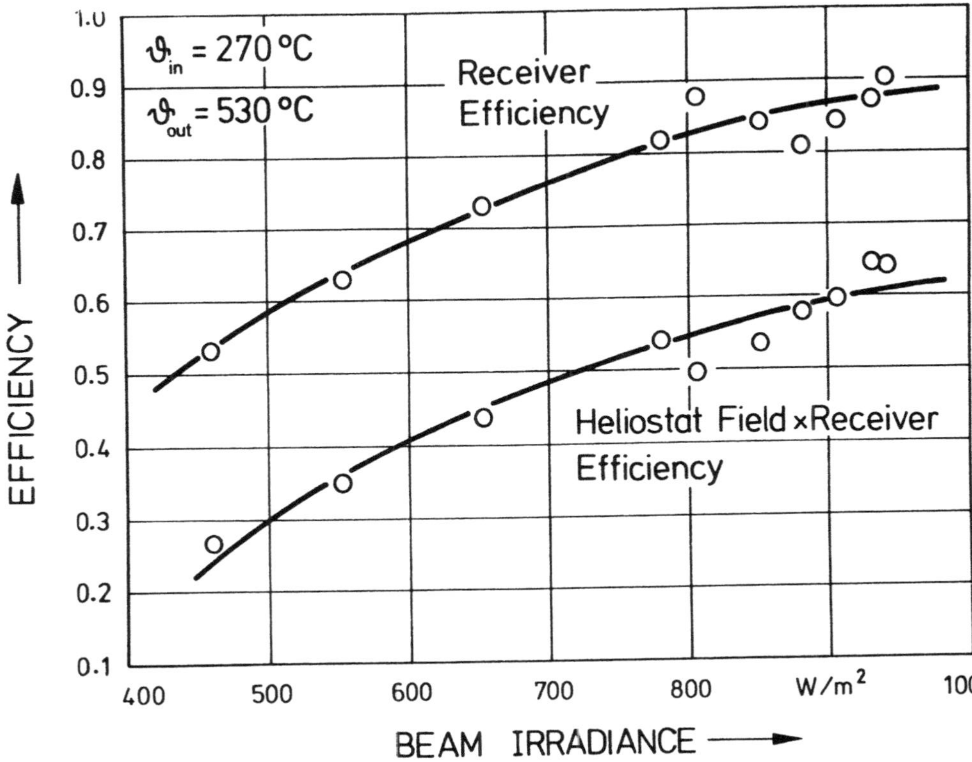

Fig. 5.5.-2: Thermal Efficiency and Receiver Efficiency versus Beam Irradiance (Sulzer Receiver)

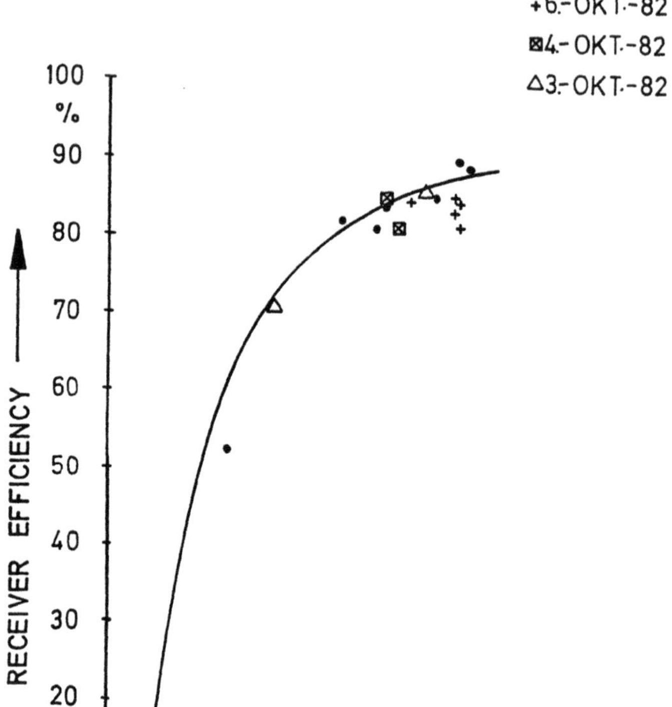

Fig. 5.5.-3: Receiver Efficiency versus Power into Cavity (Sulzer Receiver)

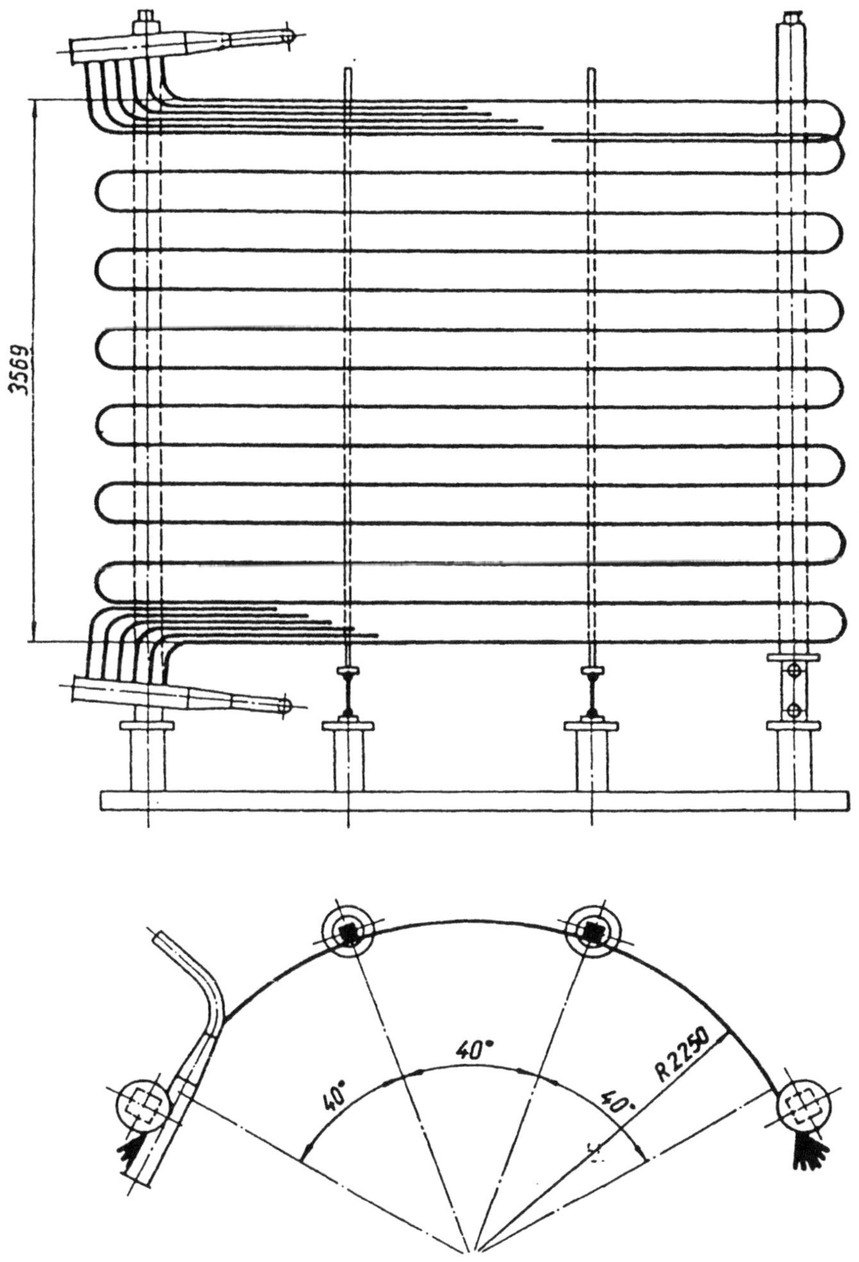

Fig. 5.5.-4: SSPS Sulzer Receiver (taken from Ref.6)

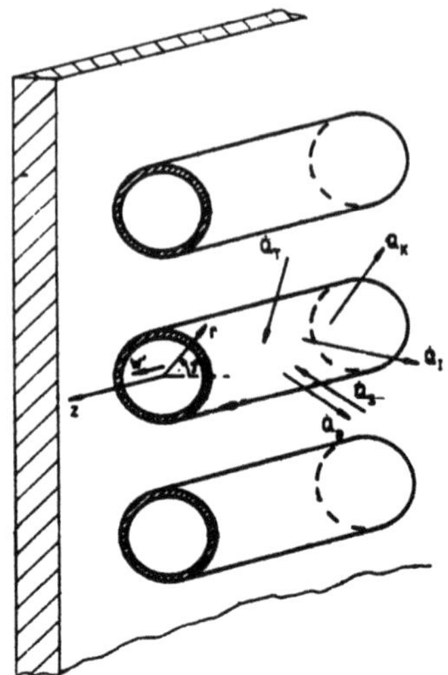

Fig. 5.5.-5: Receiver Tubes in Front of a Heat Resistant Ceramic Backwall

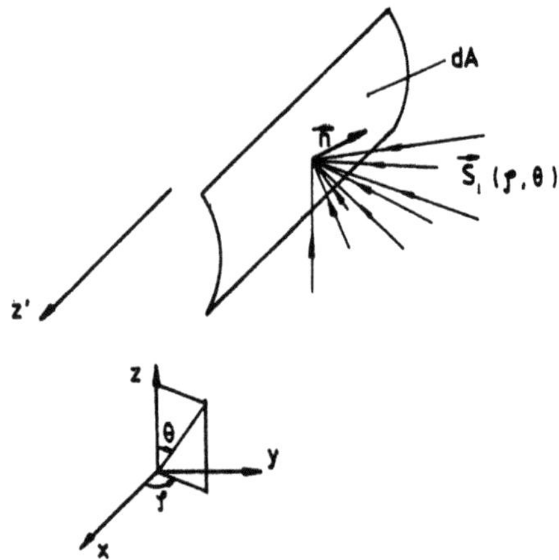

Fig. 5.5.-6: Spatial Angle Dependent Solar Radiation

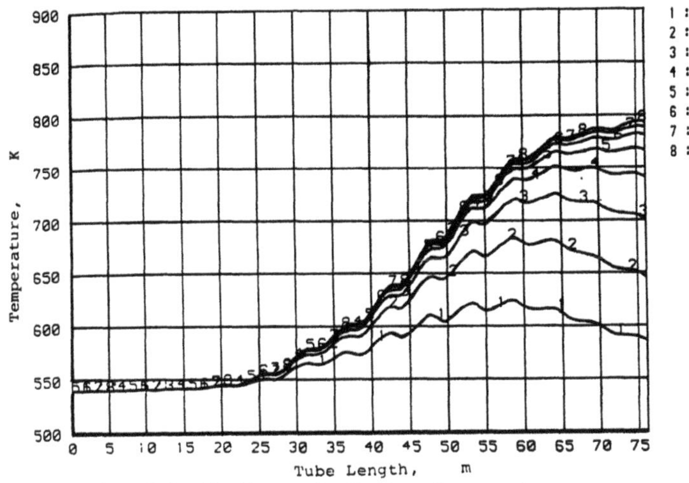

Fig. 5.5.-7: Unsteady Sodium Temperature;
Response to Step Funktion

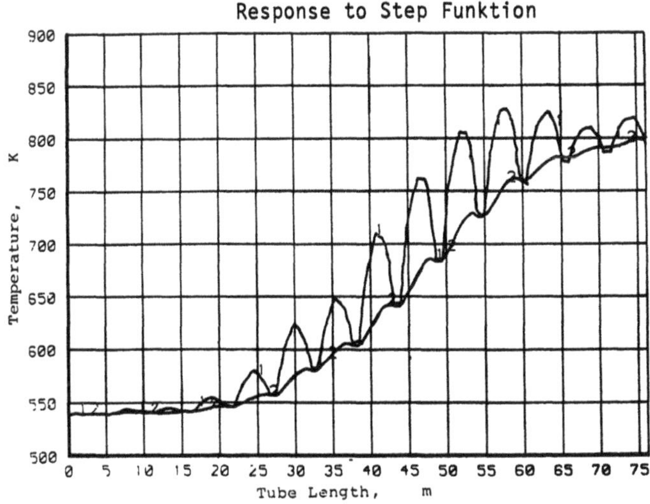

Fig. 5.5.-8: Temperature on Tube Surface;
Curve 1 - front side
Curve 2 - back side

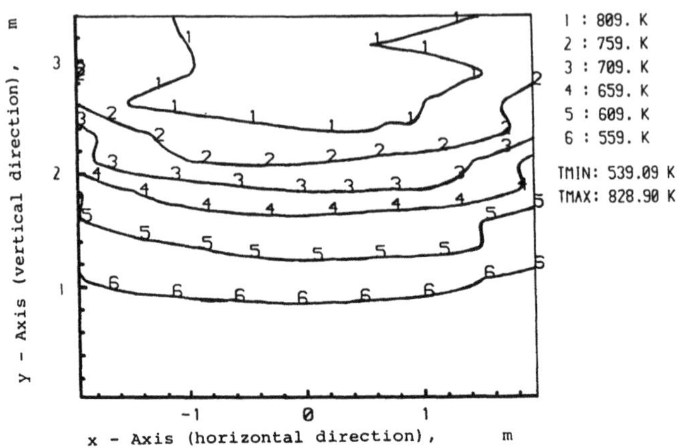

Fig. 5.5.-9: Isolines of Temperature Distribution on the Front Surface of the Tube Bundle

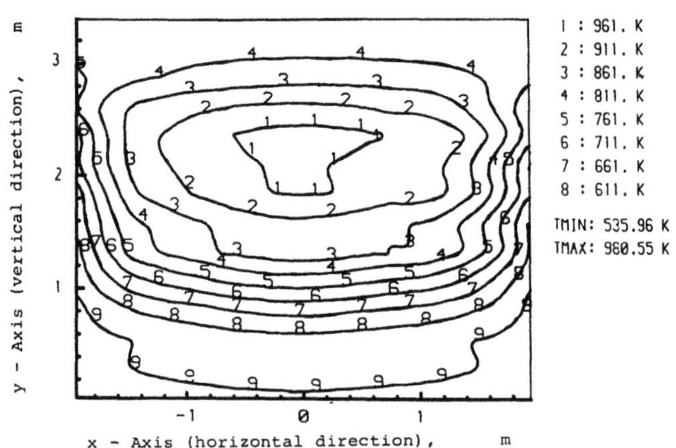

Fig. 5.5.-10: Isolines of the Temperature on the Ceramic Backwall (Steady State)

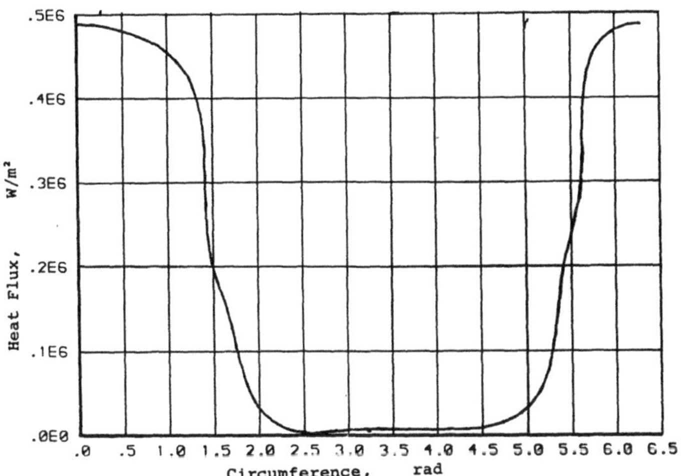

Fig. 5.5.-11: Concentrated Solar Radiation Along Circumference of One Tube

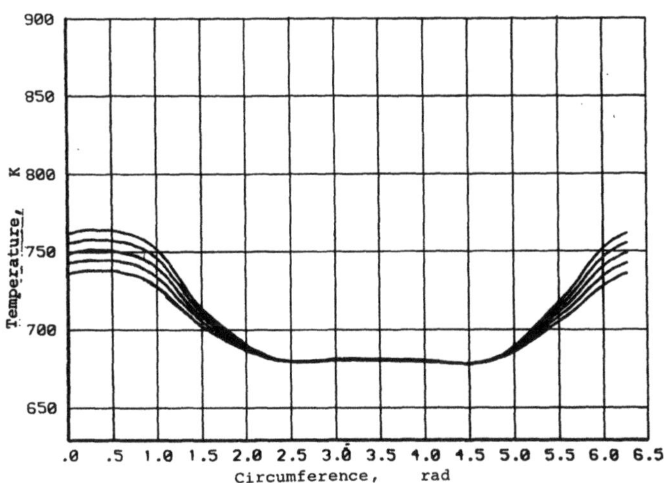

Fig. 5.5.-12: Temperatures Along Circumference of One Tube

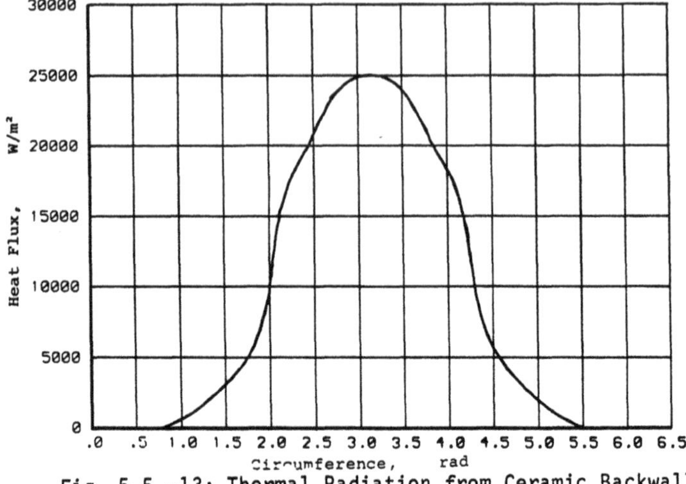

Fig. 5.5.-13: Thermal Radiation from Ceramic Backwall on Circumference of One Tube

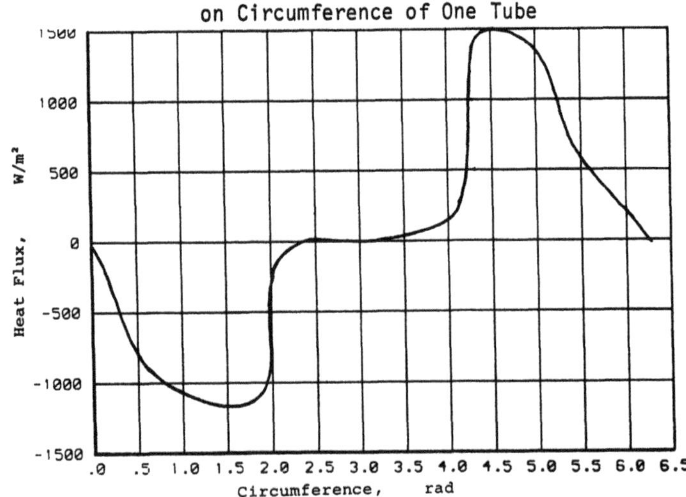

Fig. 5.5.-14: Thermal Radiation from Adjacent Along Circumference of One Tube

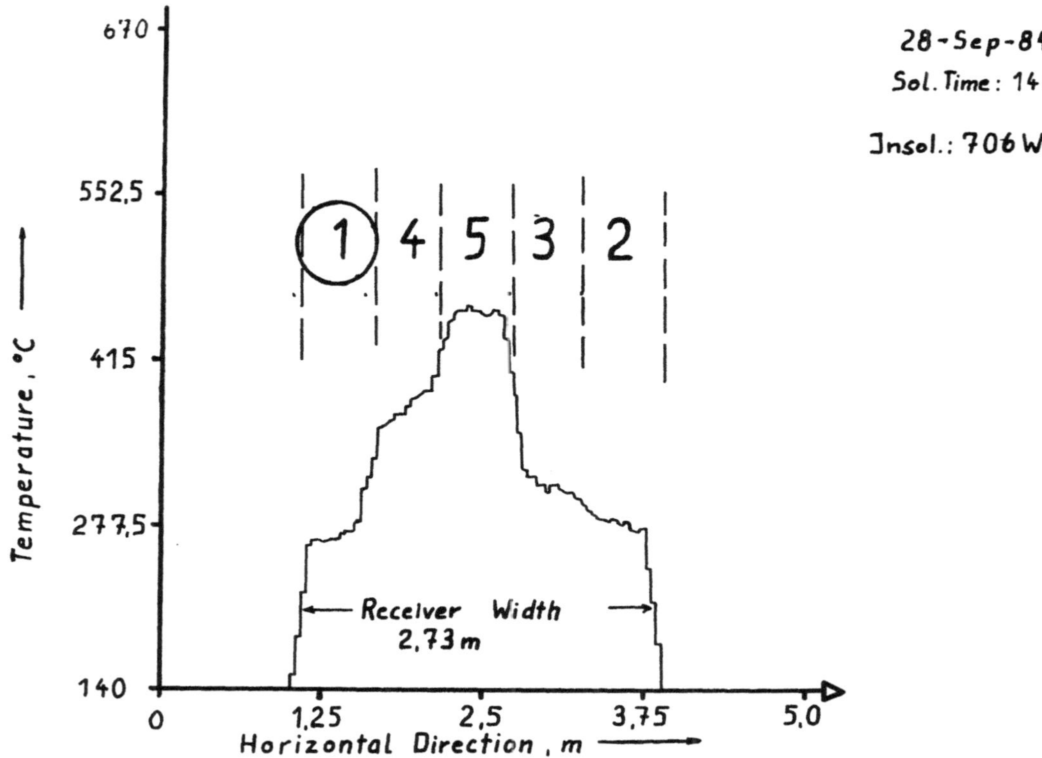

Fig. 5.5.-15: Horizontal Temperature Profile (Lower Region of ASR)

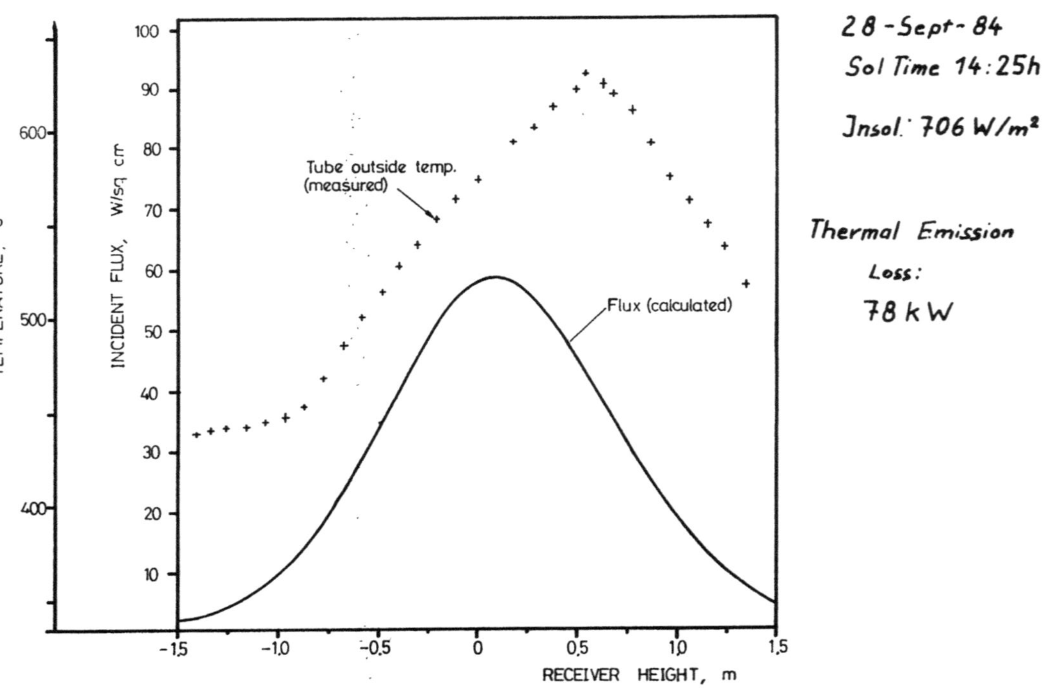

Fig. 5.5.-16: Vertical Temperature Profile of Central Panel (ASR receiver)

RECEIVER LOSSES: RESULTS OF TESTS

Heinz Jacobs, Manuel Sánchez, and Ricardo Carmona, ITET

1. INTRODUCTION

The objective is to evaluate the thermal losses from the receiver over the temperature range at which the receiver may operate. To achieve this, tests have been carried out with no concentrated power reaching the receiver and with the receiver doors open.

The tests consist in circulating sodium through the receiver at different inlet temperatures and estimating the losses by multiplying the mass flow rate times the enthalpy loss.

Sodium has been circulated through the receiver in the normal direction (i.e., flowing from the cold storage tank to the hot tank) for inlet temperatures between 200 and 300°C. To estimate the losses at higher temperatures, sodium must flow from the hot tank and therefore tests have also been conducted with sodium flowing in the reverse direction. Since the pumps do not operate 'in reverse' the latter tests have been made by controlling the argon pressure in the tanks manually to maintain a constant flow.

This paper is divided into two parts. The first refers to tests performed by the ITET on the Advanced Sodium Receiver since April 1984. A new sodium flowmeter, which has been recently calibrated, was installed by A. Brinner and F. Reich of DFVLR-Stuttgart, on April 25, 1984, and all the ASR flow data has been taken from the readings from this flowmeter.

The second part refers to tests performed on the Sulzer receiver during 1982, by J. Kraabel from Sandia, and by one of the authors of this paper (H. Jacobs) during March and April 1983. These tests were performed using the sodium flowmeter which had been installed originally on the CRS.

2. ASR TESTS

2.1 Consideration on the Flowmeter Accuracy

The flowmeter is calibrated to measure flow in the so-called forward direction: it may give wrong readings when used in reverse.

The DAS gives the temperature and the volume of the sodium in both tanks. From these data, for forward flow one may estimate how much sodium has left the cold tank and how much has been added to the hot tank during some interval.

Tab.5.6-1 lists typical results for measurements made between 7:20 and 8:45 a.m. on September 27, 1984. The cold tank has lost 18.35 tons and the hot tank has gained 18.83 tons. These numbers are within 2.5% of each other. From the flowmeter readings made during the same time interval, the total circulated sodium can also be calculated. For the example shown, the integration of the flow over time yields 18.9 tons. The difference is small (1.6%).

This kind of check may also be used for the reverse flow. In the example shown in Tab.5.6-2, for a reverse flow test performed on September 7, 1984 between 12:20 and 14:05, the cold tank gains 8.67 tons while the hot tank loses 8.8 tons -- for a difference of 1.5%. Integrating the data for the flowmeter operating during this time interval one obtains 9.09 tons, which is 4% more than the average of the two 'tank' numbers. This allows an estimate of the error from using the flowmeter in reverse. Unfortunately, this estimate is not very consistent. In the example illustrated in Tab.5.6-3, also for reverse flow, the difference between the flowmeter and the tank numbers is only 0.4%. We are using the highest observed difference (i.e., 4%) as an indication of the largest error.

2.2 The Tests

The data for the different tests are listed in Tab.5.6-2. The tests are numbered from 1 to 10. In the first four, the sodium flows in the reverse direction. For tests 6 to 10, the flow is forward.

Other input data are the flow rate, the inlet and outlet sodium temperatures, and the ambient temperature. The difference T_1 between the 'mean' sodium temperature and the ambient temperature, and the wind speed are also listed.

The results from the tests are also listed in Table 2.

2.3 Total Losses

As stated above, the total losses have been estimated by multiplying the flow rate reading from the flowmeter times the enthalpy gain in the sodium, and averaging the product over the time interval. The total loss L is listed in the ninth column of Tab.5.6-1, and is also plotted as a function of temperature in Fig.5.6-1.

The least squares fit for the data gives the straight line

$$L(kW) = -60.4 + 0.623T(°C)$$

in the interval $170 < T_o < 380°C$, with a correlation coefficient = 0.965.

2.4 Radiation Losses

The mean sodium temperature T has been assumed to be representative of the mean receiver temperature. This is justified when there is no concentrated power incident on the receiver: it has been estimated that the actual difference between the bulk medium temperature in the receiver pipe and the outside pipe temperature is only about 1°C.

The radiation losses have been estimated by multiplying the power radiated from a blackbody at the mean sodium temperature to another blackbody at the ambient temperature times the area of the receiver times the emittance of the Pyromark coating, which is taken to be 0.9, in accordance with the manufacturer's data. The shape factor has been taken to be one.

The radiated power L_r, is listed for the ten test conditions in the tenth line in Tab.5.6-2, and plotted in Fig.5.6-2. The data have been fitted by the least squares method to the second order polynominal

$$L_{r(kW)} = 21.4 - .168T + 0.000865T^2$$

in the interval $170 < T_o < 400$, with a determination factor of 0.999.

The radiated power estimated from the mean receiver sodium temperature is not equal to the power that would be estimated by adding the radiation losses from each of the five panels, computed as a function of each panel mean temperature. However, the differences are small. For example, for the conditions in Test Nº 3, the radiation losses calculated by considering each panel separately equal 68.3 kW, which differs from the result listed on the table by about 0.4%.

As noted above, the ambient temperature has been used on the reference (or 'cold body') temperature in all the calculations. The ambient temperature is different from the so-called sky temperature, but the differences are again small. For example, using the sky temperature correlation from (Ref.1), for test conditions in Test Nº 3, the difference between ambient and 'sky' temperature is only 1°C.

2.5 Convective Losses

The convective losses have been calculated by subtracting from the total losses the estimated losses due to radiation and the losses due to conduction. From work reported by Sandia (Ref.1) and by the ASR manufacturers (Ref.1), the conduction losses amount to about 5% of the total loss.

The estimates for the losses by convection L_c, are listed in Tab.5.6-2 and plotted in Fig.5.6-3. The estimates have been fit by the least squares method to the straight line

$$L_c \text{ (kW)} = -2.1.6 + 0.291T \text{ (°C)}$$

in the interval $170 < T_o < 400$°C, with a correlation coefficient of 0.982.

The last line in the Tab.5.6-2 lists the convective heat transfer coefficient estimated as the ratio between the convective losses and the product of receiver area and the difference T, between the mean sodium temperature and the ambient temperature.

The convective heat loss coefficient is affected by wind speed and direction through the Reynold's number which is very sensitive to the air speed. Unsuccessful attempts were made to quantify this effect. The existing anemometer, placed on top of the receiver tower, does not suffice to characterize the effect of the wind on the receiver which does not have cylindrical symmetry.

3. SULZER TESTS

The tests that have been described for the ASR are practically identical to tests which were initiated on the site by J. Kraabel from Sandia, in 1982. (There were differences in the way the reverse flow was driven, but these differences do not affect this discussion).

The 1982 tests are numbered from 1 to 9 in Tab.5.6-3. The 1983 tests on the Sulzer are numbered from 10 to 14.

The estimated total losses which are listed in Tab.5.6-3, are also plotted in Fig.5.6-1. For the Sulzer, the total loss can be fit into the following straight line

$$L_t \text{ (kW)} = -253 + 1.93 \text{ T°C}$$

in the interval $170 < T_o < 400$°C.

3.1 Radiation and Convection Losses

Radiation losses were estimated by using RAD1, a program written by J. Kraabel. They are listed in Tab.5.6-3 and plotted in Fig.5.6-2. The data can be fit by the polynominal

$$L_r = -17 + 0.149 \cdot T + 1.78 \cdot 10^{-4} \cdot T^2$$

in the interval $170 < T_0 < 400°C$, with a determination coefficient equal to .96.

RAD1 also estimates the convection losses (and the convective heat transfer coefficient) by subtracting the radiation losses and a correction for conduction from the total losses.

The convective losses and the convective heat loss coefficient are listed in Tab.5.6-3, and also plotted in Fig.5.6-2 and 3. They have been fit by the line

$$L_c = -225.7 + 1.62 \cdot T$$

4. REFERENCE

1. P. Berdahl and R. Fromberg, "The Thermal Radiance of Clear Skies", Solar Energy, V. 29, Nº 4, 299-315, 1982.

Reverse flow test on 14-09-84

time	Cold Storage v m3/h	T C	rho kg/m3	m tons	Hot Storage v m3/h	T C	rho kg/m3	m tons
10:25	22.38	251	891	19.94	38.75	435	847	32.82
12:00	36.74	273	886	32.75	23.72	430	849	20.13

delta m: 12.80 tons   12.69 tons
Flowmeter delta m: 12.79 tons

Reverse flow test on 09-07-84

time	Cold Storage v m3/h	T C	rho kg/m3	m tons	Hot Storage v m3/h	T C	rho kg/m3	m tons
12:20	23.23	250	891	21.59	36.67	433	848	31.09
14:05	34.20	256	890	30.26	26.25	430	849	22.28

delta m: 8.67 tons   8.80 tons
Flowmeter delta m: 9.09 tons

Normal flow test on 27-09-84

time	Cold Storage v m3/h	T C	rho kg/m3	m tons	Hot Storage v m3/h	T C	rho kg/m3	m tons
07:20	49.48	228	897	44.38	8.72	342	870	7.58
08:45	29.08	229	897	26.08	29.85	278	885	26.40

delta m: 18.35 tons   18.83 tons
Flowmeter delta m: 18.35 tons

Table 5.6.-1: Comparison between the amount of Sodium, set to the Tanks, and the Flowmeter readings.

flow : Receiver Flowrate
Tin : Inlet Temperature
Tout : Outlet Temperature
Tmea : Mean Temperature Difference
Tam : Ambient Temperature
Vel : Velocity
Ltot : total Losses
Lrad : radiative Losses
Lcon : conductive Losses
HTra : Heat Transfer Coefficient

Nr.	Date	flow m3/h	Tin degC	Tout degC	Tmea degC	Tam degC	Vel m/s	Ltot kW	Lrad kW	Lcon kW	HTra W/m2/K
1	23-05-84	6.48	411	322	345	21	5	178	65	104	38
2	14-09-84	12.10	422	380	375	26	6	154	81	66	22
3	09-08-84	6.25	414	332	349	24	8	158	68	81	30
4	14-09-84	6.60	417	340	352	26	5	156	70	78	28
5	01-07-84	7.00	262	230	218	28	11	73	26	43	25
6	01-07-84	14.00	266	247	228	28	11	87	29	53	30
7	31-07-84	7.03	215	195	172	33	6	47	17	27	20
8	31-07-84	13.98	216	205	177	33	5	51	18	30	21
9	09-07-84	6.80	219	200	174	35	1	43	18	24	16
10	09-07-84	14.80	223	212	183	35	1	51	21	29	20

Table 5.6.-2: ASR LOSSTEST

flow : Receiver Flowrate
Tin : Inlet Temperature
Tout : Outlet Temperature
Tmea : Mean Temperature Difference
Tam : Ambient Temperature
Vel : Velocity
Ltot : total Losses
Lrad : radiative Losses
Lcon : conductive Losses
HTra : Heat Transfer Coefficient

Nr.	Date	flow m3/h	Tin degC	Tout degC	Tmea degC	Tam degC	Vel m/s	Ltot kW	Lrad kW	Lcon kW	HTra W/m2/K
1	04-10-82	data not available			193	25	5	130	19	13	10
2	04-10-82	data not available			206	25	6	139	20	14	11
3	05-10-82	data not available			193	25	4	129	20	12	10
4	06-10-82	data not available			224	21	6	172	27	11	10
5	06-10-82	data not available			216	20	7	169	23	15	11
6	06-10-82	data not available			208	19	8	165	20	21	12
7	07-10-82	data not available			229	18	2	149	31	15	8
8	07-10-82	data not available			216	18	3	150	25	20	8
9	07-10-82	data not available			211	17	3	159	23	23	10
10	29-03-83	4.94	399	208	290	14	3	301	40	21	16
11	07-04-83	6.48	442	254	328	20	5	381	52	25	19
12	08-04-83	7.06	400	246	299	23	4	343	43	23	19
13	12-04-83	20.42	241	217	215	14	8	166	21	16	12
14	14-04-83	26.60	232	214	212	12	7	160	20	15	12

Table 5.6.-3: Sulzer Receiver Losstests

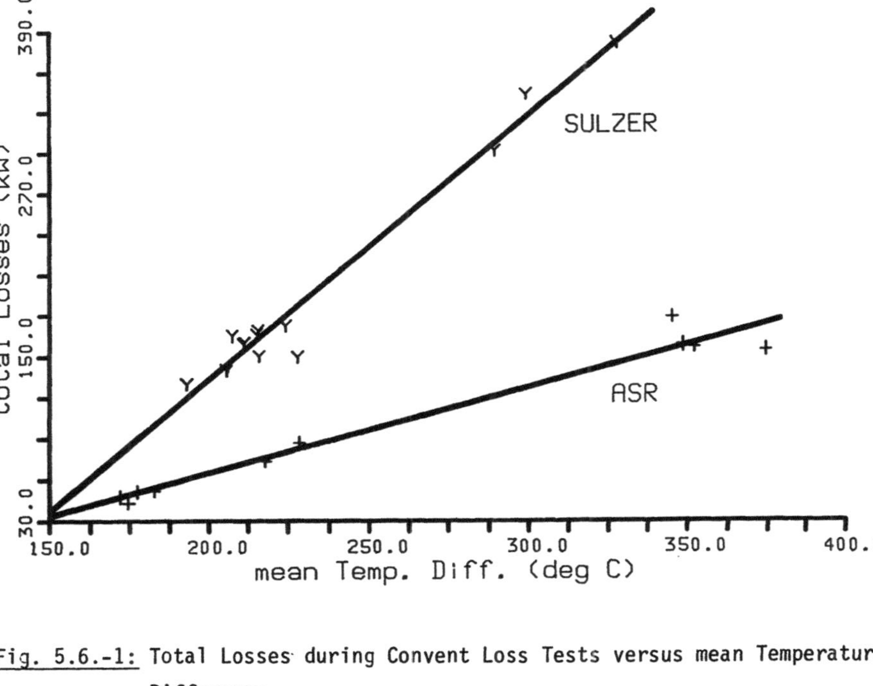

Fig. 5.6.-1: Total Losses during Convent Loss Tests versus mean Temperature Difference

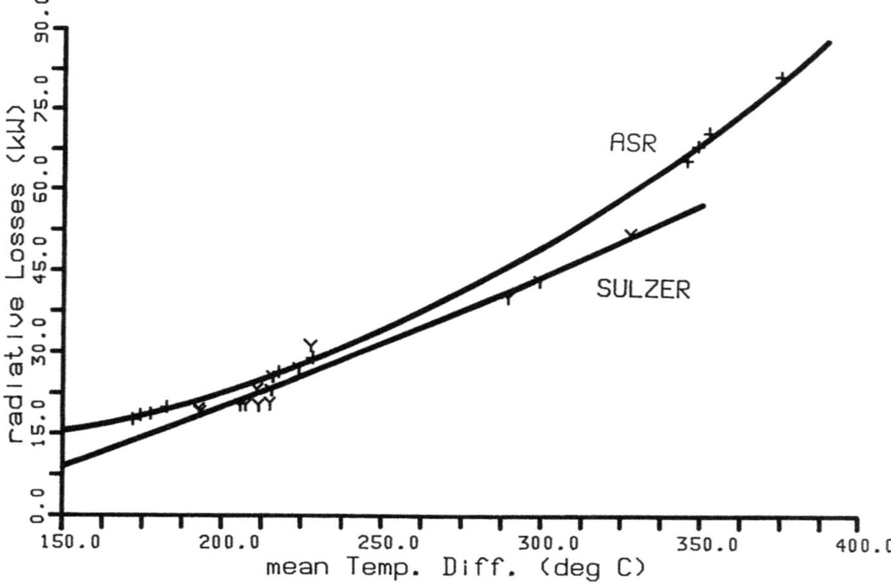

Fig. 5.6.-2: Radiative Losses versus mean Temperature Difference

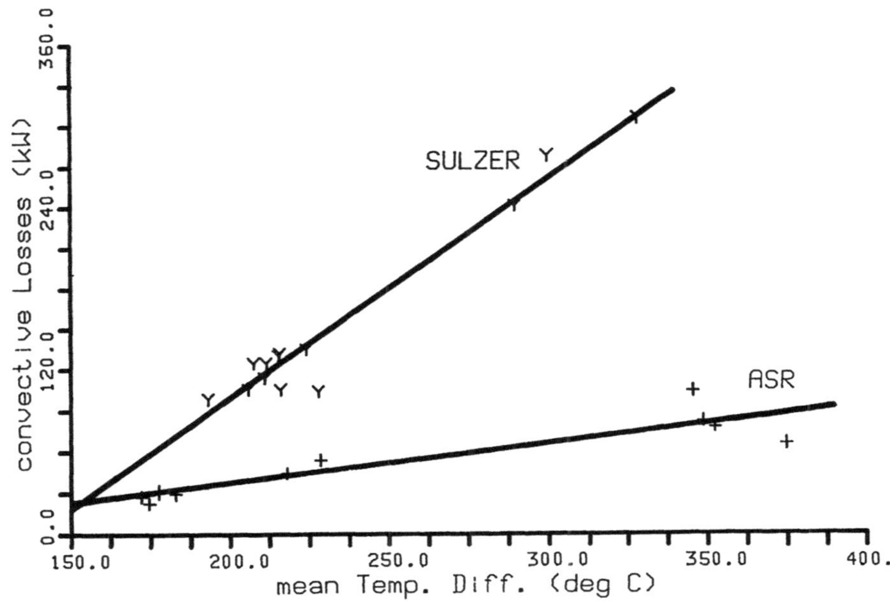

Fig. 5.6.-3: Convection Losses versus mean Temperature Difference

ASR PERFORMANCES: COMPARISON WITH SIMULATION

A.DeBenedetti , AGIP
M.Blanco , ITET

1. INTRODUCTION

During the A.S.R. operation many experimental results and data have been collected from routine operation and 'ad hoc' tests. To verify the influence of the different parameters and the validity of the hypotheses on which the design is based , those measurements must be compared with computed values. In this paper the interface between optics (incident flux distributions) and thermodynamics (temperature distributions) is analyzed and the ASR thermal performance evaluated .

Simulation and experimental results are compared under steady state conditions (full and partial load operation , losses tests) , transient conditions (start up, input power variations , pump blockage) and special situations (higher flux and temperature , etc).

2. THERMODYNAMIC MODEL DESCRIPTION

Let us consider an external receiver with tubes placed side by side as the active surface and a ceramic wall with thermal insulation as the back protection (Fig.5.7-1).

Power from the heliostat field reaches one side of the tube , while the other side faces the wall , with which it exchanges heat.

A longitudinal cross section through the tube and the backwall is illustrated inFig.5.7-2, its coordinates are x,y in the absorber plate reference system. On the tube z defines the axial position and ϑ the angular position (ϑ = 0 is the direction facing the field).
The radiation $\phi j(\vartheta)$ incident on the tube of the section j located at x,y is a function of ϑ only.

We make the assumption (which is strictly true in the case of a parallel incident beam) that :

$$\bar{\phi}j = \phi(z,\vartheta) = \phi h(x,y) * \cos(\vartheta) \qquad 0 < \vartheta < \pi/2$$

$$\bar{\phi}j = \bar{\phi}(z,\vartheta) = 0 \qquad \pi/2 < \vartheta < \pi$$

This cosine dependence is well verified for a single tube exposed to concentrated radiation (Fig.5.7-3).

Because of symmetry only one half of the tube is considered ($0 < \vartheta < \pi$) ; in turn this half is subdivided along the circumference into parts which are $\Delta\vartheta * Do/2$ wide and Δz high. Each part has an outer area $DAo = \Delta z*\Delta\vartheta*Do/2$ at temperature To.

The power incident on the tube strip i ($\vartheta_i < \vartheta < \vartheta_{i+1}$) of the element j is computed as :

$$Wi(j) = \int_{\vartheta_i}^{\vartheta_{i+1}} \phi(z_j,\vartheta) * \Delta z * \frac{Do}{2} * d\vartheta =$$

$$= [\sin(\vartheta_{i+1}) - \sin(\vartheta_i)] * \Delta z * \frac{Do}{2} * \Delta\vartheta * \phi h(x,y)$$

and the average flux $\bar{\phi}i(j)$ as :

$$\bar{\phi}i(j) = Wi(j) / (Do/2)*\Delta\vartheta * \Delta z =$$

$$= [\sin(\vartheta_{i+1}) - \sin(\vartheta_i)] * \phi h(x,y)$$

The fraction of the incident radiation which is reflected depends on the optical properties and the geometry of the surface. The side-by-side tube arrangement increases the absorptivity ; if for simplicity a 90 deg. saw-tooth profile is assumed it can be demonstrated that the actual absorption ag is (see Lecture on Receiver Theory) :

$$ag = f(a) = \frac{\sqrt{2} * a}{\sqrt{2} - (1.- a)*(\sqrt{2} - 1)}$$

with a 1.5 % increase respect to the flat surface absorptivity a.

The geometry of the surface also affects the emissivity ,

which is increased when compared with the one of the flat plate e :

$$eg = f(e) = \frac{e}{1 - (1 - e)*(1 - 1/\sqrt{2})}$$

If TF(j) (bulk fluid temperature) , Γ(j) (mass flow rate) and TB(j) (backwall surface temperature) at zj (inlet axial coordinate of the basic receiver element j) are given , the temperature distributions in the j + 1 element and then TF(j+1), Γ(j+1) and TB(j+1) can be computed .

Adimensional correlations are used to estimate the heat transfer coefficients on the outer (air) and inner (cooling fluid) tube surface. When sodium is flowing along the tube ,the following expressions for the Nusselt number NU can be used :

NU = 0.625 * (PE)**0.4 [1]
NU = 7.0 + 0.025 * (PE)**0.8 [2]
NU = 5.0 + 0.025 * (PE)**0.8 [3]

PE , the Peclet number, is the Reynolds number times the Prandtl number.

At the conditions which characterize a receiver , it is difficult to estimate the heat transfer from the tube to the environment at temperature Te ,but there are correlations to compute the heat transfer coefficient in mixed convection , he ,as a function of temperature, geometry , wind velocity , etc (see Lecture on Receiver Theory).

To heat transfer each circumferential element of the tube section presents :

-a thermal resistence RI to the sodium at bulk temperature TF

$$RI = \frac{1.}{hna * \Delta z * \Delta \theta * Di/2}$$

The inside heat transfer coefficient hna is assumed to be constant along the tube inner surface (hna = NU * kna / Di ,where kna is the sodium heat conductivity and Di the tube inside diameter).

-a thermal resistance RE towards the outside (air

and backwall). In analogy to an electrical circuit this resistance is calculated as the equivalent of convective and radiant resistances in parallel :

$$1./\ RE = 1./\ REc + 1./\ REr$$

These resistances are defined as functions of the position of the tube element along the circumference. When it is entirely in the front of the tube ($0 < \theta < \pi/2$) , one has :

$$REc = \frac{1.}{hna * DAo}$$

$$REr = \frac{To-Te}{\sigma * eg *(To**4 - Te**4) * DAo} = \frac{1.}{\sigma *eg*K*DAo}$$

where σ = K(To,Te)= (To**4 - Te**4)/(To-Te)

When the surface faces the backwall completely ($\theta > \pi/2$) , one may use :

$$REc = \frac{1.}{hb * DAo}$$

$$REr = \frac{To - TB}{\sigma *(To**4 - Te**4)*E*DAo} = \frac{1.}{\sigma * K(To,Te)*E*DAo}$$

where E = 1/ eg + 1/ eb - 1. and hb is the heat transfer coefficient through the air layer between tube and backwall.

In the case of one element lying between the two heat transfer zones ($\theta_i < \pi/2;\ \theta_{i+1} > \pi/2$) an average value is assumed.

If these resistances are assigned , the power in the cooling fluid can be computed by solving the thermal balance equation in the tube wall (see Appendix A).

The equation to solve is :

$$\frac{1}{r}\frac{\partial}{\partial r}\left(kw\ r\frac{\partial T}{\partial r}\right) + \frac{kw}{r}\frac{\partial^2 T}{\partial \theta^2} = \rho w\ Cpw\frac{\partial T}{\partial t}$$

with

$$(RE/DAo) * (Te - T) = -k * (dT/dr)\Big|_{r=Do/2}$$

and

$$(RI/DAi) * (TF - T) = -k * (dT/dr)\Big|_{r=Di/2}$$

as boundary conditions.

From the power entering the sodium it is possible to solve the continuity equation

$$\frac{\partial \Gamma}{\partial z} + \frac{\partial}{\partial t} \rho na \cdot A = 0$$

and the enthalpy balance equation

$$\frac{\partial H}{\partial z} + A \rho na \frac{\partial H}{\partial t} = Wna$$

in the fluid (see Appendix B).

3. STEADY STATE SIMULATION

3.1 NORMAL OPERATION

Incident flux measurements are available on planes (such as the HFD plane for instance) that do not coincide with the receiver plane. As heliostat aiming points lie on the receiver plane , measured flux distributions are quite different from the flux actually incident on the tubes.

To compute the thermal flux incident on the tubes as well as possible , the HELIOS program [4] is used and run to obtain first flux distributions on the measurement plane and then the flux incident on the tubes.

An incident flux map from the HFD on December 1983,

12.30 solar time is shown here(Tab.5.7-1). It will be assumed that this represents reference test conditions. A centered peak flux of 989 kW/sqm is obtained with 944 W/sqm as incident insolation. The total power collected on the HFD scanner area (5 by 4.20 m.) is 2448 kW when computed by the HP 85 which controls the bar ; it uses the simplified formula SUMi [ϕ(i)*DAr(i)].

The DAS estimates the power reflected towards the target at the same instant : 2682 kW. This value is computed considering the average instantaneous cosine factor, the number of heliostats in track , the direct solar irradiance and the actual mirror reflectivity estimated by spot measures.

To verify the HP85 result , the same distribution (in the form of a 11 by 11 matrix) is integrated applying three different mathematical techniques.

The total power on the HFD area is calculated to be :

2574 kW when integrating with a generalized spline interpolation

2616 kW when performing the integration with Simpson's formula [4]

2612 kW when performing the integration with Albrecht's formula [4]

It is assumed that 2600 kW is the most reasonable indication of the actual power incident on the HFD plane and HELIOS is used to reproduce this figure. At the same instant (December 23rd , noon) the flux distribution reflected by the 93 SSPS-CRS heliostats is computed with the input parameters set at the values used during ASR design , with the exception of direct insolation and mirror reflectivity. For insolation the actual measurements (940 W/sqm) are used ; for the average mirror reflectivity the experimental estimate is used (80%).
In these conditions HELIOS computes 2603 kW on the HFD target (a plane 5 by 4.2 m. centered at 44 m. and located 0.75 m behind the ASR tube plane) and a peak value of 1020 kW/sqm , very close to the HFD measured data (989 W/sqm) , Tab.5.7-2.

HELIOS computations , when compared with measurements taken in different days at different insolation levels , generally show a good agreement with the measurements , provided that the sky is clear , because only then the Kuiper sunshape distribution parameters used are appropriate. That is illustrated in Tab.5.7-3 where experimental measurements and HELIOS computation on the HFD plane are shown.

As the agreement between measurement and simulation is good , HELIOS is applied again to compute (with the same input stream) the flux distribution on the ASR tubes. The flux incident on the receiver tubes is higher than the one on the HFD plane as the heliostats are aimed to points on the receiver surfaces. The increase in the flux value depends on the time ; from HELIOS computations performed in different days at instants between 10.30 a.m. and noon , values in the range from 1.1 to 1.30 are obtained.

The results of this computation for December 23rd , noon are shown inTab.5.7-4:2586 kW impinge on the target with a peak of 1230 kW/sqm (1.28 the flux on HFD plane) This flux map is assumed to be the actual power distribution incident on the receiver tubes and is used as input data for the thermal analysis of the receiver performed with the thermodynamic model THERESA [5].

The result of thermal computations are compared with measurements recorded by DAS and stored on tapes (5 minutes average).
These records show at 13.03 (local time) that the receiver was operating at a mass flow rate of 30.58 cum/h (7.6 Kg/sec assuming a sodium density of 855 kg/cum) with inlet and outlet temperatures of 265 and 529 C , respectively. From a steady state thermal balance , about 2500 kW seem to enter the fluid (the DAS computed values is 2538 kW). However,this figure must be modified in order to take into account an error in the flow rate measurements (see the final report on calibration activity by A.Brinner from July 1984).

Preliminary results after the flow meter calibration suggested that the mass flow recorded values should be reduced by about 9 % in the range from 5 to 7 kg/s . That explains why , in computations, the inlet flow is reduced from 7.6 kg/sec (measured figure) to 7.0 kg/sec.

Temperatures in the receiver tubes , headers and ceramic backwall are computed with the ASR actual geometry and physical characteristics ; the incident flux distributions , the inlet temperature and the flow rate used as input are equal to measured values.

First THERESA computes the power incident on the 2.75 x 2.85 m billboard receiver , establishing how this power is divided between the 5 panels. Then the incident power distributions on 15 receiver tubes are computed (the selected tubes correspond to the instrumented tubes) and the thermal balance equations in the fluid , tube wall and backwall are solved as described at point 2 and in

Appendix A and B.

The results obtained (in no wind condition) are presented inTab.5.7-5 and summarized here :

2503 kW impinge on the receiver
2260 kW enter the sodium
90.2 % is theestimatedoverall efficiency (2260/2503)
533.5 C results the computed outlet temperature

Single panel efficiencies and losses are :

panel 1 ; 86.1 % , 22.8 kW
panel 2 : 85.7 % , 24.5 kW
panel 3 : 92.6 % , 48.9 kW
panel 4 : 91.1 % , 59.4 kW
panel 5 : 90.2 % , 81.1 kW

When the wind velocity is 6 m/s the results are :

2503 kW impinge on the receiver

2239 kW enter the receiver

89.4 % is the estimated overall efficiency (2239/2503)

531. C is the computed outlet temperature

Single panel efficiencies and losses are :

panel 1 : 84.2 % ; 26.2 kW
panel 2 : 83.7 % ; 28.0 kW
panel 3 : 92.1 % ; 52.7 kW
panel 4 : 90.5 % ; 63.4 kW
panel 5 : 89.7 % ; 85.1 kW

Convection losses increase from 58.4 at V=0 to 77.9 kW at V=6 m/s

A comparison between the computed and the measured temperatures in the panel headers is presented inFig.5.7-4; Fig.5.7-5shows the comparison for the 5th panel central tube .Fig.5.7-6presents the panel 5 central tube thermal status.
A critical point is the computation of the incident power by integration from the incident flux distributions ; small errors (about 1.5%) due to the peaked function

estimate.

All these simulations assume as a reference the 23/12 measured data , which produces an incident power of about 2500 kW , close to the full load nominal condition.

When the load decreases , the efficiency is reduced as convection and radiation losses ,which depend on temperatures , do not change. The efficiency vs. load curve is strictly connected to the receiver characteristics ;Fig.5.7-6 shows the curve for the SSPS billboard receiver.

3.2 Loss Test Simulation

To obtain convection and emission losses without the uncertainties connected with reflection evaluation , tests have been performed by circulating hot sodium through the receiver and then , when steady state is reached , estimating the escaping heat by a thermal balance calculation.

The test is simulated running THERESA without incident power and imposing at the inlet a high flow rate in order to keep the receiver temperature constant at the value at which we are going to compute losses.

Fig.5.7-7 shows emission and convection losses (curves A, B) vs. the average temperature difference , which is defined as the average temperature of the receiver minus the ambient temperature.Convection in curve A is estimated by Siebers correlation (see lecture on Receiver Theory); in Curve B the convection loss is obtained from the simple correlation

$$he = 6 + \frac{To - 270}{180}$$

Curve C shows the convection loss alone as computed by Siebers (the difference between curve A and C represents emission loss (conduction is small)).

Total losses vs. average temperature delta at different wind conditions (from 0 to 9 m/s) are plotted in Fig.5.7-8 (the line at V=0 corresponds to the curve A of the figure before).

Fig.5.7-9 shows heat transfer coefficients due to convection vs. wind velocity at different temperatures. To compute the heat transfer coefficients the exposed tube bundle surface has been considered (ASR exposed area times π/ 2).

4. TRANSIENT CONDITIONS

THERESA can compute temperatures in the fluid , tubes and backwall both in steady state and in transient condition.

An imposed time dependent behaviour can be assigned to :

- receiver inlet temperature
- receiver inlet flow rate
- flux distribution on the receiver

As example, Fig.5.7-10 shows sodium outlet temperature and ceramic backwall temperature in the case of a fast start up. That is obtained by reducing the inlet flow rate from the high value 77 kg/s to 7 kg/s , maintaining the power input at a full load condition and computing the resulting temperature transient.

4.1 Variation in Incident Power

The measurement (by R.Carmona) of the receiver outlet temperature following a small load change is presented in Fig.5.7-11 with the simulation results. Starting from a steady state situation, the receiver incident power is decreased by defocusing 6 heliostats , while the flow rate is maintained constant.

4.2 Variation in the Flow Rate

Starting from a steady state situation, the flow rate is decreased manually from 8.15 to 7.65 kg/s , keeping constant the power incident on the receiver. The outlet temperature transient from R.Carmona experiment and from simulation is presented in Fig.5.7-12.

4.3 Pump failure

In Fig.5.7-13 the simulation of the transient following a hypothetical sodium feed pump failure is shown.

At t = 2 sec. the sodium pump stops and the flow rate goes to zero in 10 sec. Four seconds after the failure (t = 6 sec) the interlocking system starts to defocus heliostats , reducing the power entering the receiver. The complete field is defocused in 10 seconds (t = 16).

The plot presents the receiver outlet temperature and the tube temperature at the hottest point.

5. SIMULATING DIFFERENT CONDITIONS

5.1 Operation at Different Temperature

Tab.5.7-8 shows the receiver condition when operated at an outlet temperature of 560 C.

The flow rate is reduced to 6.2 kg/s so that the receiver reaches 560 C with the 23/12 thermal input ;as a consequence of the increased average receiver temperature, efficiency decreases slightly.

The effect of temperature on receiver efficiency is plotted in Fig.5.7-14.

5.2 Different Panel Assembly

A simulation is presented in which a billboard receiver is assembled in a different way than ASR. The actual panel sequence is 1 , 2 , 3 , 4 , 5 with reference to the following figure.

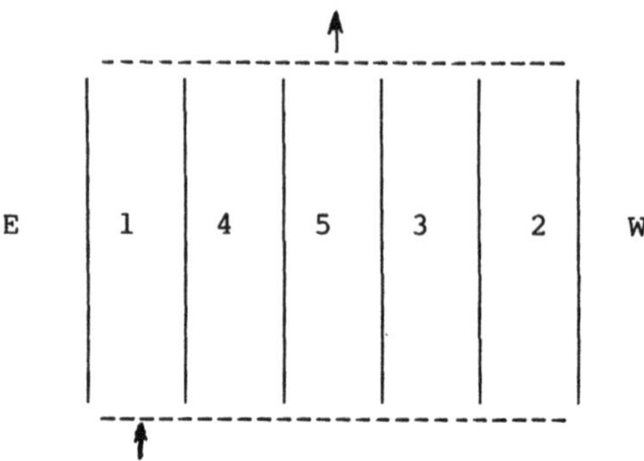

The alternative arrangement assumes the sodium inlet at the 5th panel bottom and then flowing up through panel 4 , 3 , 2 and 1 .

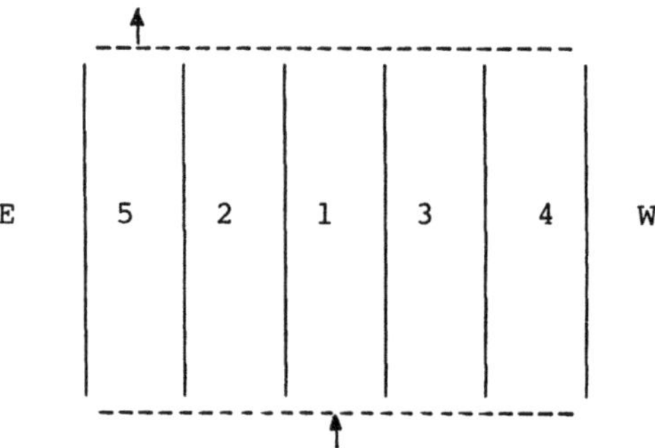

The described arrangement is less efficient because the receiver average temperature increases as power is absorbed mainly on panels 5 , 4 and 3 . Temperature regulation for a conventional loop is also more difficult , because of the longer time it takes the fluid to reach the outlet from the zone where it absorbs the heat. However stresses are alleviated in the alternative arrangement because the highest thermal flux now is absorbed by the coldest panels.

NOMENCLATURE

We define :

a = surface emissivity in infrared band
ag = equivalent overall emissivity in infrared
A = tube inner section (Di/4)** 2
Cpw = tube wall heat capacity
DAi = tube element inside surface area
DAo = tube element outside surface area
DAr = area of the HFD scanner grid associated to each measurement point
Di = tube inner diameter
Do = tube outer diameter
e = surface absorptivity in the solar band
eg = equivalent overall absorptivity in solar band
j = tube axial node
i = node along the circumference
jo = tube axial node (outside surface)
he = heat transfer coefficient with external air
hb = heat transfer coefficient with the backwall
hna = heat transfer coefficient with sodium
kna = sodium thermal conductivity
kw = tube wall thermal conductivity
H = sodium enthalpy
NAZ = number of circumferential cells
NU = Nusselt number
ri = tube inner radius (= Di/2)
ro = tube outer radius (= Do/2)
Rr = thermal resistance of the tube wall in radial direction
R = thermal resistance of the tube wall in circumferential direction
RI = convective film thermal resistance (inside)
RE = overall convective/radiative thermal resistance (outside)
s = tube wall thickness (ro - ri)
T = tube tmperature
TB = backwall temperature
Te = external air temperature
TF = fluid bulk temperature
To = outer surface tube temperature
TO = temperature at time t $-\Delta t$
t = time
z = tube axial coordinate
Wna = power entering the sodium for unit length of tube
Wi(j) = power incident on tube shell i of receiver element j

θ	=	circumferential coordinate
ϕh	=	incident flux on the receiver at x,y
ϕi	=	average flux on the tube element i
ϕna	=	thermal flux entering in the sodium
ϕ	=	flux incident on the tube wall
θi	=	$(i-1)*\Delta\theta$
Γ_{na}	=	sodium mass flow rate
Δz	=	shell axial lenght ($z_{i+1} - z_i$)
Δt	=	time step
$\Delta\theta$	=	circumferential step
ρna	=	sodium density
ρw	=	tube wall density
σ	=	Stefan-Boltzmann constant

*REFERENCES

1] B.Lubarsky
Sodium Heat Transfer
NACA Technical Note 3336 (1955)

2] R.N.Lyon
Liquid Metal Heat Transfer Coefficient
Chem.Eng.Progress Vol.47 (1951)

3] R.Seban
Trans.ASME Vol.73 (1951)

4] F.Biggs,C.Vittitoe
The HELIOS model for the optical behaviour of reflecting solar concentrators
SAND76-0347

5] THERESA : a thermal analysis code for billboard receiver
SSPS Internal Technical Report (in progress)

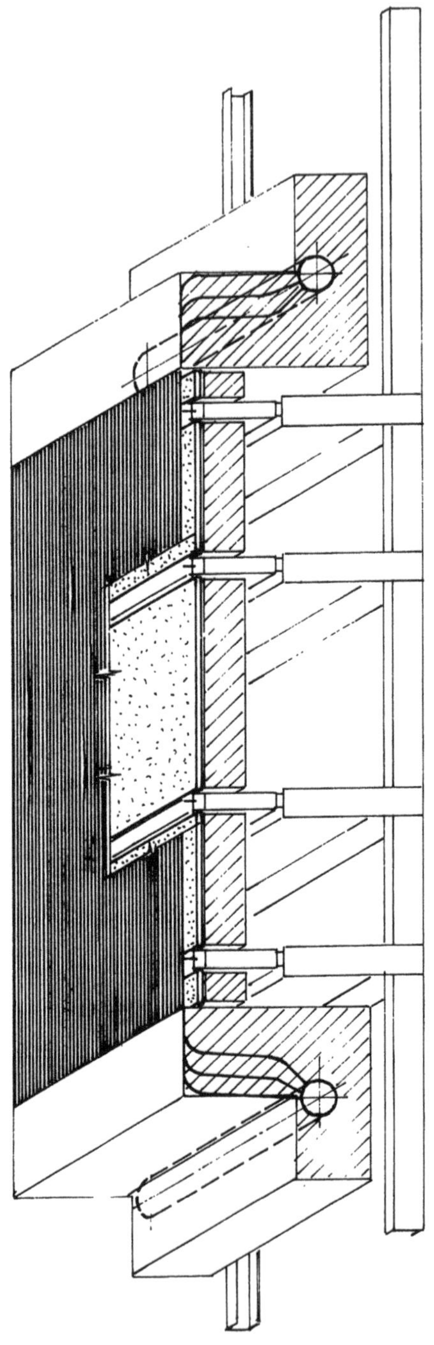

Fig. 5.7.-1: Receiver concept, assumed for theoretical considerations

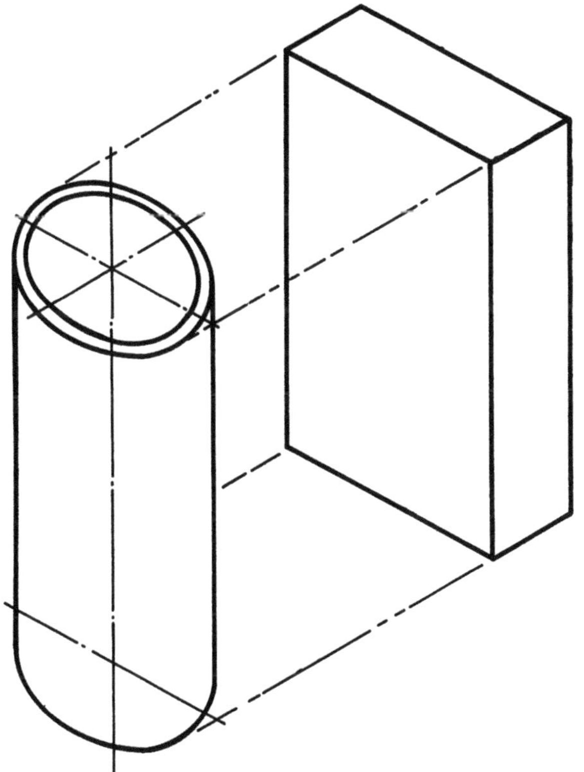

<u>Fig. 5.7.-2:</u> see text

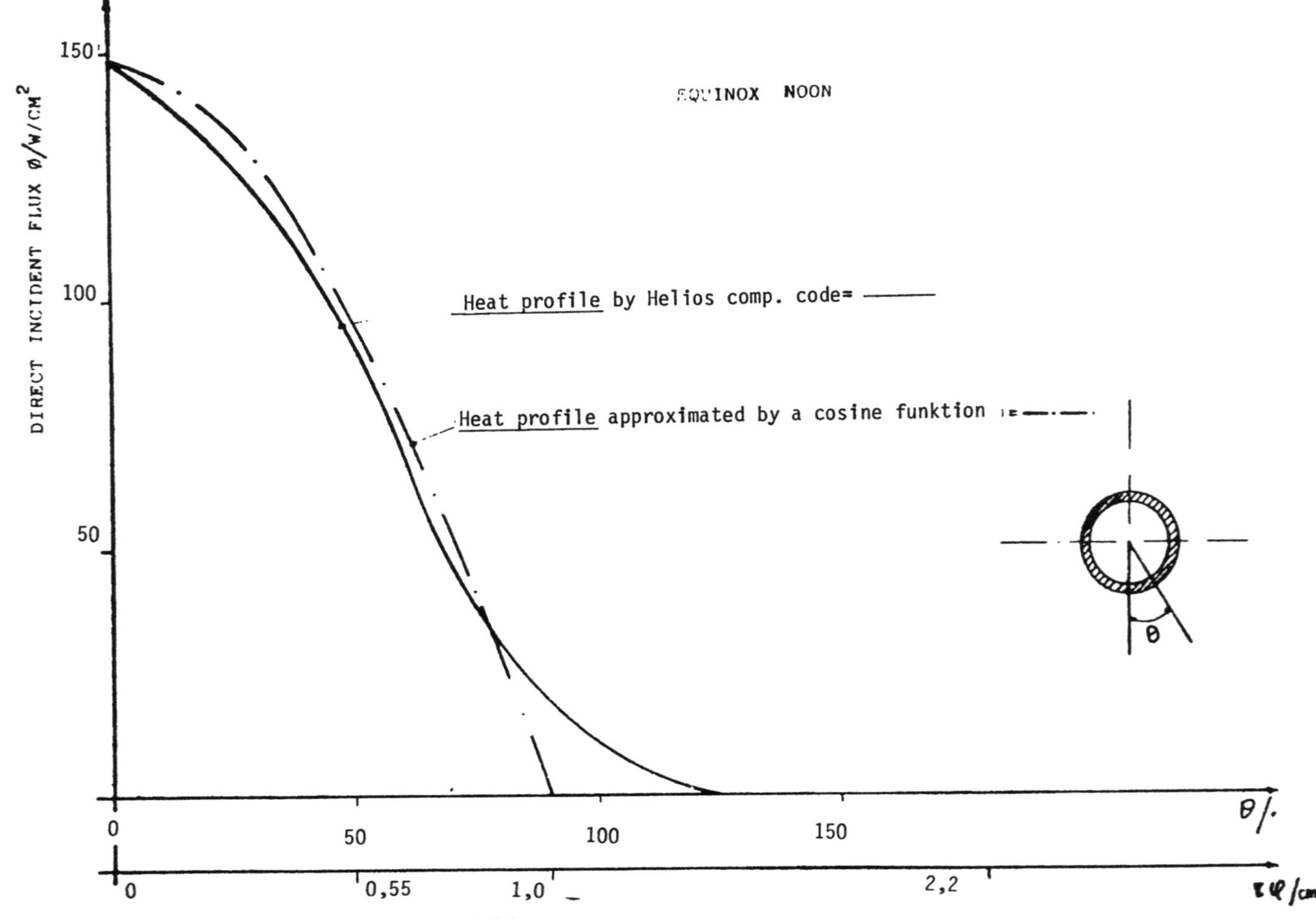

Fig. 5.7.-3: Heat Flux Profile On Isolated Tube

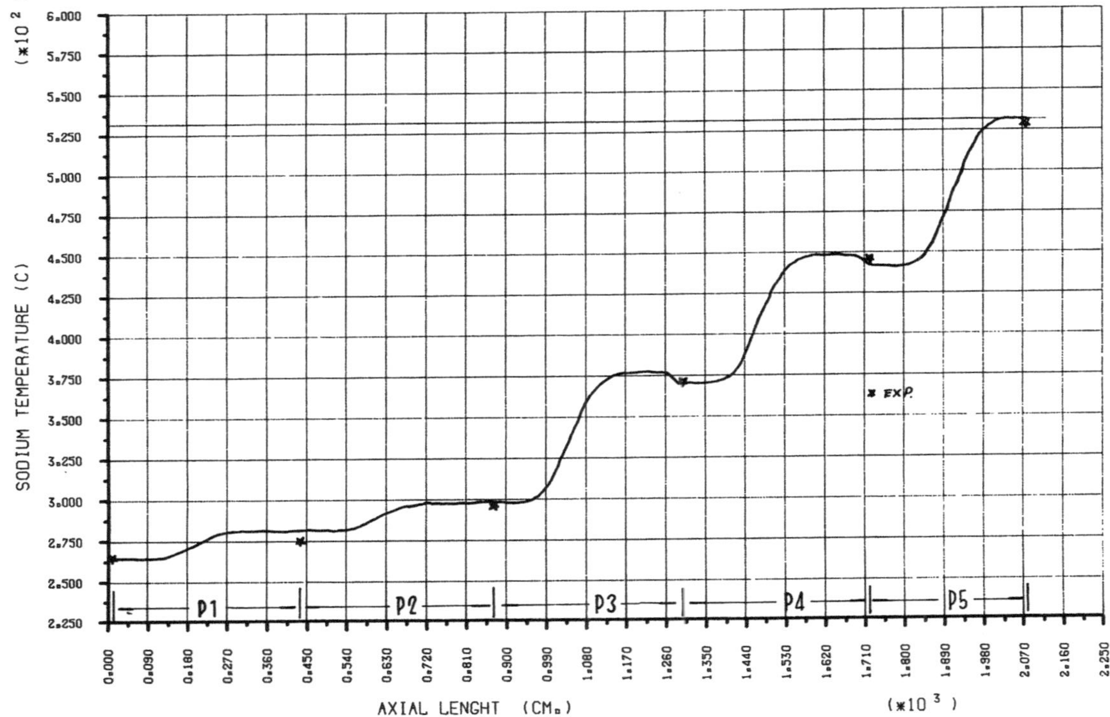

Fig. 5.7.-4: ASR Panels Central Tubes Temperature Profile

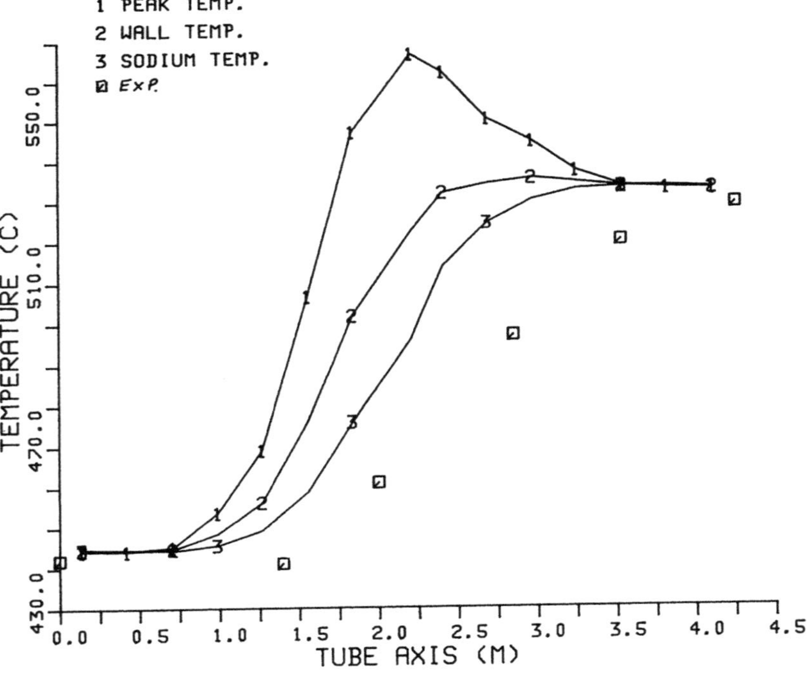

Fig. 5.7.-5: Tube Temperatures VS. Axis 5th Panels Absorber Tube

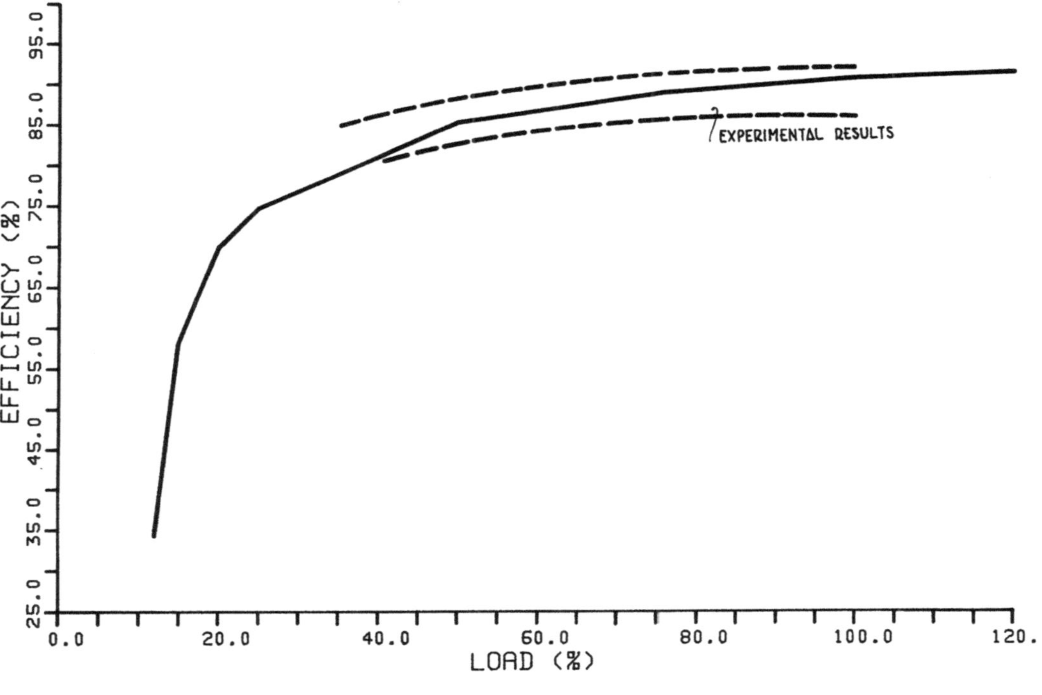

Fig. 5.7.-6: Efficiency VS. Load

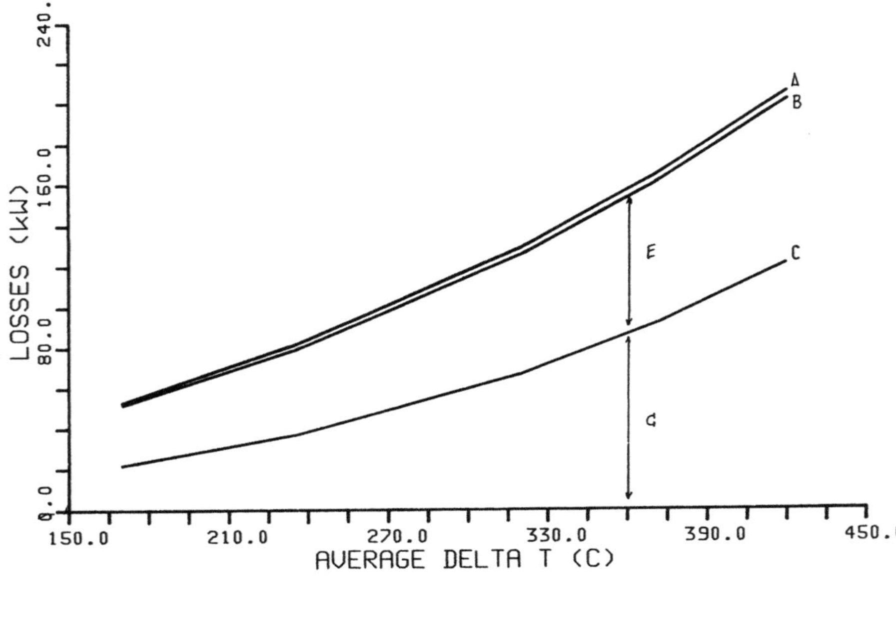

Fig. 5.7.-7: ASR Losses VS. Temp. Two Different Corr.

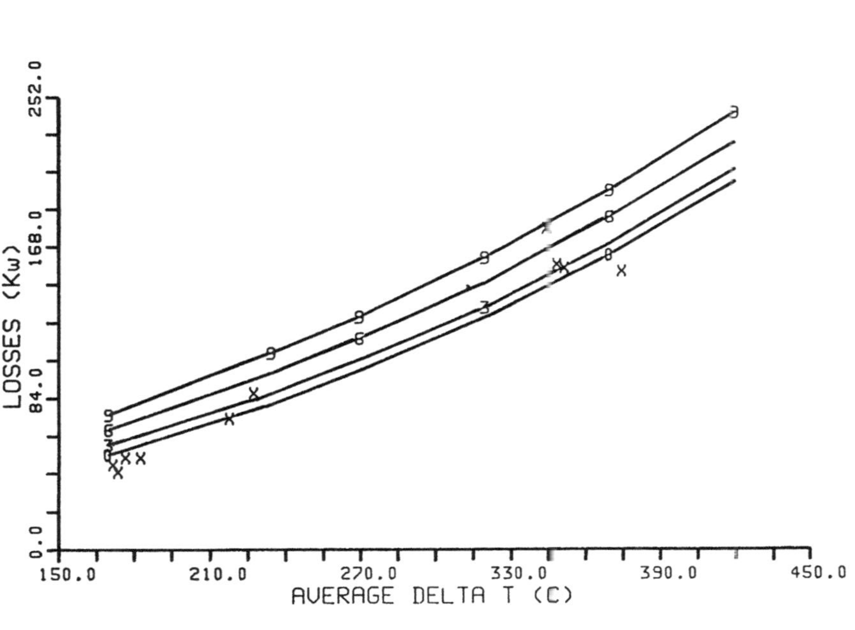

Fig. 5.7.-8: ASR Losses VS. Temperature (Measur. X) Different Wind Velocity

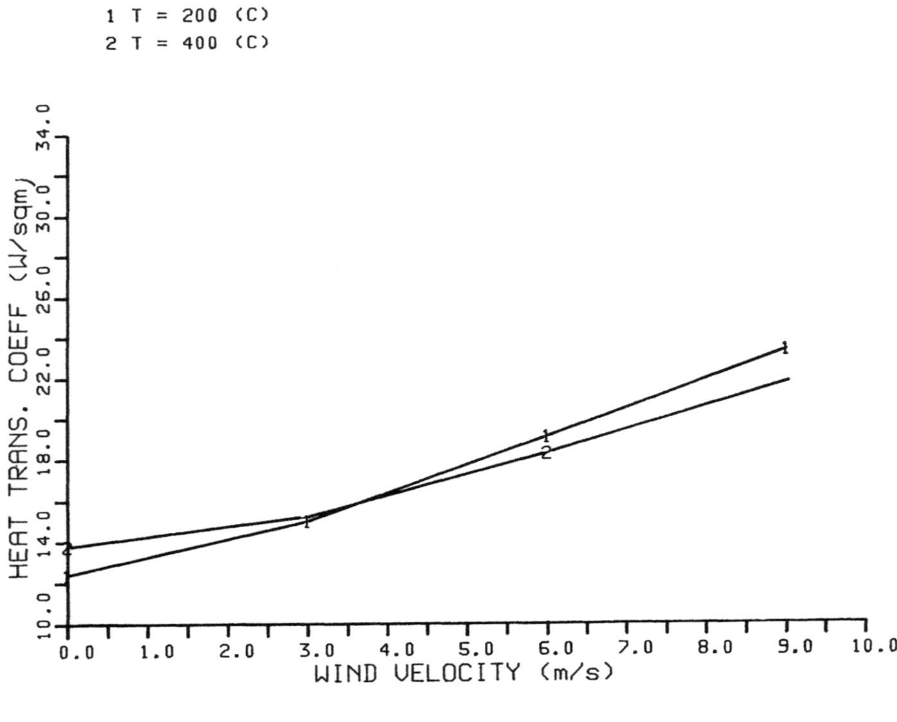

Fig. 5.7.-9:ASR H.T.C. VS. Wind Different Temp. (C)

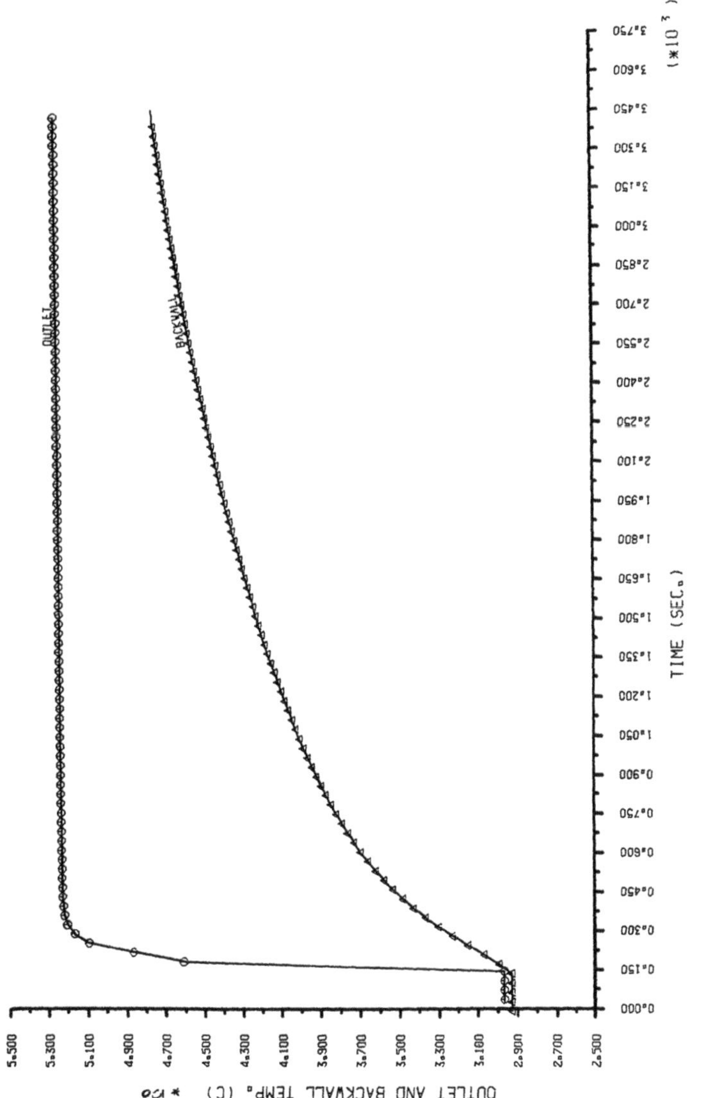

Fig. 5.7.-10: ASR Quick start-Up (With Backwall)

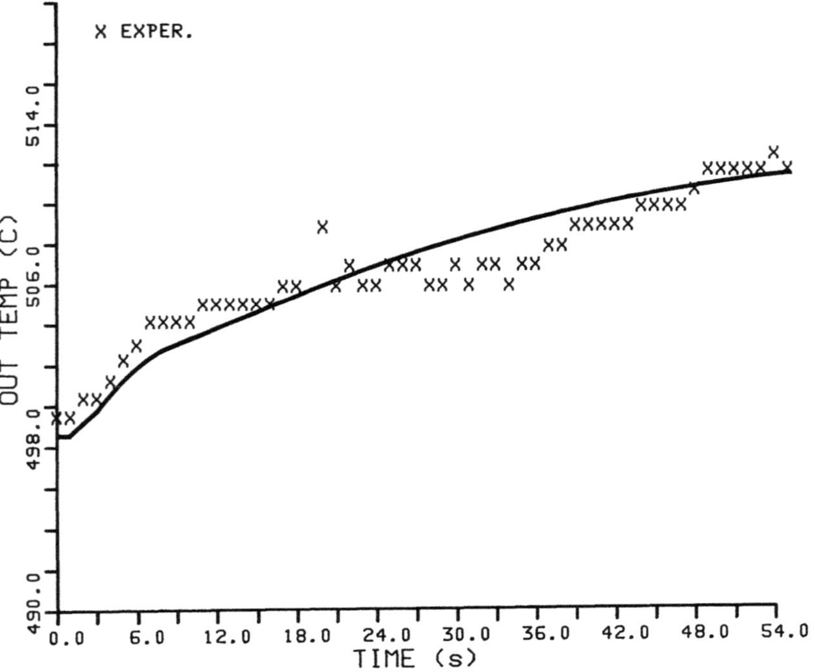

Fig. 5.7.-11: ASR Out. Temp. VS. Time After Flow Step

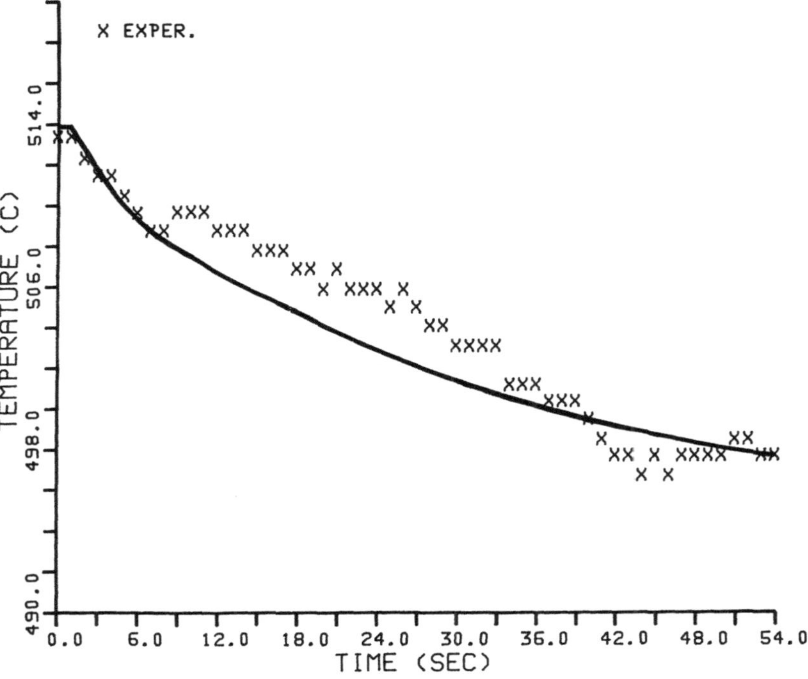

Fig. 5.7.-12: ASR Outl. Temperature After Power Reduktion

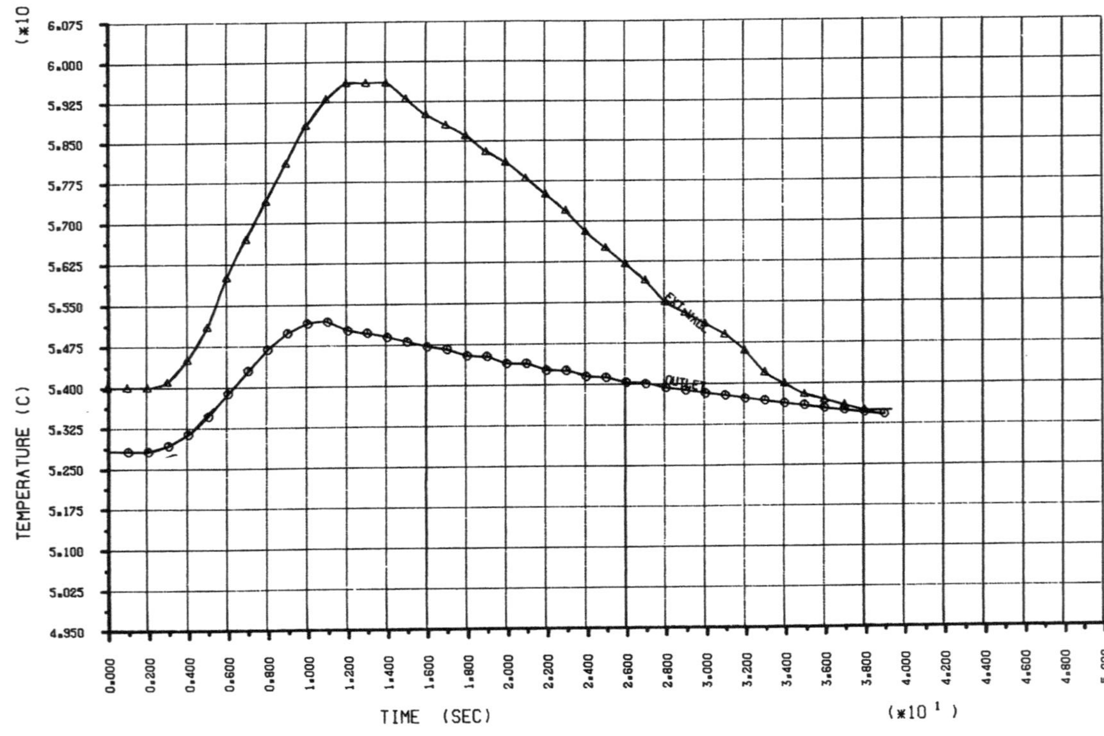

Fig. 5.7.-13:ASR Receiver Temperature VS. Time After Flow Blockage

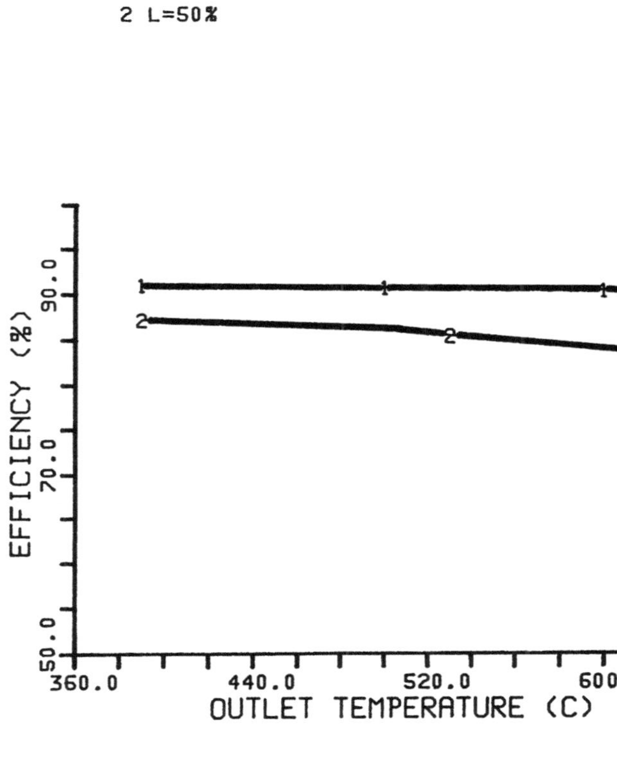

Fig. 5.7.-14: ASR Efficiency VS. Temperature At Different Loads

Table 5.7.-1: HEAT FLUX DISTRIBUTION MEASUREMENT (Corrected Values in KW/m2)	IEA-SSPS OPERATING AGENT DFVLR	Prepared by:	No.:
Test Number : ASR-17 Tracking Heliostats : 93		DATE: 23/12	

V \ H	11	10	9	8	7	6	5	4	3	2	1	P1 / P2
11	.5	.6	1.3	1.9	2.4	2.8	2.7	1.9	1.1	-.4	.3	.1 / .3
10	.3	.3	.4	1.5	5.7	13.3	10.6	3.5	.5	.2	.1	.1 / .3
9	-.8	-.4	1.0	10.6	52.1	103.2	69.5	19.1	1.2	-.5	-.6	-.7 / -.3
8	-.2	.1	7.6	65.6	271.1	469.4	313.5	89.2	7.9	.4	-.7	-.7 / -.4
7	-2.5	-1.1	17.8	182.7	560.5	908.4	658.8	214.1	16.4	-1.2	-2.8	-3.5 / -3.2
6	-8.6	-3.5	33.5	324.5	770.2	988.2	887.5	424.8	26.8	-4.5	-6.4	-7.4 / -6.9
5	-1.8	.8	49.1	406.1	747.4	565.8	604.7	349.1	36.1	-.4	-1.9	-2.6 / -2.8
4	.3	2.1	49.7	247.8	363.8	370.7	310.6	228.6	43.6	1.1	.4	-.1 / -.3
3	.3	.6	8.8	25.6	44.2	110.9	85.7	64.5	15.1	.6	.2	-.1 / 0.0
2	-.1	-.1	.1	.1	.8	2.7	2.1	2.2	.2	0.0	-.1	-.2 / -.2
1	0.0	0.0	0.0	0.0	0.0	0.0	0.0	0.0	0.0	0.0	0.0	0.0 / 0.0

SUBJECT	UNIT	VALUE
Day		357
Local Time	h:m	13:10
Solar Time	h:m	12:10
Insolation	KW/m2	.944
Tep.	Deg.C	21
Wind Speed	Km/h	1
Wind Direc	Deg.An	120
Water flow	m3/h	0.0
WP1	bar	-2.7
WP2	bar	.3
WTB1	Deg.C	22.3
WTB2	Deg.C	24.3
Water Consup	m3	0
TSPIL Max	Deg.C	187
TSPIL Min	Deg.C	81

Aver. flux KW/m2	101.1
Meas. Total Power KW	2448.0
* Thermal Power KW	2538
* PIC KW	2682
Load=Them. po. /2500=	1.0
Meas Total Pow/PIC=	.9
h. CAV=Therm. pow. /Meas. T. pow. =	1.0
h. CAV/Load=	1.0

* This is an approximate date average during five minutes taken from the D.A.S.

SSPS-ALM ASR , HELIOSTAT FIELD ERROR 3. 1 MRAD. , 23/12 NOON , HFD PLANE

SOLAR INSOLATION READ : 0. 94000E-01 WATT/CM**2

HELIOSTAT REFLECTIVITY: 80 %

W/SQ CM ON TARGET SURFACE :

X(I)= Z(J) METERS	1 7500	1 4000	1. 0500	0. 7000	0. 3500	0. 0000	-0. 3500	-0 7000	-1. 0500	-1. 4000	-1. 7500 M
2. 0700	0. 179E-05	0. 712E-04	0. 817E-03	0. 364E-02	0. 873E-02	0. 115E-01	0. 871E-02	0. 362E-02	0. 810E-03	0. 702E-04	0. 175E-05
1. 6560	0 111E-02	0. 109E-01	0. 475E-01	0. 122E+00	0. 208E+00	0. 245E+00	0. 207E+00	0. 122E+00	0. 472E-01	0. 108E-01	0. 109E-02
1. 2420	[illegible]	0. 136E+00	0. 447E+00	0. 103E+01	0. 172E+01	0. 204E+01	0. 172E+01	0. 103E+01	0. 445E+00	0. 135E+00	0. 256E-01
0. 8280	0. 165E+00	0. 715E+00	0. 224E+01	0. 522E+01	0. 885E+01	0. 106E+02	0. 883E+01	0. 520E+01	0. 223E+01	0. 710E+00	0 163E+00
0. 4140	0. 562E+00	0. 234E+01	0. 740E+01	0. 175E+02	0. 298E+02	0. 357E+02	0. 298E+02	0. 175E+02	0. 737E+01	0. 233E+01	0. 558E+00
0. 0000	0. 128E+01	0. 541E+01	0. 174E+02	0. 404E+02	0. 646E+02	0. 744E+02	0. 647E+02	0. 405E+02	0. 174E+02	0. 546E+01	0. 129E+01
-0. 4140	0. 215E+01	0 944E+01	0. 306E+02	0. 678E+02	0. 958E+02	0. 102E+03	0. 958E+02	0. 678E+02	0. 307E+02	0. 973E+01	0. 234E+01
-0. 8280	0 252E+01	0. 109E+02	0. 342E+02	0. 673E+02	0. 870E+02	0. 899E+02	0. 854E+02	0. 655E+02	0. 334E+02	0. 115E+02	0. 300E+01
-1. 2420	0. 155E+01	0. 636E+01	0. 178E+02	0 280E+02	0. 350E+02	0. 394E+02	0. 357E+02	0. 274E+02	0. 160E+02	0. 681E+01	0. 205E+01
-1 6560	0. 361E+00	0. 158E+01	0. 391E+01	0. 417E+01	0. 498E+01	0. 707E+01	0. 665E+01	0. 463E+01	0. 272E+01	0 152E+01	0. 562E+00
-2 0700	0. 200E-01	0. 158E+00	0. 410E+00	0. 266E+00	0. 273E+00	0. 560E+00	0. 522E+00	0. 272E+00	0. 131E+00	0. 931E-01	0 432E-01

TOTAL POWER ON TARGET SURFACE IS 0. 26043E+07 WATTS.

ESTIMATED ERROR IN TOTAL POWER ON TARGET SURFACE IS 0 15759E+05 WATTS.

FACET AREA = 3714. 5 M**2
REDUCED BY COS GIVES 3643. 5 M**2
FURTHER REDUCED BY SHBL GIVES 3534. 3 M**2
POWER INTERCEPTED BY MIRRORS = 0. 3425E+07 WATTS.

Table 5.7.-2: see text

TIME	IRRADIANCE	HFD MEASURE		DAS DATA		CONCENTRATION
		Flux peak	Incid. Power	Incid. Power	Power in sodium	
	W/sqm	kW/sqm	kW	kW	kW *	
23/12 11.49	950	979	/	2703	2582	1030
23/12 12.10	944	989	2448	2682	2538	1048
24/12 12.14	955	1030	/	2692	2568	1078
26/12 12.08	619	503	1274	1649	1472	810
28/12 11.34	894	931	2214	2512	2348	1040
28/12 12.06	910	985	/	2549	2350	1081
28/12 12.38	902	967	2300	2481	2375	1072
HELIOS COMPUTED DATA						
		HFD PLANE		ASR PLANE		
		Peak	Power	Peak	Power	
28/12 12.00	940	1020	2604	1230	2587	1085

* : Data affected by flow rate measure error

Table 5.7.-3: see text

FLUX DATA(KW/SQM) AND COORDINATES FROM HELIOS

Z METERS	X(I)= 1.5000	1.2000	0.9000	0.6000	0.3000	0.0000	-0.3000	-0.6000	-0.9000	-1.2000	-1.5000 M
1.5000	0.728E+00	0.233E+01	0.552E+01	0.102E+02	0.148E+02	0.167E+02	0.147E+02	0.101E+02	0.547E+01	0.230E+01	0.714E+00
1.2000	0.281E+01	0.853E+01	0.203E+02	0.383E+02	0.566E+02	0.646E+02	0.565E+02	0.381E+02	0.201E+02	0.844E+01	0.278E+01
0.9000	0.755E+01	0.235E+02	0.580E+02	0.113E+03	0.171E+03	0.198E+03	0.171E+03	0.113E+03	0.575E+02	0.232E+02	0.758E+01
0.6000	0.155E+02	0.521E+02	0.138E+03	0.273E+03	0.412E+03	0.470E+03	0.412E+03	0.276E+03	0.136E+03	0.518E+02	0.159E+02
0.3000	0.248E+02	0.945E+02	0.281E+03	0.566E+03	0.791E+03	0.855E+03	0.792E+03	0.575E+03	0.278E+03	0.951E+02	0.260E+02
0.0000	0.309E+02	0.134E+03	0.452E+03	0.935E+03	0.118E+04	0.117E+04	0.117E+04	0.937E+03	0.448E+03	0.136E+03	0.326E+02
-0.3000	0.290E+02	0.138E+03	0.502E+03	0.105E+04	0.123E+04	0.114E+04	0.121E+04	0.104E+04	0.494E+03	0.140E+03	0.308E+02
-0.6000	0.202E+02	0.969E+02	0.349E+03	0.714E+03	0.817E+03	0.749E+03	0.815E+03	0.712E+03	0.343E+03	0.986E+02	0.218E+02
-0.9000	0.101E+02	0.444E+02	0.146E+03	0.284E+03	0.334E+03	0.318E+03	0.337E+03	0.290E+03	0.144E+03	0.460E+02	0.113E+02
-1.2000	0.348E+01	0.134E+02	0.381E+02	0.696E+02	0.878E+02	0.903E+02	0.884E+02	0.711E+02	0.378E+02	0.142E+02	0.402E+01
-1.5000	0.790E+00	0.278E+01	0.709E+01	0.127E+02	0.174E+02	0.191E+02	0.173E+02	0.127E+02	0.704E+01	0.301E+01	0.927E+00

TARGET DIMENSIONS 3.00 BY 3.00

POWER ON TARGET AFTER SPLINES IS 0.25815E+04 KW AVERAGE FLUX IS 0.26862E+03 KW/SQM

TOTAL POWER ON TARGET SURFACE IS 0.25878E+07 WATTS. (SIMPSON FORMULA)

TOTAL POWER ON TARGET SURFACE IS 0.25878E+07 WATTS. (ALBRECHT FORMULA)

ERROR ESTIMATE FOR POWER MATRIX

ESTIMATED ERROR IN TOTAL POWER ON
TARGET SURFACE I 0.70656E+04

POWER INCIDENT ON PANEL N. 1 172.51KW AREA = 1.57 SQM AVE. FLUX = 110.1 KW/SQM

POWER INCIDENT ON PANEL N. 2 171.50KW AREA = 1.57 SQM AVE. FLUX = 109.4 KW/SQM

POWER INCIDENT ON PANEL N. 3 665.53KW AREA = 1.57 SQM AVE. FLUX = 424.6 KW/SQM

POWER INCIDENT ON PANEL N. 4 666.83KW AREA = 1.57 SQM AVE. FLUX = 425.4 KW/SQM

POWER INCIDENT ON PANEL N. 5 826.83KW AREA = 1.57 SQM AVE. FLUX = 527.5 KW/SQM

Table 5.7.-4: see text

ASR GLOBAL DATA (TIME= 0).

[illegible] 7.00 INCIDENT POWER (KW) = 2503.21 POWER TO FLUID (kW)= 2257.66 EFFICIENCY (%) = 90.2

ASR TEMPERATURE DISTRIBUTION (deg. Celsius):

PANEL: 1			PANEL: 4			PANEL: 5			PANEL: 3			PANEL: 2		
Tout= 282.58			Tout= 445.06			Tout= 533.56			Tout= 372.68			Tout= 299.72		
T 269.	T 281.	T 301.	T 423.	T 448.	T 462.	T 534.	T 533.	T 533.	T 389.	T 376.	T 351.	T 318.	T 298.	T 287.
F 270.	F 281.	F 301.	F 423.	F 448.	F 461.	F 534.	F 533.	F 533.	F 388.	F 375.	F 350.	F 318.	F 298.	F 287.
W 266.	W 278.	W 297	W 420.	W 445.	W 458.	W 531	W 530.	W 530.	W 385.	W 372.	W 347.	W 314	W 295	W 284.
T 270.	T 283.	T 304.	T 428.	T 455.	T 471.	T 545.	T 545.	T 545.	T 399.	T 383.	T 355.	T 321.	T 300.	T 288.
F 269.	F 280.	F 299.	F 421.	F 444.	F 456.	F 529.	F 527.	F 528.	F 383.	F 372.	F 348.	F 316.	F 297.	F 287.
W 266.	W 277.	W 296.	W 418.	W 441.	W 453.	W 526.	W 524.	W 525.	W 380.	W 368.	W 345.	W 313.	W 294.	W 283.
T 272.	T 288.	T 317.	T 445.	T 479	T 496.	T 568.	T 564.	T 567.	T 423.	T 408.	T 373.	T 334.	T 306.	T 289.
F 267.	F 273.	F 283.	F 399.	F 411.	F 415.	F 486.	F 484.	F 486.	F 342.	F 338.	F 326.	F 300.	F 290.	F 284.
W 264.	W 269.	W 280.	W 396.	W 408.	W 412.	W 483.	W 481.	W 483.	W 339.	W 335.	W 322.	W 297.	W 287.	W 281.
T 266.	T 270.	T 277.	T 388.	T 395.	T 397.	T 469.	T 468.	T 469.	T 326.	T 323.	T 316.	T 294.	T 287.	T 284.
F 265.	F 265.	F 266.	F 373.	F 374.	F 374.	F 447.	F 447.	F 447.	F 302.	F 301.	F 301.	F 283.	F 283.	F 282.
W 261.	W 262.	W 262.	W 370.	W 371.	W 371.	W 444.	W 444.	W 444.	W 298.	W 298.	W 297.	W 280.	W 279.	W 279.
Tinp= 265.00			Tinp= 372.45			Tinp= 444.81			Tinp= 299.55			Tinp= 282.38		

ON EACH PANEL:

	Panel 1			Panel 4			Panel 5			Panel 3			Panel 2		
INC. POWER (kW):	172.5			666.8			826.8			665.5			171.5		
TO FLUID (kW):	149.8			607.4			745.7			616.6			147.0		
LOSSES (kW):	22.8			59.4			81.1			48.9			24.5		
	6.19R	8.38E	8.19C	23.93R	21.87E	13.65C	29.67R	34.46E	16.95C	23.88R	14.19E	10.83C	6.15R	9.47E	8.83C
	27.2 %	36.8 %	36.0 %	40.2 %	36.8 %	23.0 %	36.6 %	42.5 %	20.9 %	48.8 %	29.0 %	22.1 %	25.2 %	38.7 %	36.1 %
EFFICIENCY (%):	86.81			91.09			90.19			92.65			85.74		

Table 5.7.-5: see text

PANEL N 5 TUBE N. 3 TUBE POSITION 0 21 (M)

IPAST= 0 TIME= 0.00 TMAX= 0.00 NPAS = 0

IC	FLUXPK	POTIN	GA	TEST	TW	TF	TBKL	CHNA	CHAIR	PTTIC	PTFIC	POIR	POCONV	POVIS	PTBKWL	RPERC
	KW/MQ	KW	KG/S	C	C	C	C	KW/MQ*C	W/MQ*C	KW	KW	KW	KW	KW	W	
1	0 0	0.000	0.179	444.4	444.6	444.7	-1 0	41.0	8.83	-0.046	-0.046	0.000	0.046	0.000	0.0	0.0
2	0.0	0.000	0.179	444.2	444.4	444.5	-1.0	41.0	8.83	-0.046	-0.046	0.000	0.046	0.000	0.0	0.0
3	26 8	0 106	0.179	445 3	444.7	444.4	441 4	41 0	7.94	0 029	0 029	0.055	0.018	0.004	0 6	27 7
4	146 3	0.580	0 179	453 8	448.5	445.6	442 5	41.0	7.97	0.484	0 484	0.057	0.018	0.021	0 6	83.4
5	331.9	1.316	0.179	469.1	456.3	449.4	446.4	40.9	8.03	1.189	1.189	0.061	0.019	0.047	0.6	90.3
6	794.4	3.151	0.179	506 6	476.0	458.9	455.9	40.7	8.18	2.946	2.946	0.072	0.020	0.113	0.6	93.5
7	1188.8	4.716	0.179	547.0	501.9	475.9	472.9	40.4	8.33	4.438	4.438	0.086	0.022	0.169	0 6	94.1
8	1178.5	4.675	0.179	566 6	522.5	496.3	493.4	39.8	8.41	4.389	4 389	0.095	0.023	0.168	0.6	93.9
9	824 3	3 270	0 179	562.7	532.1	513.7	510.8	39.2	8 40	3 034	3 034	0.096	0.023	0.117	0.6	92.8
10	439.2	1 742	0.179	550.2	534.2	524.5	521.6	38.8	8.36	1.564	1.564	0.093	0.023	0.063	0.7	89.8
11	252.0	1.000	0.179	544.5	535.6	530.2	527.3	38.6	8.35	0.850	0.850	0.091	0.023	0.036	0.7	85.0
12	90.7	0.360	0.179	537.2	534.4	532.8	529.9	38.5	8.32	0.235	0.235	0.089	0.022	0.013	0.7	65.4
13	21.2	0.084	0.179	533.4	533.3	533 3	265.2	38.4	8.31	-0.029	-0.029	0.088	0.022	0.003	0.7	0.0
14	0.0	0.000	0.179	532.8	533.0	533.1	-1.0	38.4	8.83	-0.055	-0.055	0.000	0.055	0.000	0.0	0.0
15	0.0	0 000	0.179	532.5	532.8	532.8	-1.0	38.4	8.83	-0 055	-0.055	0.000	0 055	0.000	0.0	0.0

INLET/OUTLET TUBE CONDITIONS :

	FLOW RATE (t) ,	TEMPERETURE (t) ,	FLOW RATE (t-dt)	TEMPERATURE(t-dt)
1*	0.179	444.777	0.179	444.777
16*	0.179	532.708	0.179	532.708

TUBE: INC.POWER= 21.000 LOSSES= 2.073 CONVECTION,EMISSION,REFLECTION FRACT.= 0.21 0.43 0.36 TUBE EFFIC.= 90.1

POTKT= 18.93 POTKF= 18.93 POTBK= 0.01 DP= 9854.0 DPF= 10125.4 EFFIF= 90.1 EFFIT= 90.1 REIN=(73991.6 - 82320.5)

Table 5.7.-6: see text

ASR GLOBAL DATA (TIME= 0.):

MASSFLOW (KG/S)= 6 30 INCIDENT POWER (KW) = 2503.21 POWER TO FLUID (kW)= 2247.90 EFFICIENCY (%) = 89.8

ASR TEMPERATURE DISTRIBUTION (deg. Celsius):

PANEL: 1			PANEL. 4			PANEL: 5			PANEL: 3			PANEL: 2		
Tout= 284.52			Tout= 464.64			Tout= 561.79			Tout= 384.57			Tout= 303.54		
T 270.	T 283.	T 305.	T 440.	T 467.	T 483.	T 562.	T 561.	T 562.	T 403.	T 388.	T 360	T 323.	T 302.	T 289.
F 270.	F 283.	F 305.	F 440.	F 467.	F 483.	F 562.	F 561.	F 562.	F 402.	F 388.	F 360.	F 323.	F 302.	F 289.
W 267.	W 280.	W 301.	W 437.	W 464.	W 480.	W 559.	W 558.	W 559	W 399.	W 384.	W 357	W 320.	W 299.	W 286.
T 271.	T 285.	T 308.	T 445.	T 475.	T 492.	T 573.	T 573.	T 572.	T 412.	T 395.	T 365.	T 327.	T 304.	T 290.
F 270.	F 282	F 303.	F 438.	F 464.	F 477.	F 557.	F 555.	F 556	F 396.	F 383.	F 357.	F 322.	F 301.	F 289.
W 266.	W 279.	W 299.	W 435	W 461.	W 474.	W 554.	W 552.	W 553	W 393	W 380.	W 354.	W 318.	W 298.	W 286.
T 272.	T 290	T 319.	T 461.	T 496.	T 513.	T 592.	T 588.	T 591.	T 433.	T 416.	T 381.	T 338.	T 309.	T 292.
F 267.	F 274.	F 285.	F 413.	F 427	F 432.	F 510.	F 508.	F 510	F 351.	F 346.	F 332.	F 304	F 293.	F 286.
W 264.	W 270.	W 282.	W 410	W 424.	W 429.	W 507.	W 505.	W 507	W 347.	W 343.	W 329.	W 301.	W 290.	W 283.
T 266.	T 270.	T 277.	T 400.	T 407.	T 410.	T 489.	T 488.	T 489.	T 330.	T 327.	T 320.	T 296.	T 289	T 285
F 265	F 265.	F 266.	F 385	F 386.	F 387.	F 466.	F 466.	F 466	F 306.	F 305.	F 305	F 285.	F 284.	F 284.
W 261.	W 262.	W 263.	W 382.	W 383.	W 383.	W 463.	W 463.	W 463.	W 303.	W 302.	W 301.	W 282.	W 281.	W 281
Tinp= 265.00			Tinp= 384.31			Tinp= 464.34			Tinp= 303.35			Tinp= 284.30		

ON EACH PANEL:

	Panel 1			Panel 4			Panel 5			Panel 3			Panel 2		
INC. POWER (kW):	172.5			666 8			826.8			665 5			171 5		
TO FLUID (kW)	149 6			604 7			740 3			615 5			146 7		
LOSSES (kW)	22 9			62 2			86 5			50 0			24 8		
	6 19R	8 45E	8 23C	23 93R	23 96E	14 27C	29 67R	38 94E	17 92C	23 88R	14 97E	11 13C	6 15R	9 68E	8 94C
	27.1 %	37.0 %	36 0 %	38 5 %	38 6 %	23.0 %	34 3 %	45 0 %	20 7 %	47 8 %	30 0 %	22 3 %	24 8 %	39 1 %	36 1 %
EFFICIENCY (%)	86 74			90 68			89 53			92 49			85 55		

Table 5.7.-7: see text

ASR GLOBAL DATA (TIME= 0.)

MASSFLOW (KG/S)= 7.00   INCIDENT POWER (KW) = 2503.21   POWER TO FLUID (kW)= 2195.51   EFFICIENCY (%) = 87.7

ASR TEMPERATURE DISTRIBUTION (deg. Celsius):

PANEL: 1			PANEL: 4			PANEL: 5			PANEL: 3			PANEL: 2		
Tout= 526.21			Tout= 429.04			Tout= 356.44			Tout= 499.94			Tout= 513.20		
T 513	T 524.	T 544	T 407	T 432	T 446	T 357	T 356	T 357	T 516	T 503.	T 478	T 531	T 511.	T 500.
F 514	F 525	F 545	F 407.	F 432	F 445	F 356	F 355	F 356	F 516	F 503.	F 478.	F 532	F 512	F 501
W 511	W 523	W 542	W 404.	W 428	W 442	W 353	W 352	W 352	W 513	W 500.	W 475.	W 529	W 509.	W 498.
T 514	T 527.	T 548.	T 412.	T 439	T 455.	T 368	T 368	T 367	T 526	T 510.	T 483.	T 535	T 514.	T 502
F 514	F 525.	F 544.	F 405	F 428	F 440.	F 350	F 349.	F 350	F 511	F 500	F 476.	F 531	F 512.	F 501
W 511	W 522.	W 541.	W 402	W 425	W 437.	W 347	W 346	W 347.	W 508	W 497.	W 473	W 529	W 509	W 499
T 517	T 533.	T 561	T 429	T 463	T 480	T 391	T 388.	T 390	T 550	T 535.	T 500	T 547	T 520	T 504
F 513	F 519.	F 529.	F 382.	F 395	F 399	F 307	F 305.	F 307.	F 471.	F 467	F 454	F 516	F 506.	F 500
W 510	W 516.	W 527	W 379	W 391	W 396	W 304	W 302	W 303	W 468	W 464	W 451	W 513	W 503	W 497
T 512	T 516	T 522	T 372	T 378	T 381	T 291	T 290	T 291	T 453	T 451	T 444	T 507	T 503	T 499
F 512	F 512	F 513	F 357	F 358	F 358	267	F 267	F 267	F 431	F 430	F 430	F 500	F 499	F 49[illegible]
W 509	W 509.	W 510	W 354.	W 355	W 355	W 264	W 264	W 264	W 428	W 427.	W 426	W 497	W 496	W 496
Tinp= 512.90			Tinp= 356.18			Tinp= 265.00			Tinp= 428.79			Tinp= 499.63		

ON EACH PANEL

	Panel 1			Panel 4			Panel 5	Panel 3			Panel 2		
INC POWER	172.5			666.8				665.5			171.5		
[illegible] FLUID	112.3			609.[illegible]				596.2			114.2		
[illegible]	60.2			56.9				69.3			57.3		
	6.19R	36.27E	17.75C	23.93R	19.95E	13.02C		23.88R	29.57E	15.85C	[illegible]R	33.95E	17.22C
	10.3 %	60.2 %	29.5 %	42.1 %	35.1 %	22.9 %		34.5 %	42.7 %	22.9 %	[illegible] %	59.2 %	30.0 %
EFFICIENCY	65.10			91.47				89.59			[illegible]6.58		

Table 5.7.-8: see text

APPENDIX A : Tube wall thermal balance

Let us consider an axial shell of a tube.
The thermal balance equation at the central node j is

$$\frac{T(j-1)-T(j)}{R_{\theta^-}} + \frac{T(j+1)-T(j)}{R_{\theta^+}} + \frac{T(jo)-T(j)}{Rr^+} + \frac{TF - T(j)}{RF + Rr^-} =$$

$$= C\,\frac{T(j)-T0(j)}{t}$$

At the external node jo (without thermal capacity) we can write the thermal balance equation :

$$\frac{Te - T(jo)}{RE} + \not{\emptyset} * ag * DAo + \frac{T(j) - T(jo)}{Rr^+} = 0$$

From the last equation we compute T(jo)

$$T(jo) - T(j) = \frac{RE * Rr^+}{RE + Rr^-} * \not{\emptyset} * ag * DAo + \frac{Te - T(j)}{RE}$$

and by substitution into the equation for the central node :

$$- \frac{T(j-1)}{R_{\theta^-}} + T(j* \left[\frac{1}{R_{\theta^-}} + \frac{1}{R_{\theta^+}} + \frac{1}{RE+Rr^-} + \frac{1}{RF+Rr^-} + \frac{C}{\Delta t}\right] - \frac{T(j+1)}{R_{\theta^+}}$$

$$= \frac{RE}{RE+Rr^-} * \not{\emptyset} * ag * DAo + \frac{Te}{RE} + \frac{TF}{RF+Rr^-} + \frac{C}{\Delta t}\,T0(j)$$

Defining :

$$B(j) = - \frac{1}{R_{\theta^-}}$$

$$A(j) = \frac{1}{R\theta-} + \frac{1}{R\theta+} + \frac{1}{RE+Rr+} + \frac{1}{RF+Rr-} + \frac{C}{\Delta t}$$

$$C(j) = -\frac{1}{R\theta+}$$

$$D(j) = \frac{C}{\Delta t} T0(j) + \frac{TF}{RF+Rr^-} + \frac{RE*\Phi*ag*DAo}{RE + Rr} + \frac{Te}{RE+Rr^+}$$

with the boundary conditions

$$B(1) = 0$$
$$C(NAZ) = 0$$

we can write

$$B(j)*T(j-1) + A(j)*T(j) + C(j)*T(j+1) = D(j)$$

$$1 < j < NAZ$$

which is a simple tridiagonal system of NAZ equations.

From the solution we compute

- the temperature difference DTfilm across the sodium film

$$DTfilm = \frac{RF}{Rr+RF} * (T(j) - TF)$$

- the inner wall temperature Tinside

$$Tinside = TF + DTfilm$$

- the thermal flux through the inner tube wall Φna

$$\Phi na = \frac{T(j)-TF}{Rr^- + RF}$$

Iterations are necessary to include the temperature dependence of the physical properties.

APPENDIX B : Sodium bulk mass and energy balance equations

The equations are :

$$\frac{\partial \Gamma}{\partial z} + \frac{\partial}{\partial t} A \cdot \rho na = 0 \qquad \text{continuity}$$

$$\frac{\partial H}{\partial z} + A \cdot \rho_{na} \frac{\partial H}{\partial t} = \Phi_{na} \qquad \text{energy balance}$$

By a finite difference scheme identifying the axial node by j

$$\Gamma(j+1) = \Gamma(j) - A * \Delta z * \frac{\rho_{na}(j) - \rho_{ona}(j)}{\Delta t}$$

$$H(j+1) = \frac{H(j) * \frac{\Gamma(j+1) + \Gamma(j)}{2 * \Delta z} + \frac{H0(j+1) * A * \rho_{ona}(j)}{\Delta t} + \Phi na}{\frac{\Gamma(j+1) + \Gamma(j)}{2 * \Delta z} + \frac{A * \rho_{ona}(j)}{\Delta t}}$$

It is evident that by forward steps we can solve the two equations , once the conditions at Δt - t and Φna are known.

To compute Φna, the inlet temperature at each cell is used (see appendix A) , while the correct estimate requires the average value between the inlet and the outlet. This point is solved by iterations.

A TRANSIENT RESPONSE OF THE SULZER RECEIVER

Neil Gregory, ITET, (EIR-CH)

1.0 INTRODUCTION

In an attempt to find transient data for the Sulzer receiver, three sets of tapes were read. The period chosen for such investigations, was March - April 1983. According to the general consensus of opinion, this was a period which should have contained many test days, including cloud tests, during which transient response data had been recorded.

Many difficulties were encountered in reading these tapes, as has been described in Ref. 1, and success was limited to a single day, the 6th April 1983. Further analysis of this day has revealed that no cloud test had in fact been performed, and that the transients were the result of a grid failure and a subsequent receiver trip (emergency shutdown operation).

This paper contains a discussion of the events which lead to the transients and also a description of their major features.

2.0 DATA SOURCES

The data for the 6th April 1983 were successfully read from the CRS tape and point and class summary tape set for the period : 06-04-83 to 09-05-83. Further information relating to the data manipulation and difficulties is given in Ref. 1. The explanation of these data was augmented by the comments in the operator's logbook for this period.

The data concerning heliostat tracking were taken from the point summary files at 1 minute intervals. The irradiance, receiver inlet and outlet temperatures, and flowrate data were taken from the class summary files at 10 second intervals.

3.0 SEQUENCE OF EVENTS

On the 6th April 1983, the plant was operating normally with a constant receiver outlet temperature of about 525 deg.C. It was a cloudless day, and the insolation at the time was about 770 w/m**2. At about 10:49 local time, the following events occurred:

- failure of the 25kv Sevillana grid caused the heliostats to return to the stow position.
- at 11:00, the heliostats were put into the standby position.
- at 11:11, the heliostats were manually put back into track with the receiver control system switched to automatic. In the meantime, the receiver outlet temperature had cooled to about 260 deg.C, and it was apparently the operator's intention to test the automatic receiver control system capabilities. The normal procedure for starting from such temperatures is to use manual mass flow control, starting with a relatively high flowrate to avoid high temperatures in the receiver, this procedure is particularly necessary for the ASR receiver operation.
- at 11:22, with 82 heliostats in track, a receiver trip occurred caused by a very high (HH) temperature reading of the tube bundle temperature LK01CT09. The cause of this high temperature was that the automatic receiver control system had not increased the flowrate sufficiently to correspond to a receiver power brought about by 82 heliostats in track at insolation levels of more than 800 w/m**2. This receiver trip automatically caused all heliostats to defocus.
- at 11:24, the heliostats were rapidly put into operation again with manual control, and the mass flow was increased manually according to the normal operating strategy. When the receiver outlet temperature had stabilized, the receiver control system was reset to automatic.
- by about 11:30, the receiver outlet temperature had been stabilized to approximately 525 deg.C.

The behaviour of the major parameters of the plant, that affected the operation of the receiver during this time period, are shown in Fig.5.8-1 and 2 More detailed plots of the temperature and flowrate transients are shown in Fig.5.8-3 and 4.

4.0 TRANSIENT CHARACTERISTICS

The receiver outlet temperature has been measured from the tube bundle thermocouple LK01CT09. Other thermocouple temperatures were also available on the class summary tape, but it was LK01CT09 which responded marginally faster and also caused the receiver trip.

The following tables have been included for each transient to show the maximum and minimum values obtained and also the rise/fall time measured for a 10%-90% change.

Table 5.8.-1: Grid failure transient response data

Transient	maximum	minimum	falltime (90%-10%)
Outlet temp.°C	~525	~260	5.4 min.
Flowrate m3/h	~16.5	~5	1.0 min.

Table 5.8.-2: Automatic cold startup transient response data

Transient	minimum	maximum	risetime (10%-90%)
Outlet temp.°C	~260	~566	5.3 min.
Flowrate m3/h	~5	~17.8	8.4 min.

During the automatic cold startup, the response of the flowrate started to become unstable. It was probably this instability which caused the dramatic increase in receiver outlet temperature and resulting receiver trip.

Table 5.8.-3: Receiver trip transient response data

Transient	maximum	minimum	falltime (90%-10%)
Outlet temp.°C	~566	~392	.5 min.
Flowrate m3/h	~19.7	~4.8	.5 min.

Table 5.8.-4: Manual startup transient response data			
Transient	minimum	maximum	risetime (10%-90%)
Outlet temp.°C	~392	~525	0.5 min.
Flowrate m3/h	~4.8	~28	1.2 min.

5.0 DISCUSSION

Although the 6th April 1983 was the only day during the period March - April 1983 for which transient data has been obtained, the transients which occurred do represent a situation which often happens to the CRS plant, i.e., grid failure. This day was also highlighted by relatively constant and high insolation levels and full plant operation, apart from a PCS trip.

The grid failure resulted in an automatic stowing of the heliostats, followed by an automatic reduction in flowrate. From Fig.5.8-3,it does not appear that there was any instability in the automatic flowrate control. According to a Plant Operating Authority, on this particular day, it was then decided to try and test the automatic receiver control system for a cold startup, since the receiver outlet temperature had cooled to its "cold" value. With the receiver control system in operation, the heliostats were brought back into track. However, either the instability in the flowrate (see Fig.5.8-4)or its relatively low value resulted in an inability of the system to handle the available power from the 82 heliostats at insolation levels exceeding 800 w/m**2. As a result, the outlet temperature limit of the thermocouple LK01CT09 was exceeded, which caused a receiver trip and immediate defocussing of the heliostats.

By manual operation of the flowrate it was then possible to continue with a fast restart of the plant with all heliostats in track.

6.0 CONCLUSIONS

On the basis of this occurrence, it would appear that receiver control system was, at this point in time, in need of improvement.

The transients investigated here show the importance of the man-in-the-link, although the redundancy of the system safety measures would have been sufficient to prevent any damage to the plant.

7.0 ACKNOWLEDGEMENTS

The author would like to thank both Mats Anderson and Heinz Jacobs for the invaluable help which they have given him in an attempt to read the tapes and also to Paco Ruiz, Paco Blanco and Ricardo Carmona for their assistance in deciphering the logbook to find the cause of the transients.

8.0 REFERENCES

(1) Neil Gregory, "Sulzer receiver - transient response data during the period March - April 1983", R-46/84 NG, October 1984

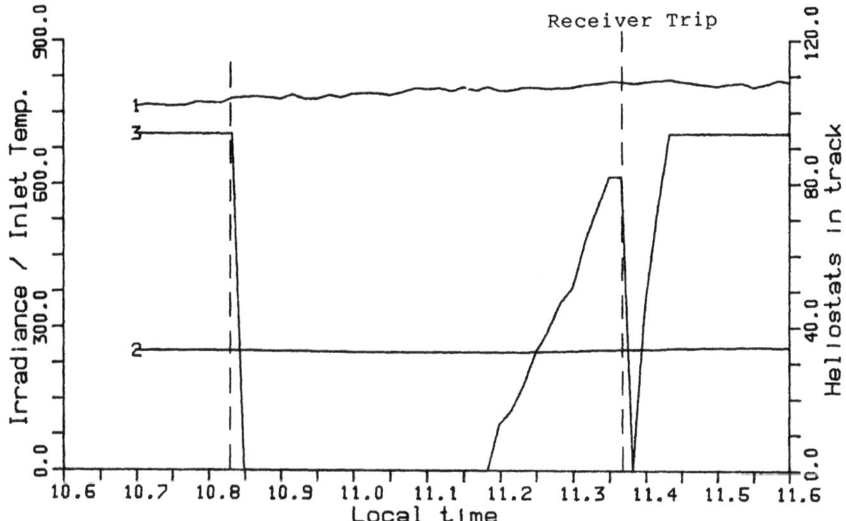

Fig.5.8-1: Mayor boundary conditions at the time of the transient

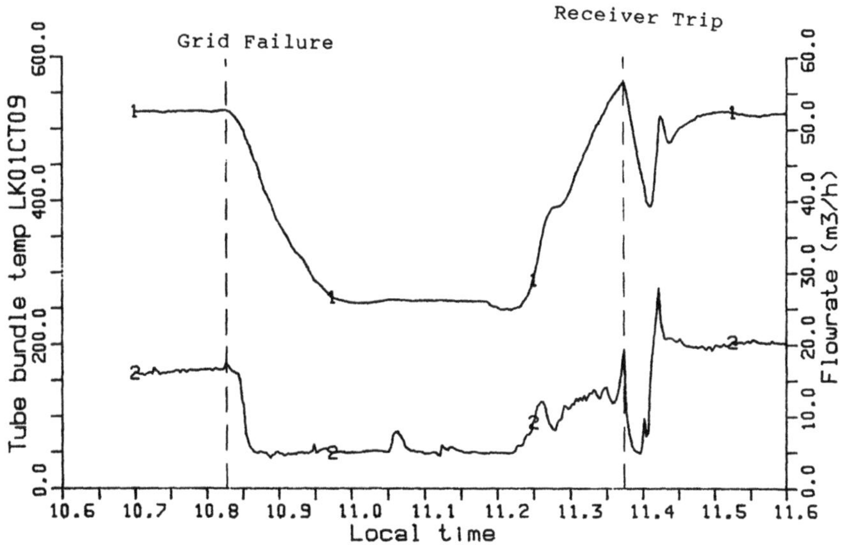

Fig.5.8-2: Transient response of flowrate and receiver outlet temperatures

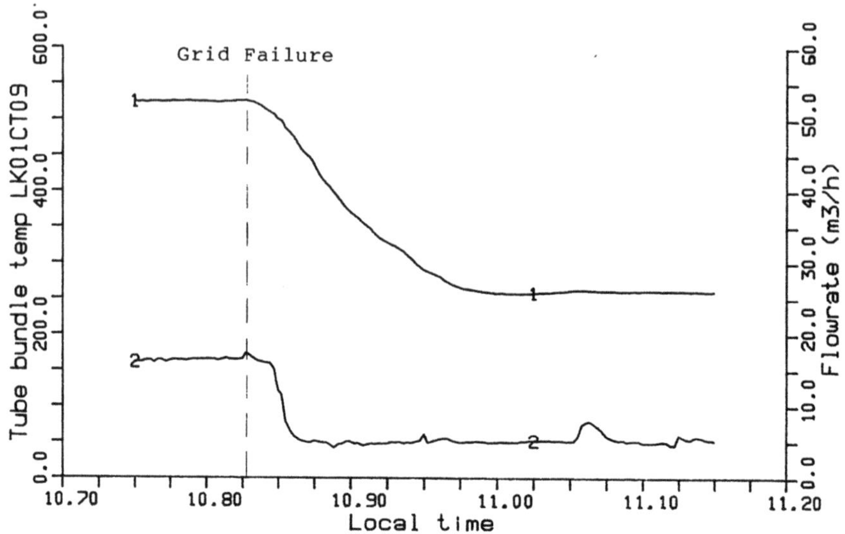

Fig.5.8-3: Detailed transient response to a grid failure

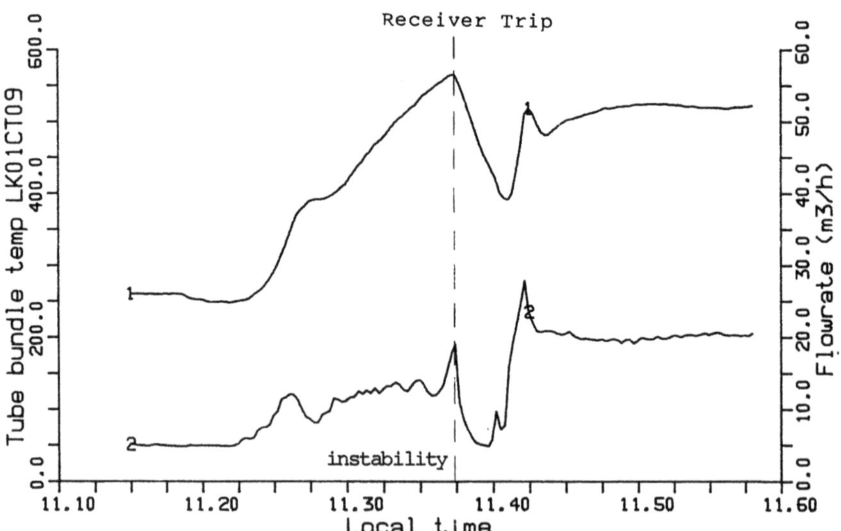

Fig.5.8-4: Detailed transient response to a receiver trip

THE ADVANCED SODIUM RECEIVER TRANSIENT RESPONSE

Ricardo Carmona and José G. Martín, ITET

1. INTRODUCTION

High efficiencies for solar energy collection ($\geqslant$ 90%) are possible with high flux, sodium-cooled receivers (Ref.1): these have low thermal inertia and short response times. Controls must keep the outlet temperature, T, constant at changing power levels. This 'quasi-stationary' state is achieved by manipulating the coolant flow rate, W. This state must be reached quickly and maintained, while keeping the temperature gradients in the receiver within tolerable limits. The Advanced Sodium Receiver (ASR) and its controls are designed to accomplish this; it is of interest to determine how close performance approaches expectations.

Variations in either W or the heat flux reaching the receiver, q, will affect T. The ASR transient response has been tested by making changes in the sodium flow rate and the heat flux reaching the receiver at different loads. The response may be roughly characterized in terms of ratios between variations in T and those in W and q, and the corresponding time constants. The results of tests performed on the ASR are evaluated and discussed in terms of these parameters. As will be shown, these results validate a dynamic computer model developed by ENEL et al. (2).

Dynamic system analysis and stability studies are facilitated if the receiver is represented in terms of a transfer function. One is suggested here, based on the actual response and the receiver configuration. The temporal response calculated from this function represents the data accurately.

Finally, this paper will present some considerations on the time required for warm-up and observations about the performance of the receiver loop (i.e., the system composed of receiver, pump, and controls) when clouds pass over the heliostat field.

2. GENERAL CHARACTERISTICS OF THE ASR RECEIVER

The ASR is a sodium-cooled receiver, 2.75 m wide by 2.85 m high, which may be considered as a representative module for a larger receiver. It is composed of five panels, all on the same plane(Fig.5.9-1).In each panel there are 39 cylindrical tubes, placed vertically and in parallel.The tubes are 2.65 m long; their outer diameter is 14 mm, and the walls are 1 mm thick. The nominal outlet temperature for the receiver is 530°C.

Sodium flows into headers above each panel and is piped along a single tube to the next panel. The thin pipes permit a large heat flux: the receiver is designed for a peak flux of 1.3 MW/m^2 under normal operation.

To flatten the flux profile, the heliostats are aimed at three points. A typical flux plot is shown in Fig.5.9-2. For operation at rated power (2840 MWth), nominal T, and an inlet temperature of 270°C, sodium flows at the rate of about 32 m^3/hr; its residence time is about 54 seconds.

3. RECEIVER TRANSIENT RESPONSE

Transient tests have been performed on the ASR. The effect on T of a step change in the flowrate W is shown in Fig.5.9-3. The insolation at the time of this test ranged between 960 and 965 W/m^2. Before the test was made, T had remained within 1° of 490°C for more than 5 minutes, and W was 5.25 ± 0.02 Kg/sec (equivalent to 21.5 ± 0.1 m^3/hr). By operating the pump under manual control, the flow was lowered to 4.82 Kg/sec (19.7 m^3/hr) in about 2 seconds, the pump response time. T rose until it reached 513°C.

As recorded by the Data Acquisition System (DAS), T has rapid fluctuations due to noise in transmission and discrete effects of the analog-to-digital convertor. To smooth out the noise, the data were passed through a filter of the second order; the figure shows the output data of the filter.

The response may be characterized in terms of the ratio between variation in T and variations in W. In general this gain has a steady-state component, G_W, which does not vary with the period of the signal, and a dynamic gain which does. For a system where the output variable, T, satisfies a first-order differential equation in time, t, a step change, ΔW, in the manipulated variable causes the change in the output, ΔT, to obey an exponential function. The receiver response permits a rough characterization in terms of G_W and the time constant, τ_W. A study by the manufacturer gives the theoretical justification for this (2). The data in Fig.5.9-3 may be characterized by G_W = 53°C/Kg sec and a time constant, τ_W = 57 seconds.

The next test consisted of raising the flow. As shown by the results in Fig.5.9-3, the gain and time constant are the same as before. The receiver is not a linear system: gains and times depend on the flow. Similar tests were carried out at high (8.06 Kg/sec) and low (2.3 Kg/sec) flow rates; the time constants, residence and dead times, and gains are listed in Tab.5.9-1. The data are given in one-second intervals: the times are approximated to the next closest digit.

The next set of tests, results of which are plotted in Fig.5.9-4, consisted of changing the heat flux at the receiver by sending heliostats out and into track. The response to these changes may be represented in terms of a constant gain, G_q, and a time constant, τ_g. At high flow (W = 8.06 Kg/sec), the flux step down consisted in sending 6 heliostats (out of 90) out of track when the insolation was 960 W/m^2. This is a step of 165 kW (down from 2645 kW, close to design conditions).

Similar tests (Fig.5.9-5) were performed at medium flows with 6 heliostats sent out of track (out of a total of 61), and at low flow with 3 heliostats sent out of track (out of 29). The gains are listed in Tab.5.9-1. The time constants were found to be the same as those for the flow changes. In general, τ_w should be smaller than τ_q because of the delay induced by the transfer of heat across the tube walls. Here, the difference is not noticeable because the heat transfer is fast.

Gains and time constants vary approximately inversely with the flow rate. They can be approximated by: τ = 250/W seconds, G_W = 218/W °/Kg sec, and G_q = 0.73/W °C/kw, with less than 6% error when W is given in Kg/sec. The difference between G_W and G_q is due to the different units used; equal percent variations in input give rise to approximately equal percent variations T.

There are thermocouples at each header, so it is possible to compute gains and time constants for each panel for all tests. As an example, the time constant is about 5.5 seconds for all of the panels at high flow. From Panel 1 to Panel 5 the gains are 2, 1, 7.3, 7, and 9.4 °/Kg sec, respectively. The panels are connected in series; the temperature increase and gain are proportional to the total energy absorbed. Because of the flux profile, the central panel (#5) has the highest gain.

4. COMPARISON WITH SIMULATED RESULTS

ENEL has developed a dynamic simulation model that is applicable to the ASR (2), and this program has been used to simulate the flow tests. The model considers the effect of a constant wind on convection losses, but not gusts or effects on heliostat tracking. Fig.5.9-4/5 show the simulated and actual response. During the test, the wind ranged between 20 and 25 km/hr, and there were small variations in the insolation (about 5 W/m^2). The actual response curve is not as smooth as the simulation.

The actual flux shape also introduces errors. We used the standard design profile: work is in progress to use actual flux distribution data from the Heat Flux Distribution (HFD) sensing bar. Aside from these considerations, the fit between actual and simulated responses does validate the model.

5. RECEIVER TRANSFER FUNCTION

It is of interest to represent the receiver response in terms of a Laplace transfer function. To facilitate the following discussion, the simulated results will be used as reference. The receiver cannot be represented in terms of linear differential equations over the whole range of operation. It is necessary to make linear approximations about equilibrium points.

Considerations of the receiver geometry and the receiver response itself suggest a general form for the transfer function. Most of the flux reaching the receiver is absorbed by the three central panels (there is a delay between the time sodium leaves one of these panels and reaches the outlet thermocouple).

Fig.5.9-6a shows the response when the flow is changed from 5.42 Kg/sec to 4.78 Kg/sec in one step (a second ramp). The shape of the curve suggests that it may be approximated by three main contributions, with different time delays (Fig.5.9-6b). The first is an 'S' shape which appears about at t ~ 4 sec. This 'S' shape is of the type which characterize a second-order system. The first contribution comes from the panel which is closest to the outlet (i.e., Panel 5). The second contribution appears at t ~ 21 sec. This is the contribution from the next to the last panel (i.e., Panel 4): sodium leaving Panel 4 reaches the sensor point at the outlet in approximately 21 sec. For simplicity, we take it to be a contribution from a first-order system. Finally, about 43 sec after the perturbation (the time for the sodium leaving Panel 3 to reach the outlet), there is another component, also approximated as the response from a first-order system. This component is related to the contribution from the first three panels.

Taking the three contributions together, the overall transfer function may then be represented as

$$\frac{\Delta T(s)}{\Delta W(s)} = \left(\frac{b_1 s + b_2}{s^2 + a_1 s + a_2} e^{-t_{d_1} s} + \frac{b_3}{s + a_3} e^{-t_{d_2} s} + \frac{b_4}{s + a_4} e^{-t_{d_3} s}\right) \frac{0.5}{s + 0.5}$$

This function has three time delays and eight coefficients. The factor 0.5/(5+0.5) is the response of a thermocouple with a time constant of 2 sec--the one at the outlet. The lags are inversely proportional to the flow rate; they may be approximated by td_1 = 16/W, td_2 = 100/W, and td_3 = 210/W.

The system is nonlinear and the coefficients must be identified at the different flows. They were identified by using the Powell algorithm (3), minimizing the sum of the mean square errors with a program developed at the University of Seville. The coefficients are listed in Tab.5.9-2.

From these coefficients, a general simulation program also developed in Seville has been used to obtain the temporal response in time domain for low, medium, and high flows. A response is shown by the dotted line in Fig.5.9-6,while the solid line represents the results from the ENEL model.

6. CONTROLS

The control system for the receiver must manipulate the flow so as to maintain T at some reference value, now chosen to be 530°C. Fig.5.9-7 shows the block diagram for the controls. The inlet temperature may vary; however, the cold storage limits these variations, which are slow compared with the receiver response. The present ASR control system assumes that the inlet temperature is constant. Small oscillations in T are eliminated in the hot storage tank. The most important variations are those in the absorbed heat flux. These may be large amplitude variations with high frequency components. For example, after a cloud passes the heat reaching the receiver rises quickly. The controls must avoid high temperature peaks after such passages.

The control system consists of feedforward and feedback loops. The inner loop manipulates the flow. This loop must be fast enough so that it does not interact with the outer ones. To speed up the control response, the intermediate loop detects the variations in the temperature differences across all panels. The outer loop, where the response is slower than the intermediate one, maintains T constant.

The flow rate ranges from 8 to 40 m^3/hr; the regulator parameters must vary to maintain stability over the operating range. The feedforward action depends on the heat flux: it limits temperature peaks when the flux rises sharply. This action is more important at low flow rates, where response times are longer. Feedforward is imprecise because heat flux is not estimated accurately. The average of the output from ten photovoltaic sensors, mounted on heliostats in the field is the signal used. There are several problems with this arrangement. First, the cells are mounted horizontally, and therefore the gain is not constant, but depends on sun angle. Second, they measure global insolation; thus a cloud may actually increase the horizontal insolation. Lastly, shading of the sensors, particularly in winter, causes the inaccuracies.

Because of the imprecise estimation of the solar flux, the action is the relevant one: its main purpose is to protect the receiver from overheating when the insolation rises.

The response of the control system in the automatic mode to changes in the power incident on the receiver is illustrated in Fig.5.9-8. The number of heliostats in track is 86 before the test; the insolation is 965 W/m^2. The test consists of taking nine or more heliostats out of track and returning them to track after eight minutes. (These changes are almost instantaneous, i.e., they are much faster than can be expected from a cloud, although the relative insolation change is not as large. The purpose of this test is to verify the effectiveness of the cloud loop (i.e., the feedback action), because there is no change in the actual insolation and therefore no change in the feedforward signal).

The outlet temperature is plotted in (b) and the flowrate in (c). When the heliostats are taken out of track, the temperature drops about 10°C in about 30 seconds, but this drop is arrested by the drop in the flowrate initiated by the feedback signal which started as soon as the temperature started to drop, and the temperature rises towards the normal operating value. These are oscillations which are very apparent in the graph because of its large scale. When the heliostats are placed in track, the temperature rises to about 4°C above the set point before going down again. The test, therefore, verifies that the feedback action is adequate.

Warm-Up

In the topic reports by the contracting companies, 'warm start-up' is the operation by which the receiver is heated from about 270 to 530°C. The low temperature is, of course, the design inlet temperature for the receiver. 'Cold start-up' is the operation by which the receiver is filled. Cold start-up requires careful attention -- at the moment, it is certainly not performed routinely. Presently, when the receiver is warmed up at high insolation (for example, when a day 'clears'), the operation is performed in about ten minutes.

Because of its short time constant, the receiver could be heated from 270 to 530°C in less than three minutes at noon. For early morning warm start-up, when the heat flux and the sodium flow rates are smaller, the time required would be several times longer.

Based on the calculations on the stresses induced on the structure by rapid temperature changes, the designers recommend that the temperature variation should not exceed 8°/sec anywhere, which corresponds to a warming period of less than three minutes.

Response time and structural limitations are not the only factors which limit the start-up time. It is also affected by the amount of power incident on the receiver, which depends on the heliostat field accuracy and the receiver size and spillage limitations.

If all the present heliostat field is focused in the early morning as soon as the insolation is above 300 W/m^2, the lateral protection of the relatively small receiver overheats. It overheats because of large heliostat image sizes and tracking accuracy. Because the walls overheat, not all of the heliostats are put into track. Therefore less power reaches the receiver, and the operating outlet temperature cannot be reached at the minimum flow.

If a quicker actual start-up is desired, there are several solutions. The lateral walls can be replaced by others which can tolerate the higher incident power. Tracking could be improved -- the field is not tracking as accurately as it should. Some measure could be taken to improve heliostat images in the morning -- particularly that of the east side heliostats. Finally, the minimum flow rate could be lowered.

The receiver was designed for a minimum flowrate as low as 1/10 of the nominal value (i.e., 3 m^3/hr), but the present pumping system does not permit operation below 6 m^3/hr. During the acceptance tests, the control system designer noted that at lower rates the flow oscillated. This oscillation was related to the slow response of the pump controls: there was no problem with the first receiver, a high thermal inertia cavity type. However, the pump controls had to be modified to increase the response for the ASR. The limitation on the minimum flow raises the minimum irradiance at which the receiver operates.

Because the operating outlet temperature cannot presently be reached quickly in the morning, a common operating strategy is to continue circulating heated sodium back to the cold tank. As the sodium is circulated overnight, the cold tank cools down and the present early morning operating strategy is to heat the receiver up to the design temperature of 270°C. This strategy is sensible in that energy is collected early. However, the period between when the heliostat field is first put in track and when the operating temperature is reached in the receiver is no real indication of the warm-up time required for this receiver.

Cloud Passages

Outlet temperature and flow rate during a cloud passage around noon are plotted in Fig.5.9-9. Two clouds of short duration (about 50 seconds each) make the insolation drop to about 400 W/m^2 (the irradiance plot represents the reading from a pyrheliometer installed atop the main SSPS building). The flow drops to about $8m^3/hr$, and there is a slight decrease in T of about 10°K. Note that the flow drop ('effect') seems to anticipate the irradiation drop ('cause'). The feedforward action is initiated by sensors installed in the field rather than by the pyrheliometer, and irradiance does not fall simultaneously at both locations.

The passage of a cloud of longer duration (about 100 seconds) is illustrated in Fig.5.9-10.The insolation drops to about 200 W/m^2: W drops to its minimum value (6 m^3/hr) and remains there for about 50 seconds. When the cloud passes, the feedforward signal makes the pump operate at maximum power (maximum W is about 40 m^3/hr). T begins to rise, but the induced flow surge stops and reverses the rise: T drops to about 420°C before it rises again. Thus, the control system does keep the receiver from overheating in spite of large and fast insolation rises. However, the outlet temperature drops too much. This undesirable effect could be eliminated through a better evaluation of the heat flux reaching the receiver.

7. CONCLUSIONS

- The compact receiver responds quickly to variation in the heat flux. The controls successfully protect the receiver from overheating and from large temperature gradients.

- The dynamic simulation model developed by the manufacturer has been validated by the actual receiver response.

- A transfer function has been suggested; this function represents the receiver response accurately. It is now possible to use this function to predict what T will be for arbitrary changes in W and in flux. It is also possible to study the stability of the receiver loop.

- A more accurate estimate of the heat flux at the receiver is desirable in order to improve the response.

  The time required to warm the receiver early in the morning is not limited by the receiver response time or structural integrity. The field tracking accuracy and image quality are not appropriate to the receiver dimension and lateral protection and therefore the whole field cannot be put into track early. This, and the limitation imposed by a relatively high minimum flow, delays the time at which the operating temperature can be reached in the receiver.

- At high insolation, the POA now heats the receiver in about ten minutes. Because of its short response time, the receiver could be heated in less than three minutes, and it appears that this fast warm-up can be accomplished without exceeding maximum design temperature variations.

- The 'safety margin' between the actual and possible warm-up times costs about 150 kWh of thermal energy per start-up.

8. ACKNOWLEDGMENT

The simulation program was implemented on the SSPS computer by J. Magnani. The discussion on the actual time required for start-up was helped by lengthy discussions with the POA staff. To identify the coefficients for the transfer function, the authors relied on support from Eduardo Fernández Camacho and Francisco Rodriguez Rubio of the Engineering School of the University of Seville. The authors are grateful for this help.

9. REFERENCES

1. JACOBS, H. (1984). Performance of the Sulzer Cavity Receiver and the Franco-Tosi External Receiver....Paper presented at this workshop.
2. AGIP NUCLEARE and FRANCO-TOSI INDUSTRIALE, (1983). The Advanced Sodium Receiver (ASR). SSPS Technical Report Nº 3/83. Topic Report Nº 3.
3. POWELL, H. B. (1964). An Efficient Method for Finding the Minimum of a Function of Several Variables...... Computer Journal, 7. p.155.

FLOW W, (Kg/s)	TIME CONSTANT (sec)	RESIDENCE TIME (sec)	DEAD TIME td (sec)	GAIN G_w, (°C/Kg/s)	GAIN G_q, °C/kW
HIGH (8.1)	31	50	2	29	0.10
MEDIUM (5.1)	48	80	3	40	0.14
LOW (2.3)	115	172	7	90	0.28
FLOW DEPENDENCE	250/W	400/W	16/W	218/W	0.73/W

Table 5.9.-1: Response for Different Flow Rates

	a1	a2	a3	a5	b1	b2	b3	b5
HIGH	1.48	1.33	0.12	0.037	-0.735	-11.57	-1.11	-0.42
MEDIUM	0.82	0.30	0.062	0.026	-0.744	- 3.98	-0.98	-0.38
LOW	0.47	0.07	0.032	0.013	-0.733	- 2.13	-1.04	-0.32

Table 5.9.-2: Coefficient Identification

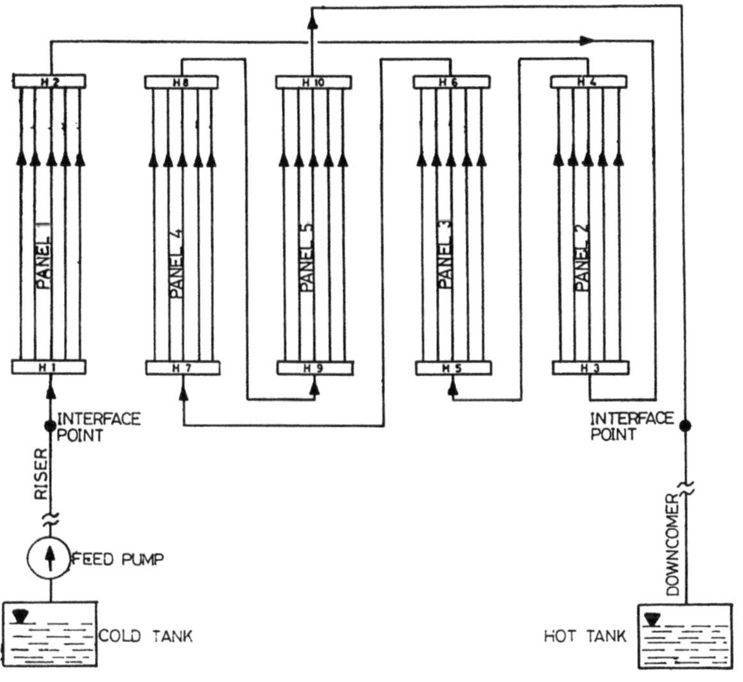

Fig. 5.9.-1: ASR Flow Diagram (Ref.2)

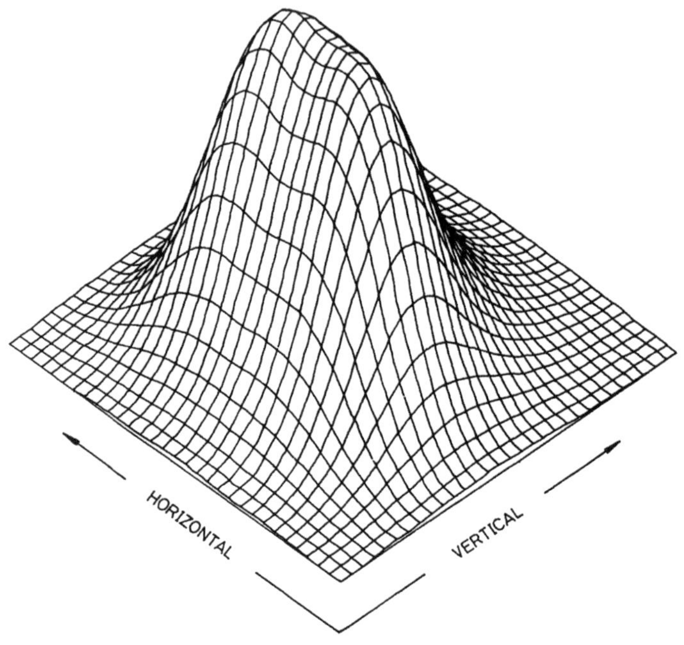

Fig. 5.9.-2: Typical Flux Plot at the Receiver Plane (Ref.2)

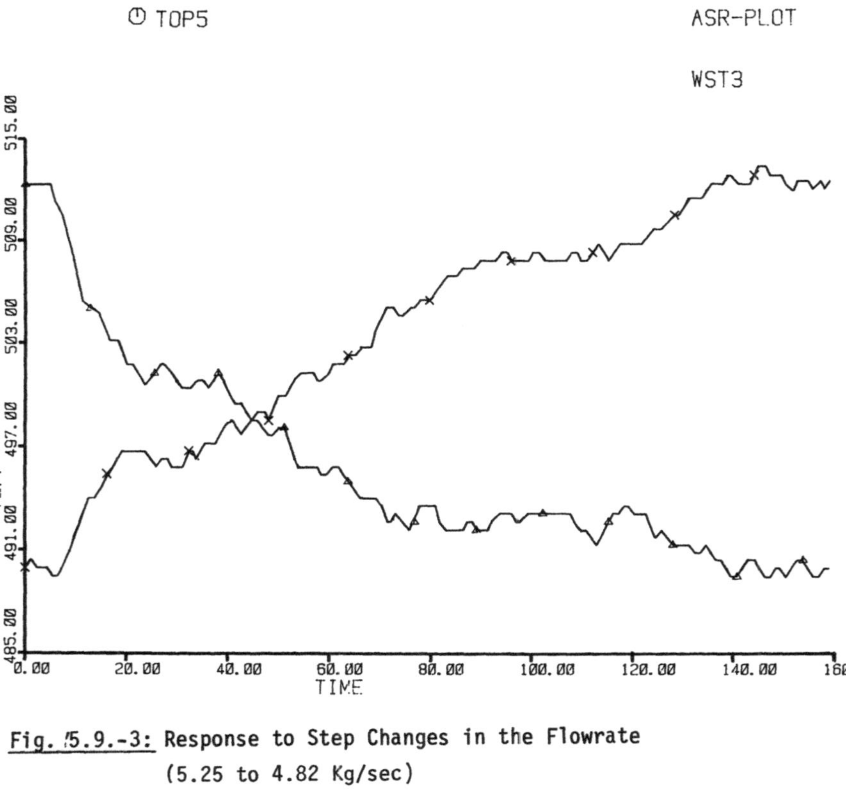

<u>Fig. 5.9.-3:</u> **Response to Step Changes in the Flowrate (5.25 to 4.82 Kg/sec)**

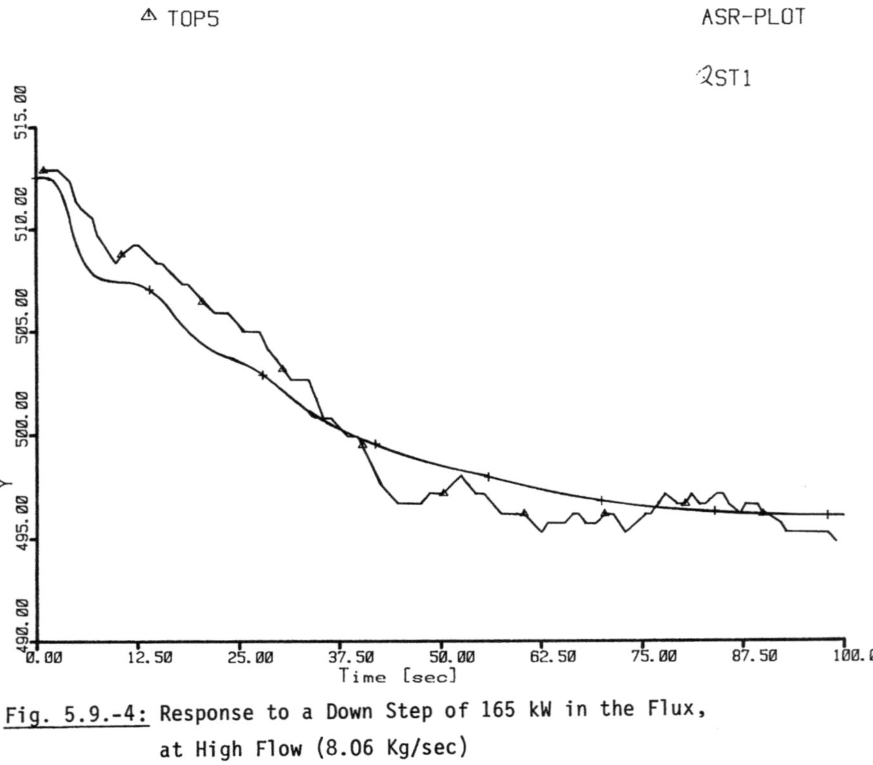

Fig. 5.9.-4: Response to a Down Step of 165 kW in the Flux, at High Flow (8.06 Kg/sec)

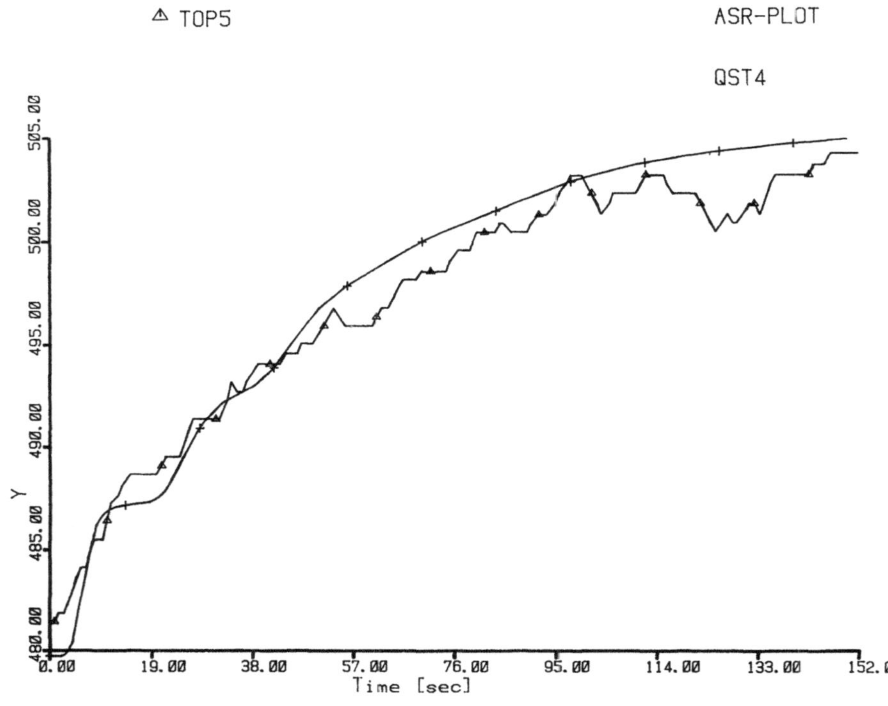

Fig. 5.9.-5: Response to an Upward Step of 165 kW in the Flux, at Medium Flow

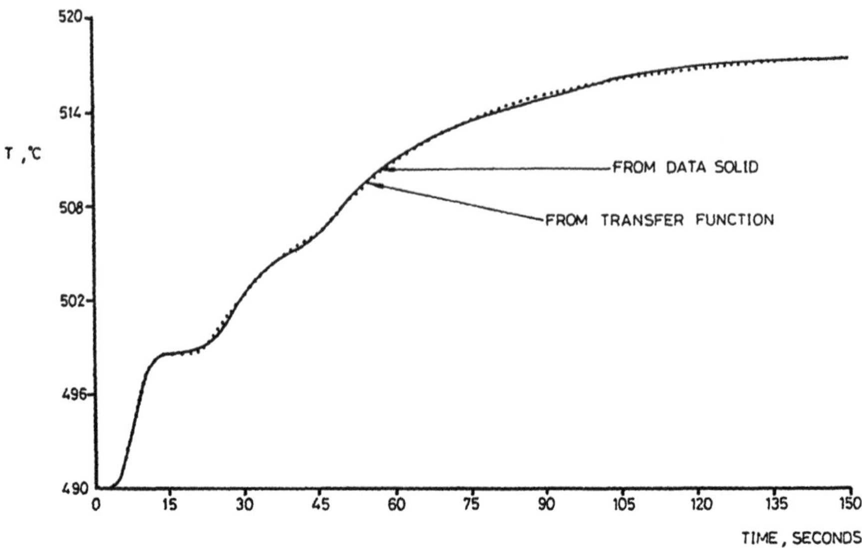

Fig. 5.9.-6a: Response

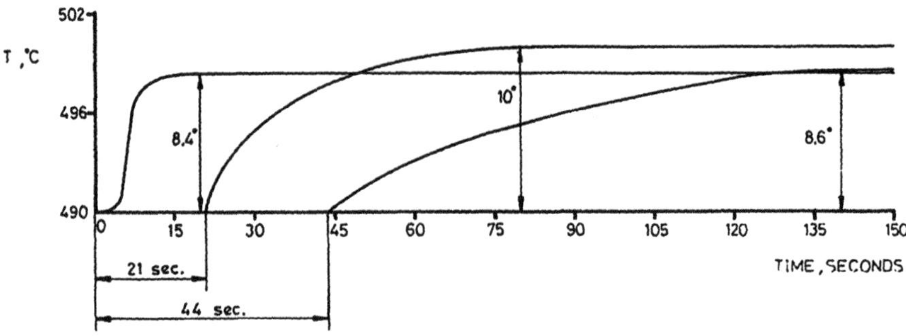

Fig. 5.9.-6b: Contributions to Response

Fig. 5.9.-6: Effect of Change from 5.42 to 4.78 Kg/sec

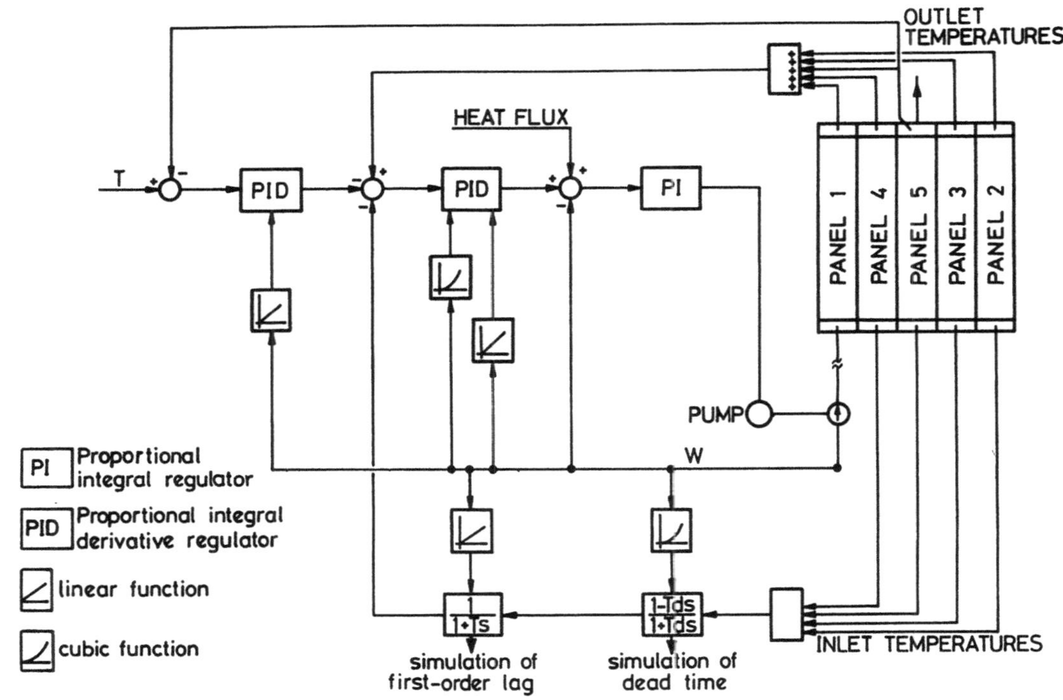

Fig. 5.9.-7: Control System (Ref.)

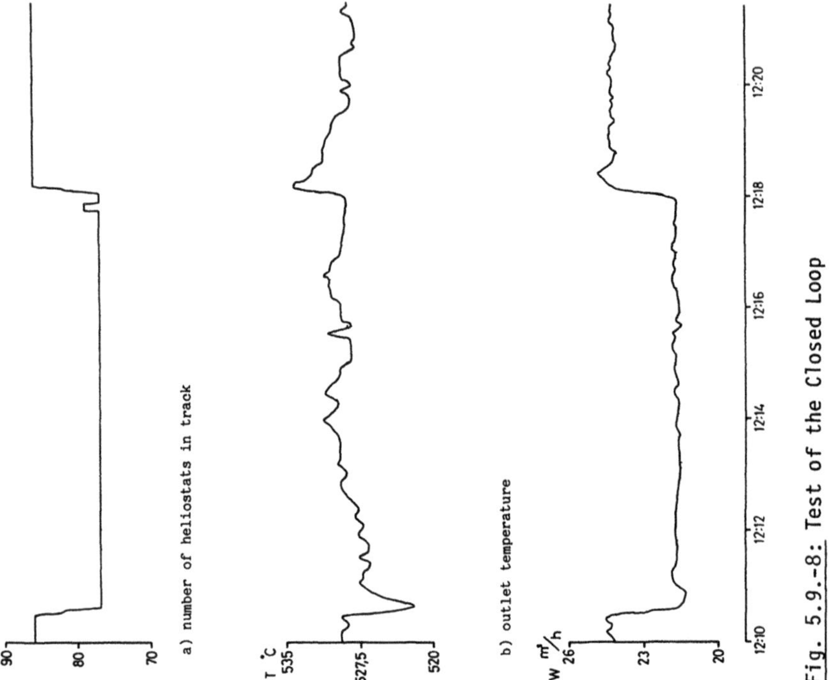

Fig. 5.9.-8: Test of the Closed Loop

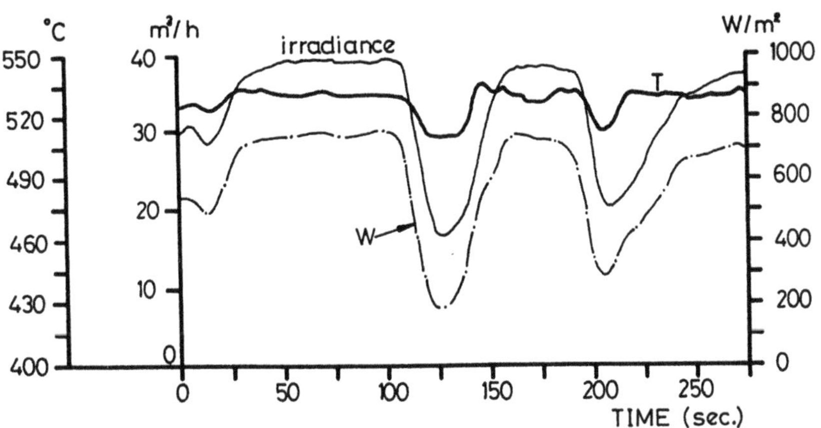

Fig. 5.9.-9: Effect of Passage of Clouds of Short Duration

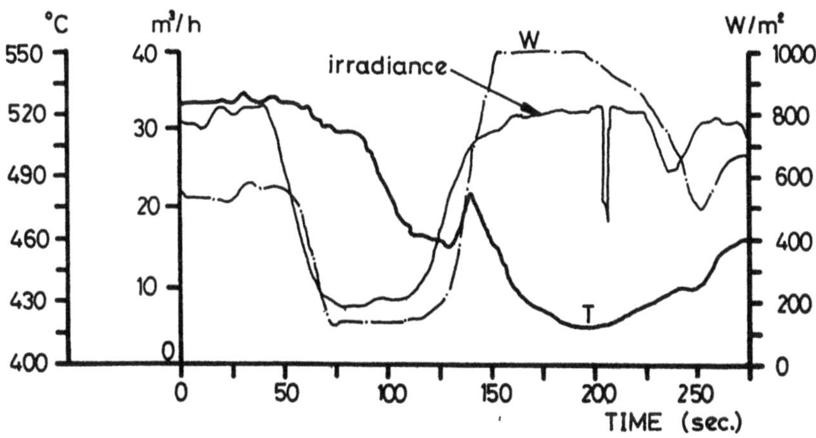

Fig. 5.9.-10: Effect of Passage of Clouds of Long Duration

DIFFERENCES BETWEEN FILLING STRATEGIES FOR THE SULZER AND ASR RECEIVERS

Francisco Ruiz and Antonio Cuadrado, Sevillana

1. INTRODUCTION

One point to be considered in the comparison of two types of receivers is the ease of operation and the impact of the operational modes on the overall behavior of the system. Though little difference exists between the ASR and Sulzer receivers in most operation modes, there is a point where differences are relevant, and a change in the operation strategy is needed.

These operational differences are mainly due to the type of receiver (external in the case of the ASR and cavity in the case of the Sulzer) as well as the position of the tubes in each receiver (five vertical panels through which sodium circulates up and down in the ASR, and one horizontal tube bundle in the Sulzer receiver). As a result of this variation, the filling of the Sulzer receiver is a fast and routine procedure (approximately one hour), easily repeated every morning so that draining the tube bundle at night is feasible; the ASR filling is a complicated and slow operation (approximately eight hours), which makes any periodic draining and filling unfeasible.

2. FILLING PROCEDURE

2.1 Sulzer Receiver

After preheating the receiver by focusing some heliostats on the aperture's center, the heliostats are defocused, and the receiver doors are closed to achieve a uniform inside temperature. Once this has been obtained, the sodium ascent begins through the normal inlet tube. The sodium is moved by the pressure difference (3/4.5 bar) previously established between the cold tank and regeneration tank (the latter communicates with the upper side through the venting line), and as well as the cold sodium pump which works during all the process. When the sodium begins to exit, the venting line is closed, and sodium is sent to the tanks (cold and hot). This marks the end of the filling procedure.

2.2 ASR

For a detailed explanation of the procedure, including the problems encountered during the last filling, see Appendix 1. Generally speaking, the filling process is as follows.

The receiver is preheated by focusing a series of heliostats on distinct parts of the aperture (Appendix 1; figure A) as well as using trace-heating of the non-irradiated parts. Upon reaching a temperature as homogeneous as possible in the panels and back ceramic wall, filling begins without defocusing the heliostats. Panel #1 is filled through the inlet pipe, while the rest of the panels is filled using the draining tubes which communicate with the outlet tube of the receiver.

The filling is carried out using only pressure differences in the tanks, and without the help of the cold pump as in the case of the Sulzer receiver.

Synchronization is an essential part of the filling process to avoid temperature differences among the panels and the consequent thermal shock.

Once the receiver is filled, the procedure is similar to the one followed in the case of Sulzer. The required time for this process is approximately three hours.

3. DIFFERENCES BETWEEN THE TWO FILLING PROCEDURES

3.1 Preheating

Sulzer

The preheating of the Sulzer receiver is carried out with a group of heliostats focusing on the aperture center to achieve a homogeneous temperature in the entire tube bundle. This procedure reduces the possibility of thermal shock during filling. Furthermore, the large thermal inertia of the receiver reduces the cooling rate, allows the filling procedure to take place when the heliostats are out of focus and the doors are closed.

ASR

During filling, some heliostats have to be refocused to obtain uniform energy and temperature distributions. In spite of these efforts, temperature differences of more than 100°C do exist, which causes significant thermal shocks during filling.

Tests have shown that defocusing the heliostats and closing the doors does not equalize the temperature, as in the case of the Sulzer receiver; these temperatures decrease very quickly, such that there is not enough time to perform the filling. Therefore, the filling procedure must be carried out with the heliostats in track, which introduces the risk of emergency defocusing during the procedure. In addition, stable sunshine is required during this process.

3.2 Filling

Sulzer

The filling is carried out quickly through only one tube and with the help of the cold pump. This decreases the risk of any possible plugs and the existience of argon bays. The procedure involves no risk and does not change the normal operation; therefore no special procedure needs to be established.

ASR

The filling must be done simultaneously via the inlet and drain tubes. There is a risk of filling one panel before the others due to the existence of cold points or valve blockages. This would imply that the sodium passes from one panel to another through the upper venting tubes, producing strong temperature differences among the tubes of the one panel, causing deformation. (This incident occurred during the second filling of the receiver.)

The filling synchronization, as well as the following safety venting, requires the individual operation of the venting valves. This point was not foreseen in the initial design -- it was implemented later.

4. CONCLUSIONS

In the Sulzer receiver, the filling process is quick, easy, and can be performed by one person in one hour. This implies that the receiver can be drained daily with no need to maintain the loop in operation during the night, weekend, or non-operational periods, resulting in a savings of personnel.

In the case of the ASR, the process is a delicate and very long operation. At least three people and a time period of three hours under stable sunshine conditions are necessary. This implies the impossibility of periodic draining; consequently it is necessary to maintain 24-hour operation, resulting in costly personnel shifts.

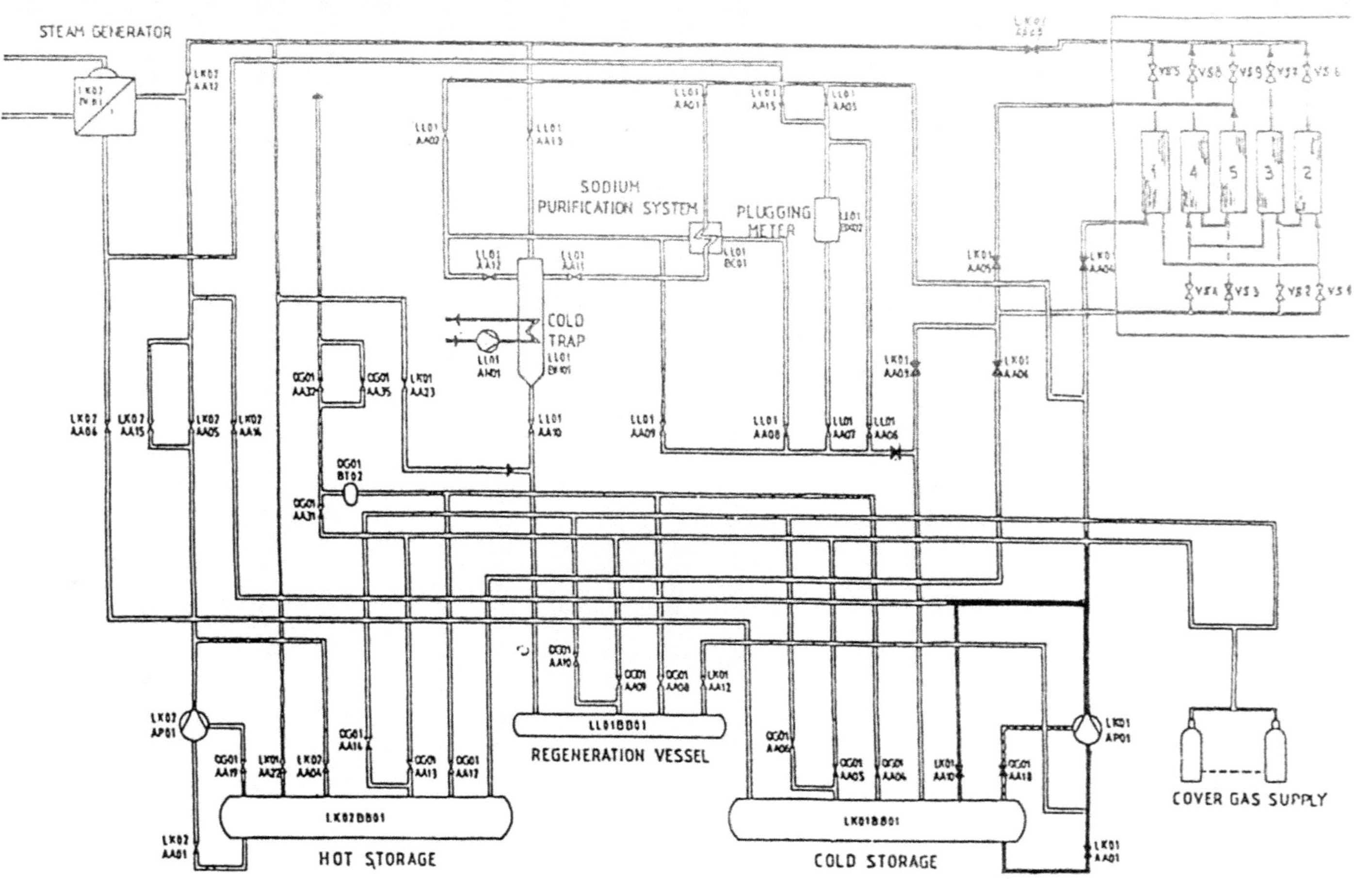

Fig. 5.10.-1: ASR Sodium Heat Transfer System

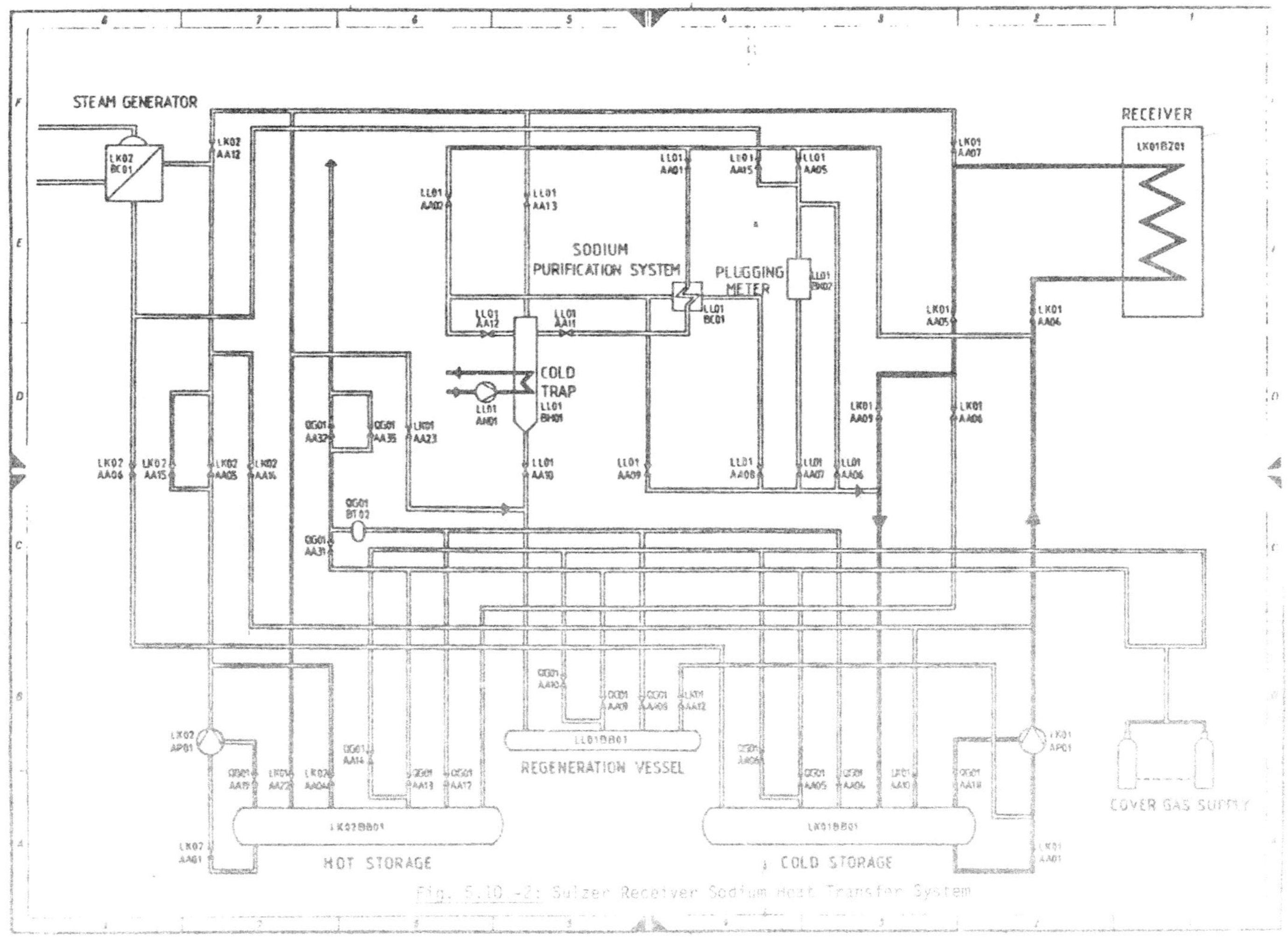

Fig. 5.10.-2: Sulzer Receiver Sodium Heat Transfer System

APPENDIX 1

ASR FILLING PROCESS

The last ASR filling was realized on April 18, 1984. The available data were analyzed in order to obtain experience and successful results concerning this delicate operation.

The plant status before the start of the filling process was as follows:
- sodium plugging temperature - 100°C (measured on April 10, 1984 before draining)
- sodium temperature in the cold tank - 242°C
- sodium level in the cold tank - 1.6 m
- argon gas pressure in the hot and cold tanks - 5.9 bars
- insolation - between 680 and 715 W/m^2 during the operation time
- all the temperatures of the trace-heating system were correct

At 12:37, the preheating of the receiver was started using heliostats with the focussing points modified beforehand so that all the surface was homogeneously covered.

At 15:10, the filling started; sodium was detected in the receiver at 15:21. Twenty minutes later the filling was performed. Figures A to J show the process in detail using as a reference the changes suffered by the temperatures of the different parts of the receiver when sodium arrived.

Main Problems

There were two main problems when filling the receiver. The first one is the low temperature observed in the lower headers of Panels #2, #3, #4, and #5, in spite of the trace-heating. At the start of the filling procedure, the inlet temperatures in the collectors were P1: 171°C; P4: 133°C; P5: 118°C; P3: 130°C; and P2: 124°C. The heating electrical consumption was measured and found to be correct. The reason for this could be the presence of water in the insulation due to rain the week before.

The second problem was detected later. The existence of some cold points in the filling tubes of Panels #2,...#5 (draining pipes) caused a strong temperature drop in the sodium going up. The sodium at 242°C into the tank went down to 144°C in some points of the panels, and to 170°C in the thermocouples of the draining pipes. This rapid temperature descent could cause thermal shock in the receiver.

Comments and Explanations of Figures

Figure A. The position of the preheating aiming points is chosen so that a homogeneous heating of all the surface is assured. Obviously, all the heliostats were not focussed at the same time, only those necessary to maintain a uniform temperature between 250°C and 300°C.

The lower table shows the coordinates X,Y,Z of the focussing point.

Figure B. Draining tubes. Thermocouples were installed in the draining tubes. These were connected provisionally to the microprocessor through the inputs corresponding to TC 38. 39., 40, and 41.

Consequent to the cooling of the sodium as mentioned before, the temperature of the thermocouples went down suddenly until 170°C was reached in the coldest one. The differences observed in the reaction time are due to the thermocouple being more or less separated from the draining collectors.

Different behavior is shown in the last figure representing the final collector of the first panel, where the sodium arrival temperature is adequate.

Figure C. Lower headers temperature. While the temperature increased suddenly in Panel #1 from 174 to 245°C, in the others this increase was more gradual.

The maximum daily referenced to Panel #1 is 28 minutes and it is shown on the furthest panel of the filling line, Panel #2.

Figure D. Down corners and down level of the center panel. Sodium reaches Panel #1 at 15:21:18, decreasing the temperature from 260 to 164°C. After ten to fifteen minutes, and after having heating all the headers connections, it goes up again to 250°C.

At 15:23:45, sodium reaches Panel #2, two hours and seventeen minutes after having detected sodium presence in the first panel.

Afterwards, a moderate increase is observed in Panel #5 as a result of the circulating sodium (60 cm) through the tubes; it goes from 300°C to 332°C, then later drops to 144°C.

During this process, there is a stop in the filling of nearly two minutes due to the irregularities abovementioned. This stop is clearly observed in the figure of Panel #1 where sodium temperature drops for a second time at 15:23:08.

Figure E. Central level (Panels #3, #4, #5). Considering that the thermal transient is different in each panel due to the different temperatures of the panels and the different radiation received by the sodium circulating through each tube, the important point is that in this level and at these panels , sodium reaches all nine thermal couples at the same time (15:23:48).

Figure F. Upper level. The thermocouples of Panels #2, #3, #4, and #5 clearly detected the presence of sodium at 15:23:56. The thermocouples of Panel #1 did not detect it so clearly because the sodium enters at a higher temperature: a significant change can hardly be appreciated. This happens at approximatley 15:23:50.

Figure G. Upper Headers. The first collector slightly detected the sodium at 15:24:20 -- twenty seconds before the others.

The changes are less evident in the lower levels due to the small difference between the sodium and collector temperatures.

Figure H. Filling process in Panel #5. No abnormalities were observed in the filling.

The decrease in the change of temperatures in the corresponding thermocouples is clearly noted as one goes up in the receiver.

Figure J. Filling times. Hours, minutes, and seconds in which the presence of sodium is detected for the first time in the different areas of the receiver.

Conclusions

Plugging temperature must be as low as possible and hot tank temperature as high as possible, so that if a cold point exists in the filling tubes the effects would be minimum.

It is not convenient to stop the process once the sodium is going up as this could cause larger cooling effects. The sodium flow prevents any plugging.

Once the sodium is detected in the lower part of the panels (not yet in the upper headers) it can be convenient to stop the process momentarily to equalize the sodium levels in the different panels.

The preheating of the lower collectors by the electrical tracing must be observed and the necessary solutions should be studied to increase the temperatures in the collectors.

The sodium cooling problem in its way up through the filling tube seems to be due to insufficient trace-heating in one or more sections of the pipe used for filling.

The filling procedure requires a total of three hours from the beginning of the preheating. This long period of time, the difficulties, and the personnel requirements, as well as thermal shock which is produced, prevent routine filling and draining procedure.

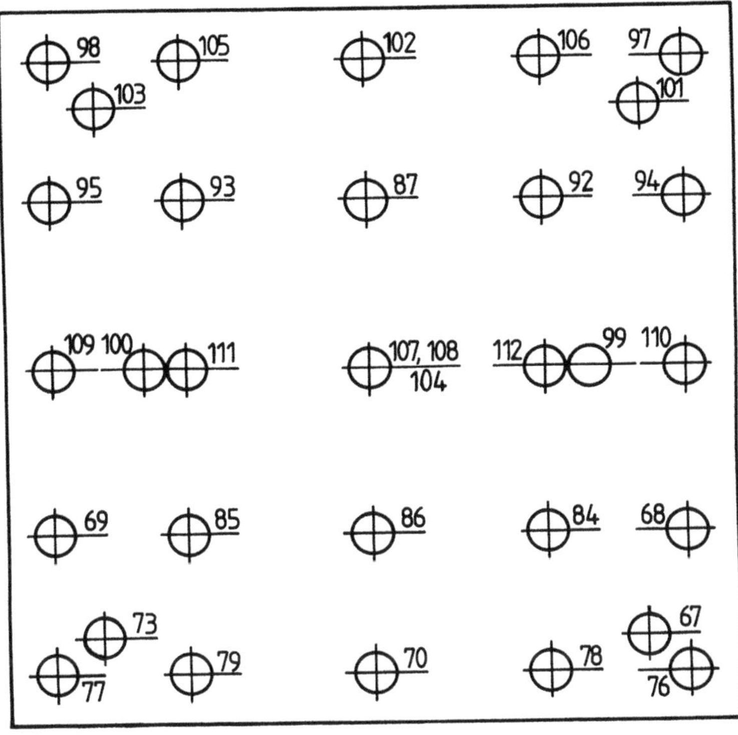

101.- -0.900, 0, 45.0
73.- 0.900, 0, 43.0
67.- -0.900, 0, 43.0
71.- -0.900, 0, 43.0
105.- 0.500, 0, 45.0
106.- -0.500, 0, 45.0
95.- 1.000, 0, 44.5
93.- 0.500, 0, 44.5
87.- 0,0, 44.5
92.- -0.500, 0, 44.5
94.- -1.000, 0, 44.5

109.- 1.000, 0, 44.0
111.- 0.500, 0, 44.0
112.- -0.500, 0, 44.0
110.- -1.000, 0, 44.0
69.- 1.000, 0, 43.5
85.- 0.500, 0, 43.5
86.- 0,0, 43.5
84.- -0.500, 0, 43.5
68.- -1.000, 0, 43.5
77.- 1.150, 0, 42.75
79.- 0.500, 0, 43.0

70.- 0,0, 42.750
78.- -0.500, 0, 43.0
76.- -1.150, 0, 42.750
97.- -1.150, 0, 45 250
98.- 1.150, 0, 45 250
100.- 0.900, 0, 44.000
99.- 0.900, 0, 44.000
107.- 0,0, 44.0
108.- 0,0, 44.0
104.- 0,0, 44.0
103.- 0.980, 0, 45.0

Fig. A: Preheating Aiming Points

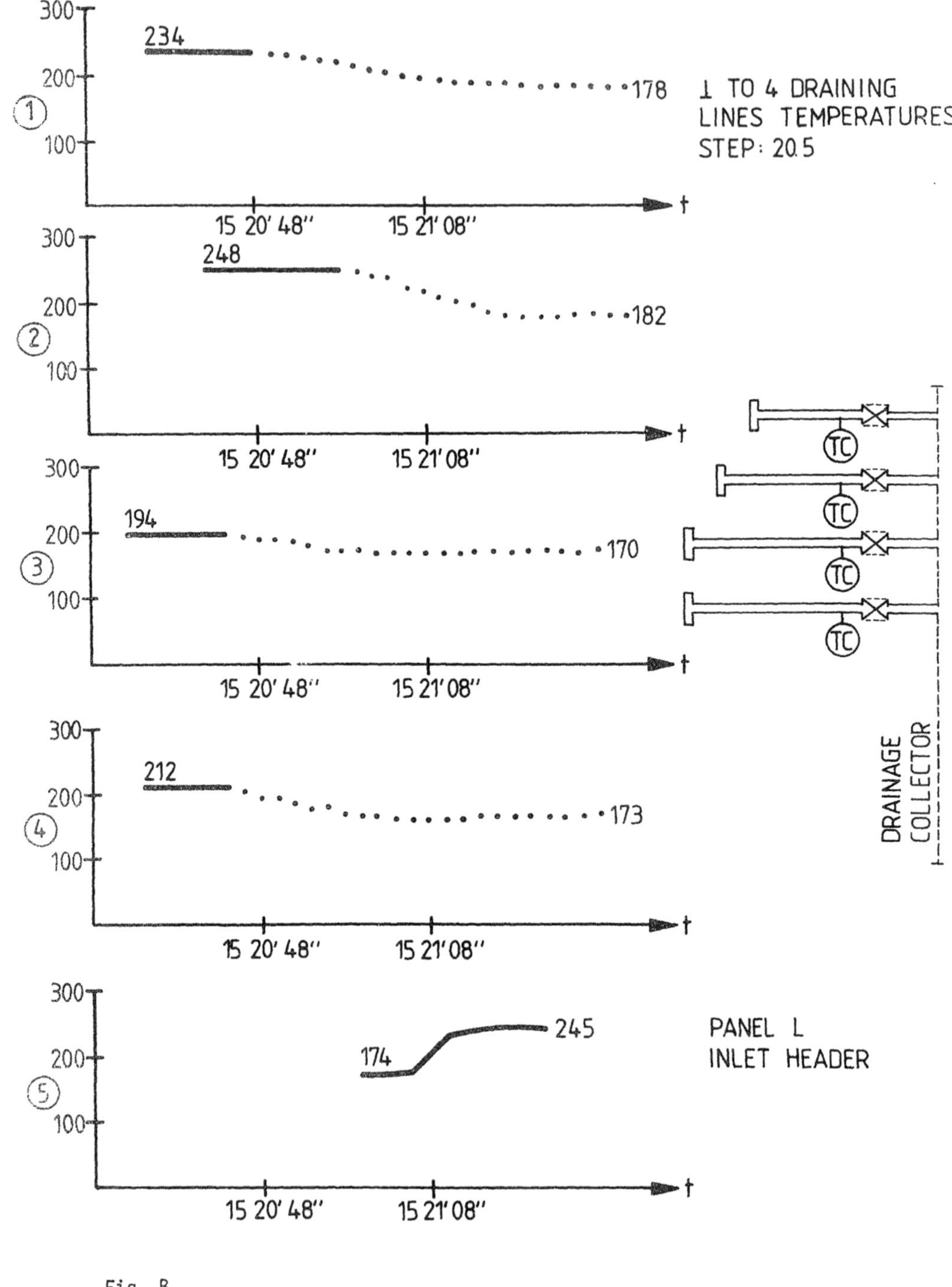
300
200
100
①
234
178
1 TO 4 DRAINING
LINES TEMPERATURES
STEP: 20.5
t
15 20' 48''
15 21' 08''
②
248
182
③
194
170
④
212
173
⑤
174
245
PANEL L
INLET HEADER
TC
DRAINAGE
COLLECTOR

Fig. B

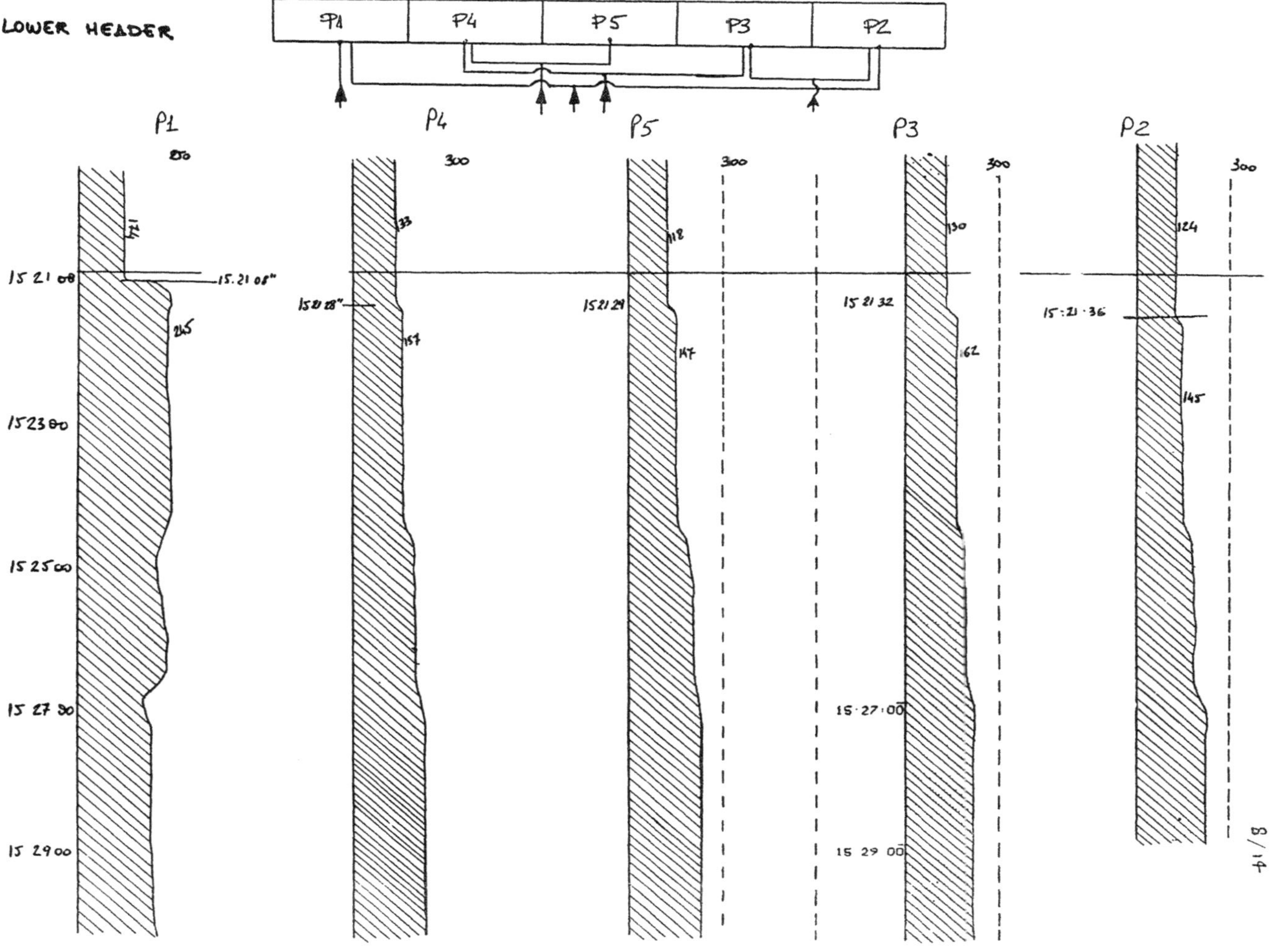
LOWER HEADER
P1
P4
P5
P3
P2
300
124
130
118
133
174
15 21 08
15.21 08"
15 21 28"
15 21 29
15 21 32
15:21·36
245
157
147
162
145
15 23 00
15 25 00
15 27 30
15·27·00
15 29 00
8/14

Fig. C

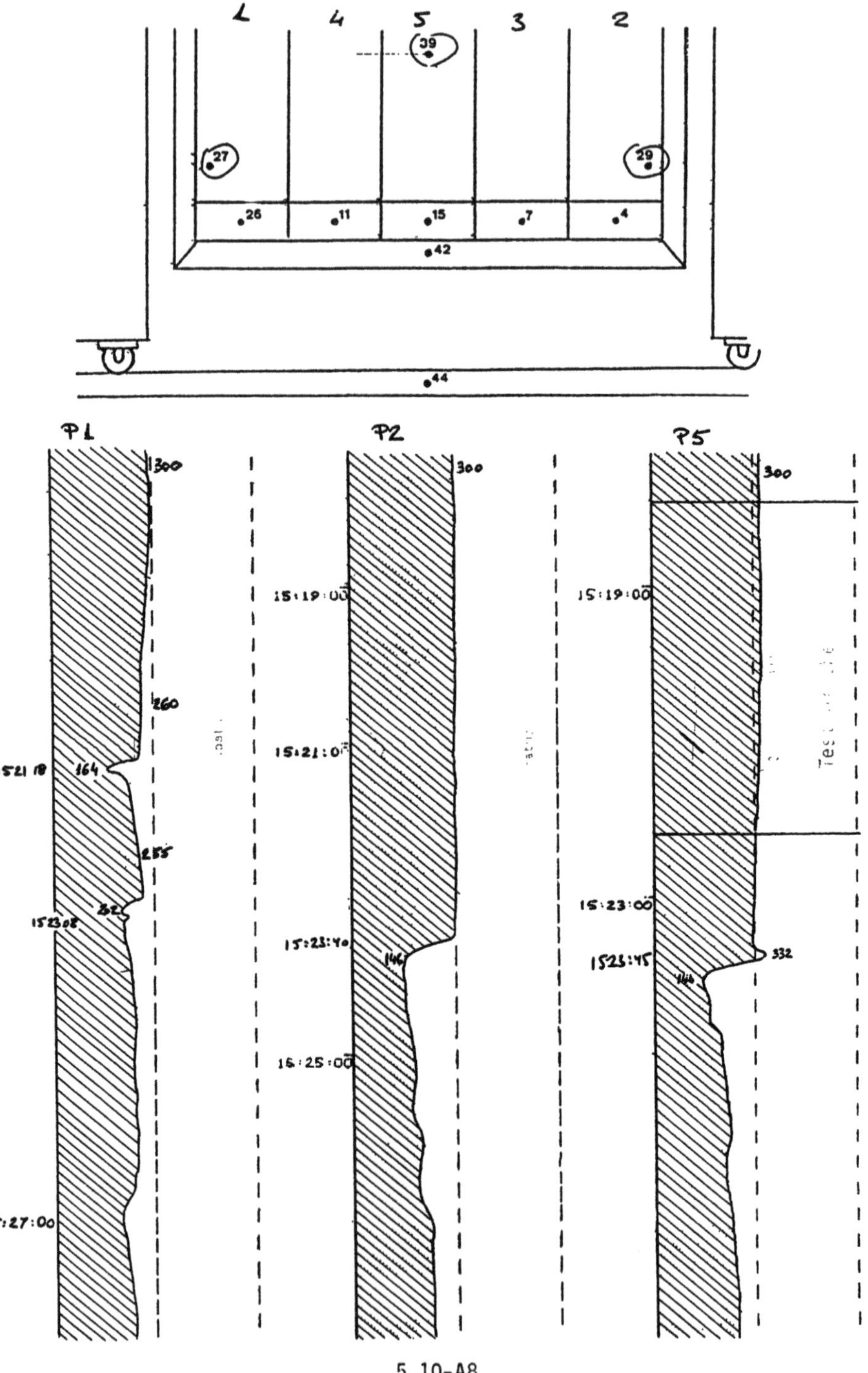
1
4
5
3
2
39
27
29
26
11
15
7
4
42
44
P1
P2
P5
300
260
152118
164
255
262
152308
15:27:00
15:19:00
15:21:00
15:23:40
146
15:25:00
15:23:00
152345
332
144

5.10-A8

Fig. D

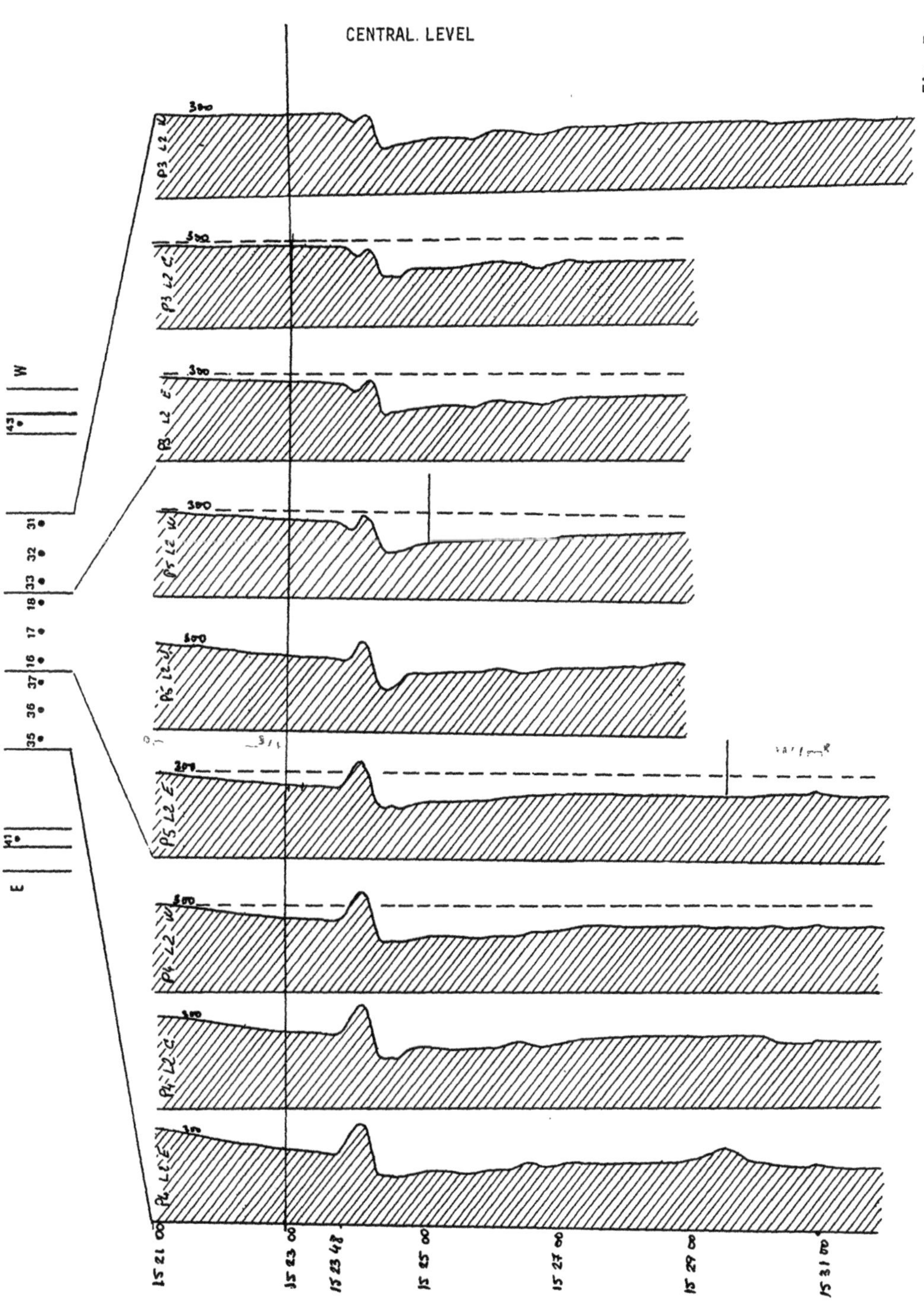
CENTRAL. LEVEL
Fig. E
W
E
31
32
33
18
17
16
37
36
35
15 21 00
15 23 00
15 23 48
15 25 00
15 27 00
15 29 00
15 31 00

UPPER LEVEL

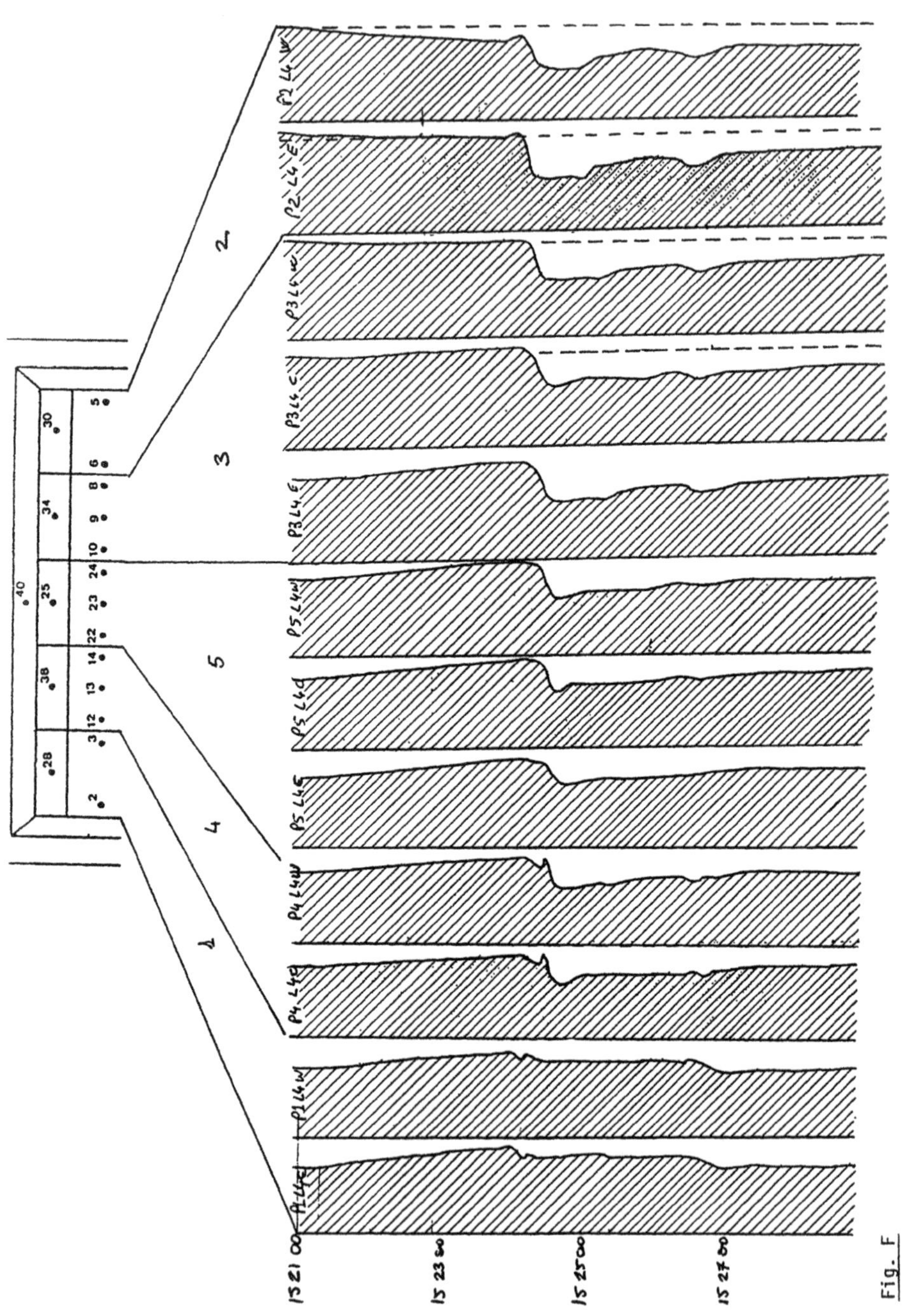

Fig. F

TOP HEADERS

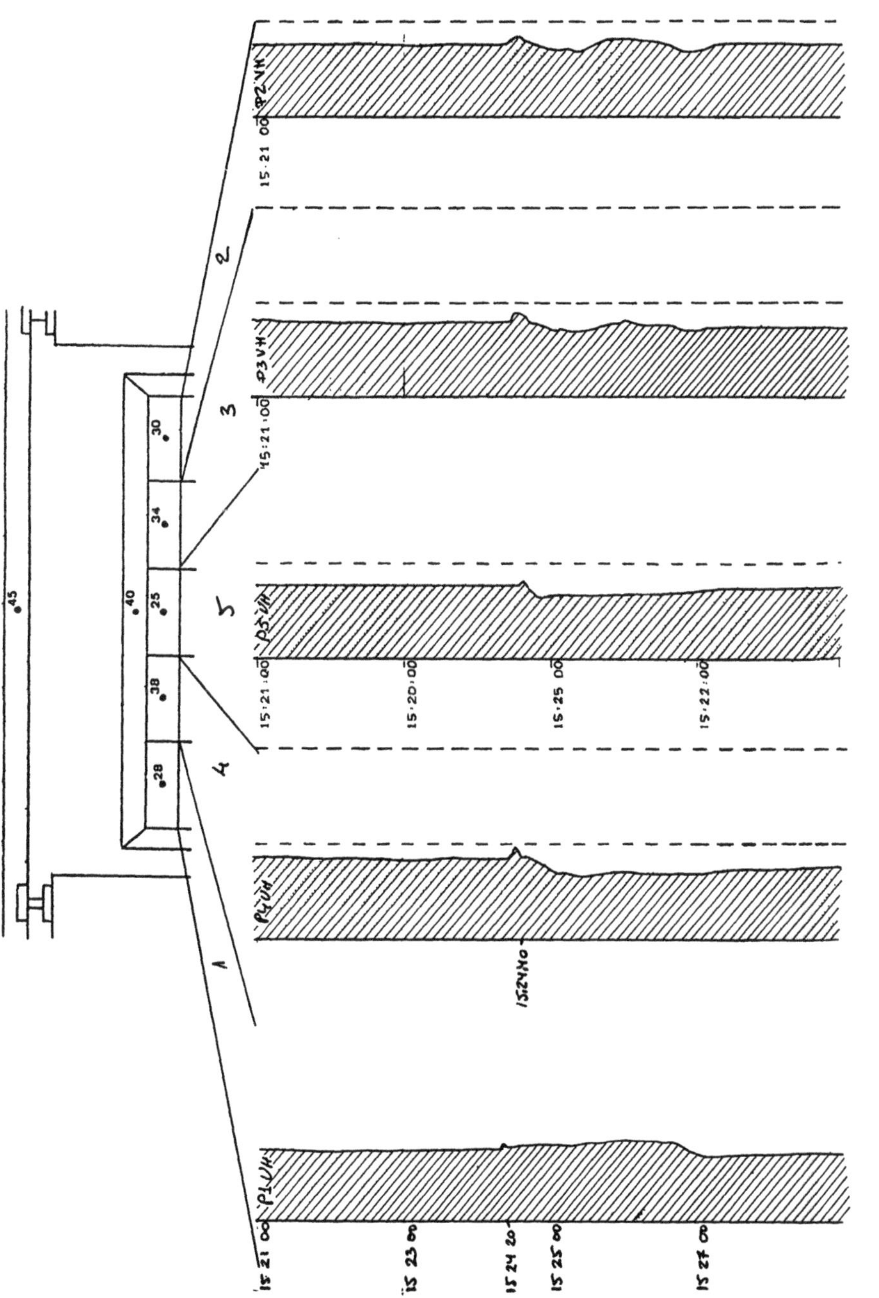

Fig. G

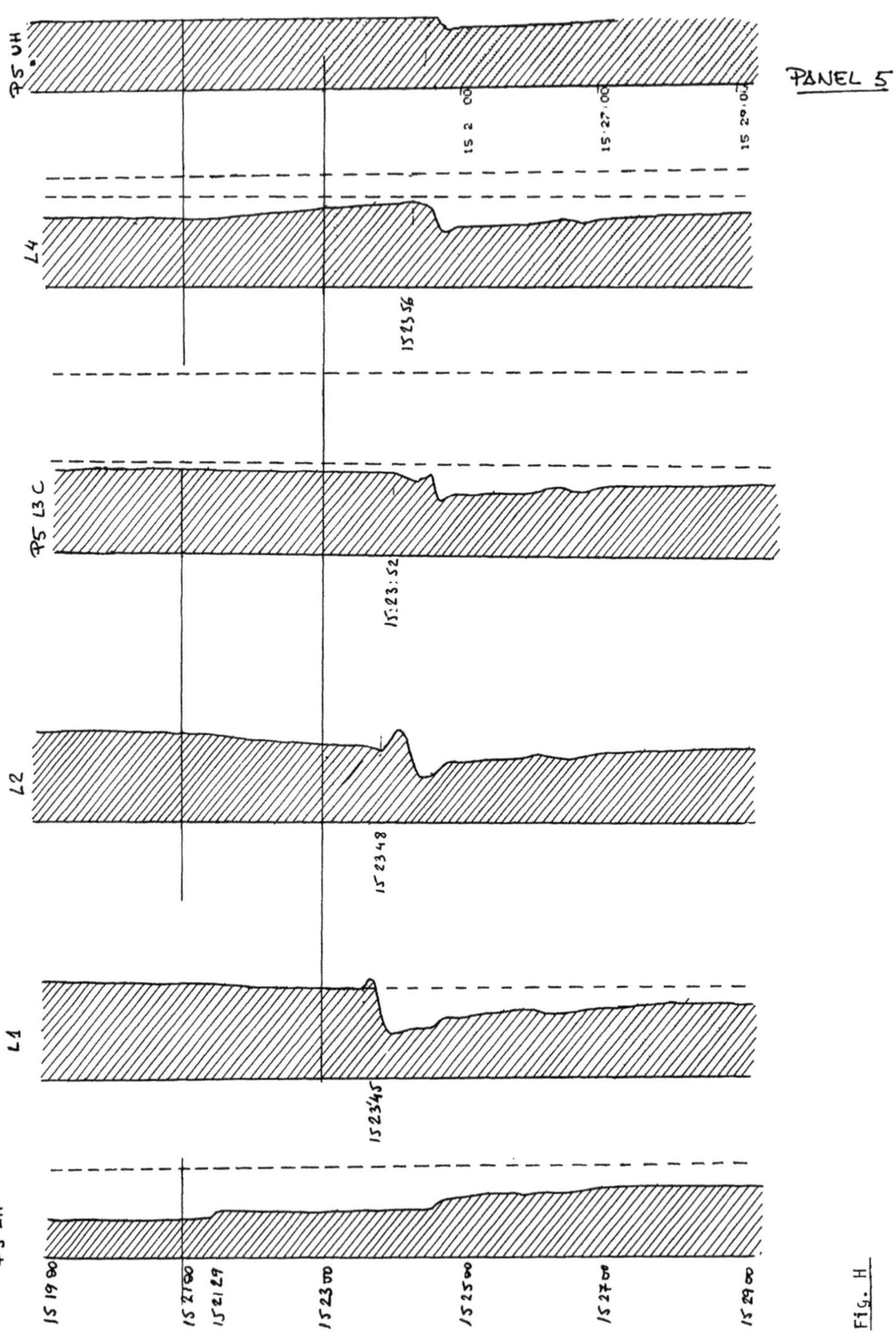

PANEL 5
PS UH
L4
PS L3 C
L2
L1
PS LH
15 19 00
15 21 00
15 21 29
15 23 00
15 25 00
15 27 00
15 29 00
15 23 56
15:23:52
15 23 48
15 23 45
Fig. H

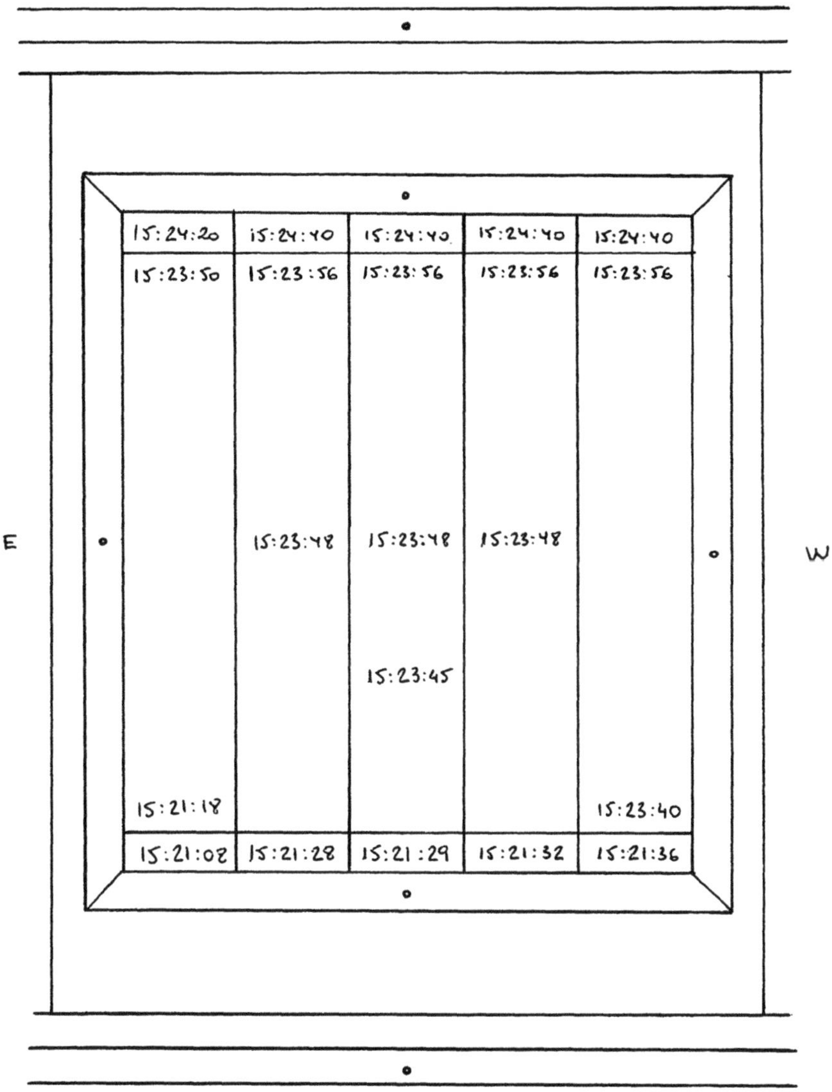

E

W

15:24:20	15:24:40	15:24:40	15:24:40	15:24:40
15:23:50	15:23:56	15:23:56	15:23:56	15:23:56
	15:23:48	15:23:48	15:23:48	
		15:23:45		
15:21:18				15:23:40
15:21:08	15:21:28	15:21:29	15:21:32	15:21:36

Fig. I

RESULTS ON THE PERFORMANCE OF THE SULZER CAVITY RECEIVER AND THE FRANCO-TOSI EXTERNAL RECEIVER

Heinz Jacobs and Clifford S. Selvage, ITET

Summary

This paper presents some of the results of evaluation work performed on the cavity receiver (Sulzer) and the external receiver (ASR) installed at the IEA/SSPS Project.

Operational data are combined with data obtained in loss tests which were performed on both receivers.

Several different methods of evaluation produce the same results, which are that the cavity receiver has a lower efficiency than the external receiver, and the losses of the cavity receiver are considerably higher than those of the external receiver.

The figures presented in this evaluation have an error band of 5%.

1. INTRODUCTION

Two different receiver concepts were manufactured and tested in the 500 kW Central Receiver System experiment at the Small Solar Power Systems Project in Almería, Spain (Tab.5.11-1 and 2 (Ref.6)).

One is a cavity receiver and the other is an external receiver. Both receivers use sodium as the heat transfer fluid.

The basic design and thermal analysis of the cavity receiver was carried out by INTERATOM. Detailed design with stress analysis, the manufacture, and erection of the receiver on site were performed by Sulzer with acceptance testing in the Spring of 1981. The contract companies responsible for the design, manufacture, and erection of the external receiver were AGIP and Franco-Tosi Industriale. Acceptance testing took place in the Autumn of 1983.

The work concerning the evaluation of the receivers was performed on site in Almería and considers two topics: 1. evaluation of operational data, and 2. evaluation of the loss tests.

The results of this evaluation are presented in this paper. It is desirable to continue performing convection loss tests of the ASR to verify the evaluation performed to date, and to improve the accuracy of that variation. Additional work on the influence of wind on ASR convective losses and an evaluation of the energy needed for start-up of the ASR have not been performed to date, but might be completed before the end of 1984.

2. CONVECTION LOSS TEST

The temperature drop and the flow of sodium pumped through the receiver with no solar power delivered from the heliostat field provides an indication of the convection and radiation losses (Ref.1). The receiver conditions during this reverse flow test are slightly different than normal because the surface temperature is lower and the temperature profile is not the same as when the receiver is collecting energy. Two different types of convective loss tests have been performed; a normal flow, using warm sodium from the cold tank into the inlet of the receiver with outlet returning to the cold tank, and a reverse flow, in which hot sodium is taken from the hot tank, introduced into the receiver at the normal outlet of the receiver with the sodium exiting the receiver at the normal inlet of the receiver into the cold tank. The advantage of the normal flow test is that it does not require any special operational changes, however the temperatures are well below normal operating temperatures. This test has been performed several times following normal operation, the only difference is that the heliostats are out of track.

The reverse flow test is more complicated. For this reason it has been performed only a few times.

The procedure of the reverse flow test for the cavity receiver is

- the receiver door is closed
- the valves are switched to reverse flow
- sodium flows until temperature is at steady-state (at least 15 minutes).

For the external receiver (ASR), this procedure is

- preheat the receiver back wall with a few heliostats (this avoids a long wait to achieve steady-state temperature)
- switch valve to reverse flow (flow is a result of a difference in gas pressure between the two tanks)
- steady-state temperature after 20 minutes.

Results

The results of the loss tests are presented in Ref.1.

To interpret these results, it is necessary to look carefully into the difference between test status and operation status. There are several arguments which indicate that operational losses are different than test losses.

First, the operational surface temperature during normal operation of the absorber tubes is considerably higher than during the reverse flow test. Therefore, we reason that the operational losses must be higher than these reverse flow tests indicate, at least for the ASR.

Second, the temperature profile for the ASR during operation is different than for the reverse flow loss test. During the reverse flow test, the profile of temperature decrease is more or less linear, from the receiver normal outlet to the receiver normal inlet. However, under normal energy collecting operation, the temperature profile shows a rather small temperature for the edge panels, panels 1 and 2, then a significantly larger temperature rise in panels 3 and 4, with the largest temperature rise in the last and central panel, panel 5. To attempt to correct for the profile influence, the receiver is divided in five parts for each panel.

Since the inlet and outlet temperature of each panel is recorded on CRS data tapes, a detailed loss calculation was performed. Each panel has its own loss figure, based on the empirical equations in Ref.7. The Sulzer receiver was subdivided into seven loops for the same purpose.

To compare loss test efficiency figures with direct measurement numbers, it seems reasonable to take care of the absorptivity. An average figure for K is given in Ref.9 as 0.94. Finally the efficiency η was calculated from

$$\eta = \alpha \cdot \left(1 - \frac{P_{loss}}{P_{absorbed} + P_{loss}}\right)$$

3. RECEIVER EFFICIENCY DURING OPERATION

Another method to calculate receiver losses is to compare incoming power from the field with the absorbed power. The incoming power is calculated using the HELIOS computer code. The calculated figures are then used to construct a look-up table which gives power as a function of day and solar time under design conditions (2). Additional measurements were also performed using HERMES (3) and FAS (4), and the results are used to support this table.

The figures from the look-up table (matrix) need to be corrected by current data of insolation, reflectivity, and the number of heliostats in track. The insolation and the number of heliostats in track are recorded automatically by the Data Acquisition System (DAS). The reflectivity is measured on site periodically. A theoretical analysis of the data which allows a limited forecast is given in Ref.5. In addition it is necessary to correct the flowrate of the receiver circuit as describd in Ref.8.

The absorbed power is calculated using enthalpy and mass flow. The estimated uncertainties of HELIOS calculation (5%), operating with an undesignated number of heliostats (5%), facets' soiling (3%), and calculation of absorbed power (4%), produced an overall margin of error of 5% (Ref.6).

Results

In the following figures the results presented are daily energy averages for a day which had at least 2 hours of uninterrupted receiver operation.

The processed data are read directly from the DAS data tapes, containing five-minute average values of all measurement points.

Tab.5.11-2 and 3 give examples of the output file of the data processing for a steady-state period of both receivers. Besides the meteo data and the input data for efficiency calculation (flow, temperatures), the loss and efficiency figures for both methods are found:

- calculated using the HELIOS look-up table
- calculated using loss test results

The averages are presented in Tab.5.11-4 and 5. Fig.5.11-1 and 2 show the efficiency of both receivers vs. distributed power. Each point is the average of one day as in Tab.5.11-4 and 5.

The efficiency is calculated by comparing absorbed power with the HELIOS ouput "power on target".

Fig.5.11-3 and 4 show the efficiency calculated from the loss test results.

The highest observed efficiency for both receivers can be expressed as:

	measured	calculated	difference
Sulzer	76.0%	77.4%	1.4%
ASR	91.9%	89.8%	2.1%
difference	15.9%	12.4%	

One should also know that these efficiency figures are based on power into cavity representing power on target, which is already reduced by the spillage losses.

Fig.5.11-5 to 8 contain the losses (measured using HELIOS, and calculated during loss tests) vs. absorbed power.

Fig.5.11-9 and 10 show the efficiency as a function of the irradiation. If irradiation is low, the efficiency will be lower than normal because of the relatively constant losses. The influence of wind speed and direction on the losses cannot be separated by using data from five-minute averages.Fig.5.11-11 to 14 show no tendency in any case.

4. DISTRIBUTION IN THE RECEIVERS

The power is absorbed in seven loops of the Sulzer receiver and five panels of the ASR. An average distribution is as follows:

Sulzer receiver	% of absorbed power	
loop #7	1.2	top
" 6	19.4	
" 5	40.2	
" 4	28.8	
" 3	10.5	
" 2	1.3	
" 1	-1.8	bottom

ASR	% of absorbed power	
panel #1	8.2	edge
" 2	7.5	edge
" 3	27.0	
" 4	27.1	
" 5	30.2	center

Fig.5.11-15 and 16show the measured percentage of energy received by each panel vs. absorbed power.

It is obvious that in the Sulzer receiver it is the four center loops that are most involved in gaining energy. Loop #1 (bottom) loses more energy than it receives.

5. LOSSES DURING THE NIGHT

The losses during the night for the cavity receiver are 6 - 9 kW, and 16 - 18 kW for the external receiver. The reasons for the low standby losses of the cavity receiver are the fact that the doors fit the cavity better, and the fact that the high thermal inertia caused by the ceramic wall reduces the temperature drop in the sodium. However, it is necessary to heat up the ceramic wall in the beginning of the next operational period.

6. CONCLUSIONS

The external receiver (ASR), with a high heat flux, has higher efficiency than the cavity receiver (Sulzer). This is proven by comparing the results of the loss tests: the losses of the cavity receiver appear to be approximately twice those of the external receiver. One reason for this is that the smaller heat transfer area of the external receiver produces smaller losses. The actual area with high temperatures of the external receiver is reduced by the two edge panels (Ref.1,2). This results in only 3 of the 5 panels having really high losses.

7. REFERENCES

1. John S. Kraabel, "Convection Testing of Central Receiver System - Fall 1982, and Receiver Efficiency, Fall Measurement Campaign, 1982", SSPS - CRS Midterm Workshop Proceedings, SSPS Technical Report No 4/83 (1983).

2. M. Becker, H. Ellgering, and D. Stahl, "CRS-Construction Report", SSPS SR2 (1982).

3. W. Schiel, "HERMES Measurements", Proceedings op. cit. (1983).

4. G. Von Tobel, Ch. Shelden, and M. Real, "Concentrated Solar Flux Measurement at the IEA/SSPS Plant", SSPS Technical Report № 2/82 (1982).

5. M. Sánchez, "A Mathematical Model for Estimating the Average Heliostat Field Reflectivity", SSPS Internal Report (1984).

6. M. Pescatore, "Comparison Between the Sulzer Cavity Receiver and the Franco-Tosi External Receiver", SSPS Internal Report (1984).

7. R. Carmona and H. Jacobs, "Convection Loss Tests", Presentation at the IEA/SSPS Deliverable Review, October 15-18, 1984 in Almería, Spain (1984).

8. A. Brinner, "Calibration of Relevant Measuring Sensors", SSPS Technical Report (1984).

9. T. van Steenberghe, "Solar Absorptance Measurements on the First CRS Receiver Tubes", SSPS Internal Report (1982).

DESIGN DATA	CAVITY RECEIVER (first built sodium)	EXTERNAL RECEIVER (advanced sodium)
HEAT EXCHANGER	1 vertical 120° cylinder tube bundle with 6 parallel sodium-carrying tubes directed in a serpentine from bottom to top with 14 passes	5 vertical tube panels with 39 tubes, each connected serially
DIMENSIONS OF HEAT EXCHANGER	Radius: 2.25 m Height: 3.61 m Arc Length: 4.71 (active)	Height: 2.85 m (= active tube length) Width: 2.75 m
TOTAL WEIGHT	approx. 25 T	approx. 20 T
NORTH ORIENTED APERTURE	9.0 m^2, octagonal shape	9.0 m^2, rectangular shape
ABSORBING SURFACE	17 m^2	7.8 m^2
APPROX. TUBE LENGTH	87 m (66 m active)	50 m (14.3 m active)
TUBE GEOMETRY	Ø 38 mm X 1.5 mm	Ø 14 mm X 1 mm
TUBE PAINT (black)	Pyromark 2500	Pyromark 2500

Table 5.11.-1a: Comparison of Design Data for both Receivers

OPERATING CONDITIONS	CAVITY RECEIVER	EXTERNAL RECEIVER
THERMAL POWER - Input	2,840 kW	2,840 kW
- Output	2,508 kW	2,525 kW
SODIUM - Mass Flow	7.34 Kg/s	7.34 Kg/s
- Inlet/Outlet	270°C/530°C	270°C/530°C
- Pressure Drop	0.5 bar	1.5 bar
- Flow Velocity	1.5 m/s	1.85 m/s
HEAT FLUX - Peak	0.6 MW/m^2	1.38 MW/m^2
- Average	0.16 MW/m^2	0.36 MW/m^2
- Aiming	single point	three points

Table 5.11.-1b: Comparison of Operating Conditions for both Receivers

SULZER RECEIVER OPERATION   testdate 18-03-83

(01)	#154,1	Temperature 8m	[deg C]
(02)	#151,1	Wind Speed	[km/h]
(03)	#152,1	Wind Direction	[deg]
(04)	#150,1	Direct Insolation CRS	[W/m2]
(05)	NHTRACK	Number of Heliostats in Track	
(06)	LK01CT03	Receiver Inlet Temperature	[deg C]
(07)	LK01CT25	Receiver Outlet Temperature	[deg C]
(08)	LK01CF01	Receiver Inlet Flow	[m3/h]
(09)	calculated:	Power into Cavity (HELIOS)	[kW]
(10)	calculated:	Absorbed Power by Sodium	[kW]
(11)	calculate d	Losses during Operation	[kW]
(12)	calculated:	Efficiency (HELIOS)	[%]
(13)	calculate d	Losses due to Losstests	[kW]
(14)	calculated:	Efficiency (Losstest)	[%]

LocTm	SolTm	(01)	(02)	(03)	(04)	(05)	(06)	(07)	(08)	(09)	(10)	(11)	(12)	(13)	(14)
12:15	10:57	16.8	19.7	74.5	911.2	91.0	256.7	527.8	22.17	2481.	1903.	578.	76.7	511.	78.8
12:20	11:02	16.9	18.2	81.4	909.7	91.0	256.5	526.2	22.31	2481.	1905.	577.	76.8	508.	78.9
12:25	11:07	16.8	19.9	74.4	910.2	91.0	256.4	527.0	22.45	2483.	1923.	560.	77.4	509.	79.1
12:30	11:12	17.0	17.1	65.9	911.4	91.0	256.3	527.6	22.37	2488.	1921.	568.	77.2	509.	79.1
12:35	11:17	17.2	14.4	96.1	910.3	91.0	256.2	526.8	22.64	2486.	1940.	546.	78.0	506.	79.3
12:40	11:22	17.2	17.1	85.0	909.8	91.0	256.1	527.3	22.57	2487.	1938.	549.	77.9	506.	79.3
12:45	11:27	17.3	12.9	113.6	909.1	91.0	256.1	526.3	22.65	2486.	1938.	547.	78.0	503.	79.4
12:50	11:32	18.2	11.0	100.9	911.4	91.0	256.0	526.9	22.68	2494.	1946.	548.	78.0	501.	79.5
12:55	11:37	18.3	8.1	91.3	911.2	91.0	255.9	526.4	22.54	2494.	1930.	564.	77.4	499.	79.5
13:00	11:42	18.0	10.7	87.6	907.9	91.0	255.8	527.1	22.59	2487.	1941.	546.	78.1	498.	79.6
13:05	11:47	18.3	7.5	103.6	907.5	91.0	255.7	525.4	22.20	2486.	1896.	590.	76.3	494.	79.3
13:10	11:52	18.1	10.9	113.1	908.9	91.0	255.8	527.0	22.55	2492.	1937.	555.	77.7	494.	79.7
13:15	11:57	18.2	17.0	104.7	910.1	91.0	256.5	526.9	22.82	2496.	1954.	543.	78.3	495.	79.8
13:20	12:02	17.7	16.1	97.9	908.9	91.0	257.2	525.8	22.79	2494.	1938.	556.	77.7	496.	79.6
13:25	12:07	17.7	14.5	103.5	910.6	91.0	257.8	527.3	22.83	2497.	1947.	550.	78.0	498.	79.6
13:30	12:12	18.4	17.2	99.8	908.5	91.0	258.4	526.7	22.82	2490.	1937.	553.	77.8	496.	79.6
13:35	12:17	18.1	16.9	94.8	909.5	91.0	258.8	526.3	22.98	2491.	1945.	546.	78.1	496.	79.7
13:40	12:22	17.9	17.6	84.2	909.6	91.0	259.1	526.8	22.82	2490.	1932.	558.	77.6	498.	79.5
13:45	12:27	18.4	11.8	122.0	908.8	91.0	259.5	526.9	22.99	2486.	1944.	542.	78.2	496.	79.7
13:50	12:32	18.1	11.5	104.2	907.1	91.0	259.8	526.6	22.92	2481.	1934.	547.	78.0	498.	79.5
13:55	12:37	18.5	14.6	84.7	904.8	91.0	260.3	526.6	22.86	2473.	1924.	549.	77.8	496.	79.5
14:00	12:42	18.4	13.9	75.8	905.8	90.7	261.1	527.7	22.61	2465.	1905.	559.	77.3	497.	79.3
14:05	12:47	18.3	15.0	102.1	905.6	90.9	261.7	527.7	22.81	2469.	1918.	551.	77.7	497.	79.4
14:10	12:52	18.4	13.7	100.0	904.4	91.0	262.4	527.1	22.51	2468.	1882.	585.	76.3	496.	79.2
14:15	12:57	18.8	14.7	107.5	904.5	91.0	263.5	529.0	22.89	2467.	1919.	548.	77.8	498.	79.4
14:20	13:02	18.5	15.5	110.0	901.4	91.0	264.2	527.1	22.84	2455.	1896.	560.	77.2	497.	79.2

Table 5.11.-2: Major Sulzer Receiver Parameters at around Solar Noon

Date 18-Mar-1983

ASR RECEIVER OPERATION testdate 10-12-83

(01)	#154,1	Temperature 8m	[deg C]
(02)	#151,1	Wind Speed	[km/h]
(03)	#152,1	Wind Direction	[deg]
(04)	#150,1	Direct Insolation CRS	[W/m2]
(05)	NHTRACK	Number of Heliostats in Track	
(06)	LK01CT01 R	Receiver Inlet Temperature	[deg C]
(07)	LK01CT36	Receiver Outlet Temperature	[deg C]
(08)	LK01CF01	Receiver Inlet Flow	[m3/h]
(09)	calculated:	Power into Cavity (HELIOS)	[kW]
(10)	calculated:	Absorbed Power by Sodium	[kW]
(11)	calculate d	Losses during Operation	[kW]
(12)	calculated:	Efficiency (HELIOS)	[%]
(13)	calculate d	Losses due to Losstests	[kW]
(14)	calculated:	Efficiency (Losstest)	[%]

LocTm	SolTm	(01)	(02)	(03)	(04)	(05)	(06)	(07)	(08)	(09)	(10)	(11)	(12)	(13)	(14)
12:00	10:58	13.7	0.2	114.8	813.1	93.0	262.5	529.3	23.14	2168.	1950.	218.	89.9	243.	88.9
12:05	11:03	14.6	1.0	109.3	818.6	93.0	263.3	529.4	23.38	2190.	1965.	225.	89.7	243.	89.0
12:10	11:08	13.7	2.3	120.5	817.1	93.0	264.2	529.4	23.62	2190.	1978.	213.	90.3	245.	89.0
12:15	11:13	14.1	0.4	142.5	817.1	93.0	265.7	529.0	23.71	2195.	1970.	224.	89.8	245.	89.0
12:20	11:18	14.3	0.2	126.0	823.3	93.0	267.3	529.2	23.86	2215.	1971.	245.	89.0	245.	88.9
12:25	11:23	14.5	0.2	167.7	822.6	93.0	268.3	529.6	23.95	2218.	1974.	243.	89.0	246.	88.9
12:30	11:28	14.3	2.3	177.0	819.4	93.0	268.2	529.1	24.08	2213.	1981.	232.	89.5	246.	88.9
12:35	11:33	14.5	0.9	188.2	820.4	93.0	268.1	529.4	24.25	2220.	1999.	222.	90.0	247.	89.0
12:40	11:38	14.5	0.4	175.2	820.1	93.0	268.0	529.3	24.32	2223.	2004.	219.	90.1	247.	89.0
12:45	11:43	15.0	0.4	123.9	821.5	93.0	267.9	529.2	24.19	2231.	1993.	238.	89.3	246.	89.0
12:50	11:48	14.6	2.5	208.5	823.6	93.0	267.8	528.9	24.22	2241.	1995.	246.	89.0	247.	89.0
12:55	11:53	14.2	3.0	205.5	819.6	93.0	267.7	529.5	24.22	2234.	2000.	234.	89.5	248.	89.0
13:00	11:58	14.9	0.5	178.7	822.8	93.0	267.7	529.4	24.23	2247.	2000.	247.	89.0	246.	89.0
13:05	12:03	15.4	0.2	70.9	824.6	93.0	267.8	529.1	24.04	2251.	1982.	269.	88.0	245.	89.0
13:10	12:08	15.6	0.2	136.2	818.9	93.0	267.7	529.2	24.02	2231.	1981.	251.	88.8	245.	89.0
13:15	12:13	15.8	0.2	194.6	813.7	93.0	267.7	529.2	23.96	2213.	1976.	237.	89.3	245.	89.0
13:20	12:18	16.0	2.2	153.4	814.4	93.0	267.7	528.8	23.98	2211.	1975.	237.	89.3	245.	89.0
13:25	12:23	15.0	3.2	189.6	814.2	93.0	267.7	529.2	24.08	2206.	1986.	220.	90.0	247.	89.0
13:30	12:28	15.1	1.7	167.2	820.6	93.0	267.7	529.4	24.25	2220.	2002.	218.	90.2	247.	89.0
13:35	12:33	16.3	0.2	103.9	818.1	93.0	267.7	529.1	23.91	2209.	1971.	238.	89.2	245.	88.9
13:40	12:38	16.5	0.6	159.9	811.8	93.0	267.7	529.1	23.61	2188.	1947.	241.	89.0	244.	88.9
13:45	12:43	15.7	1.1	202.3	803.8	93.0	268.0	528.9	23.46	2162.	1931.	231.	89.3	244.	88.8
13:50	12:48	15.1	3.9	226.1	793.4	93.0	268.1	529.2	23.29	2130.	1918.	212.	90.1	244.	88.7
13:55	12:53	16.0	0.2	215.4	793.9	93.0	269.1	529.0	23.44	2128.	1921.	207.	90.3	243.	88.8
14:00	12:58	16.7	0.2	159.7	789.4	93.0	272.2	528.8	23.51	2112.	1900.	212.	90.0	243.	88.7
14:05	13:03	16.4	0.8	216.4	781.1	93.0	274.5	529.4	23.26	2081.	1866.	215.	89.7	243.	88.5

Table 5.11.-3: Major ASR Parameters around Solar Noon

Date 10-Dec-1983

SULZER RECEIVER OPERATION

(00)	Number of 5 minutes averages measured		
(01)	#154,1	Temperature 8m	[deg C]
(02)	#151,1	Wind Speed	[km/h]
(03)	#152,1	Wind Direction	[deg]
(04)	#150,1	Direct Insolation CRS	[W/m2]
(05)	NHTRACK	Number of Heliostats in Track	
(06)	LK01CT03	Receiver Inlet Temperature	[deg C]
(07)	LK01CT25	Receiver Outlet Temperature	[deg C]
(08)	LK01CF01	Receiver Inlet Flow	[m3/h]
(09)	calculated:	Power into Cavity (HELIOS)	[kW]
(10)	calculated:	Absorbed Power by Sodium	[kW]
(11)	calculate.d	Losses during Operation	[kW]
(12)	calculated:	Efficiency (HELIOS)	[%]
(13)	calculate d	Losses due to Losstests	[kW]
(14)	calculated:	Efficiency (Losstest)	[%]

date	(00)	(01)	(02)	(03)	(04)	(05)	(06)	(07)	(08)	(09)	(10)	(11)	(12)	(13)	(14)
29 08 82	20	26.8	28.1	85.4	763.4	87.7	209.4	521.4	14.09	1886.	1415.	471.	74.9	410.	77.4
30 08 82	53	26.3	34.8	90.0	645.1	81.4	249.5	523.2	11.61	1489.	1009.	480.	67.5	446.	69.0
02 09 82	18	25.6	17.7	135.8	797.1	87.0	231.5	526.7	14.99	1903.	1416.	488.	74.2	433.	76.3
03 09 82	53	24.6	22.5	86.0	799.6	86.4	272.9	528.1	17.32	1905.	1392.	514.	72.7	498.	73.3
05 09 82	74	26.3	13.7	204.9	763.2	86.0	264.2	529.1	15.12	1768.	1263.	505.	71.0	476.	72.1
06 09 82	102	30.5	18.8	222.9	822.5	84.4	268.1	527.1	16.60	1871.	1357.	514.	71.9	473.	73.4
07 09 82	56	24.4	30.4	82.1	690.0	82.4	262.8	528.1	13.41	1655.	1125.	531.	67.4	466.	70.2
23 09 82	12	26.8	22.3	236.6	840.4	84.6	253.7	520.0	18.74	2173.	1588.	585.	72.8	452.	77.5
24 09 82	89	26.8	12.8	210.3	873.7	85.9	253.4	526.5	18.94	2170.	1602.	568.	73.2	476.	76.3
25 09 82	98	27.0	33.9	199.8	800.9	87.0	265.4	527.5	17.83	2049.	1469.	580.	71.2	477.	74.9
26 09 82	13	24.8	16.0	213.5	809.2	87.0	263.5	523.8	17.70	1981.	1427.	554.	71.7	476.	74.6
27 09 82	34	24.9	11.7	123.7	681.5	87.0	255.1	527.9	13.66	1667.	1176.	491.	69.7	457.	71.1
01 10 82	46	23.2	18.8	281.1	798.6	87.0	262.2	529.0	18.59	2090.	1565.	525.	74.8	482.	76.4
02 10 82	86	25.5	13.5	71.4	832.6	85.9	253.4	527.2	18.30	2087.	1552.	535.	74.1	474.	76.3
03 10 82	83	27.1	13.1	167.2	796.1	86.3	263.2	527.0	17.76	2017.	1483.	534.	72.9	474.	75.1
04 10 82	20	25.4	24.1	74.0	556.7	87.0	247.0	523.7	10.91	1475.	963.	512.	64.0	433.	67.6
06 10 82	24	21.0	30.5	251.2	849.1	86.7	262.9	525.0	20.19	2198.	1674.	524.	76.1	481.	77.6
07 10 82	70	19.8	16.5	213.9	867.0	86.2	255.6	526.9	18.29	2137.	1573.	564.	72.9	487.	75.5
16 03 83	13	17.3	17.3	268.9	931.0	77.9	262.4	522.4	19.62	2184.	1608.	576.	73.1	502.	75.5
17 03 83	65	15.8	14.3	215.7	854.3	90.5	235.8	524.6	18.71	2249.	1701.	548.	75.3	485.	77.4
18 03 83	69	16.6	16.3	93.2	871.1	90.2	255.3	526.1	19.98	2270.	1709.	562.	74.6	505.	76.5
19 03 83	70	21.0	9.8	186.6	883.1	89.0	262.5	527.7	20.53	2303.	1715.	588.	74.1	498.	77.1
21 03 83	35	20.2	12.3	209.7	767.3	89.5	221.4	522.0	14.50	1934.	1396.	539.	71.5	466.	74.3
25 03 83	50	13.1	31.7	282.4	864.2	93.0	264.1	523.8	21.82	2393.	1796.	597.	73.8	518.	76.2
26 03 83	75	19.2	30.1	221.0	860.9	92.3	273.4	526.3	22.30	2325.	1770.	555.	75.8	521.	76.8
27 03 83	45	17.3	22.0	220.7	803.0	92.1	263.9	525.5	18.77	2110.	1550.	561.	72.6	521.	74.1
28 03 83	15	13.7	20.7	285.3	836.9	89.2	275.8	522.5	21.59	2298.	1677.	621.	72.8	520.	76.0
30 03 83	73	13.4	18.4	176.6	770.4	91.3	267.3	526.1	18.62	2084.	1521.	562.	72.8	517.	74.4
31 03 83	52	17.4	12.0	207.8	758.3	92.0	257.6	524.5	16.49	1959.	1396.	563.	70.2	503.	72.4
02 04 83	30	14.9	22.5	223.5	858.0	93.0	259.9	526.0	21.44	2411.	1803.	608.	74.7	518.	77.6
04 04 83	89	16.5	16.1	204.6	817.5	91.5	265.4	527.4	19.31	2173.	1601.	572.	72.9	514.	74.9
05 04 83	42	18.9	25.9	240.8	801.4	92.7	247.0	527.0	17.95	2143.	1604.	538.	74.5	478.	76.5
06 04 83	84	19.7	14.1	210.2	825.7	93.3	264.5	528.1	20.00	2256.	1664.	593.	73.2	507.	76.1
09 04 83	96	24.5	21.7	232.2	850.1	91.4	259.8	528.4	19.77	2267.	1682.	585.	72.9	494.	75.9
10 04 83	5	26.5	28.1	215.7	607.0	91.4	254.3	529.0	12.50	1654.	1088.	566.	65.0	466.	69.0
11 04 83	36	28.9	29.0	255.4	596.8	83.6	247.5	524.9	10.87	1468.	958.	509.	64.9	436.	68.3
12 04 83	29	16.4	23.6	85.9	719.4	93.0	258.1	526.8	16.88	2013.	1429.	584.	69.7	496.	72.8
14 04 83	32	15.2	33.0	92.9	688.5	94.0	240.2	526.2	14.89	1946.	1365.	581.	69.4	477.	73.2

Table 5.11.-4: Daily Average of the Major Receiver Parameters

ASR RECEIVER OPERATION AVERAGE DATA
==================================

(00)	Number of Measurement Points		
(01)	#154,1	Temperature 8m	[deg C]
(02)	#151,1	Wind Speed	[km/h]
(03)	#152,1	Wind Direction	[deg]
(04)	#150,1	Direct Insolation CRS	[W/m2]
(05)	NHTRACK	Number of Heliostats in Track	
(06)	LK01CT01 R	Receiver Inlet Temperature	[deg C]
(07)	LK01CT36	Receiver Outlet Temperature	[deg C]
(08)	LK01CF01	Receiver Inlet Flow	[m3/h]
(09)	calculated:	Power into Cavity (HELIOS)	[kW]
(10)	calculated:	Absorbed Power by Sodium	[kW]
(11)	calculateed	Losses during Operation	[kW]
(12)	calculated:	Efficiency (HELIOS)	[%]
(13)	calculateed	Losses due to Losstests	[kW]
(14)	calculated:	Efficiency (Losstest)	[%]

date	(00)	(01)	(02)	(03)	(04)	(05)	(06)	(07)	(08)	(09)	(10)	(11)	(12)	(13)	(14)
01 12 83	59	18.6	17.0	94.6	828.4	92.6	279.0	528.1	24.57	2125.	1922.	204.	90.4	241.	88.8
08 12 83	40	14.0	2.4	214.0	822.4	85.2	269.8	521.8	22.05	1955.	1753.	202.	89.7	233.	88.1
09 12 83	22	12.3	2.4	255.5	805.1	92.0	280.2	519.2	23.01	1928.	1724.	204.	89.4	237.	87.8
10 12 83	45	14.7	2.0	164.9	794.5	93.0	267.9	529.1	22.98	2106.	1893.	213.	89.9	242.	88.6
11 12 83	13	12.2	7.0	224.2	806.0	93.0	260.9	515.2	23.56	2083.	1892.	190.	90.9	235.	88.9
23 12 83	48	18.9	1.7	197.5	919.0	92.7	268.7	527.3	26.75	2404.	2180.	223.	90.7	248.	89.7
24 12 83	53	18.0	12.9	98.5	921.3	92.8	270.2	528.7	26.45	2386.	2157.	229.	90.4	249.	89.6
27 12 83	59	14.5	18.5	101.6	726.9	92.8	261.2	528.5	19.90	1885.	1683.	202.	89.3	228.	88.0
28 12 83	32	15.6	25.0	99.5	865.4	91.4	254.4	529.1	23.17	2264.	2018.	246.	89.2	239.	89.4
29 12 83	34	12.3	15.5	103.1	796.2	91.9	273.4	528.7	22.61	2034.	1816.	218.	89.3	240.	88.3
30 12 83	60	13.7	1.8	208.5	842.3	91.5	270.5	529.0	23.51	2148.	1915.	233.	89.2	242.	88.7
05 01 84	29	10.3	7.1	222.9	862.8	90.9	261.5	527.2	23.76	2223.	1994.	229.	89.7	244.	89.1
10 01 84	10	8.4	7.2	115.7	883.0	91.0	267.4	529.5	25.19	2325.	2082.	243.	89.6	251.	89.2
12 01 84	27	14.4	0.8	204.8	853.8	89.0	275.7	528.1	24.54	2191.	1952.	239.	88.8	245.	88.6
10 02 84	28	11.5	18.2	110.1	909.9	86.9	258.1	529.1	24.16	2325.	2070.	255.	89.1	244.	89.4
01 03 84	15	10.0	8.6	240.1	739.1	86.0	262.8	526.4	17.67	1691.	1480.	212.	87.3	225.	86.5
14 03 84	21	11.0	26.1	236.6	823.0	80.5	268.2	524.6	19.48	1821.	1580.	241.	86.8	231.	87.0
15 03 84	17	9.3	51.1	274.4	859.5	56.5	269.2	527.0	13.30	1274.	1082.	192.	84.9	215.	83.4
16 03 84	12	11.8	27.6	277.0	871.0	83.9	272.9	522.4	23.35	2118.	1837.	281.	86.7	242.	88.3
27 03 84	25	13.4	2.0	111.9	736.7	73.3	256.1	524.0	13.66	1347.	1149.	198.	85.2	207.	84.4
22 05 84	14	25.0	0.0	0.0	919.7	61.8	266.7	503.5	18.63	1594.	1394.	199.	87.5	212.	86.8
24 05 84	13	25.0	0.0	0.0	891.3	74.3	276.4	523.6	20.94	1842.	1625.	217.	88.2	230.	87.6
11 06 84	36	25.0	0.0	0.0	621.9	89.0	263.2	526.7	15.82	1511.	1318.	193.	86.7	213.	85.7
12 06 84	31	25.0	0.0	0.0	695.7	76.6	264.8	527.1	15.10	1422.	1249.	173.	87.0	211.	84.7
13 06 84	77	25.0	0.0	0.0	689.1	88.0	266.2	529.0	16.93	1599.	1408.	191.	87.5	218.	86.1
14 06 84	103	25.0	0.0	0.0	686.4	78.8	268.0	529.0	14.93	1418.	1229.	189.	86.3	211.	84.9
19 06 84	10	25.0	0.0	0.0	534.8	84.8	266.2	513.7	9.42	927.	736.	191.	79.3	186.	79.7
20 06 84	21	25.0	0.0	0.0	727.1	65.1	261.3	529.1	9.94	1013.	840.	173.	82.9	191.	81.3
02 07 84	39	25.0	0.0	0.0	590.6	80.1	243.4	529.0	11.77	1230.	1068.	162.	86.7	195.	84.4
26 07 84	53	32.5	0.0	98.4	520.8	86.9	236.4	527.7	9.70	1064.	900.	163.	84.6	177.	83.5
01 08 84	77	27.9	44.6	93.3	789.7	85.6	262.1	529.4	15.87	1528.	1340.	188.	87.7	210.	86.4
02 08 84	16	31.3	22.5	104.0	908.8	90.6	267.4	528.5	21.58	1944.	1776.	167.	91.4	227.	88.7
03 08 84	29	32.8	13.0	158.6	718.6	91.3	265.8	529.3	16.35	1515.	1359.	156.	89.6	209.	86.6
06 08 84	27	26.4	40.3	101.5	683.2	84.2	269.9	519.6	14.20	1280.	1106.	174.	85.6	202.	83.6
07 08 84	22	27.0	19.6	108.7	754.7	89.9	266.2	528.4	16.13	1490.	1334.	156.	89.4	211.	86.2
23 08 84	14	28.8	25.5	230.6	861.9	75.7	262.3	528.7	20.46	1929.	1722.	207.	89.3	223.	88.4
24 08 84	17	25.5	8.1	131.8	745.7	80.4	249.7	526.8	16.33	1655.	1432.	223.	86.2	210.	86.6

Table 5.11.-5: Daily Average of the ASR Receiver Parameters

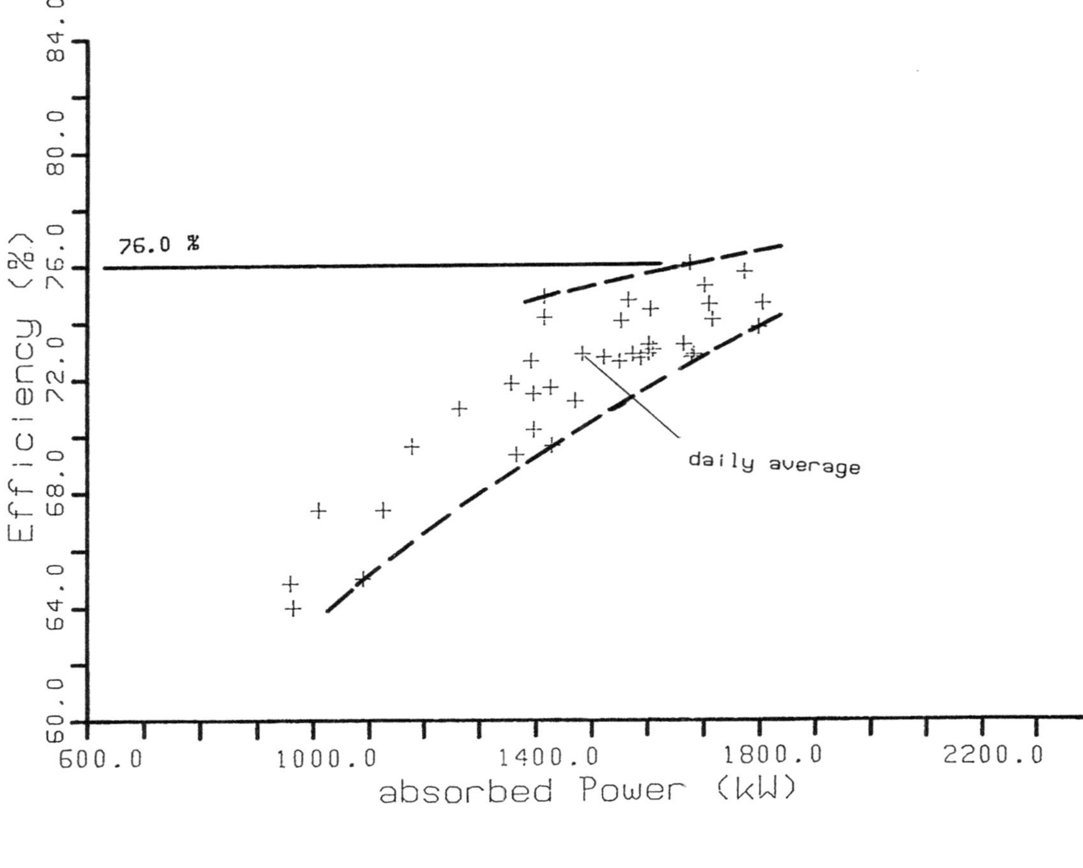

Fig. 5.11.-1: Measured Efficiency of the Cavity Receiver versus Absorbed Power

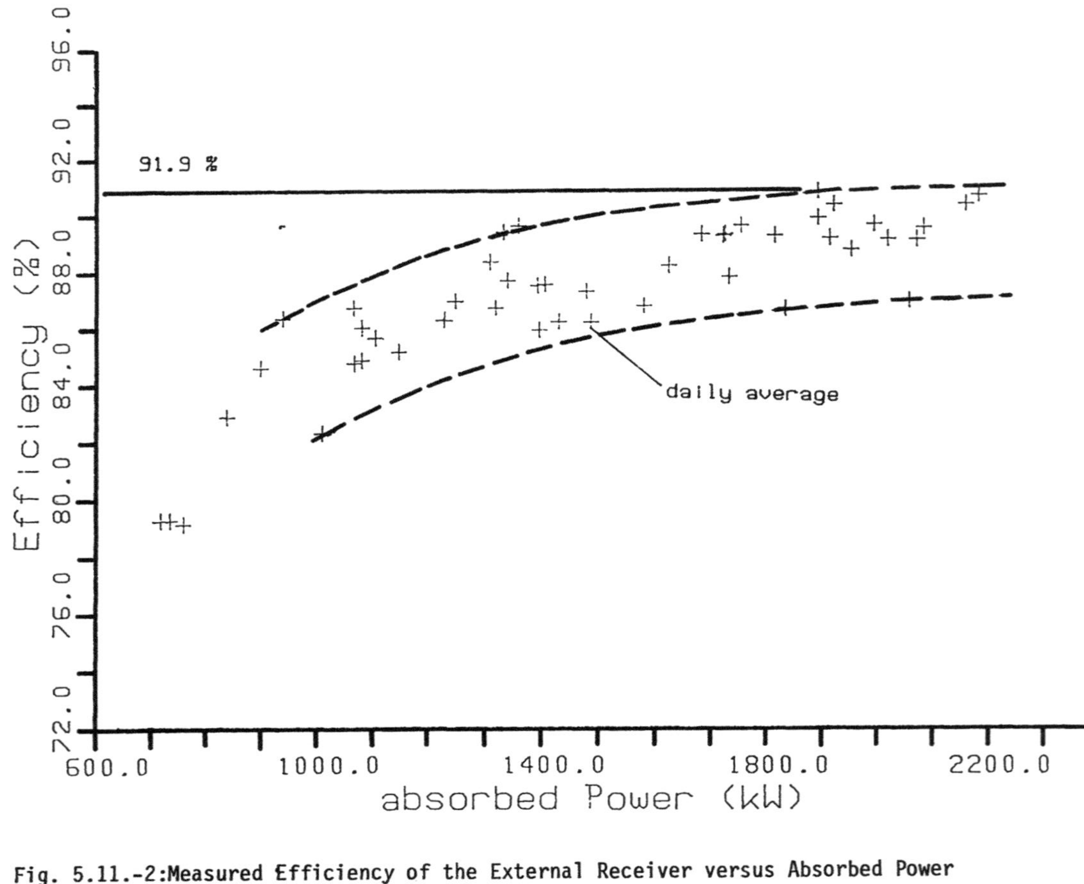

<u>Fig. 5.11.-2:</u>Measured Efficiency of the External Receiver versus Absorbed Power

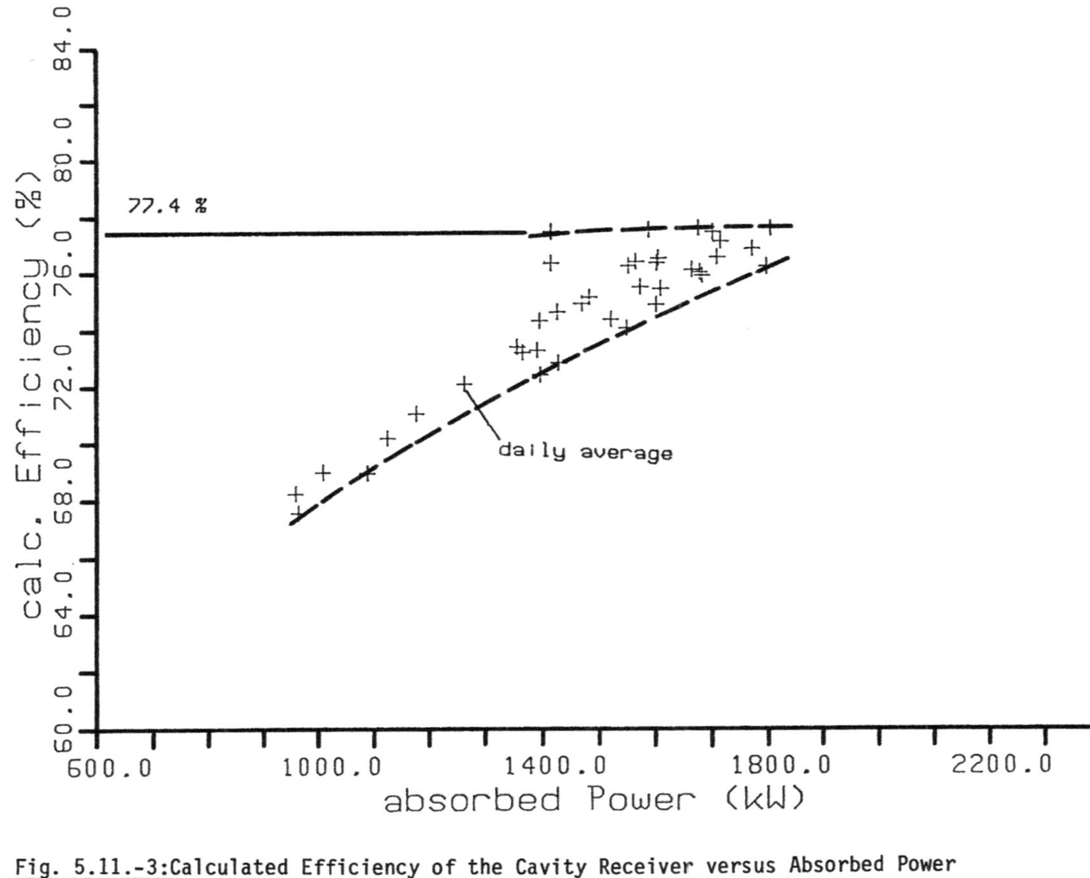

Fig. 5.11.-3:Calculated Efficiency of the Cavity Receiver versus Absorbed Power

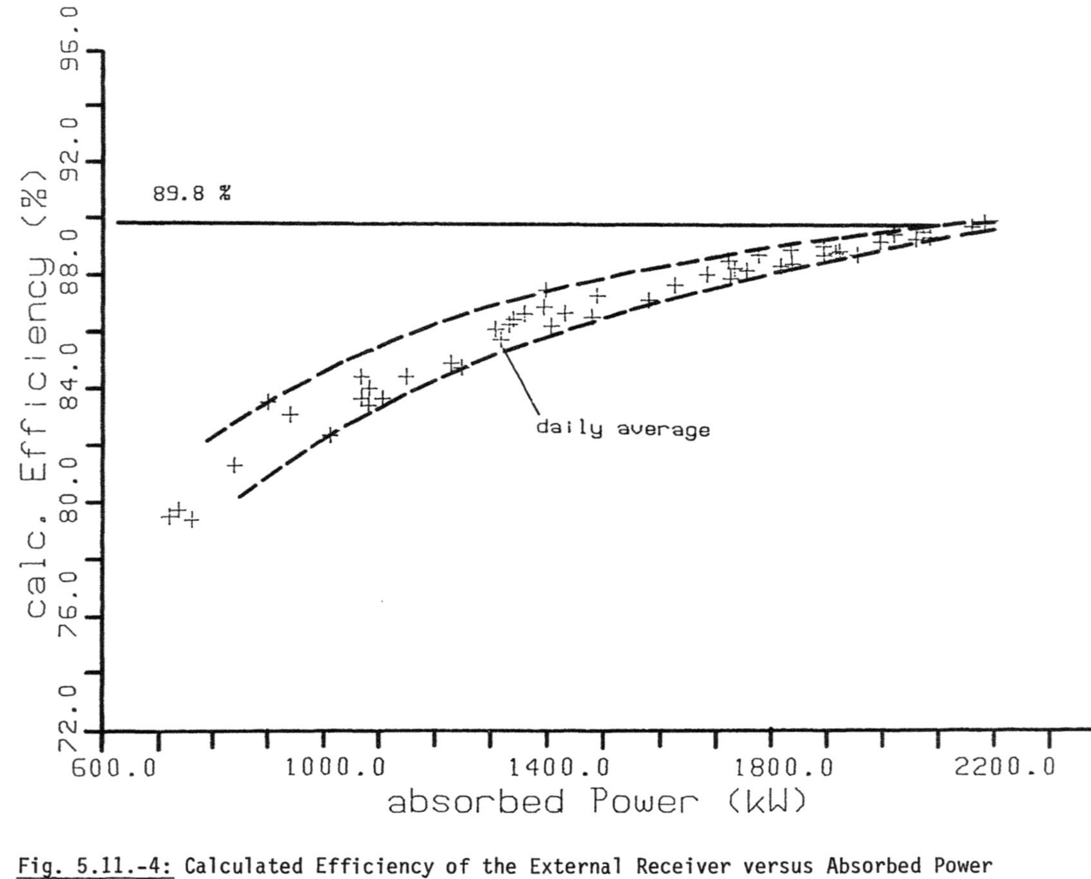

<u>Fig. 5.11.-4:</u> Calculated Efficiency of the External Receiver versus Absorbed Power

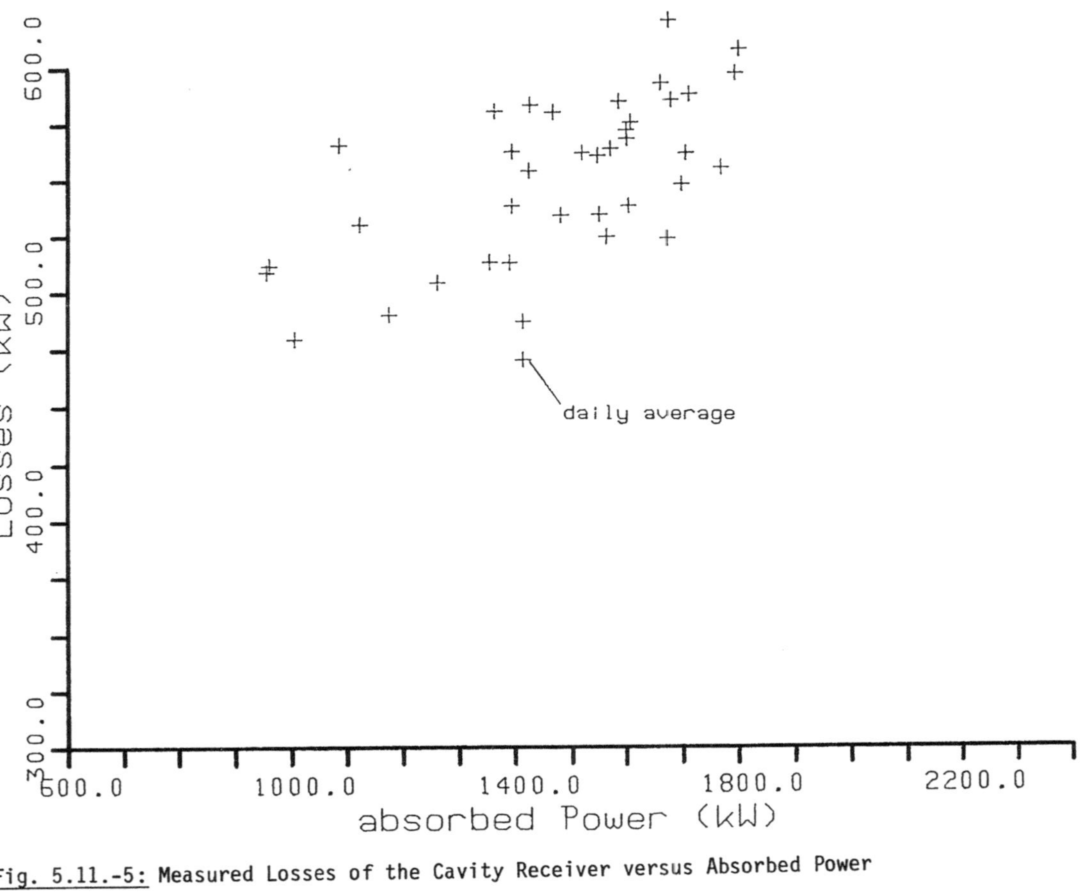

Fig. 5.11.-5: Measured Losses of the Cavity Receiver versus Absorbed Power

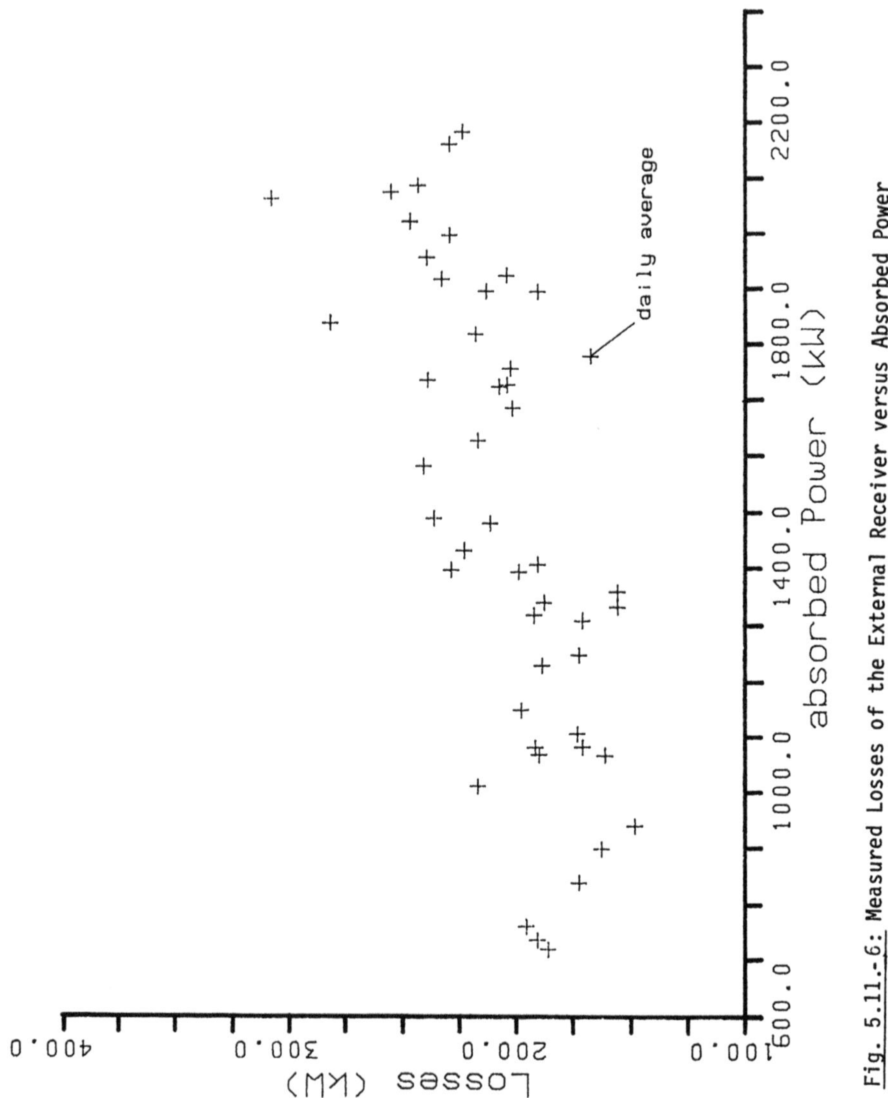

Fig. 5.11.-6: Measured Losses of the External Receiver versus Absorbed Power

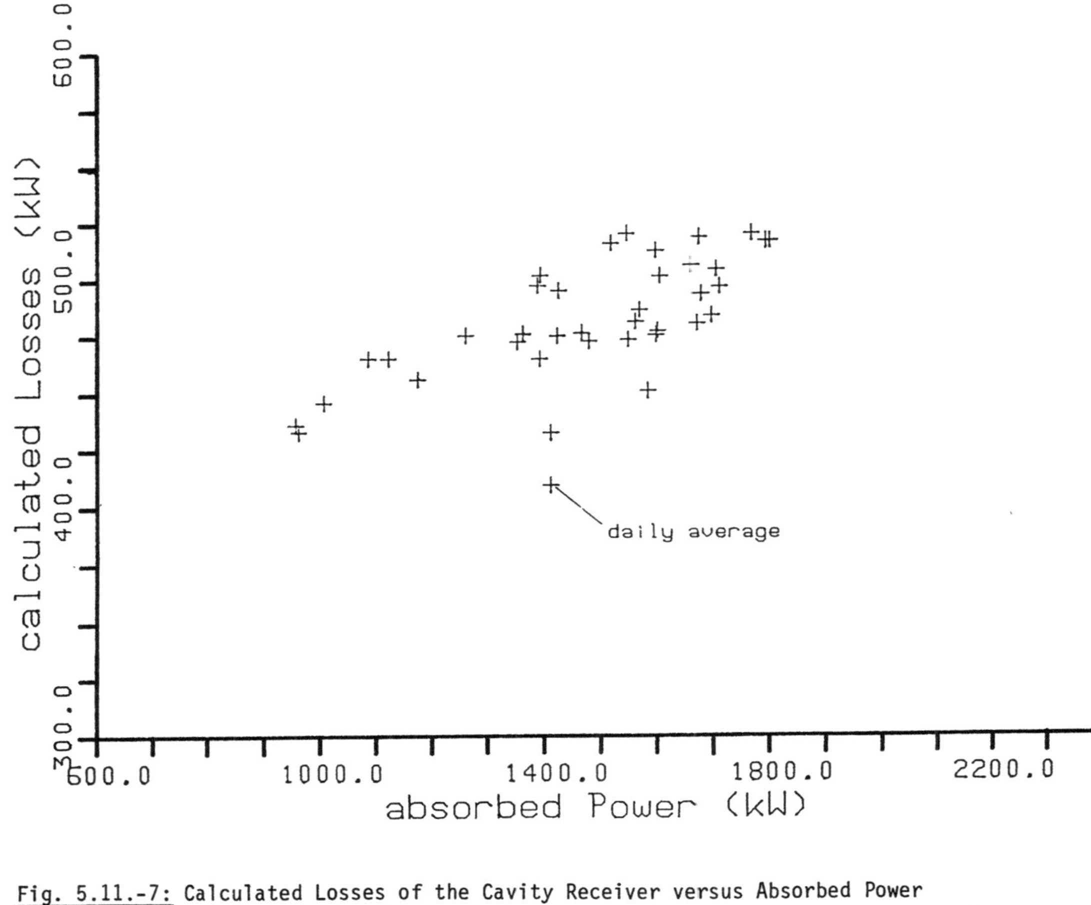

Fig. 5.11.-7: Calculated Losses of the Cavity Receiver versus Absorbed Power

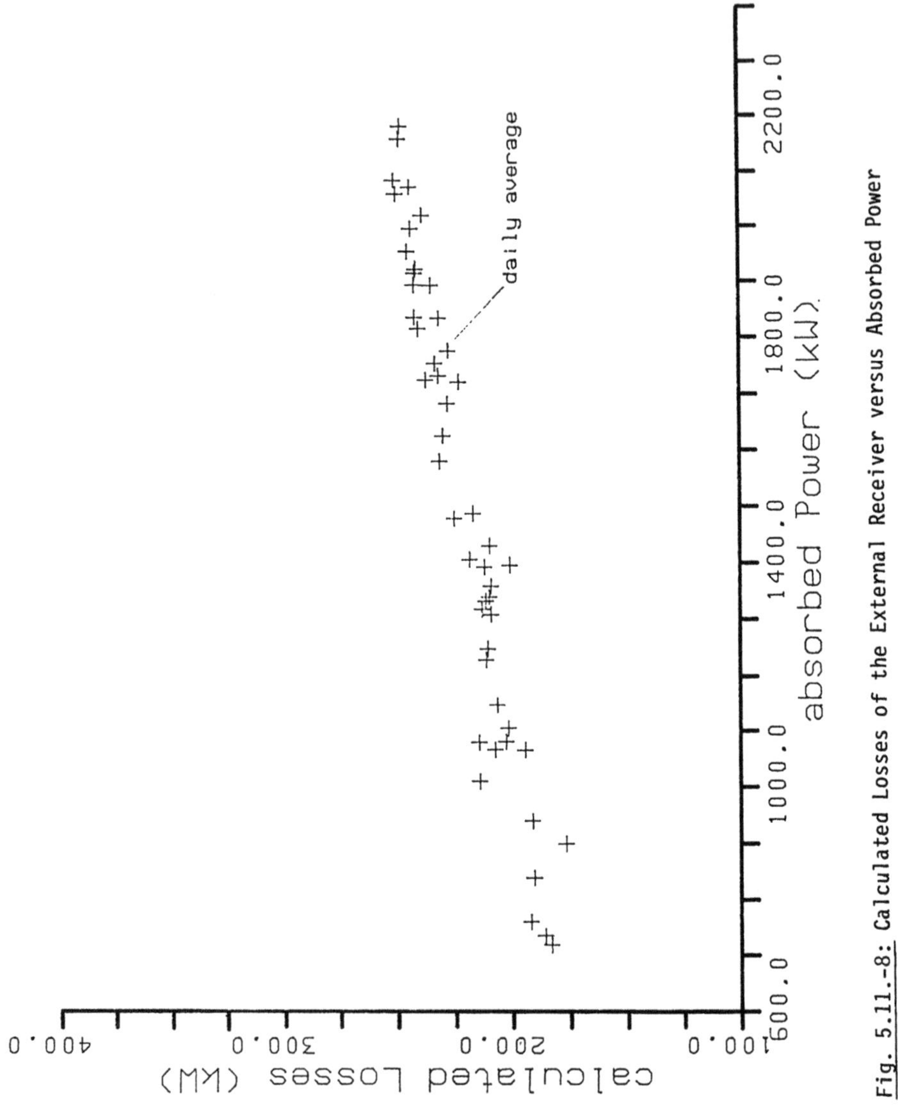

Fig. 5.11.-8: Calculated Losses of the External Receiver versus Absorbed Power

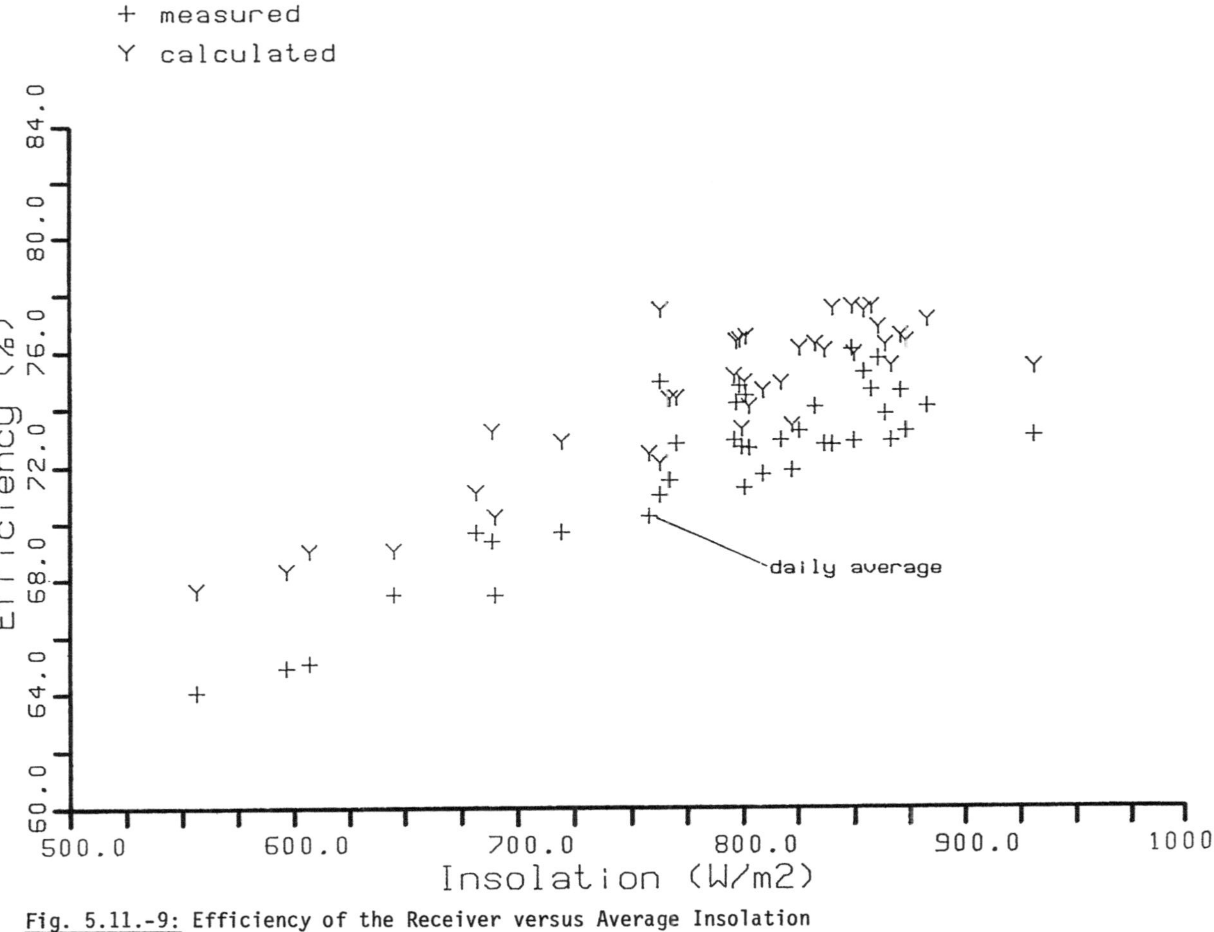

Fig. 5.11.-9: Efficiency of the Receiver versus Average Insolation

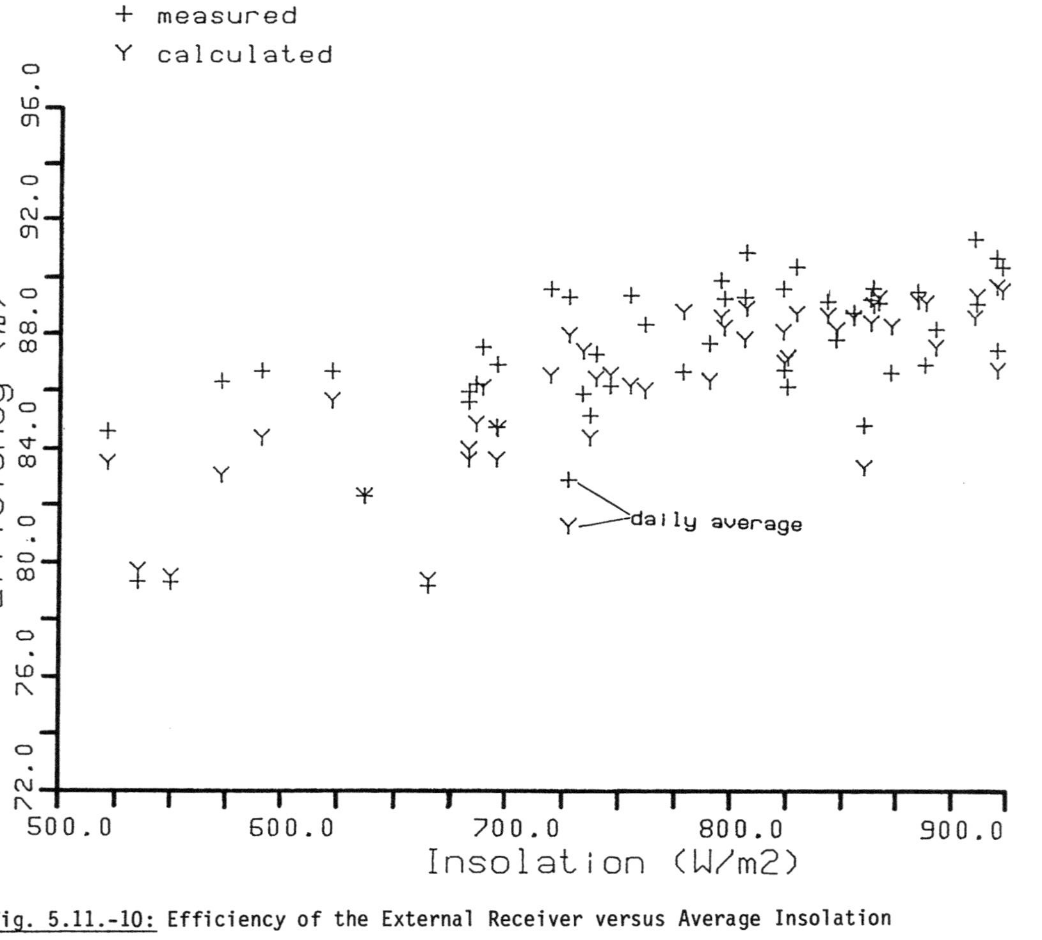

Fig. 5.11.-10: Efficiency of the External Receiver versus Average Insolation

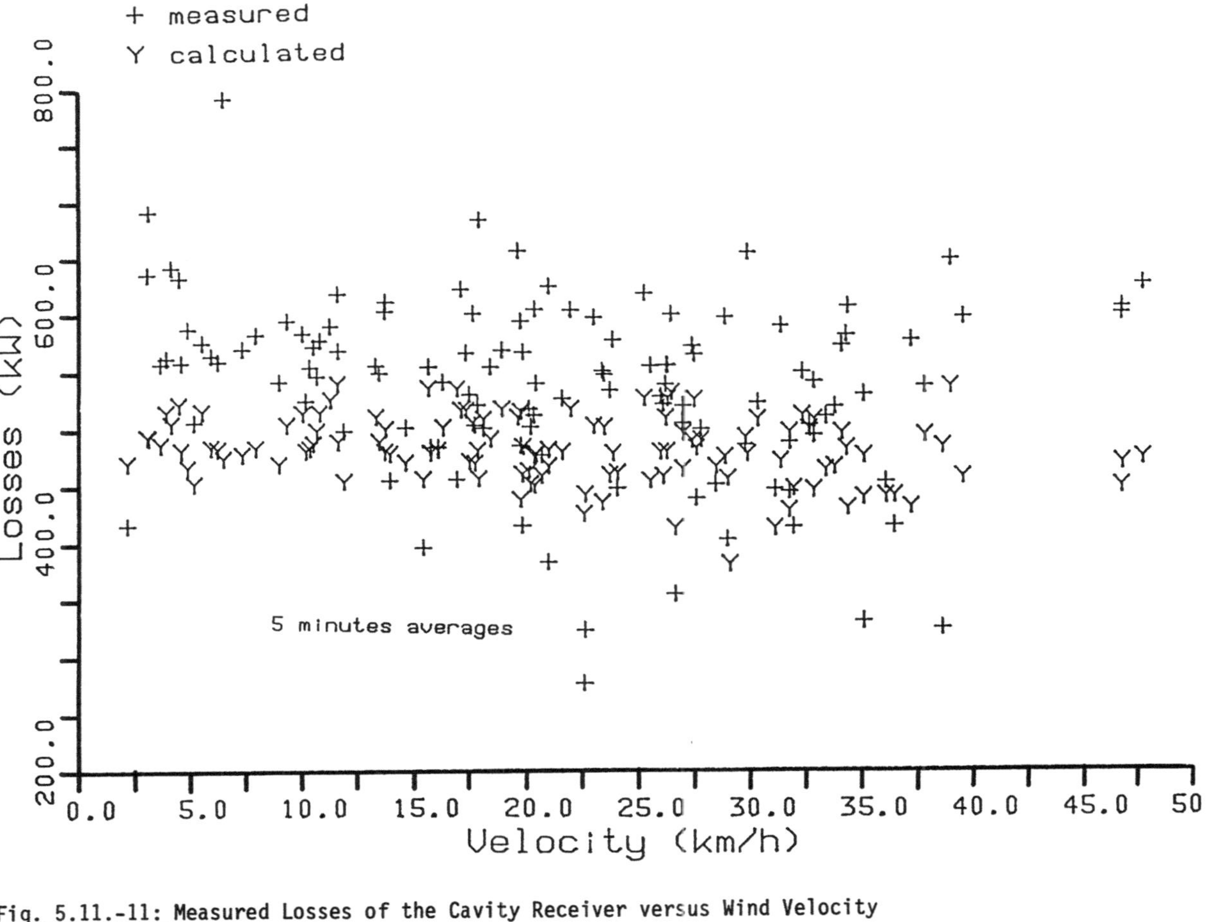

Fig. 5.11.-11: Measured Losses of the Cavity Receiver versus Wind Velocity

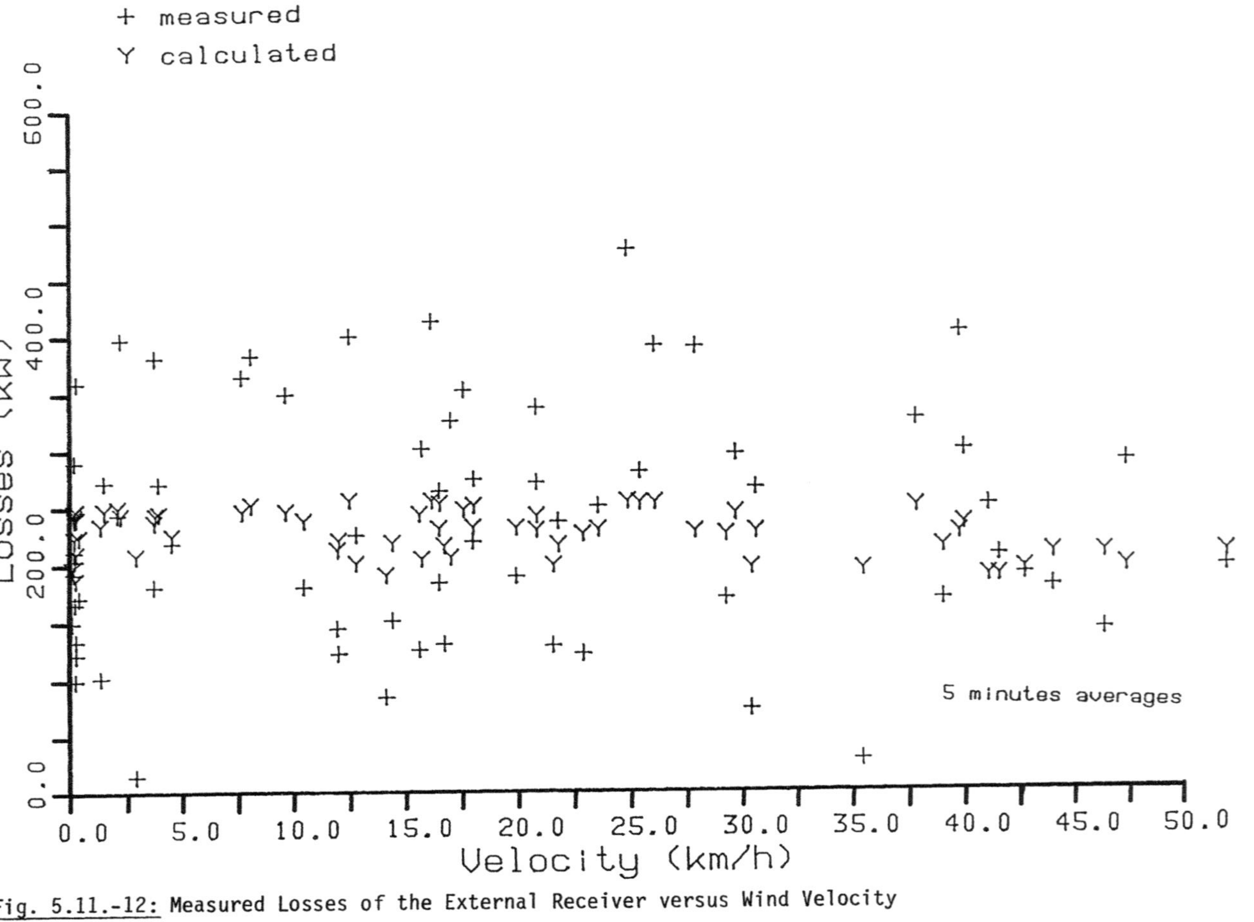

Fig. 5.11.-12: Measured Losses of the External Receiver versus Wind Velocity

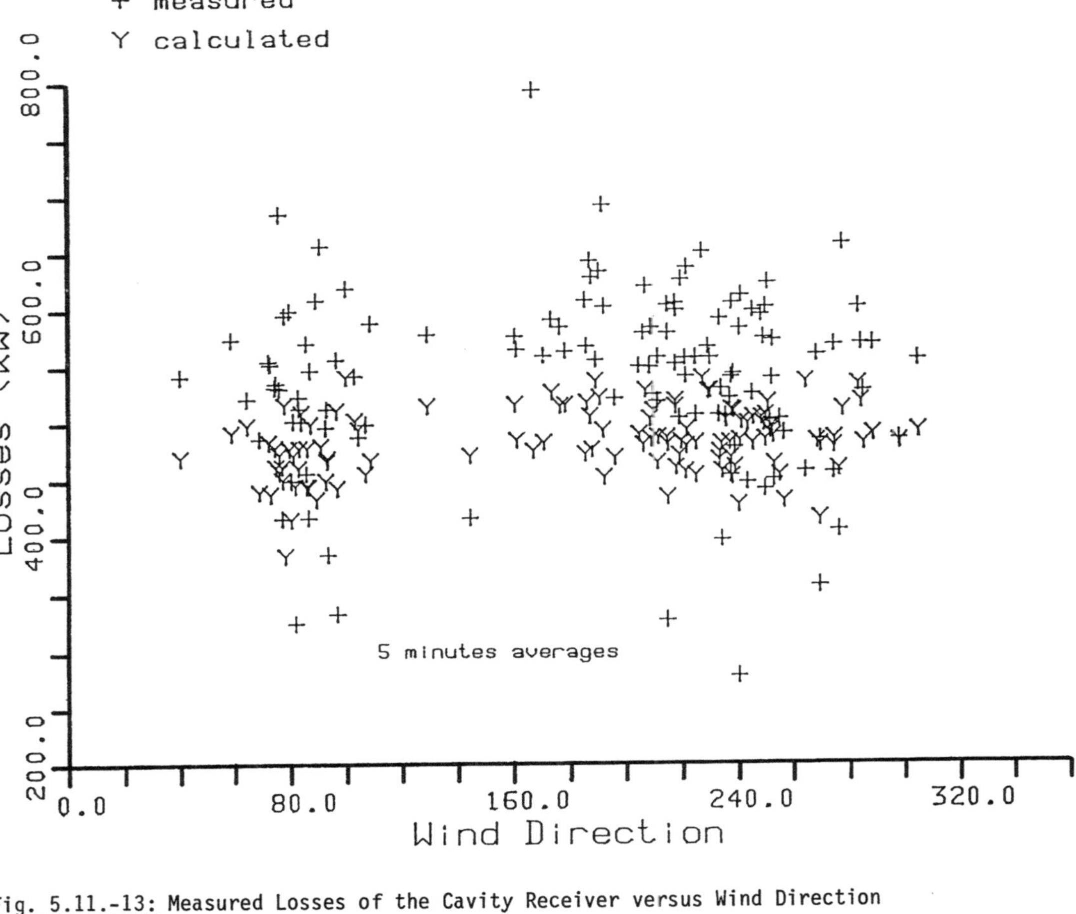

Fig. 5.11.-13: Measured Losses of the Cavity Receiver versus Wind Direction

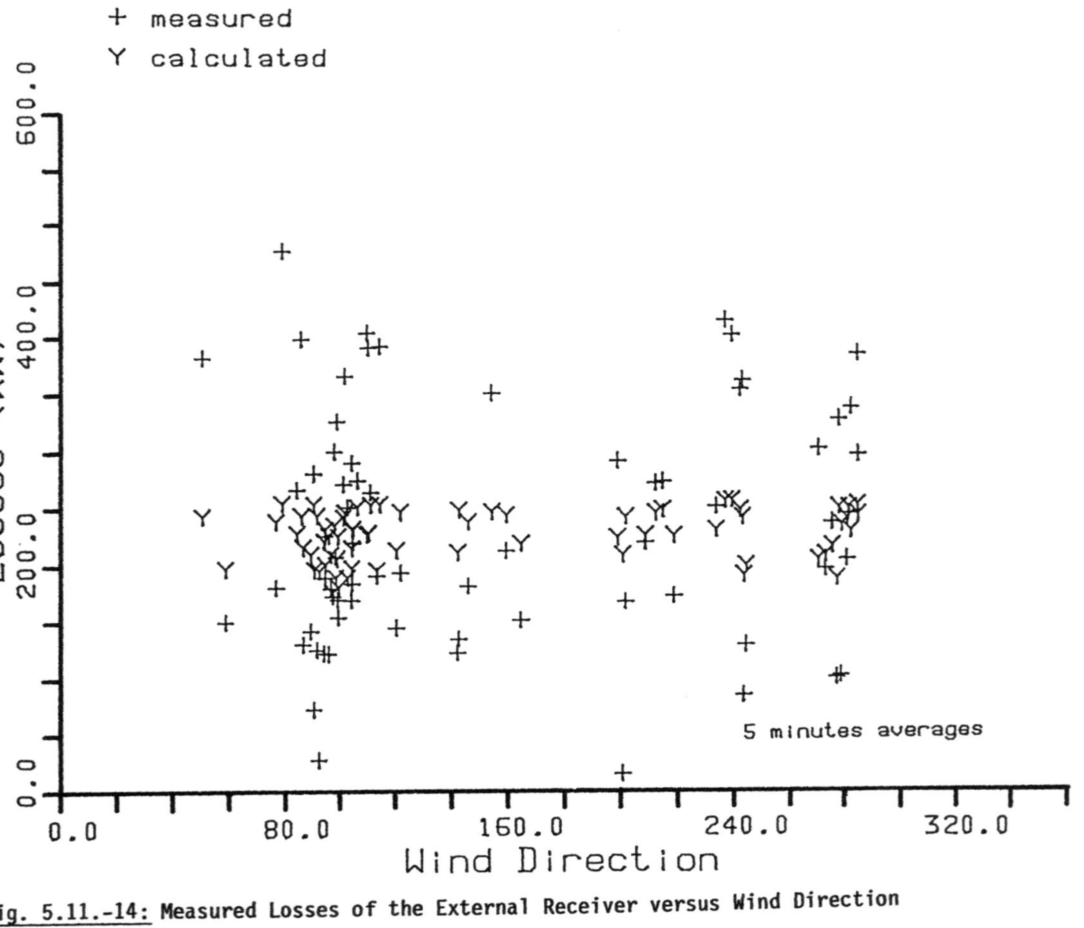

<u>Fig. 5.11.-14:</u> **Measured Losses of the External Receiver versus Wind Direction**

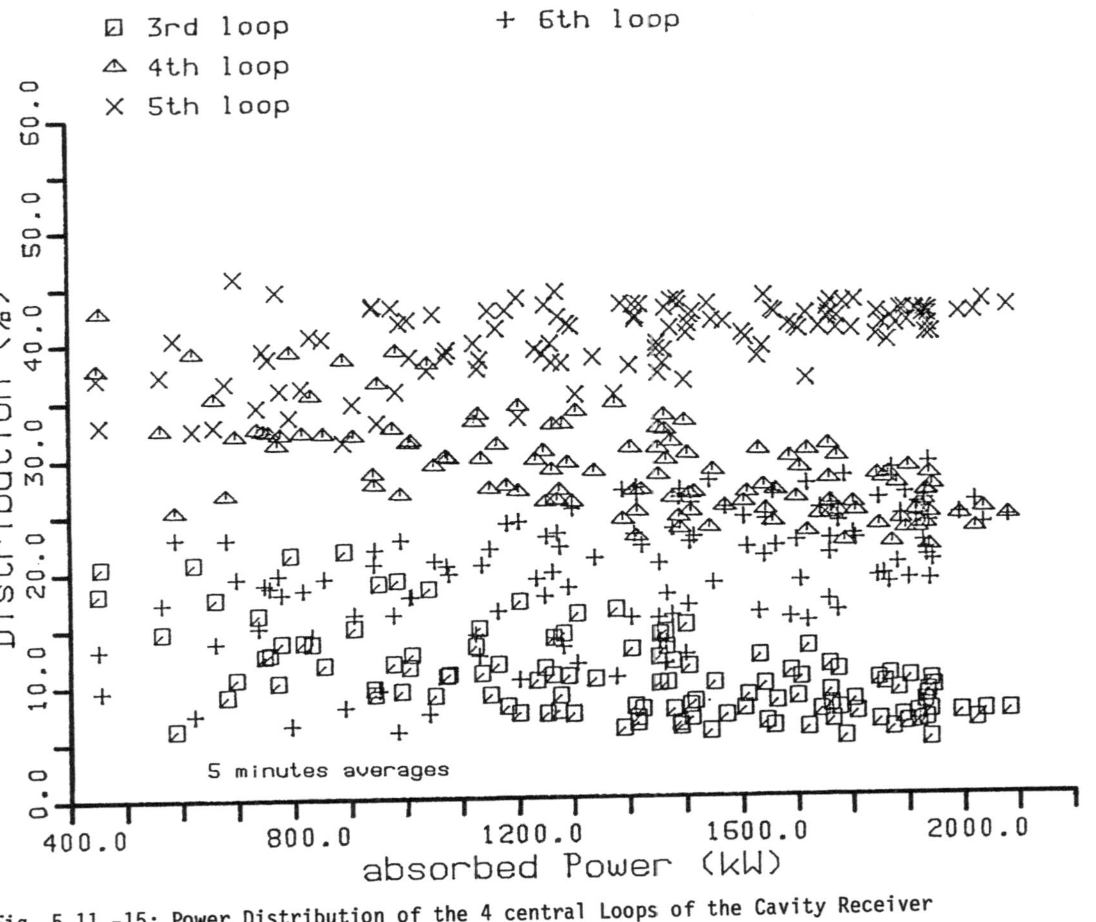

Fig. 5.11.-15: Power Distribution of the 4 central Loops of the Cavity Receiver

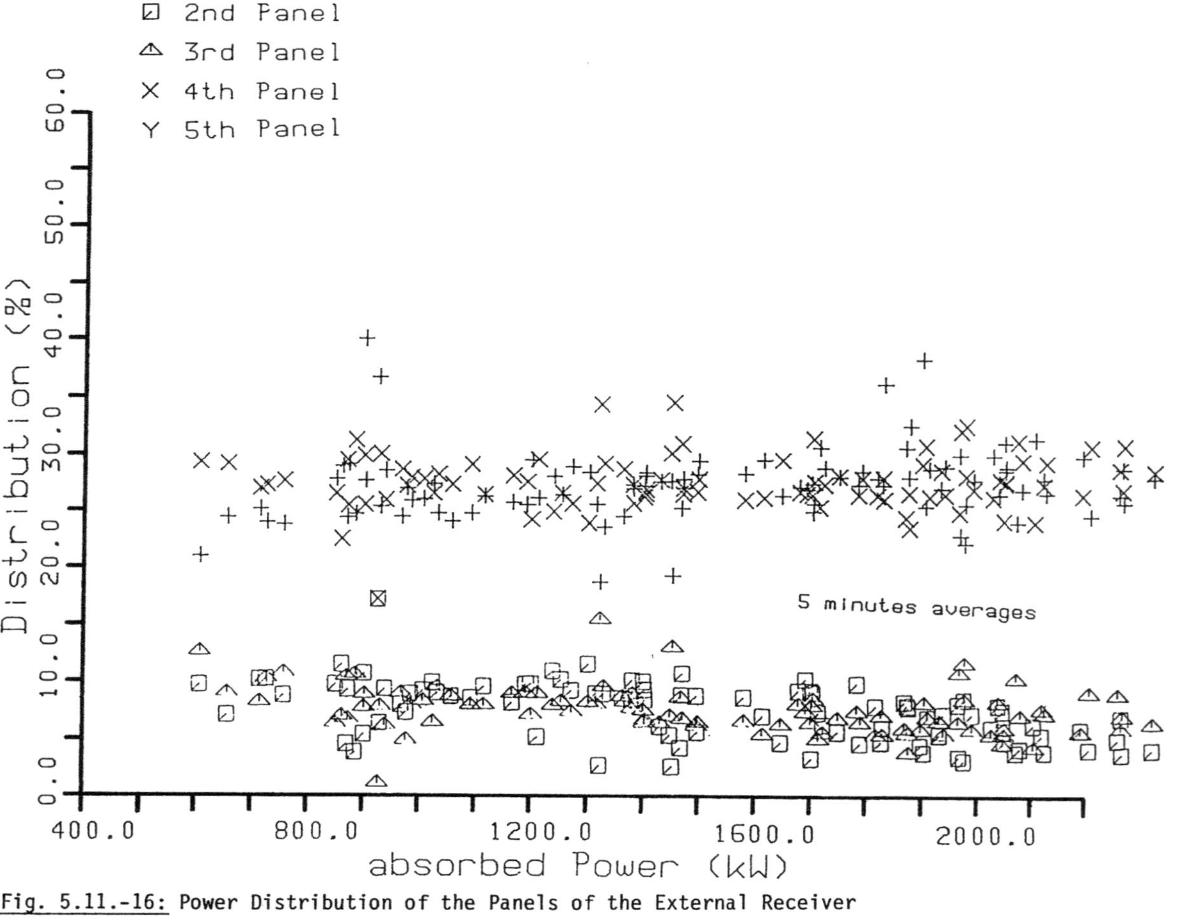

Fig. 5.11.-16: Power Distribution of the Panels of the External Receiver

THERMAL LOSSES/THERMAL INERTIA

THERMAL LOSSES/THERMAL INERTIA

INTRODUCTION

Quantifying the loss stair-step has been one of the important goals of the evaluation team at the SSPS project. This evaluation area addresses each of the loss elements as they were evaluated, and also discusses one of the interesting realizations that resulted from this evaluation - thermal inertia, a major hinderance to successful solar thermal systems application.

H. Jacobs and M. Andersson analyzed the different operating modes of the over-all system and evaluated the piping and storage tank losses over the various operating modes. The report on this evaluation is in the report LOSSES OF THE PIPING AND TANKS. The normal cool down characteristics of the hot and the cold sodium storage tanks is presented. The energy loss, when sodium is kept flowing through the receiver during non solar hours, is also discussed. We are introduced to the subject of thermal inertia.

The conclusions are:

1 - Losses and thermal inertia in the pipes and tanks limit plant operation.
2 - The difference in normal receiver outlet temperature and the PCS operating temperature is small, therefore losses in the storage tanks are very important.

Looking next at the Power Conversion System (PCS), the losses due to low sodium temperature, start-up losses, and losses during normal operation, are evaluated by H. Jacobs and R. Carmona and reported on in the document PCS LOSSES.

The conclusions of this report are:

1 - The Power Conversion System requires more energy than the rest of the system can collect.
2 - The amount of thermal losses observed is a function of the time since the PCS was last operated.

Although it would seem appropriate for a report on the receiver losses to be included in the preceding evaluation area, Receiver Performance, the report REMARKS ON RECEIVER BEHAVIOR is included in the loss evaluation area in order to include all loss areas in one section. C. Selvage used this report to formalize several internal reports addressing the area of receiver losses. Work by several previous ITET members where receiver absorptivity, reflectivity and conductivity were measured and reported on, are referenced in this report with each referenced report duplicated for future reference.

The conclusions are:

1 - Each of the loss elements in the system are important and most have been measured.
2 - The accuracy of these measurements could be improved.

A very serious loss area, that remains a serious problem, is the requirement to trace heat to all of the containment metal for the sodium. This is not a problem unique to a sodium cooled system but any system whose coolant solidifies at temperatures above ambient, such as molten salt. P. Wattiez and A. Cuadrado surveyed the electrical power consumed by the SSPS/CRS trace-heating system and provided the report CRS PARASITIC CONSUMPTION: THE TRACE HEATING.

The conclusions are:

1 - Trace heating uses much of the energy generated.
2 - Trace heating of the SSPS plant can be reduced by careful plant operational management.

The intermittency of the solar energy source makes imperative the consideration of time dependent effects in the evaluation of solar thermal plants. At the SSPS project, the "loss stair-step" concept, which guided most of the early evaluation work, has gradually been complemented by an "inertial ladder" concept in which different subcomponents have characteristic time responses and delays which have an important effect on the actual plant performance. This "paradigm shift" is discussed in some detail by C. Selvage in the report IMPLICATIONS FOR DESIGN AND OPERATION.

The conclusions are:

1 - Thermal inertia is a major problem in the SSPS/CRS.
2 - Thermal inertia can be designed around.

PIPING AND TANK LOSSES

Heinz Jacobs and Mats Andersson, ITET

1. INTRODUCTION

In this paper, the thermal losses for the main piping and tanks in the CRS-system are presented. The main objective is to provide the necessary data for calculating the overall losses in the system at normal operation conditions. The calculations are based mainly on data from the CRS-DAS (Data Acquisition System) collected during the period May to September 1984, after the calibration phase which was finished at the end of April.

2. SYSTEM DESCRIPTION

Fig.6.1-1 shows a simplified drawing of the system with the main pipes and the two tanks; those treated here are marked with numbers. These pipes are those that are used during normal operation: filling the hot storage tank from the receiver and circulating hot sodium through the steam generator producing steam. The pipes are insulated with mineral wool mats, 160 mm thick for the hot (530°) and 130 mm for the cold (275°) pipes.

3. CALCULATION METHOD FOR PIPE LOSSES

Of the four pipes that are listed in the following table, only three have been treated with accurate methods. As there does not exist any temperature sensor in the inlet to the cold storage, pipe 4 had to be approximated using results from the other pipes.

№	code	description	diameter (mm)	length (m)	supports (numbers)	temp. (°C)
1	LK01-BR01/02	cold storage-receiver	80	91.7	25	275
2	LK01-BR03	receiver-hot storage	80	86.4	30	530
3	LK02-BR01/02	hot storage-steam generator	80	40.3	19	530
4	LK02-BR03	steam generator-cold storage	50	25.6	17	275

Table 6.1.-1: Description of the Pipes

Except for pipe 2, all results are obtained by using normal operation data stored as five-minute averages on tapes. For pipe 2, calculations have also been done using data from two reverse flow tests.

The losses have been calculated using the enthalpy difference between inlet and outlet temperatures of each pipe. The well known relations for density and specific heat for sodium are not presented here.

3.1 Theoretical Calculations

In order to get some approximate results for comparison, the thermal losses have also been calculated theoretically using the basic equation:

$$\dot{Q} = k . A . \Delta \vartheta \; ; \quad \frac{s}{\lambda} \gg \frac{l}{\alpha} \quad \Longrightarrow$$

$$\frac{\dot{Q}}{l} = \frac{\lambda}{s} . \frac{\pi . (Do - Di) . \Delta \vartheta}{\ln \left(\frac{Do}{Di} \right)}$$

The losses through supports and hangers, etc. are not calculated.

3.2 Steady State Criteria

All calculations have been done using data selected by criteria chosen to ensure steady state conditions. The temperature difference has been limited to a value 5 to 7° with a maximum variation of 0.3 to 0.5° during at least half an hour. Higher ΔT's would be preferable but this would conflict with the limiting of the flow rates (low accuracy at low flows).

4. RESULTS OF PIPE LOSSES

The results from both the theoretical calculations and those obtained by using measured data are presented in Table 6.1.-2.

Calculations Based on Measurements				Theoretical Calculations
Pipe No	Total Loss (kW)	$T_{av}-T_{amb}$ (°C)	k (W/m K)	(only for pipes)
1	16.3	240	0.74	0.24
2	29.3	403	0.84	0.40
3	23.2	480	1.20	0.49

Table 6.1.-2: Result of Pipe Losses,measured and calculated

5. LOSSES OF THE SODIUM STORAGE TANKS

The thermal behavior of the storage tanks has an important influence on the way to operate the whole CRS plant. In particular, the cooling down of the hot storage limits the amount of electrical energy which one can produce in one day.

It is not only the energy loss but also the fact that one cannot use the sodium with temperatures under 480°C. Since it is not foreseen to recirculate this sodium from the hot storage to the receiver, one needs to get rid of it by producing minor quality steam. The energy finally goes to the cooling tower.

The cooling down of the tanks is determined by the mass of steel, quality of insulation, and mass of sodium in the tank.

If one keeps the hot sodium pump running (in bypass from the hot storage, through the pump, and back into the tank) during the night, it cools down faster than without pump operation.

To avoid the 'waste of energy due to low sodium temperature', it is possible to increase the mass of sodium in the tank during the night. Since the losses are supposed to be constant (Ref.1), the temperature drop would become a lower value.

6. THEORETICAL CALCULATION OF THE THERMAL LOSSES

To aid in this calculation, the thermal losses are divided into several groups:

Q_1 - Losses through the insulation
Q_2 - Losses through the supports
Q_3 - Losses through the inlet and outlet tubes
Q_4 - Losses through the insulation of the repair-caps (only with the cold storage).

The calculations are based on the following measured data, taken in October 1982:

- sodium temperature in the tank
- surface temperature
- temperature of the support
- ambient temperature in the sodium hall

The calculations were performed using the heat transfer equations described in Ref.1. The results are the heat transfer coefficients and the loss distributions for both tanks (Tab.6.1.-3).

	Cold Storage	Hot Storage
Heat Transfer $W/m^2/K$	0.588	0.453
through insulation (%)	57.3	60.4
through supports (%)	36.8	35.8
through tubes (%)	3.1	3.8
through repair-caps (%)	2.7	---

Table 6.1.-3: Results of Theoretical Calculations of Tank Losses

7. UNDISTURBED TANK LOSSES

To verify the calculated loss figures, the cooling down behavior of the tanks was investigated. Undisturbed means that the sodium pump was not running during the test. The calculation method presupposes that the cooling down follows the theoretical equations.

The cooling down behavior can be influenced by the electrical trace-heating. Due to the higher tank surface temperature during the trace-heating operation, the losses are slightly higher. For the cold storage, only the 'trace-heating' figure is available.

Tab.6.1-4 presents the results of this test. The dimension is watts per square meter of tank surface and per Kelvin degree.

Heat Transfer Number $W/m^2/K$	Cold Storage	Hot Storage
without trace-heating	---	0.452
with trace-heating	0.644	0.468

Table 6.1.-4: Heat Transfer Figure for Undisturbed Cooling Down of Tanks

8. DISTURBED COOLING DOWN OF THE HOT STORAGE

Disturbed cooling down means that the hot sodium pump keeps running in bypass. Additional thermal losses of pipes and pump casing produce a greater temperature drop.

The cold storage provides the receiver and steam generator with sodium to keep the components warm during non-operational time. These additional losses cause additional temperature drops in the tank.

Fig. 6.1.-2/3 show the difference in temperature drop between disturbed and undisturbed cooling down.

The temperature of the cold storage has a high influence on plant efficiency because the PCS needs at least 480°C inlet temperature. By keeping more sodium in the tank during night, the desired temperature can be achieved earlier in the day.

Tab.6.1-5 shows how the tank cools down, depending on level and inlet temperature.

T_{start}	level = 0.4 m	1 m	2 m
525°C ----►	450	485	498
500°C ----►	429	462	474
475°C ----►	406	439	451
450°C ----►	386	416	427

Table 6.1.-5: Temperature of the Cold Storage Tank After Cooling Down 12 Hours

9. ACCUMULATION OF RESULTS

In order to present the achieved results in a way that one can compare the different losses to each other, a summary has been made for the two most common modes of operation:

- energy collecting (energy to receiver and energy to steam generator)
- recirculation of sodium (during nights and nonoperational time)

Tab.6.1.-6 presents the loss figures for both operational modes.

	Losses Measured During Test		Losses While Collecting Energy		Losses During Recirculation	
	Temp K	Loss kW	Temp K	Loss kW	Temp K	Loss kW
Pipe 1	240	16.3	240	16	240	16
Pipe 2	403	29.3	480	35	240	17
Pipe 3	480	23.3	480	23	240	12
Pipe 4	-	-	-	-	240	(12)
Pipe 5	-	-	-	-	240	(8)
Hot Storage	500	23	500	23	500	23
Hot Bypass	-	-	-	-	500	(10)
Cold Storage	270	17	270	17	270	17
Cold Bypass	-	-	-	-	270	(5)
Receiver	500	230/500	500	230/500	270	7/17
Steam Generator	-	-	500	(3)	270	(2)
TOTAL				344/614		129/139

() = estimated
/ = Sulzer/ASR

Table 6.1.-6: Summary of Losses

10. REFERENCES

1. H. Jacobs, "Thermal Losses of the Sodium Storage Vessels of the CRS", SSPS Technical Report Nº 7/83, November 1983.

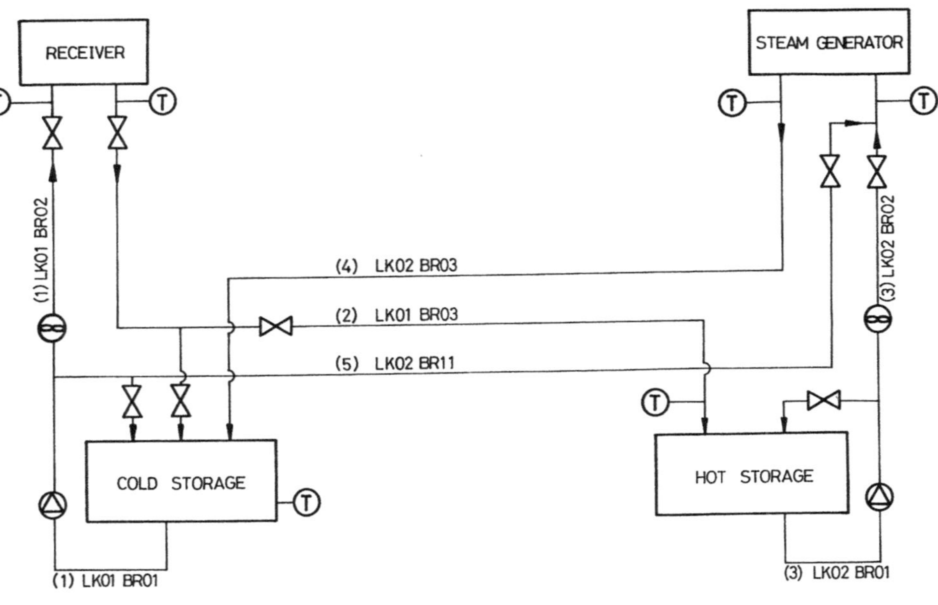

Fig. 6.1.-1: CRS-Flowchart

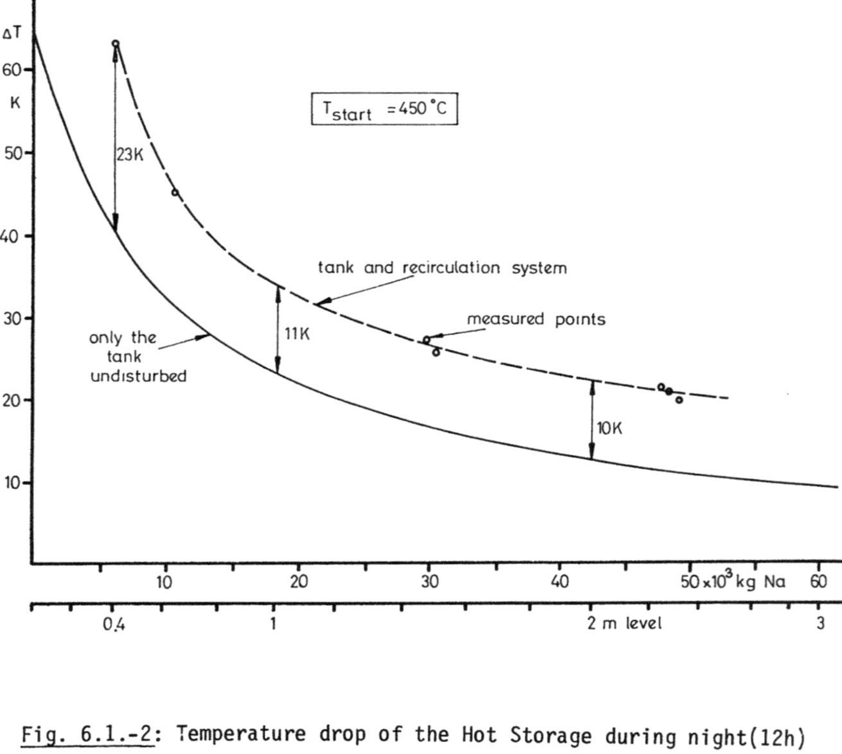

Fig. 6.1.-2: Temperature drop of the Hot Storage during night(12h)

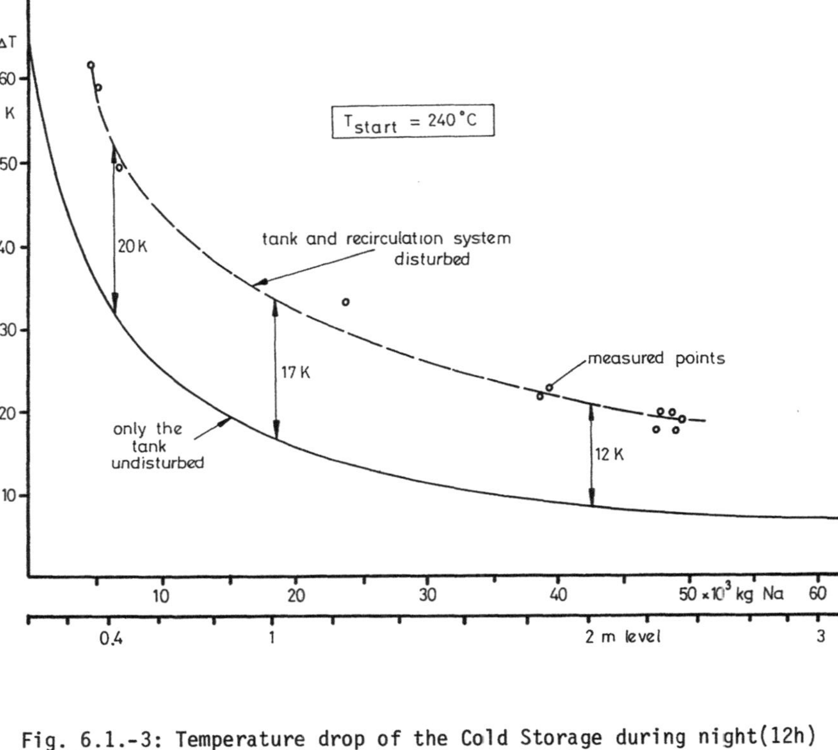

Fig. 6.1.-3: Temperature drop of the Cold Storage during night(12h)

POWER CONVERSION SYSTEM

Heinz Jacobs and Ricardo Carmona, ITET

1. DESCRIPTION OF THE POWER CONVERSION SYSTEM

The Power Conversion System (PCS) (Fig.6.2-1)consists of a Sulzer, sodium heated, steam generator that supplies a Spilling power conversion system based on a six-cylinder steam reciprocating engine.

Steam Generator

The steam generator has a major influence on the Spilling system design. This once-through-type heat exchanger must be operated within strict limits to minimize thermal shock due to load changes. The change should not exceed 5%/minute. In addition, since the primary fluid is sodium, the temperature of the feedwater must be controlled very closely to avoid water temperatures below the sodium melting point.

By-Pass System

To avoid thermal shocks to the steam generator, an automatic steam expansion system has been provided to by-pass the engine. The by-pass system allows the PCS to be started, preheated, and loaded, independent of the steam motor. Once the by-pass is in full operation, the steam motor can be run as desired. The by-pass has proved invaluable -- it allowed a means of unloading the hot sodium storage tank when the temperature was too low for operation of the engine.

Steam Engine

The six-cylinder Spilling engine is basically a standard machine that expands the steam generator output from 100 bar to 0.3 bar in five stages.

Regulation circuit

The regulation circuit maintains the flowrate of the sodium to around 10 times the flowrate of the water. During normal operation of the generator, the flowrate of the sodium is controlled by the flowrate of the feedwater and the sodium outlet temperature (Fig.6.2-2). By design, the sodium outlet temperature affects the regulation circuit very gradually. Slow changes ensure that stable sodium conditions exist, which are more important than stable steam temperatures. A regulation mechanism for constant steam temperatures would be much more complicated.

If the PCS works in the bypass mode, the load is determined by the operator in the control room. In motor-generator operation, the load is defined by the need of the generator. At the steam side of the generator, the design pressure is 100 bar. Since the end of March 1983, this pressure has been limited to 90 bar; in March 1984 it was further reduced to 80 bar.

2. START-UP PROCEDURE

Before start-up can be initiated, sodium from the cold storage tank is pumped through the steam generator until the sodium inlet temperature reaches 230°C. The start-up procedure of the PCS then begins.

The feedwater is heated by the electric auxiliary heater until the temperature of the feedwater is greater than 200°C. The feedwater pump starts and pumps steam through the water inlet pipe in the steam generator. When the water inlet temperature reaches the set-point (150-180°C), the signal "Ready for SODIUM HEAT TRANSFER SYSTEM" (SHTS) is displayed on the panel in the control room. At this point, the mixing procedure can start. During this procedure, the sodium inlet temperature increases from the temperature of the cold storage to that of the hot sodium.

To start the steam engine requires that the steam temperature be at least 480°C. Therefore, one must get rid of the low temperature sodium in the hot storage tank via the steam generator. At the same time, sodium from the receiver is increasing the mixing temperature in the storage tank. If the temperature conditions are achieved, the steam engine is able to start. The motor needs around 25 minutes for warm-up, after which the alternator is connected to the grid.

3. PCS LOSSES

The losses of the PCS can be divided into: (1) losses due to low sodium temperature; (2) start-up losses; and (3) losses during normal operation. The low temperature sodium losses contain the energy wasted during start-up. These losses are in conjunction with the status of the hot storage. The start-up losses are associated with warm-up losses. The work to determine these two kinds of loss is not yet finished; the results will be included in the final version of this paper.

Start-Up Losses

The DAS reports the sodium temperature and flowrate at the inlet to the steam generator. Using data, one may estimate the power sent to the generator. Integrating the power between the time at which the power conversion system is 'started' until the time at which the steam motor starts gives the energy required to start the steam generator. Data from 12 days chosen at random in 1984 shows that the start-up losses ranged between 1200 and 1600 kWh.

From the same sodium data at the steam generator inlet, integrating between the time when the steam motor starts and the time when power is first fed to the grid, gives the engine start-up losses. For the same 12 days, these losses ranged from about 600 to 800 kWh. Therefore, the whole PCS start-up losses average about 2100 kWh, or about 20% of all the energy fed to the PCS during a good operational day.

Losses During Normal Operation

Fig.6.2-3 and Tab.6.2-1show an overview of the energy efficiency. One cross contains the figure for one day. The numbers are all under the 22% limit. Due to load change, the energy efficiency becomes a lower value. Figure 4 shows the total efficiency for one day; the start-up is included. All figures are under the limit-line, which has the following equation:

Total Efficiency = less than 6.5% + 9.3 * electrical energy (MWh)

The difference between the limit-lines in Fig.6.2-3/4 is related to the losses due to start-up. The actual power efficiency is plotted in Fig.6.2-5 Each plotted point refers to one five-minute average. Table 2 gives an example for one day (22/8/84). All efficiency values are under the limit-line.

Obviously, the losses are subdivided into constant losses, losses due to the load (or gross power), and additional losses. The constant losses are independent of the load. Together with the losses due to the load, they produce the limit-line in Fig.6.2-4 The additional losses are the differences between this line and the actual measured values. The mathematical analysis of the limit-line produces 425 kW for constant losses and 2.5 * gross (kW) for load dependent losses.

The additional losses are in the range of 200 kW (average). One reason for these losses can be the load changing. Another possibility that should be investigated is that the steam valve in the bypass may not be constantly closed during steam motor operation.

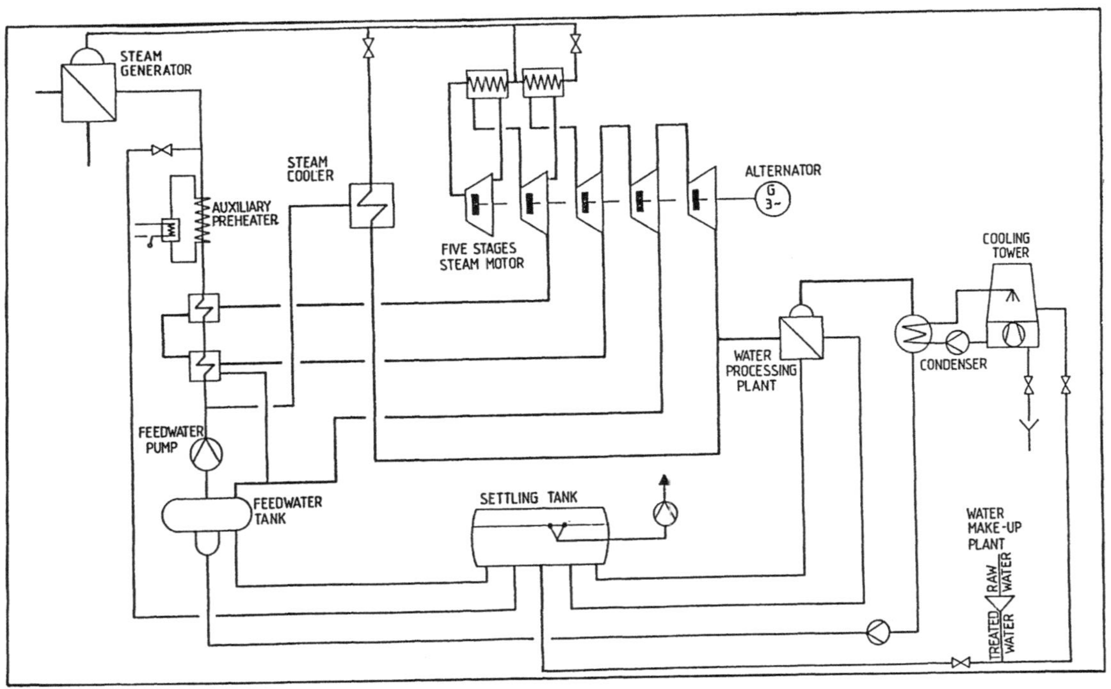

Fig. 6.2.-1: CRS-PCS Flowchart

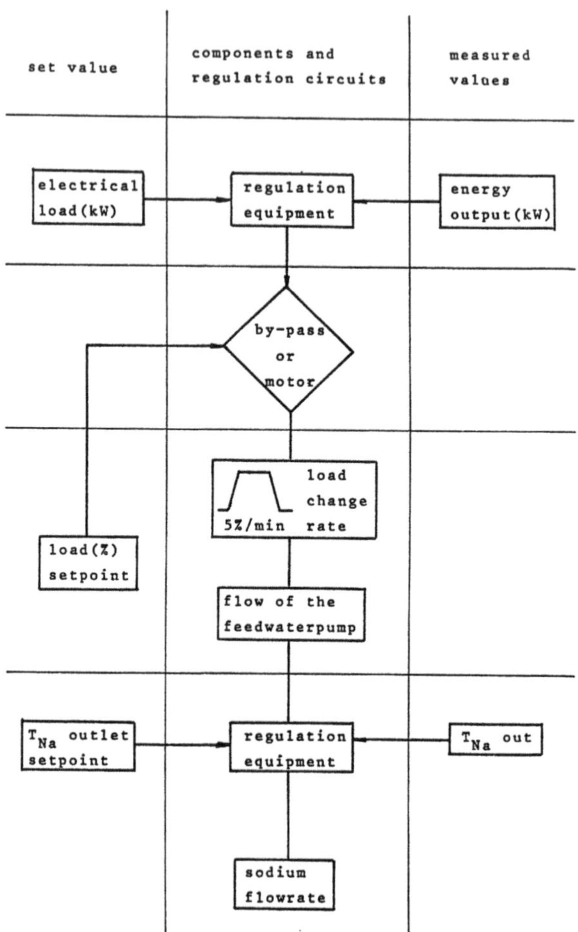

Fig. 6.2.-2: CRS-PCS Regulation System

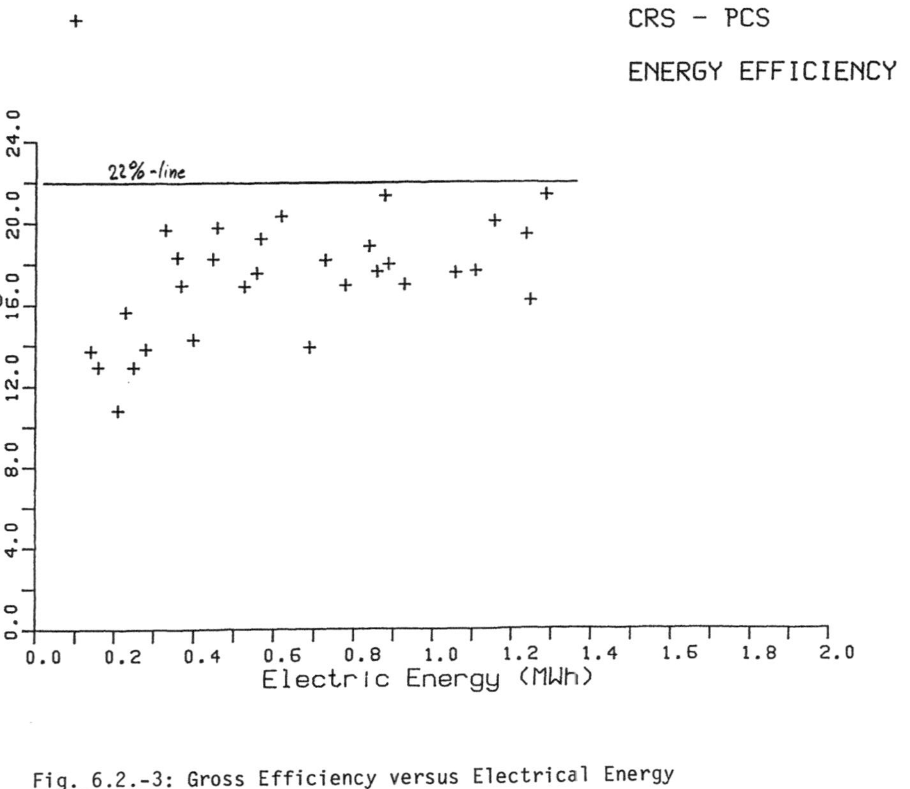

Fig. 6.2.-3: Gross Efficiency versus Electrical Energy

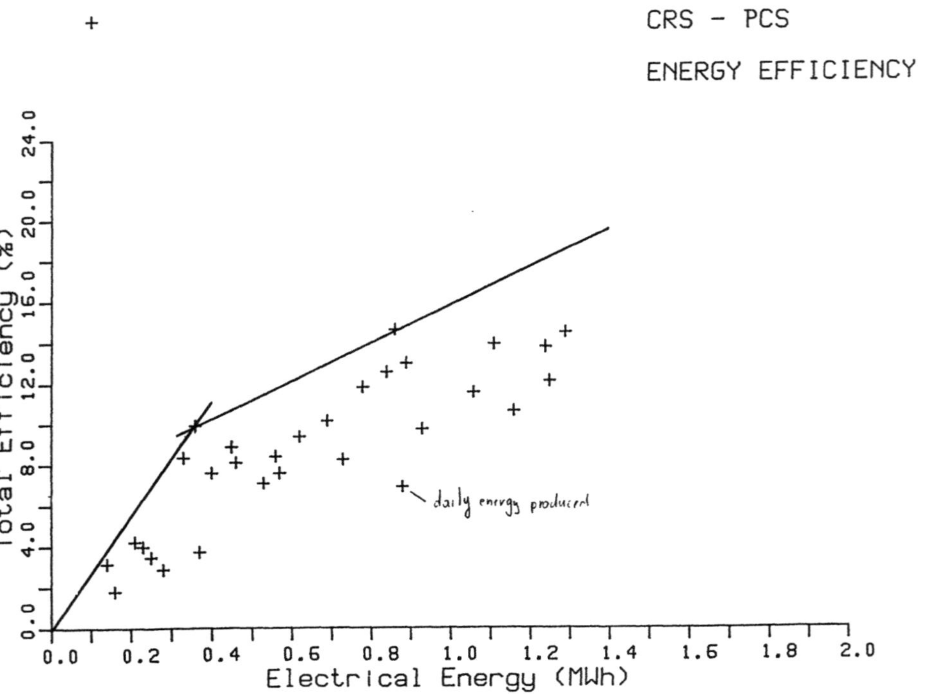

Fig. 6.2.-4: Total Efficiency versus Electrical Energy

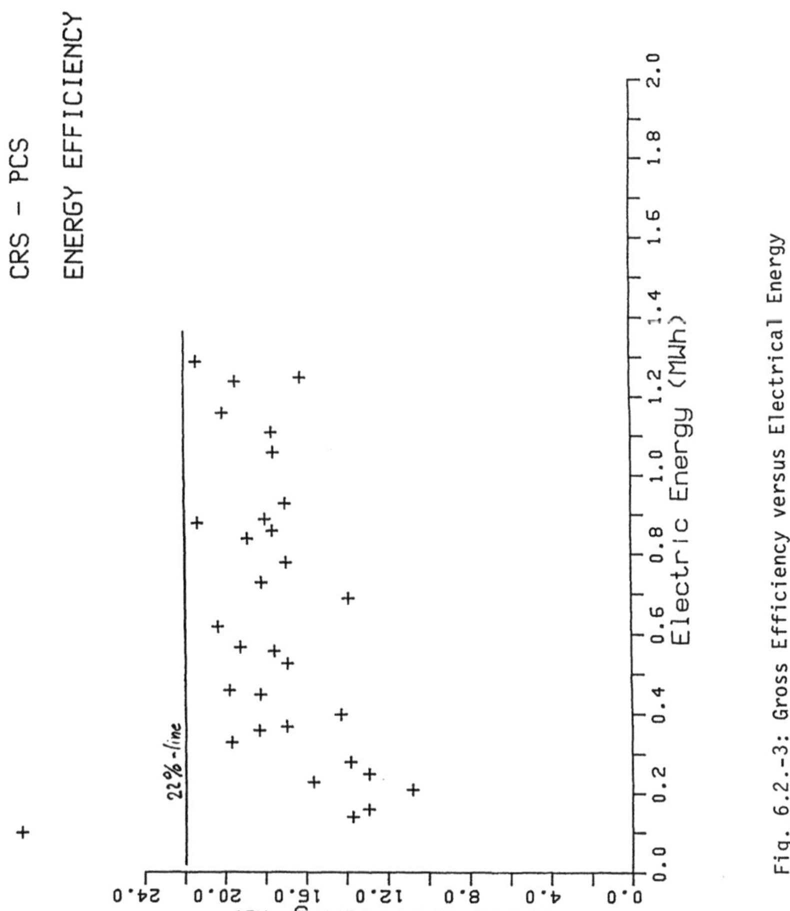

Fig. 6.2.-3: Gross Efficiency versus Electrical Energy

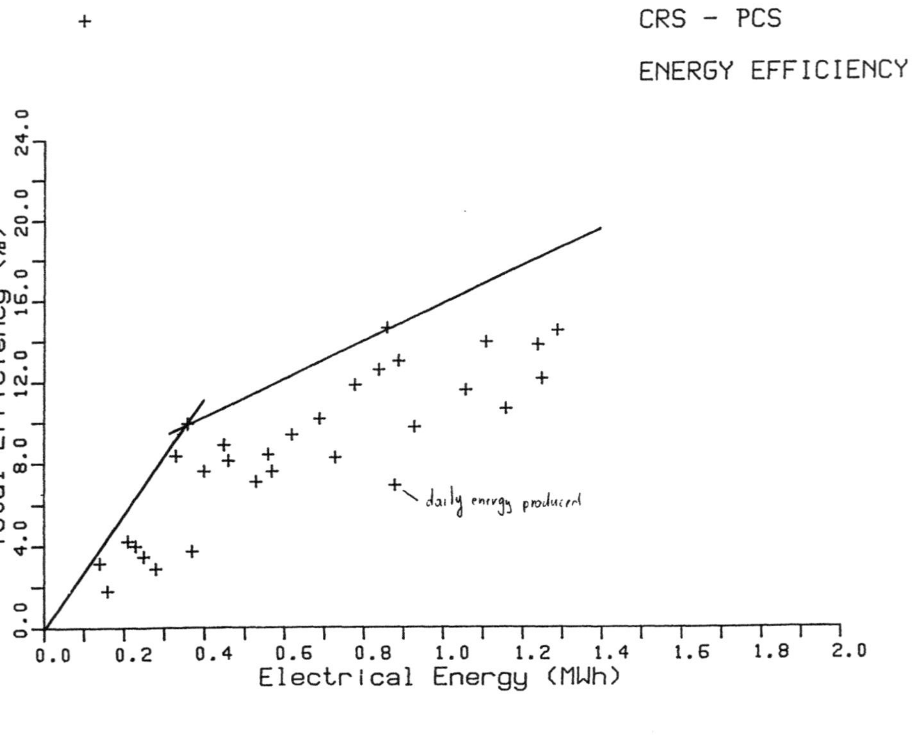

Fig. 6.2.-4: Total Efficiency versus Electrical Energy

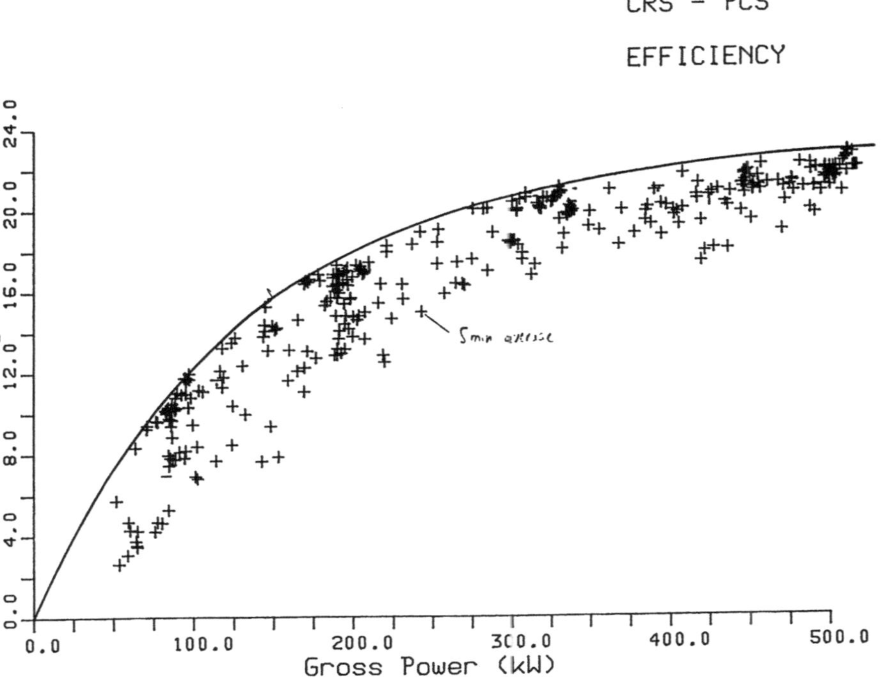

Fig. 6.2.-5: Efficiency versus Gross Power

PCS - OPERATION

(1)	Start Time Flow to Steam Generator	[hh:mm]
(2)	Time when Steam Conditions Reached	[hh:mm]
(3)	Start Time for Electric Production	[hh:mm]
(4)	Stop Time for Electric Production	[hh:mm]
(5)	Maximal Steam Pressure	[Bar]
(6)	Maximal Steam Temperature	[Deg C]
(7)	Energy to Steam Generator (total)	[MWh]
(8)	Energy to SG before Electric Production	[MWh]
(9)	Energy to SG during Electric Production	[MWh]
(10)	Electric Energy Produced	[MWh]
(11)	Max Electric Power	[kW]
(12)	Efficiency (Elenergy/Steam Gen (10/7))	[%]
(13)	Efficiency (Elenergy/Steam Gen (10/9))	[%]

DATE	NUM	(1)	(2)	(3)	(4)	(5)	(6)	(7)	(8)	(9)	(10)	(11)	(12)	(13)
10-12-83	344	11:40	12:15	15:00	15:50	83.43	504.44	8.84	3.97	4.03	0.73	518.00	8.27	18.16
23-12-83	357	11:10	13:25	14:05	17:50	83.16	506.06	10.85	4.66	5.77	1.16	504.81	10.66	20.04
27-12-83	361	10:15	12:10	14:55	17:50	85.36	506.84	8.84	5.78	1.22	0.16	204.06	1.78	12.93
28-12-83	362	11:25	13:05	14:35	17:50	85.31	506.75	5.85	3.52	1.49	0.23	472.44	3.99	15.65
30-12-83	364	09:55	11:55	12:20	17:55	85.02	510.78	7.48	2.04	2.95	0.57	508.44	7.60	19.23
10-01-84	10	10:00	11:35	13:25	18:05	82.37	501.50	9.84	5.99	2.05	0.28	403.03	2.87	13.81
01-03-84	61	10:20	12:00	13:35	18:55	84.63	500.50	9.59	4.06	5.51	0.93	507.06	9.75	16.97
06-03-84	66	13:10	14:20	15:00	19:00	84.28	496.09	7.14	2.85	1.92	0.25	210.53	3.47	12.91
14-03-84	74	10:45	14:00	15:00	19:10	85.23	492.44	9.88	4.07	2.17	0.37	490.81	3.73	16.95
24-05-84	145	07:25	14:30	15:05	16:25	85.24	470.66	4.43	2.64	1.02	0.14	349.47	3.15	13.71
28-05-84	149	11:52	16:52	17:17	21:12	84.03	480.94	5.11	2.52	2.49	0.45	501.75	8.89	18.23
31-05-84	152	13:52	16:02	16:37	20:57	83.93	474.50	3.61	1.54	1.96	0.36	476.53	9.94	18.31
06-06-84	158	08:58	12:43	13:08	21:18	83.72	504.56	8.91	2.29	6.05	1.29	483.88	14.51	21.37
07-06-84	159	08:13	09:08	10:08	21:18	83.72	511.72	15.48	1.60	13.79	2.73	338.25	17.65	19.81
08-06-84	160	11:04	11:59	12:24	21:19	83.66	499.28	5.88	0.87	4.88	0.86	341.41	14.62	17.61
11-06-84	163	10:29	13:19	13:54	21:24	83.69	488.22	6.79	1.74	4.97	0.69	402.13	10.17	13.88
12-06-84	164	09:59	13:34	14:29	21:24	83.92	493.28	3.89	1.83	1.66	0.33	403.34	8.36	19.66
13-06-84	165	08:10	08:40	09:05	21:25	83.76	502.56	10.33	0.85	7.76	1.25	210.69	12.14	16.18
18-07-84	200	11:30	15:30	16:50	21:20	83.58	492.88	7.46	4.20	3.14	0.53	468.06	7.10	16.90
20-07-84	202	11:11	14:01	14:26	21:21	83.34	499.19	6.57	1.90	4.59	0.78	493.03	11.82	16.94
01-08-84	214	11:06	15:01	15:51	21:11	83.59	494.09	6.63	3.40	3.06	0.62	484.75	9.38	20.32
02-08-84	215	14:31	16:31	16:56	21:11	83.51	480.31	5.18	2.17	2.77	0.40	250.25	7.63	14.26
03-08-84	216	10:21	13:36	14:46	21:06	83.65	494.78	6.85	1.78	4.95	0.89	404.00	12.99	17.96
07-08-84	220	10:55	14:25	15:55	21:05	83.63	494.38	6.69	3.31	3.21	0.56	302.94	8.43	17.55
16-08-84	229	10:54	13:59	14:29	20:54	83.57	496.06	9.02	2.57	6.42	1.24	405.13	13.80	19.40
17-08-84	230	10:44	14:19	15:09	20:54	83.59	498.13	6.68	2.15	4.45	0.84	307.75	12.56	18.84
21-08-84	234	10:08	11:53	16:38	20:48	83.55	479.69	5.08	3.05	1.99	0.21	191.91	4.22	10.81
22-08-84	235	10:02	13:02	13:32	20:47	83.63	499.72	7.98	1.59	6.32	1.11	510.53	13.95	17.63
24-08-84	237	08:07	10:12	14:52	20:42	83.87	511.47	12.83	6.47	4.15	0.88	510.75	6.90	21.32
30-08-84	243	11:50	13:05	14:45	20:35	83.51	500.56	9.19	3.08	6.07	1.06	508.53	11.58	17.53
31-08-84	244	11:05	14:20	15:25	20:35	83.52	488.78	5.65	3.21	2.32	0.46	500.38	8.12	19.77

Table 6.2.-1: Summary of PCS Operational Data

program: PCS job: 8-OCT-1984 15:22:25 1330

PCS - OPERATION testdate 22-08-84
===============

(1)	LK02CF01	Steam Generator Flowrate	[m3/h]
(2)	LK02CT03	Steam Generator Inlet Temp.	[deg C]
(3)	LK02CT05	Steam Generator Outlet Temp.	[deg C]
(4)	#114,1	CX01CE05 Gross Electrical Power	[kW]
(5)	LA01CP02	Steam Pressure	[bar]
(6)	calculated:	Power to Steam Generator	[kW]
(7)	calculated:	PCS total Losses	[kW]
(8)	calculated:	PCS total Efficiency	[%]
(9)	calculated:	additional Losses	[kW]

LocTm	SolTm	(1)	(2)	(3)	(4)	(5)	(6)	(7)	(8)	(9)
13:32	11:20	18.55	484.1	262.3	69.9	84.1	1227.9	1158.0	5.70	558.2
13:37	11:25	16.47	483.9	260.8	101.8	81.7	1096.5	994.7	9.29	315.1
13:42	11:30	20.06	485.5	263.3	190.8	82.6	1329.0	1138.2	14.36	236.2
13:47	11:35	19.56	486.8	263.5	198.8	83.4	1301.6	1102.8	15.28	180.7
13:52	11:40	18.81	487.6	262.8	197.0	82.8	1260.2	1063.2	15.63	145.7
13:57	11:45	18.19	488.3	263.9	192.3	82.9	1215.9	1023.6	15.82	117.7
14:02	11:50	18.20	489.1	262.3	197.7	82.7	1229.4	1031.7	16.08	112.4
14:07	11:55	17.99	489.8	264.1	195.5	83.0	1209.2	1013.7	16.17	99.9
14:12	12:00	17.88	490.7	262.4	193.9	82.7	1214.7	1020.7	15.97	110.9
14:17	12:05	17.77	491.4	262.5	195.1	82.8	1210.8	1015.7	16.12	102.9
14:22	12:10	17.68	492.2	262.8	193.7	82.8	1206.5	1012.7	16.06	103.5
14:27	12:15	18.98	492.8	263.8	232.9	82.2	1291.8	1058.8	18.03	51.5
14:32	12:20	25.70	494.4	266.5	364.3	85.3	1740.8	1376.5	20.93	40.8
14:37	12:25	32.32	496.1	269.0	472.4	88.4	2179.3	1706.8	21.68	100.7
14:42	12:30	34.48	497.2	269.8	501.7	90.2	2326.8	1825.1	21.56	145.8
14:47	12:35	34.97	497.9	269.1	505.7	90.6	2373.3	1867.6	21.31	178.5
14:52	12:40	34.89	498.5	268.5	508.0	90.5	2380.7	1872.7	21.34	177.6
14:57	12:45	35.03	499.2	267.7	510.5	90.5	2405.0	1894.4	21.23	193.1
15:02	12:50	35.05	499.7	268.2	508.3	90.6	2405.4	1897.2	21.13	201.5
15:07	12:55	33.86	500.5	269.2	483.3	90.9	2321.3	1838.0	20.82	204.6
15:12	13:00	24.59	500.3	267.7	312.3	88.3	1696.0	1383.7	18.41	178.1
15:17	13:05	17.50	499.2	263.4	194.5	83.0	1224.2	1029.6	15.89	118.3
15:22	13:10	17.23	499.5	262.6	194.4	82.5	1210.9	1016.5	16.06	105.4
15:27	13:15	17.12	499.7	264.3	194.2	82.5	1195.5	1001.3	16.24	90.9
15:32	13:20	17.19	500.3	265.3	195.2	82.8	1197.7	1002.5	16.30	89.4
15:37	13:25	17.34	500.9	267.8	197.4	82.8	1197.7	1000.2	16.49	81.6
15:42	13:30	17.16	501.2	269.9	197.6	82.9	1176.1	978.6	16.80	59.7
15:47	13:35	17.13	501.8	269.5	197.8	83.0	1178.0	980.1	16.80	60.5
15:52	13:40	17.00	502.2	270.1	195.5	82.5	1168.0	972.5	16.74	58.9
15:57	13:45	17.00	502.3	270.5	196.3	82.8	1166.9	970.6	16.82	54.9
16:02	13:50	17.03	502.8	270.9	197.6	82.7	1169.0	971.4	16.90	52.5
16:07	13:55	16.87	503.4	271.4	197.0	82.7	1157.9	960.9	17.01	43.5
16:12	14:00	16.99	503.5	271.7	198.8	83.0	1165.8	967.0	17.05	44.9
16:17	14:05	17.05	504.0	273.2	197.3	82.6	1164.3	967.0	16.95	48.8
16:22	14:10	16.83	504.5	272.4	197.4	82.8	1155.0	957.7	17.09	39.2
16:27	14:15	16.93	504.7	273.3	199.1	82.9	1158.7	959.6	17.18	37.0
16:32	14:20	16.86	505.1	273.1	198.3	83.0	1156.4	958.0	17.15	37.2
16:37	14:25	16.82	505.4	273.3	198.2	83.0	1154.4	956.2	17.17	35.8
16:42	14:30	16.72	505.8	273.8	198.1	82.9	1147.0	948.9	17.27	28.6
16:47	14:35	16.63	505.9	274.3	196.2	82.7	1138.5	942.3	17.23	26.7
16:52	14:40	16.73	505.9	274.0	197.8	82.8	1146.7	948.9	17.25	29.5
16:57	14:45	16.75	506.2	274.1	197.3	82.8	1149.5	952.2	17.16	34.0
17:02	14:50	16.58	506.5	274.9	196.3	82.7	1134.7	938.4	17.30	22.5
17:07	14:55	16.68	506.7	275.2	198.3	82.8	1141.1	942.8	17.38	22.0
17:12	15:00	16.71	506.9	275.4	198.2	82.8	1142.5	944.3	17.35	23.8
17:17	15:05	16.54	507.1	275.2	196.3	82.8	1132.9	936.6	17.32	21.0
17:22	15:10	16.87	507.1	275.3	198.5	82.7	1155.8	957.2	17.18	35.9
17:27	15:15	16.63	507.1	275.3	197.7	82.8	1139.3	941.6	17.35	22.3
17:32	15:20	16.67	507.1	275.8	197.3	82.7	1138.7	941.5	17.32	23.4
17:37	15:25	16.71	507.1	276.4	197.5	82.7	1138.6	941.1	17.34	22.5
17:42	15:30	16.45	507.1	277.3	196.2	82.8	1116.9	920.7	17.57	5.2
17:47	15:35	16.61	507.1	275.5	197.4	82.8	1136.5	939.2	17.37	20.7
17:52	15:40	16.48	507.1	276.4	196.4	82.6	1123.2	926.8	17.49	10.7
17:57	15:45	16.49	507.1	275.7	196.1	82.8	1127.7	931.6	17.39	16.3
18:02	15:50	15.90	507.1	276.3	174.0	83.5	1084.7	910.7	16.04	50.8
18:07	15:55	11.72	505.9	272.2	87.0	81.1	809.8	722.8	10.75	80.2
18:12	16:00	11.96	505.3	272.1	92.1	80.9	824.5	732.3	11.17	77.0
18:17	16:05	11.81	504.8	272.5	91.2	80.8	811.9	720.8	11.23	67.9
18:22	16:10	11.68	504.5	273.3	88.3	81.0	799.1	710.8	11.04	65.2

Table 6.2.-2: Example of PCS Operation

REMARKS ON RECEIVER LOSSES

Clifford S. Selvage, ITET

1. INTRODUCTION

Since the early days of solar central receiver system design (1973 - 1975) there has been extensive discussion and theoretical analysis about receiver design, usually having to do with receiver losses. The advocates of cavity receivers developed extensive three-dimensional mathematical models of the convection flow in and around the cavity and several large experiments were designed and constructed that provided experimental results modifying the theoretical models. I refer specifically to the work of our French colleagues from CNRS in Poitiers, P. LeQuere, B. Dutry, et al, and J.S. Kraabel and D.L. Sabers from Sandia National Laboratories, Livermore. These works have helped all of the solar thermal community to clarify our thinking and solidify our position either for the cavity or the "billboard", the external receiver. Cavity receivers are in use at the French CRS Themis, the Spanish CRS CESA-UNO, and were used at the Japanese CRS "Sunshine", the European Community's Eurelios, and the IEA/SSPS. External receivers are in service at the 10 MWh_e CRS at Barstow and at the IEA/SSPS CRS.

Clearly there are some differences in system performance related to the choice of cavity or external type of receivers. We see some indications, which are not fully resolved, of shorter start-up times with the external receivers and some rather solid evidence of convective losses that are far less than expected for the external receiver. Clearly these surprises are not entirely related to the difference of cavity or external, but are also influenced by the heat transfer medium or receiver coolant used.

At the IEA/SSPS project, both a cavity receiver and an external receiver of the billboard type, used the same receiver coolant -- sodium -- and so some acceptable and reasonable comparisons should be possible. To that end extensive loss experiments have been conducted on both receivers and much data has been recorded which should provide a solid foundation to resolve the argument. However, very little of this data has received detailed analysis. Recently, a calibration study has resulted in some modification of the preliminary conclusions which were reported at the 2nd International Central Receiver Workshop in Varese, Italy in June

1984. The extensive calibration effort has shown that the flow meters in the receiver circuit gave large errors in flow readings when operated in normal forward flow direction, thus making the cavity receiver look more loss-prone than earlier reported.

2. CONSIDERATION OF THE LOSSES

A receiver collects and converts radiant energy to thermal energy and then transfers the thermal energy to the receiver coolant. The potential loss areas are: (1) reradiation, where radiant energy is reflected off of the receiver absorber surfaces, (2) convection, where air flows around the receiver tubes that contain the receiver coolant and therefore carry thermal energy away, and (3) conduction wherein thermal energy is carried away from the coolant tubes through the structure that supports the tubes and the receiver.

The following sections comment on each of these loss areas and the efforts taken at SSPS to determine the magnitude of each loss area.

3. RERADIATION

As usual, there are several ways to quantify the loss. In light of the fact that reradiation is largely the result of reflection off of the absorbing surfaces, a measure of this loss should be the reflectance of the absorber surface. Absorptivity (reflectivity) measurements of the absorber surfaces of both the cavity and external receivers were performed several times at the SSPS, usually confirming the expected absorptivity numbers as published by the manufacture of the Pyromark 2500, the black paint used on both receivers.

The first of these works was reported by T. van Steenberghe of the ITET, at the SSPS Central Receiver Workshop held in Tabernas in April 1983. This investigation was on the cavity receiver manufactured by Sulzer and Mr. van Steenberghe used a solar spectrum reflectometer, made by Devices and Services, Co. of Dallas, Texas, to measure the hemispherical directional reflectance averaged over a simulated air mass 2 solar spectrum. This instrument can provide a resolution of 0.001, a repeatability of 0.003, and an accuracy of 0.02. The results from Mr. van Steenberghe's investigation are summarized as follows:

- New uncured coating: = 0.90
- New cured coating: = 0.96
- Irradiated coating: = 0.92 to 0.96

The 0.92 value was obtained in the highest flux (discolored) region of the receiver panel. The maximum absorptance loss after 1000 hours of receiver exposure is thus limited to 4%.

The complete paper by Mr. van Steenberghe describing this investigation is in Appendix A.

Radiation-absorptivity measurements were also made on the Advanced Sodium Receiver, the external-billboard receiver, which also uses Pyromark on its absorbing surfaces. However, that data, which contained "no surprises", is not available at this time.

A real-time system for measuring the reradiated power from either receiver was designed, constructed, and mounted on the receiver side of the Heat Flux Distribution measuring system -- the HFD bar. Descriptions of both the HFD system and the reradiation measurement system are covered in Appendix B in the paper by Mr. van Steenberghe as presented in the April 1983 CRS Workshop. Although the reradiation system was put together and operated, the difficulty of recording the data, competition with the need for HFD measurements, and the unrealiability of the HFD system prevented acquiring any meaningful results.

4. CONVECTION

Convective heat transfer is the arena in which the highly active discussions of external versus cavity have taken place. As stated earlier, considerable analytical work has been performed by some of the best heat transfer and technical mathematicians in an effort to relate knowledge from normal heat transfer systems and small experimental efforts to these very large heat transfer surfaces. These surfaces under consideration are in the size range that results in the normal factors in convection calculations, such as Reynolds number and Grashof numbers, being several orders of magnitude larger than previously considered. Some of the work that was specific to the IEA/SSPS project was accomplished by D.L. Siebers and J.S. Kraabel, and a recent publication titled "Estimating Convective Energy Losses from Solar Central Receivers", SAND84-8717 published in April 1984 is worthy of note. Dr. Kraabel participated in a measurement campaign at the SSPS site and defined several convective experimental possibilities that have been refined and implemented at this facility. These experiments are of a calorimetric nature using the receiver coolant (sodium) to provide the energy that is lost by convection while flowing through the receiver with no other energy input. Kraabel's paper, which defines these tests and relates the data acquired to statements of convective loss was presented in the SSPS/CRS Workshop of April 1983 and is included as Appendix C. Kraabel concluded that convection of the IEA/SSPS cavity receiver was of reasonable size and of a magnitude similar to pretest prediction. Further, that the test method devised would give accurate conductive loss data.

The method described by Kraabel and expanded on by the International Test and Evaluation Team (ITET) and the Project Operating Authority (POA) at the SSPS plant resulted in some tests on each of the two receivers using both the normal flow calorimetric test and reverse flow calorimetric tests. Results of these tests with the cavity receiver are reported by H. Jacobs of the ITET in the report included as Appendix D.

After the work by Mr. Jacobs was completed, a comprehensive calibration effort was conducted which revealed some very important sodium flow signal errors. Mr. Jacobs redid the data analysis using the correction factors provided from the calibration, and the conclusions on losses are modified. In addition to this recalculation on the cavity receiver a larger number of convective loss tests were performed with the external receiver. Analysis of this data has been accomplished by R. Carmona, H. Jacobs, and M. Sánchez of ITET in the recent months and is presented in this book as a paper under the title Receiver Losses - Results of Tests. Their paper describes experiments on both receivers with very recent results.

5. CONDUCTION

The energy loss through the supporting structure is usually much smaller than the convective losses. Conductive losses calculations, although complex, are usually rather dependable, given that dimensions and material properties of the supporting structure are known. Kraabel states that conduction may be determined by a combination of temperature measurements and the thermal conductivity or by observing the behavior of the receiver when the coolant is being circulated through it with the doors closed. It is clear that with no incident flux to the receiver and with no radiation and convection from the receiver the power loss from the coolant must be equal to the conduction.

H. Jacobs used this reasonable position in measurements of the receiver during normal flow and reverse flow tests with the doors closed and has reported on conductive losses in Appendix D.

APPENDIX A

SOLAR HEMISPHERICAL ABSORPTANCE MEASUREMENTS ON THE TUBES OF THE FIRST SSPS-CRS SODIUM RECEIVER

By

T. van Steenberghe
IEA/SSPS Project - ITET

CONTENTS

1. ABSTRACT

Absorptive paints are generally used for high-temperature solar receiver surfaces since more selective coatings have proved to degrade quickly under such conditions.
This paper presents the results of solar averaged hemispherical absorptance measurements carried out on the first CRS receiver and compares them with those obtained for new coatings. The aging of the absorptive coating under concentrated solar radiation is evidenced: a loss of absorptance limited to 4% was detected.

2. INTRODUCTION

Special coatings are in general use in various types of solar collector system receivers. Selective coatings are used for those receivers which are intended for low or medium temperatures; for high temperatures, such selective coatings are not presently available. Temperature is one of the factors that most adversely affects the durability of these coatings (Reference 1).
However, absorptive coatings with few or no selective characteristics are generally used to improve the behaviour of high-temperature receivers, such as those found in tower systems. It is interesting to observe the behaviour of such coatings when they are exposed to high fluxes for extended periods of time.

The first (Sulzer) CRS receiver is a cavity receiver. The heat exchange surface is made of six parallel tubes running in fourteen passes horizontally from the lower left corner to the upper left corner along a portion of a cylinder (Figure 1). These tubes are made of stainless steel that is coated into a special absorptive black paint (Pyromark 2500). A solar spectrum reflectometer was used to measure the solar averaged hemispherical absorptance of the tube coating at various locations on the receiver panel. The results of these measurements are presented here. In addition, some measurements of the solar averaged hemispherical absorptance of other parts of the receiver have been performed. They are deemed to be valuable, e.g.. as input for detailed radiative modelling of the receiver cavity. These data are also included.

3. MEASUREMENT METHOD

From Kirchoff's law and the law of conservation of energy, it can be shown (Reference 2) that the monochromatic directional emittance $\varepsilon_\lambda(\mu,\phi)$ and the monochromatic directional absorptance $\alpha_\lambda(\mu,\phi)$ can be calculated for opaque bodies from the monochromatic directional-hemispherical reflectance $\rho_\lambda(\mu_i,\phi_i)$:

$$\varepsilon_\lambda(\mu,\phi) = \alpha_\lambda(\mu,\phi) = 1 - \rho_\lambda(\mu_i,\phi_i)$$

where: $\mu = \cos\theta$, with θ the polar angle,
ϕ = the azimuthal angle,
λ = the wavelength,
i = incident radiation

If the incident radiation is diffuse, then the directional hemispherical reflectance is equal to the hemispherical-directional reflectance:

$$\rho_\lambda(\mu_i,\phi_i) = \rho_\lambda(\mu_r,\phi_r)$$

where r = reflected radiation. This reciprocity relationship is illustrated in Figure 2a.
Integrating over the hemisphere and over the solar spectrum wavelengths, the solar averaged hemispherical absorptance of the surface is obtained as:

$$\alpha_{solar} = 1 - \rho_{solar}(\mu_r,\phi_r)$$

3.1 Measurement instrumentation

The instrument used on site is a Solar Spectrum Reflectometer, from Devices & Services Co., Dallas,Texas. It measures the hemispherical directional reflectance averaged over a simulated Air Mass 2 solar spectrum.

The light provided by the measurement head is made diffuse by a white coating on its inside surface, while the detectors are collimated to the measuring port with an angle of 20º (Figure 2B).

The solar spectrum averaging is realized using a single tungsten-halogen light source and four detectors with filters. Figure 3 shows the response of the individual detectors, designated as UV, Blue, Red,and IR.

Figure 4 shows the electronically obtained linear combination which closely approximates the Thekaekara Air Mass 2 spectrum (Reference 3) in the energy (area) sense.

The essential specifications of the instrument are listed below:

- Resolution 0.001
- Repeatability 0.003
- Absolute accuracy 0.02

3.2 Measurement technique

The instrument calibration depends on the type of measurements to be made.

First the basic instrument calibration was checked and adjusted. Next, the specific calibration procedure, which was devised by SANDIA, was used for measurement on the first and last tubes of the CRS receiver panel.

The standard calibration procedure was used for all measurements on flat surfaces. It was also used for measurements on all intermediate tubes, which are too close together to allow correct positioning of the measurement head with the tube jig installed.

The SANDIA- devised and standard calibration procedures were compared on the most easily accessible tubes to check measurement consistency. This proved to be satisfactory.

Repeated measurements were taken on all surfaces to insure repeatability.

3.3 Measurement locations

Measurements were taken on several types of Pyromark absorptive coating:

- new, uncured
- new (unexposed), cured
- exposed, cured

They were also taken on some non-active surfaces.

The measurements of the absorptance of the new, uncured coating were taken on the CESA-1 receiver superheater tubes before their installation. The aspect of the coating is matt black.

Then, measurements on the lowest tube of the SSPS-CRS receiver were taken; this region of the receiver is essentially unirradiated, but as the receiver had already operated for some 3000 hours, the coating there could be considered well-cured though still new. This new, cured coating has the same aspect as the new coating, i.e., black and matt.

Finally, measurements were taken on other SSPS-CRS receiver tubes, as indicated on Figure 5, passing through the most irradiated region of the panel. In this peak flux region, a slight discoloration of the coating can be seen (grey to violet to pink). This discoloration is especially visible on those tubes which are bent inwards, toward the concave (hottest) side of the receiver panel. These measurements were made after 1000 hours of exposure to concentrate flux.

The measurements on the non-active surfaces were taken on:

- uncoated parts of the receiver tubes (headers, bends),
- the cavity floor,
- the cavity walls, including the back of the doors,
- the cavity entrance frame, horizontal and vertical,
- the octagonal aperture frame corners,
- some white (reflective) coated parts such as door fronts and aperture corners.

Figure 6 shows schematically the measurement locations.

4. RESULTS

4.1 Absorptive coating

The solar averaged hemispherical absorptance results are summarized in the table below:

Coating	Absorptance min.	Absorptance average	Absorptance max.
New, uncured	.886	.90	.904
New, cured	.956	.96	.963
Irradiated	.924	.94/.96	.962

It can be seen that the curing process makes the coating significantly more absorptive. The absorptance increases from the original 0.90 to 0.96, a 7% rise. The 0.96 absorptance found for the new/cured coating is in perfect agreement with the value obtained by Martin Marietta Corporation (Reference 4).

For both the uncured and cured coatings, a fairly uniform absorptance can be observed, at least for clean coatings. Dust has an adverse effect, resulting readings as low as 0.85.

For irradiated regions, the measured absorptance values range from values typical of unexposed regions downwards. The absorptance in the highest flux region, where the discoloration is visible, is lowered to 0.94 and even locally to 0.92, which means a 2% to 4% loss.

4.2 Non-active surfaces

The table below gives the obtained results of solar averaged hemispherical absorptance on non-active surfaces.

Location	Aspect	Absorptance
Tubes, headers, bends	Uncoated S.S. oxidized (yellow)	0.54 ... 0.68
Cavity		
floor	S.S., white bright to satin	0.66 ... 0.68
wall (left)	SS, grey, satin	0.50
entrance floor	SS, white/grey, satin	0.62 ... 0.65
entrance wall (left)	SS, oxydized slightly	0.49 ... 0.55
	strongly	0.72
door inner wall	SS, white	0.45
aperture corners inner face	steel, grey	0.78
White coatings		
door outer wall	white, dirty	0.32
aperture corners outer face	white, clean	0.25

5. CONCLUSIONS

Measurements of solar averaged hemispherical absorptance were carried out on the CRS receiver tube absorptive black coating.

The results are summarized as follows:

- New uncured coating: α = 0.90
- New cured coating: α = 0.96
- Irradiated coating: α = 0.92 to 0.96

The 0.92 value was obtained in the highest flux (discoloured) region of the receiver panel. The maximum absorptance loss after about 1000 hours of receiver exposure is thus limited to 4%.

Some measurements were also taken on non-active surfaces around the receiver. These measurements might be useful for detailed radiative modelling of the cavity.

6. REFERENCES

(1) T. van Steenberghe and A. Sevilla
Solar hemispherical absorptance measurements on SSPS-DCS receiver tube selective coatings.
IEA-SSPS - DCS Workshop - December 9-10, 1982

(2) J.A. Duffie and W. A. Beckman
Solar Engineering of Thermal Processes.
John Wiley - N.Y. - 1980

(3) M.P. Thekaekara
Suppl. to the Proc. 20th Annual Meeting on Inst. for Environmental Sci. 21, "Data on Incident Solar Energy" - 1974

(4) Pyromark 2500 - Measurements by MMC - 1979
Private communication 1982

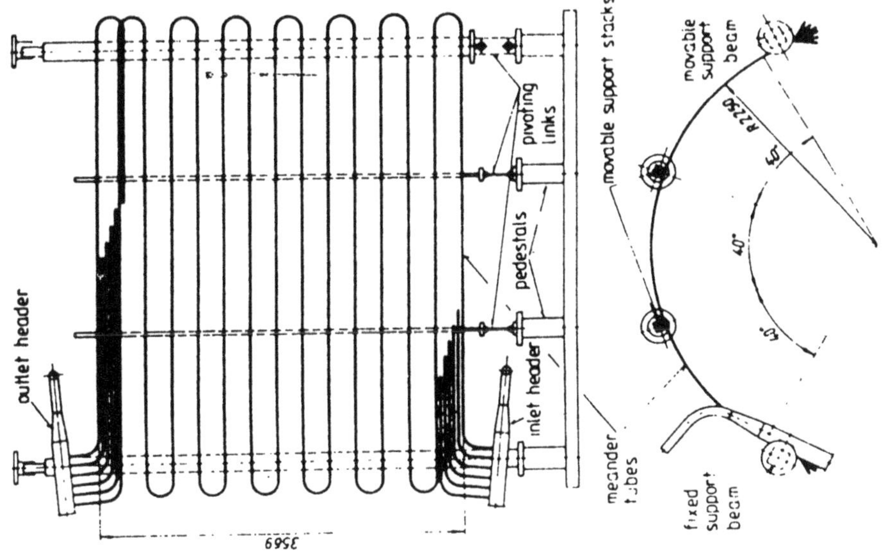

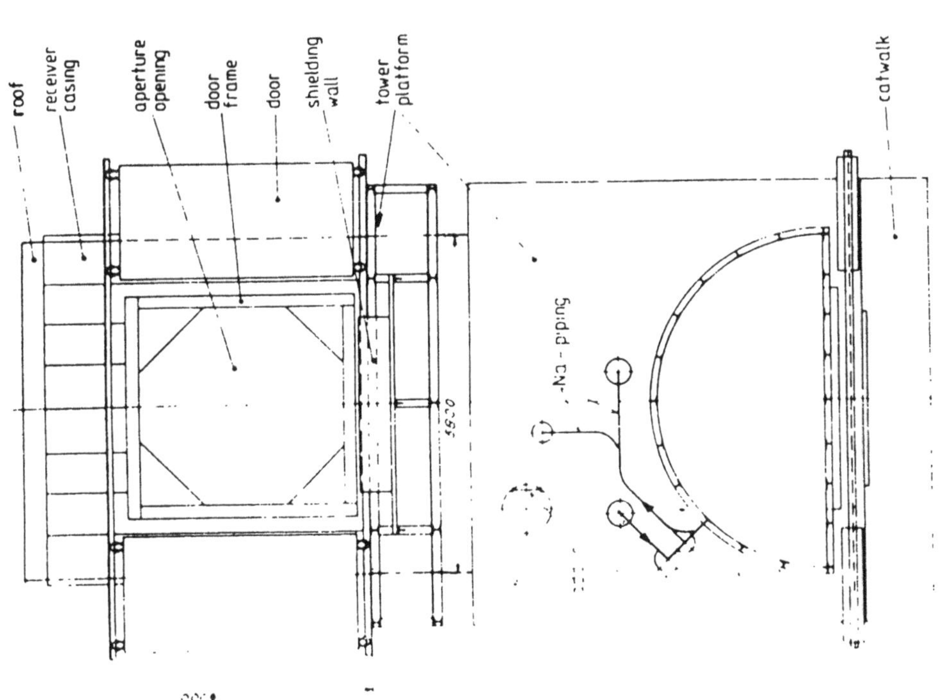

Figure 1
Receiver

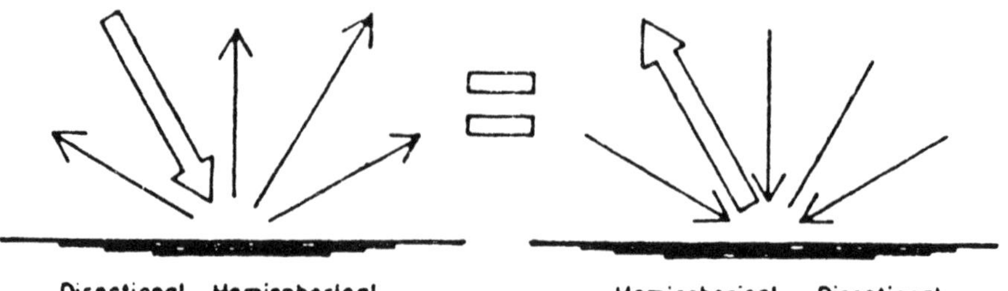

Figure 2a

Reciprocity Relationship for Hemispherical-Directional and Directional-Hemisperical Reflectivities

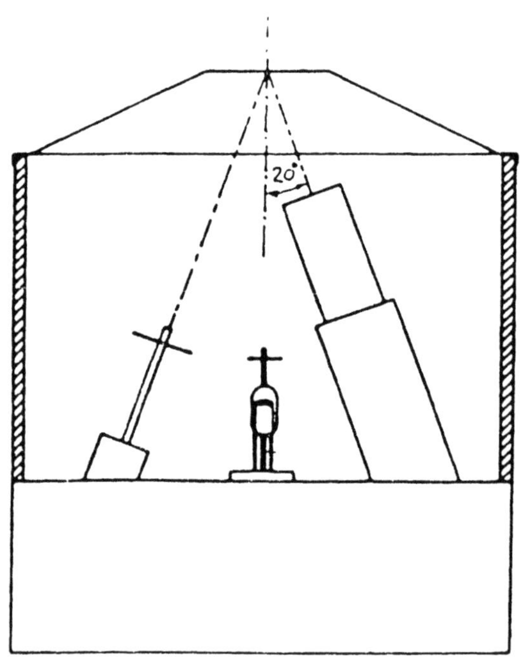

Figure 2b

Cross Sectional View of Measurement Head

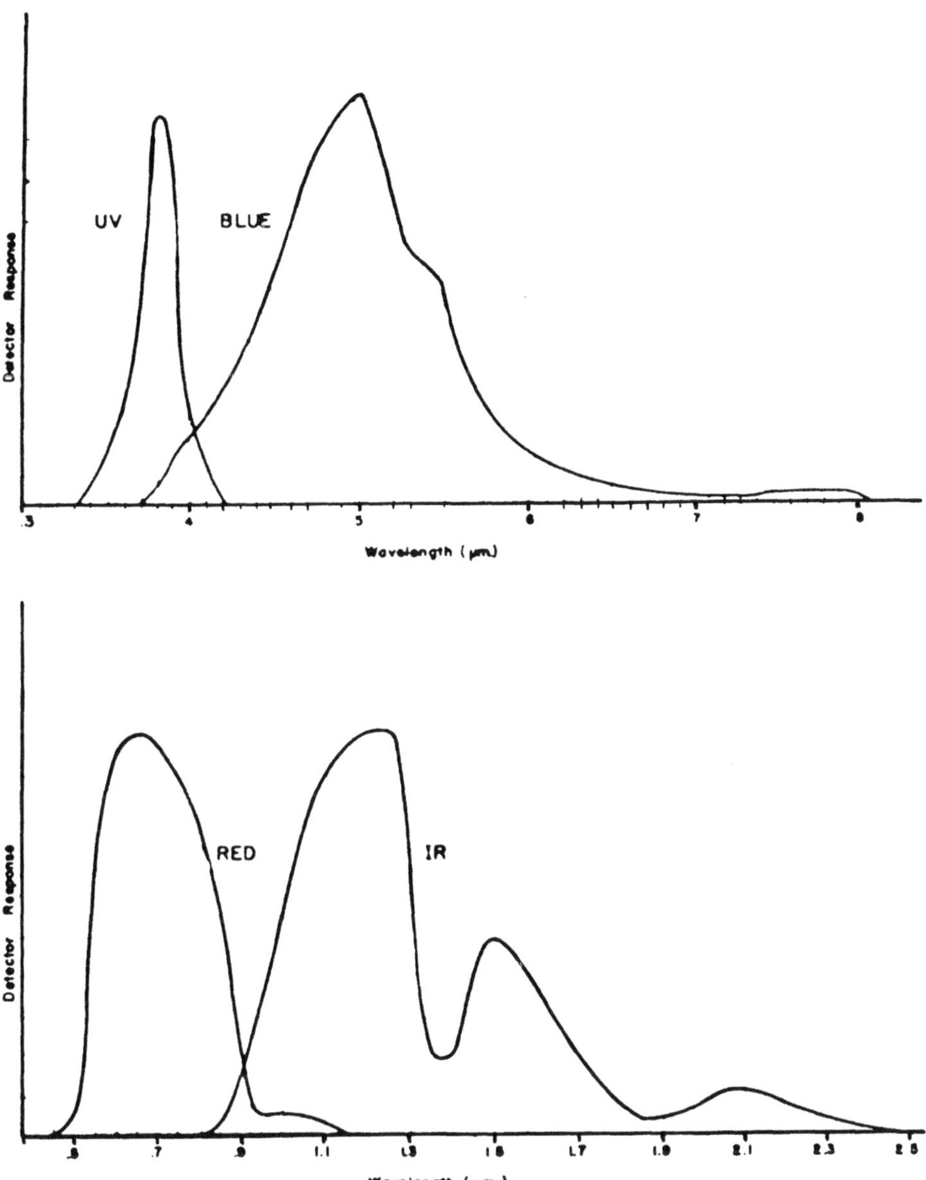

Figure 3
Response Spectra of Detectors

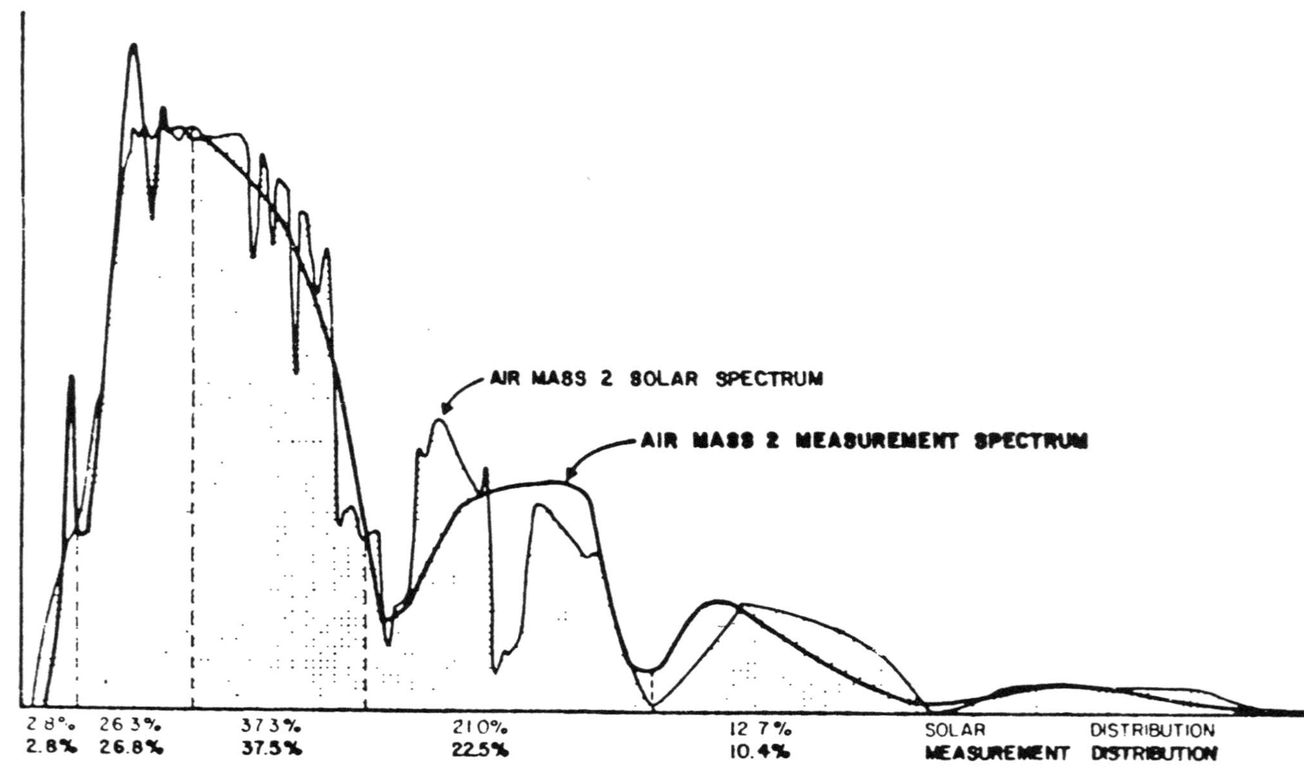

Figure 4

Comparison of the Air Mass 2 Solar Spectrum and the Air Mass 2 Measurement Spectrum

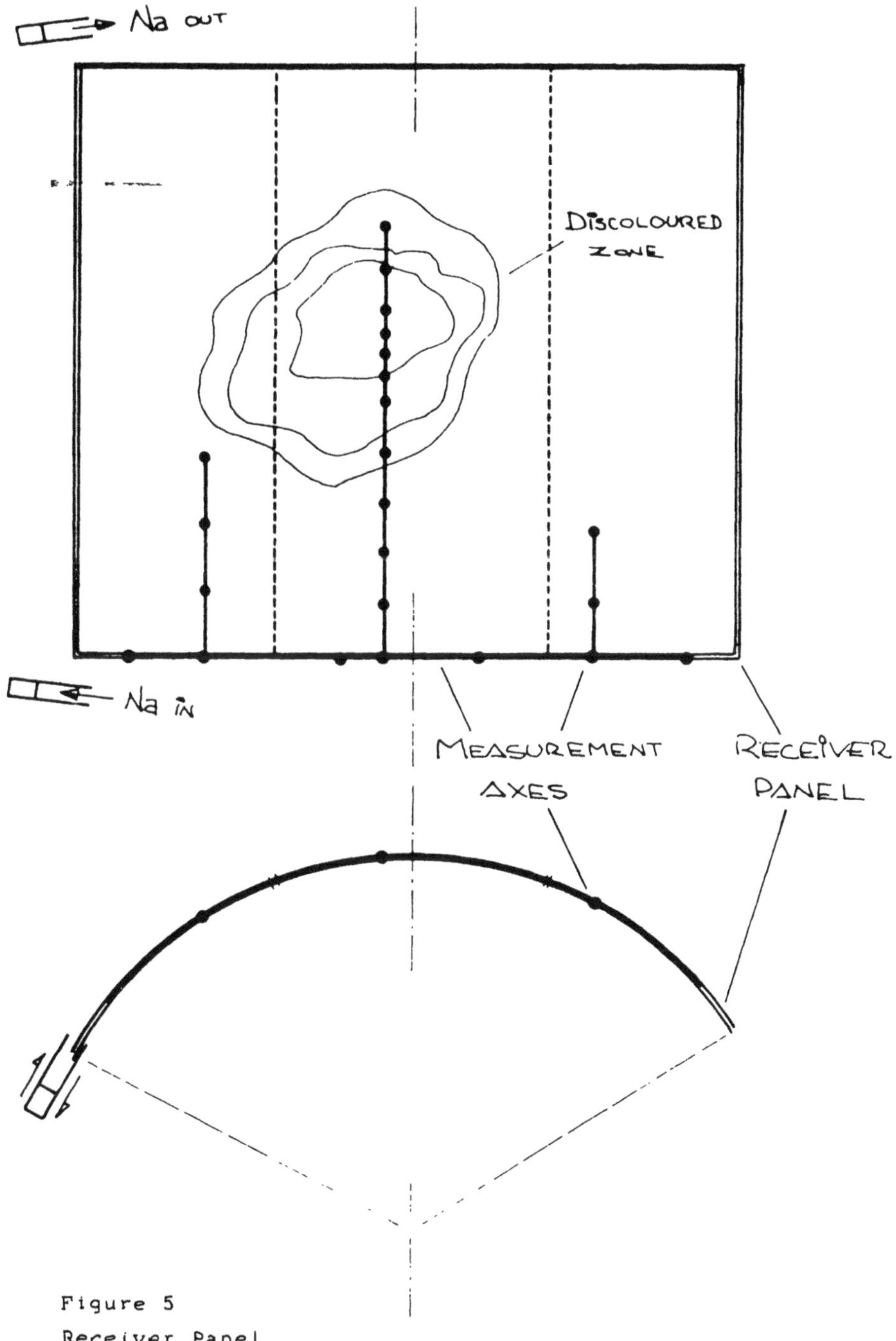

Figure 5
Receiver Panel
Absorptivity Measurement Locations

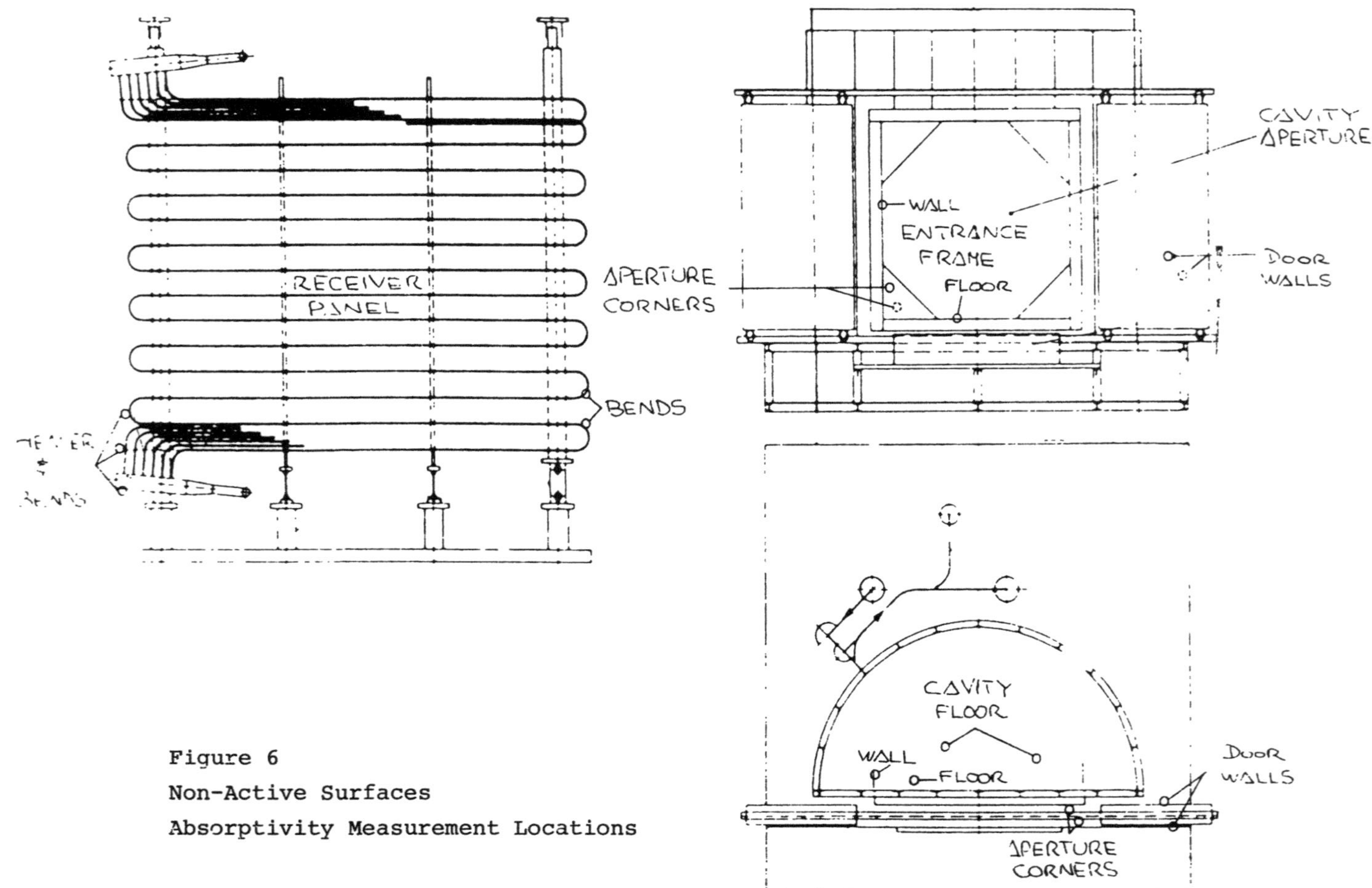

Figure 6
Non-Active Surfaces
Absorptivity Measurement Locations

APPENDIX B

HEAT FLUX DISTRIBUTION MEASUREMENT SYSTEMS

By

T. van Steenberghe
IEA/SSPS Project - ITET

CONTENTS

1. INTRODUCTION

Critical information for a tower solar plant such as the SSPS-CRS is the power reflected by the heliostat mirror field to the receiver.
This value is needed to evaluate both the heliostat field performance and the performance of the receiver and downstream power plant.

In the case of the SSPS-CRS, this information had additional importance at the time of acceptance tests. This power represented the interface between the two main subcontractors: Martin Marietta Company which delivered the heliostat field with its computer-based control (see Table 1 for field specifications), and Interatom, which delivered the sodium heat transfer systems (receiver, heat storage, and steam generator) and the power conversion system (steam motor/alternator set, condenser, and cooling tower).

The main nominal data of the SSPS-CRS are summarized in Table 2.

A heat flux distribution (HFD) measurement system was therefore installed at the SSPS-CRS to experimentally determine the power delivered by the heliostat field to the sodium receiver.

This paper describes this measurement system. Results are given and compared with predictions made by the HELIOS computer code (Reference 1), in another paper (Reference 2) presented to this workshop.

Finally, how the concept was extended to measure the radiative heat losses (reradiated power) from the receiver is explained.

2. INCOMING HEAT FLUX DISTRIBUTION MEASUREMENT SYSTEM

The measurement system is composed of:

- measuring equipment comprising a frame that has a moving traverse bar with flux transducers and auxiliary equipment,
- a dedicated data acquisition and control system under the control of two microcomputers.

The DFVLR (FRG), Operating Agent for the SSPS project under contract with the IEA, nave taken responsibility for the measuring equipment and for the choice of the data acquisition and control system.

Belgonucleaire (Belgium), under contract with the Operating Agent, provided the data acquisition and control system and the associated software program to operate the complete HFD measuring system.

2.1 The measurement equipment

The measurement equipment (Reference 3) is essentially composed of a traverse bar with flux transducers and auxiliary equipment, a rectangular frame through which the bar is sweeping, and a stepping motor and associated electronics driving the bar. This equipment is much like that at the Central Receiver Test Facility, Albuquerque, New Mexico (USA).

General layout

Figure 1 shows the arrangement of the equipment around the receiver cavity on top of the CRS tower.

The frame around the cavity is about 7 m high by 8 m wide. The vertical measuring bar is maintained by the top drive spindle and rolls on a lower rail. There is a standby 'Parking' position on the right side, where the bar is shadowed by a protective cover.
The front face of the water-cooled bar is equipped with radiometers measuring the incoming flux; the corresponding amplifiers are accomodated inside the bar.

Functioning

The bar is driven by the steppping motor to sweep the measurement plane. The motor is programmed to stop so measurements can be taken at eleven positions equally spaced along the bar path.

The stopping time at every measurement position (less than 1 sec) is determined by the measurement control system. The travel time between successive positions is determined by the stepping motor characteristics; although it was designed to be about 4 sec, it is actually longer. The total bar travel time is about 3 minutes. After the measurements are taken at the eleventh position, the bar is driven back to its parking position at a higher velocity and without stops.

The measurement plane is parallel to and 0.85 m in front of the receiver aperture plane. The radiation reflected by the heliostat field reaches the measurement plane from a horizontal angular range of about 80° and a vertical angular range of about 30°.

Cooling

The traverse bar is cooled by water. There are two separate cooling circuits: one for the bar itself, and another for the radiometers.

The bar cooling circuit leads downward along the frontface and returns upward along the backface. This ensures the refrigeration of the bar structure, especially around the radiometers on the frontface, and provides an adequate buffering for the amplifiers.

The radiometer cooling circuit, on the other hand, provides parallel flow to every radiometer. This water is the heat sink.

Instrumentation

The radiometers are calorimeters from HYCAL Engineering, USA, with five different full-scale ranges from 170 kW/m2 to 3400 kW/m2. They are stated to have $\pm$ 3.0% accuracy and $\pm$ 0.5% repeatability. They can operate to temperatures up to 200°C, with a linear ($\pm$ 2%) response. Ten of them are distributed along the bar with their ranges chosen according to predictions by the HELIOS computer code.

Each radiometer is connected to an amplifier with automatic range switching as a function of the input signal.

Some radiometers which were originally planned to be installed at fixed positions on the frame in order to measure the spillage flux were not installed; the eleventh radiometer at the lower end of the bar was not installed either, because it would have been shaded most of the time by the lower platform structure.

In addition to this flux instrumentation, there is the so-called 'housekeeping' instumentation which takes temperature measurements, cooling water flowrate, and pressure measurements. Bar position is encoded by a potentiometer.

2.2 Data acquisition and control system

The data acquisition and control system HP 304A is composed of the following equipment (Reference 4):

- the controller, an HP 85A microcomputer, with 32K memory and interface cards;
- the measuring instrument, a high-resolution digital voltmeter HP 3456A;
- the scanning control unit, HP 3497A, with:
    - o system clock/calender
    - o 60 analog input channels
    - o 16 digital input channels
    - o 16 digital output channels

This system is interconnected by means of the HP-IB bus system (IEEE-488) and is installed in the CRS control room. Figure 2 is a schematic of the system.

The associated software program, Solar Flux Automatic Measurement Program (SOFAP), was developed to allow the computer to perform the measurement of the reflected solar heat flux distribution, including he control of the complete measurement system and the processing of the data (Reference 3).
The task of the SOFAP program can be divided as follows (Figure 3):

- Measurement system check and set-up
- Control of the traverse bar movement
- Measurement sequences trigger and control
- Handling and processing of measurement results
- Outputting of measurements results

The auxiliary programs TSOFAP and RSOFAP shown in Figure 3 are a slowed-down test version of SOFAP and a program for data retrieval from magnetic tape, respectively. These programs will not be further described here. The remainder of this section is devoted to a brief functional description of the main software.

Table 3 lists the input and output signals to and from the data acquisition and control system.

Besides the analog signals, the system needs two digital signals sent by the stepping motor electronics and gives to the computer the status of operation of the bar: bar inside/outside of parking or bar moving/staying. This allows the computer to take the necessary actions in case of any incident. The two digital outputs are provided by the controller to the stepping motor as clearance to move to the next position (go) or as an emergency command.

The program architecture is modular. The main blocks and subroutines are:

- clock and identification
- initialization
- data read
- system and instruments check and set-up
- measurement sequencing and bar movement control
- alarms
- data decoding and processing, calculations
- editing preparation
- storage of data on magnetic tape
- printing
- plotting
- transfer of data to main CRS-DAS

The program works automatically after the operator has given the necessary inputs, defined the output, and given the clearance to initiate the measurements..

Measurements are first taken in the parking position to monitor the ancillary 'housekeeping' equipment and to take account of the possible offset of the radiometers.

The bar is then allowed to move to the first measurement position. As soon as the bar has stopped, the solar heat flux measurements are taken and the system outputs the command to drive the bar to the next position, and so on through the eleven positions.

After measurements have been taken in position eleven, the bar is driven back rapidly to the parking position to measure any offset of the radiometers due to temperature increases during the experiment and to perform the housekeeping measurements again.

During the measurements, the computer CRT is used to indicate to the operator the progress of the test, i.e., bar movement, taking of measurements, etc.

Immediately after the measurements, the data are processed by the computer. Processing involves the following steps:

- data decoding
- conversion of measured voltages to physical units
- correction for temperature rise offset
- calculations of average heat flux, integrated power in the measurement plane, peak flux, etc.

The output is then prepared and printed as previously required:

- printed tables (voltages and/or uncorrected/corrected physical units)
- pseudo 3-D plots

Finally, the data are stored on magnetic tape and/or sent to the main CRS/DAS if required.

3. RERADIATED HEAT FLUX DISTRIBUTION MEASUREMENT SYSTEM

The global receiver instantaneous efficiency can be determined using either the results from the incoming HFD measurements or the HELIOS simulation code predictions for the input power. Available sodium flowrate and temperature measurements can be used to calculate the sodium enthalpy gain for the output power. However, it would be interesting to be able to divide the receiver losses among those resulting from convection, reradiation (reflection and emissive radiation), conduction, and geometry (spillage).

It appeared that the HFD system could lend itself to the measurement of the reradiation losses from the receiver through the aperture, provided the appropriate radiometers were added to the back of the moving traverse bar (Reference 5).

A simple model (Reference 6) of the radiative behaviour of the receiver was devised in order to evaluate the heat flux reaching the back of the bar and to define the required radiometers. As can be seen from Figure 4, the peak flux is predicted to be less than 10 kW/m2.

Suitable radiometers from Thermogage Inc., USA, with battery-operated amplifiers were borrowed from Sandia National Laboratories. Since no cooling water could be diverted from the main HFD bar, and since the back radiometers would be in the shade of the bar, an experimental structure that requires no cooling was designed.

This structure consists of a double radiation shield made of brightly polished stainless steel and contains both the radiometers and their amplifiers. Three thermocouples are installed to monitor the temperature transient inside the structure during the bar sweep. The structure is secured to the back of the bar by means of welded bolts. Ten radiometers are distributed along the bar as indicated in Figure 5.

Since the scanner inputs available within the data acquisition and control system are saturated, the measurement cannot be taken simultaneously with the HFD measurements. An external analog switch, driven by four digital output channels of the scanner, is used to direct the signals to the voltmeter, under control of the computer. A special software program written for this purpose works similarly to SOFAP and has the same capabilities.

Measurements have now been carried out for some weeks. The results could not be evaluated for this workshop but will be analyzed and reported in the future.

4. REFERENCES

(1) The HELIOS model for optical behaviour of reflecting solar concentrators.
SANDIA REPORT - SAND 76-0347 (1979)
F. Biggs - C. Vihitoe

(2) Comparison of SSPS-CRS heat flux measurements with theoretical HELIOS predictions.
Paper presented to this SSPS-CRS workshop (1983)
Dr. Kiera

(3) Device for the measurement of heat flux distribution.
SSPS Technical Report 5/81 (1981) - DFVLR

(4) CRS heat flux distribution data acquisition system software documentation.
Belgonucleaire Report BN 8103.03 (9181)
T. van Steenberghe - R. Verbiest

(5) CRS receiver reradiation measurement system.
SSPS Report R-58/82 (1982) and other internal documents
T. van Steenberghe

(6) CRS receiver reradiation and its measurement.
SSPS Report R-56/82 (1982)
T. van Steenberghe

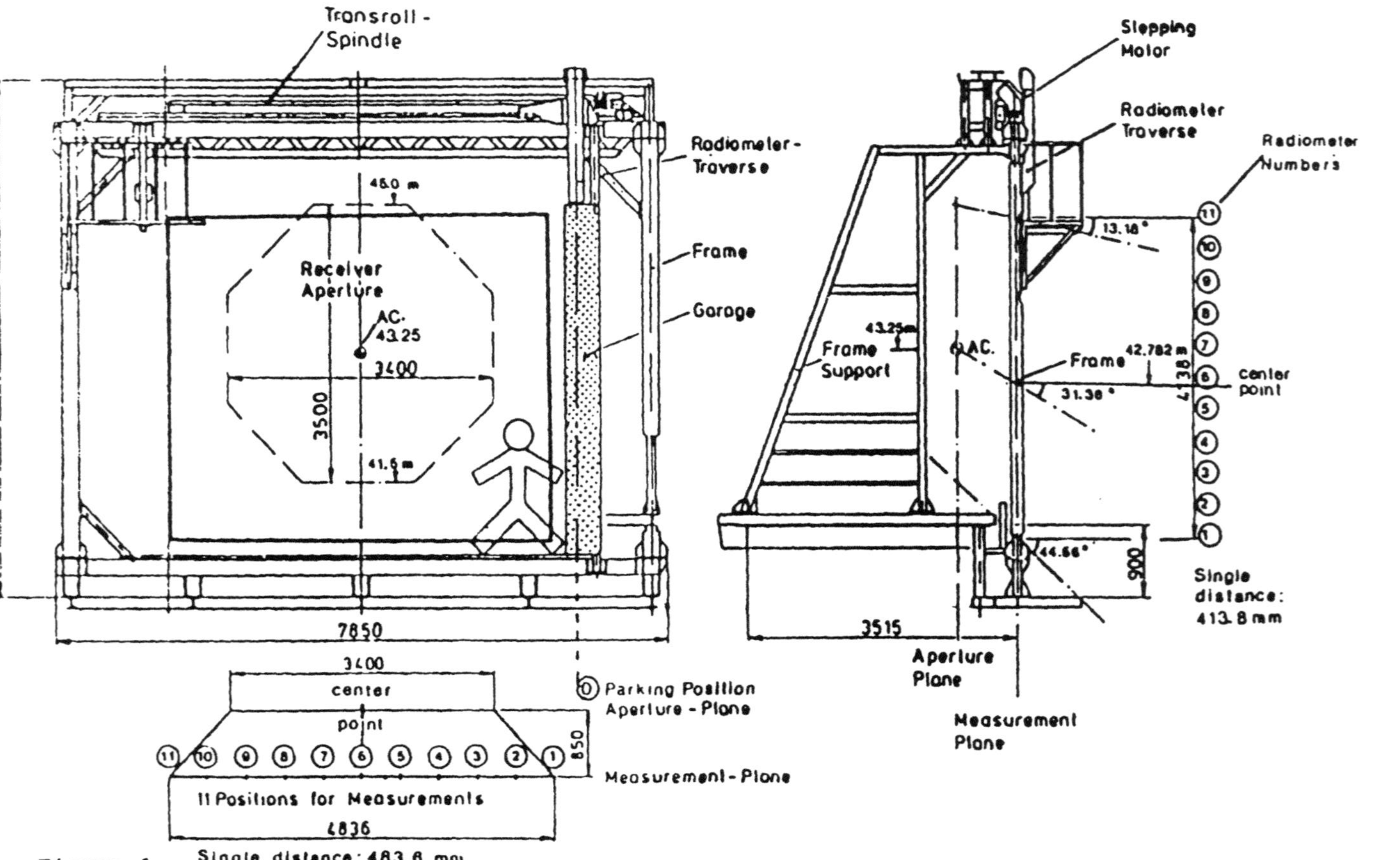

Figure 1

Arrangement of Heat Flux Distribution Measurement System

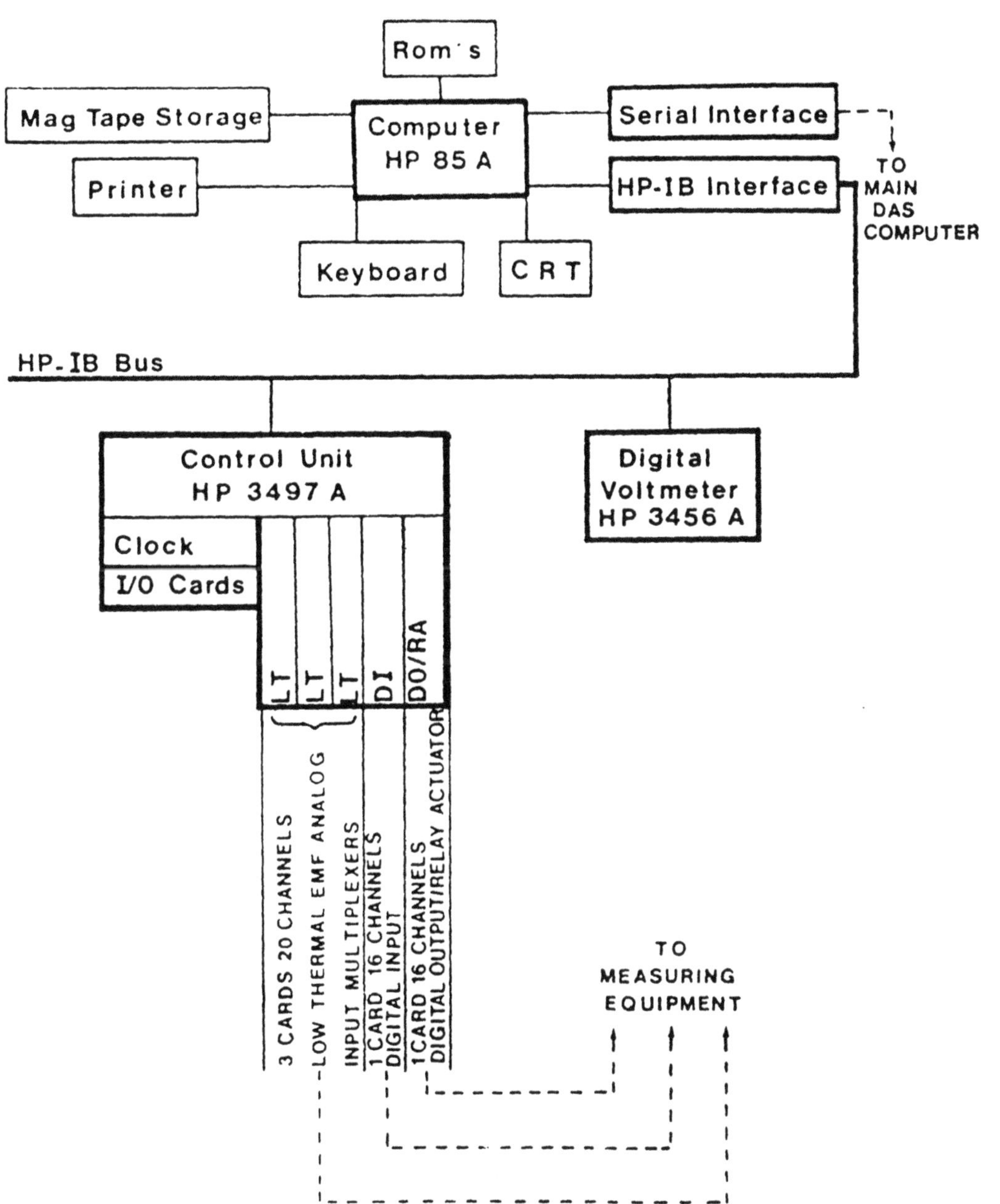

Figure 2
Data Acquisition Control System HP 3054 A

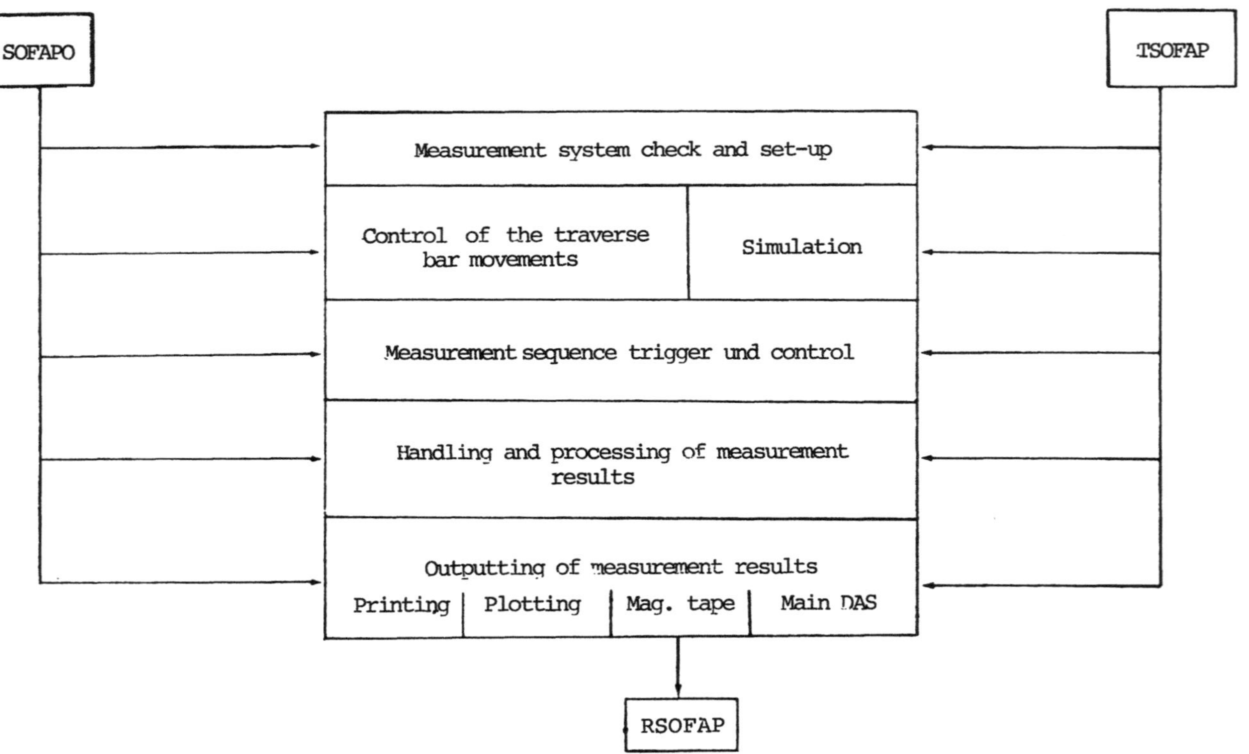

Figure 3
SSPS CRS HFD Measurement Software

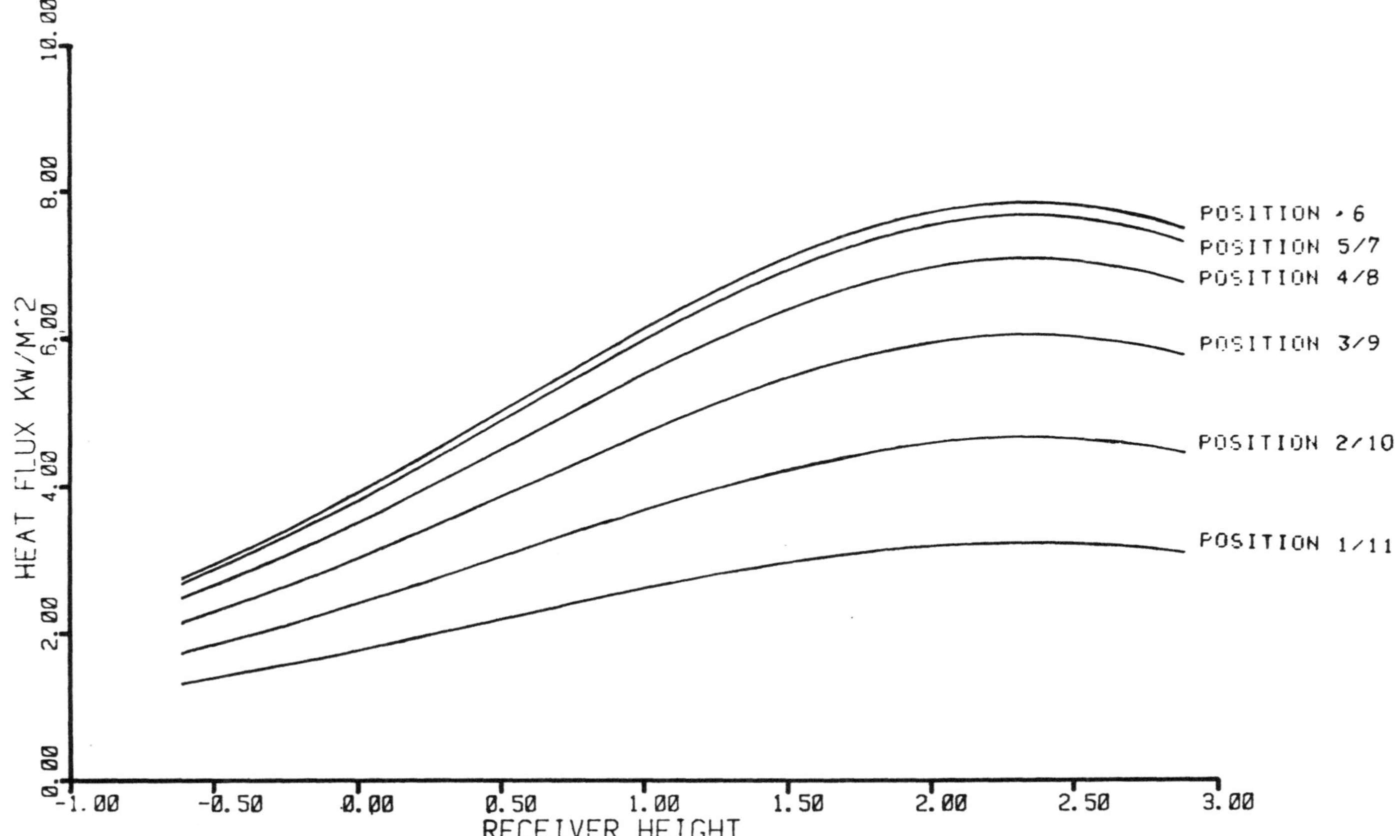
10.00
8.00
6.00
4.00
2.00
0.00
HEAT FLUX KW/M^2
POSITION ,6
POSITION 5/7
POSITION 4/8
POSITION 3/9
POSITION 2/10
POSITION 1/11
-1.00
-0.50
0.00
0.50
1.00
1.50
2.00
2.50
3.00
RECEIVER HEIGHT

Figure 4

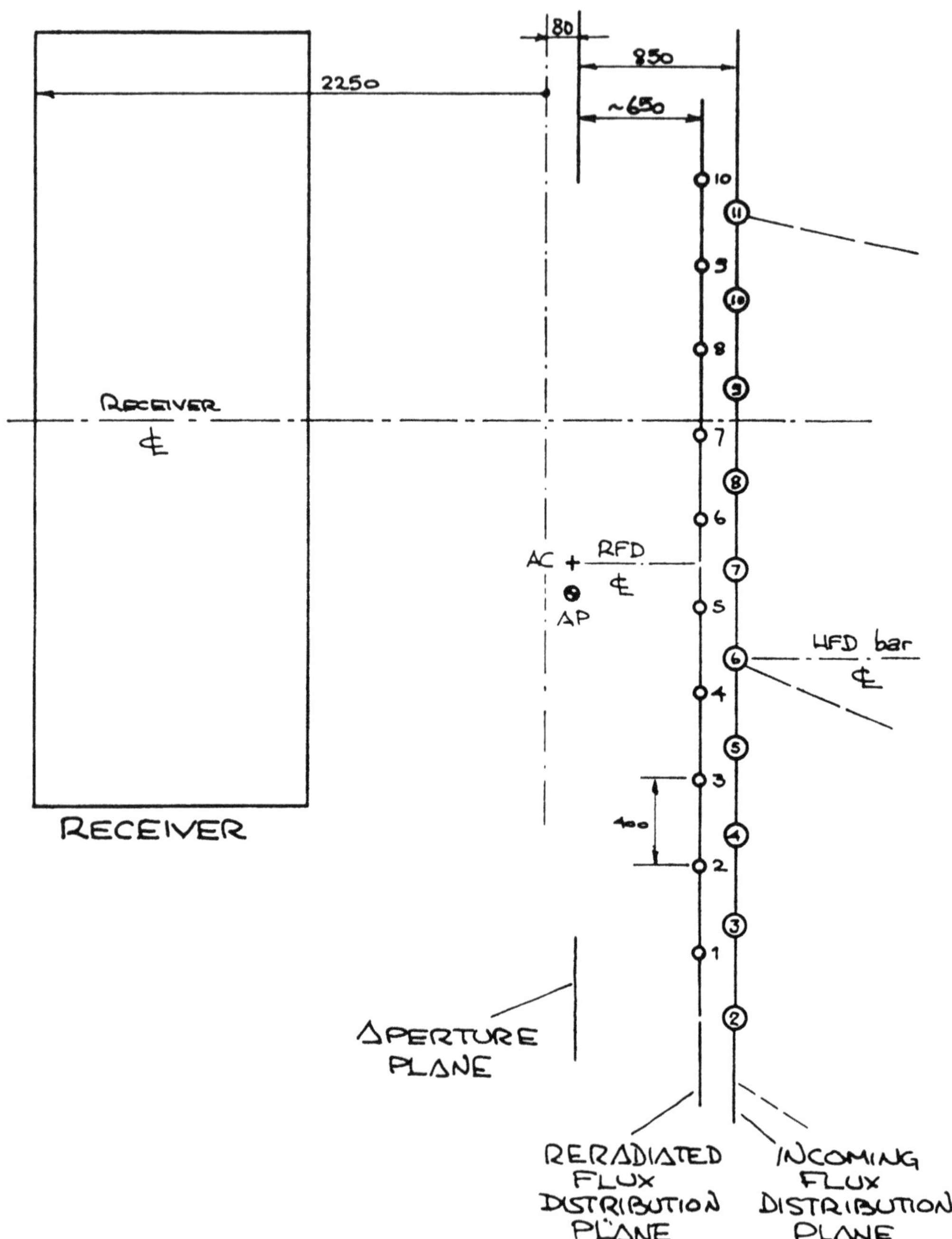
80
850
2250
~650
10
11
9
10
8
9
RECEIVER
℄
7
8
6
AC
RFD
℄
7
AP
5
HFD bar
℄
6
4
5
3
400
4
2
3
1
2
APERTURE
PLANE
RERADIATED
FLUX
DISTRIBUTION
PLANE
INCOMING
FLUX
DISTRIBUTION
PLANE
RECEIVER

Figure 5

	Continue Operation Nominal Output	Continue Operation 50% of Nominal Output	Survival
Direct Insolation	1100 w/m^2	700 w/m^2	-
Wind Velocity	$\leq$ 13 Km/hr	$\leq$ 50 Km/hr	$\leq$ 144 Km/hr
Seismic Activity (Ground Acceleration)	after $\leq$ .3 m/s^2	after $\leq$ .3 m/s^2	$\leq$ .6 m/s^2
Various	after lightning strikes one component		19 mm hail at 20 m/s sandstorms thunderstorms rain lightning

Table 1

SSPS CRS Heliostat Field Baseline Specifications

Central Receiver System

Design	day 80, 12:00 (equinox noon)	
	insolation	0,92 kW/m²
Heliostat field:	total reflective surface area	3655 m²
	concentration ratio	377
	solar multiple	1,11
	land-use-factor	0,22
Cavity receiver:	heat transfer medium	Sodium
	aperture size	9,69 m²
	active heat transfer surface	16,9 m²
	inlet temperature	270°C
	outlet temperature	530°C
Thermal storage	two-tank-system, storage medium	Sodium
	capacity equivalent to	1,0 MWh_e
	hot storage temperature	530°C
	cold storage temperature	270°C
Steam generator	sodium inlet temperature	525°C
	sodium outlet temperature	275°C
	steam outlet temperature	500°C
	steam pressure	100 bar
Power (at design point):	insolation	3362 kW
	radiation into cavity	2840 kW
	thermal	2203 kW
	gross electric	599 kW
	net electric	500 kW
Efficiencies (at design point)	insolation/aperture plane	84,5%
	thermal/gross electric	27,2%
	thermal/net electric	22,7%
	insolation/net electric (excluding correction for solar multiple)	14,9%
	insolation/net electric (including correction for solar multiple)	16,5%

Table 2

Main System Design Data for CRS

1. Input Signals.

1.1. Analog signals.

Measuring instruments :

14	Radiometers	R1 to R14
14	Amplifiers ranges	A1 to A14
14	Temperatures in amplifier boxes	TA1 to TA14
2	Power supply monitoring	M1 and M2

Cooling system :

2	Water pressure	WP1 and WP2
3	Water temperature (radiometers)	WTR1 to WTR3
1	Flow rate	WF
2	Water temperature (traverse bar)	WTB1 and WTB2

Traverse bar :

4	Surface temperature	STB1 to STB4

Stepping motor :

1	Number of rotations	N

Total : 57 analog input signals

1.2. Digital Signals.

Stepping motor :

2 status of operation — a and b

Total : 2 digital input signals

2. Output Signals.

2.1. Analog Signals.

None.

2.2. Digital Signals.

1 Stepping motor 'GO' command.
1 Stepping motor emergency 'Back to Parking' command.

Total : 2 digital output signals.

Table 3
Input and Output Signals

APPENDIX C

CONVECTION TESTING OF THE CENTRAL RECEIVER SYSTEM, FALL 1982*

By

John J. Kraabel
Sandia National Laboratories

CONTENTS

*This work was supported by the U. S. Department of Energy

CONVECTION TESTING OF THE CENTRAL RECEIVER SYSTEM, FALL, 1982

John S. Kraabel

ABSTRACT

A set of strategies for measuring the convective heat losses from the IEA SSPS central receiver system was evaluated during the Fall Measurement Campaign, 1982. The test method, chosen for its accuracy, utilized a zero incident solar flux technique. The heat into the system is provided solely by the hot sodium; the losses from the receiver to the environment may then be measured from the mass flow rate and the temperature drop in the receiver. The emitted radiation and the conduction are calculated and subtracted from the total, the remainder being the convective heat losses. An analysis that determines the most accurate test conditions is presented along with the results of the tests. The test results are limited due to the short time available to perform the tests. Suggestions are given for future convection testing of central receivers.

NOMENCLATURE

A = Area of the aperture, 9.69 m^2
A_c = Effective cavity interior area, 67.9 m^2
c_p = Specific heat
h = Heat transfer coefficient
$\dot{m}$ = Mass flow rate
q_i = Incident power into the receiver
q_a = Power absorbed by the heat exchanger
q_r = Reflected radiant energy out of the aperture
q_e = Emitted radiant energy out of the aperture
q_c = Convective heat losses from the receiver
q_{cd} = Conductive heat losses from the receiver
q_s = Stored heat
q_t = Total power from the receiver, $q_a + q_r + q_e + q_c + q_{cd}$
T_i = Receiver sodium inlet temperature
T_∞ = Free-stream temperature
T_o = Receiver outlet temperature
T_w = 1/2 (T_i+T_o), average receiver wall temperature
ΔT = T_i-T_o
ϵ_{eff} = Effective cavity emissivity
σ = Stefan-Boltzmann constant, 5.73 x 10^{-8} W/m^2K^4

INTRODUCTION

It is necessary to either measure or to estimate the thermal losses from a solar central receiver in order to make the most accurate determination of the efficiency of the receiver[1]. The efficiency is found by dividing the total power into the receiver into the power absorbed by the receiver. The total power can be found by measuring the incident power or by adding the thermal losses to the measured absorbed power. Similarly, the absorbed power may be found by direct measurement or by subtracting the thermal losses from the measured incident power. The efficiency may then be calculated if either (or both) the incident power or the absorbed power are known. The most accurate of the three methods uses either the incident power or the absorbed power measurements combined with the thermal losses. The efficiency, when determined by dividing the absorbed power by the incident power, is highly sensitive to inaccuracies in those measurements and therefore has a high inherent uncertainty.

It is important, therefore, to try to measure the thermal losses and in particular the convective losses from the CRS. Estimation of the reflected and the emitted radiative losses and of the conductive losses from the receiver will be accomplished analytically. The convective losses are measured directly.

A variety of techniques have been used or suggested for measuring the convective losses from central receivers. Most approaches are not sufficiently accurate to produce meaningful results. This is not surprising since the goal is to make scientific measurements using a very large apparatus with industrial quality instrumentation in a relatively uncontrolled environment.

A technique is used here, similar to that first suggested by Siebers[2] and utilized by Martin-Marietta at the CRTF at Sandia Laboratories[3]. In addition, the receiver conditions for testing are optimized to try to achieve the highest possible accuracy.

POSSIBLE TEST METHODS

A review of possible techniques for testing the receiver is presented as a review of the approach taken in determining the most accurate method for measuring the convective losses

1. Normal Operation. Testing during normal operation requires solving the energy balance equation

$$q_c = q_i - q_r - q_e - q_a - q_{cd} \tag{1}$$

The results of an uncertainty analysis are shown ir Figure 1, which shows accuracy versus incident power with the accuracy of the incident flux measurement as a parameter. The large errors or uncertainties arise, because in this method, the convection is the difference between two large numbers. The uncertainties in each are large compared to the convection. Note that the greatest accuracy is the special case of zero incident power or Reverse Operation.

2. Reverse Operation. Because of the greater accuracy, the favored method is to flow hot sodium through the receiver with no heliostats in operation. The sole source of heat is the heat exchanger panel; the heat input may be easily measured. This method will be discussed in more detail later.

3. Full Power Transient. While running at full available power, the heliostats removed. The receiver is allowed to come to steady-state and the convective losses are measured as:

$$q_c = q_a - q_e - q_r - q_{cd} - q_s \tag{2}$$

The major source of uncertainty in this method is the steady-state assumption. Much of the receiver is at a high temperature due to the reflected incident radiation. When the heliostats are removed, many of these hot internal cavity surfaces contribute to the energy balance. Insufficient time is available to reach a true steady-state. To properly account for the complex unsteady heat transfer in the receiver would require a sophisticated thermal model which would account for the convection, conduction, and radiation exchange inside the CRS. The level of effort required is probably unwarranted and certainly neither the time nor the facilities to accomplish this task exist on site.

4. Step Power Input. The incident power from the heliostat field is reduced in steps. At some point, the receiver output is equal to the input and the total thermal losses are determined. This technique has essentially the same accuracy problems as the first technique.

5. No-Sodium, Steady. In the event of a failure of the sodium system, some measurements could be made by simply heating the tubes. Accuracics would be simiiar to number one with the exception that the heat exchanger temperatures would be much less certain because the thermocouples are located on the ends rather than distributed throughout. This method is probably not a practical technique.

6. No-Sodium, Unsteady. Similar to 5 except the doors are closed to allow the receiver temperatures to equilibrate. When uniform, the doors are opened and the transient cool-down is observed. This is the better of the two no-sodium techniques but still has the problems of number 3.

The conclusion of this brief review is that the best method and the one chosen for this study is the second or Reverse Flow method. It may be seen from the analysis of the first method that the highest accuracy is at zero flux. Should a marked improvement be made in measuring the incident flux into the receiver, method 1 may become more useful. The data will be available for a method 1 calculation as a part of normal operation; comparison of the results may prove interesting. Method 4 is only a special case of 1, and the transient methods and no-sodium techniques are clearly beyond our capabilities at the present time.

REVERSE OPERATION TESTS

The convection tests were run by simply opening the doors of the receiver without providing any incident power from the heliostat field. The mass flow rate of the sodium was adjusted to the pre-test optimum value and the entire system was allowed to come to steady-state.

The convection may then be determined by

$$q_c = q_a - q_e - q_{cd} - q_s \tag{3}$$

The storage term, q_s, was kept to a minimum by running the test after the receiver has had a thorough thermal soak at close to the test temperature. Ideally, if the test is to be run using sodium from cold storage, then it should be run prior to morning start-up after the receiver has been at the cold storage temperature most of the night. Similarly, if the test can be run from hot storage, then it should be run at the end of the day. The tests during the Fall Measurement Campaign were all run from cold storage and generally at times of convenience such as when cloud cover precluded normal operation.

The heat input to the receiver, q_a is measured by measuring $\dot{m}$, T_i, T_o , and evaluating c_p as a function of temperature. The conversion of m from units of m^3/hr to kg/s requires evaluating the sodium density at the inlet temperature where m is measured. The measurements of $\dot{m}$, T_i and T_o were steady to within one percent and therefore did not require time averaging.

The conduction may be determined by combination of the available temperature measurements and the thermal conductivity or by observing the behavior of the receiver when it is operating with the doors closed. From equation 3, it may be seen that with no incident flux to the receiver and with no radiation and convection from the receiver, the absorbed power must be equal to the conduction. The conductive losses are small compared to the convection, and although the uncertainties in the conduction may be large, their influence on the accuracy of the convection measurements will be small.

The emitted radiation is a complex problem, and it had been intended to use two methods to evaluate the losses. A computer program was used to solve a simplified thermal radiation model of the receiver. Interior surface temperatures and surface radiation properties were necessary data [4]. The second method was not finished in time for the Fall Measurement Campaign. It was to use calorimeters facing into the cavity to measure the emitted radiation and, when operating normally, the reflected radiation. The calorimeters would probably have given a higher degree of accuracy than the calculations and consideration should be given to including them in any future studies. Care must be taken in selecting the instrumentation because of the large difference in the magnitude of the radiation during convection tests and normal operation. It may not be possible to obtain accurate measurements for both cases with only one set of calorimeters.

Relevant data not directly involved in the measurements included wind velocity and direction atmospheric pressure and temperature. At some future date, some measure of the free-stream turbulence should be obtained. It may be possible to make crude estimates from the available data.

OPTIMIZATION ANALYSIS

Prior to running the tests, an analysis was performed to determine the most accurate method of running the test. The primary sources of uncertainty in the tests come from the uncertainties in measuring the mass flow rate and the temperatures of the sodium as it enters and leaves the receiver. The uncertainties in these measurements are expressed in terms of a fixed percentage of the full-scale capability of the transducers. As a result, if the test is run with the mass flow rate very low, then it would become the largest contributor to the overall uncertainty. Similarly, if the temperature difference were small, then the uncertainty in the mass flow rate would be small while the contribution to the total uncertainty from the temperature difference would be the largest single source of uncertainty. This concept is quantified in the following analysis.

A crude form of equation (3) is:

$$q_c = \dot{m}c_p(T_i - T_o) - \epsilon_{\text{eff}}\sigma A(T_w^4 - T_\infty^4) \tag{4}$$

Here the conduction is considered to be negligible and the storage term is assumed to be zero. The terms which contribute to the uncertainties are m, T_o, T_i, ϵ, T_w, and T_∞.

The uncertainty in q_c comes from differentiating equation 4 in terms of the contributors to the uncertainty. Thus,

$$\frac{\partial q_c}{\partial \dot{m}}\delta \dot{m} = c_p(T_i - T_o)\delta \dot{m} \tag{5}$$

$$\frac{\partial q_c}{\partial T_i}\delta T_i = \dot{m}c_p\delta T_i \tag{6}$$

$$\frac{\partial q_c}{\partial T_o}\delta T_o = -\dot{m}c_p\delta T_o \tag{7}$$

$$\frac{\partial q_c}{\partial T_w}\delta T_w = -4\epsilon\sigma A T_w^3\delta T_w \tag{8}$$

$$\frac{\partial q_c}{\partial T_\infty}\delta T_\infty = 4\epsilon\sigma A T_\infty^3\delta T_\infty \tag{9}$$

$$\frac{\partial q_c}{\partial \epsilon}\delta\epsilon = \sigma A_{ap}(T_w^4 - T_\infty^4) \tag{10}$$

It can be shown by using the data in the sample calculation for the convective loss measurements described later on that the uncertainties introduced by the last three expressions are negligible compared to the magnitude of the first three. Thus, we will ignore the influence of the uncertainties described by equations 8, 9, and 10. The uncertainty in the both temperature measurements are equal, the total uncertainty is then

$$\delta q_c = \sqrt{\left[c_p(T_i - T_o)\delta\dot{m}\right]^2 + 2\left[(\dot{m}c_p)\delta T\right]^2} \tag{11}$$

This value will be minimized when the two terms under the square root sign are equal. That is when

$$c_p(T_i - T_o)\delta\dot{m} = \sqrt{2}\dot{m}c_p\delta T \tag{12}$$

solving for m, we obtain the optimum value

$$\dot{m}_{\text{opt}} = (T_i - T_o)\frac{\delta\dot{m}}{\sqrt{2}\delta T} \tag{13}$$

This formula gives us the relation between the mass flow rate and the difference between the inlet and outlet temperatures. This is one equation in two unknowns; on a given day, the inlet temperature is unknown. If enough time were available, it would be possible to adjust the mass flow rate to the receiver, monitor the resultant temperature difference and then readjust as necessary until the proper ratio between the mass flow rate and the temperature difference was achieved. An alternative is to make a guess at the losses and solve for the unknown variable. We may write the convection as

$$q_c = hA_c(T_w - T_\infty) \tag{14}$$

where as a first approximation [7],

$$h = 0.8055(T_w - T_\infty)^{0.426} \tag{15}$$

Substituting equations (13), (14), and (15) into equation (4), we obtain

$$0.8055A_c(T_w - T_\infty)^{1.426} = c_p(T_i - T_o)^2\frac{\delta\dot{m}}{\sqrt{2}\delta T} - \epsilon\sigma A(T_w^4 - T_\infty^4) \tag{16}$$

It may be assumed for the convection tests that

$$T_w = 0.5(T_o + T_i)$$

or, $T_w = T_o + \frac{1}{2}(T_i - T_o) = T_o + \frac{1}{2}\Delta T$.

We may then write

$$0.8055 A_c (T_w - T_\infty)^{1.426} = c_p \frac{\delta \dot{m}}{\sqrt{2}\delta T} \Delta T^2 - \epsilon\sigma A(T_w^4 - T_\infty^4) \quad (17)$$

The solution to this equation was obtained numerically on an HP-41 calculator by selecting a value of ΔT and solving for T_w. The solution was then used to determine the mass flow rate that corresponded to the temperature difference. If we solve for the approximate value of the ratio in equation 13, we find

$$\dot{m} = 0.19(T_i - T_o) \quad (18)$$

where $\dot{m}$ is in m^3/hr and T_i-T_o is in C.

During the convection tests, the minimum uncertainties will be attained if the optimum ratio between m and △T is maintained. It should also be noted that the higher the value of the inlet temperature of the sodium, the higher the convective losses; subsequently, the values of m and △T are greater and the accuracy of the measurement increases.

TEST RESULTS

Convection testing during the Fall Measurement Campaign was limited for a variety of reasons, including plant down-time and insufficient thermal capacity in the cold storage tank. Operations from hot storage would have alleviated some of these problems because of the greater thermal capacity. It is necessary to have sufficient heat stored in the sodium to run a full steady-state test without lowering the temperature in the system to too low a value. For these tests, all run from cold storage, 200 C was used as a minimum value for the outlet sodium temperature. This value was chosen because the Data Acquisition System did not print out temperatures below 200 C. A slightly lower temperature limit could have been used but there seemed to be little advantage in doing so.

The data are presented in Table 1. The nine data points actually are derived from five experimental runs. The last six data points were obtained in two sets of three; each set was obtained in a single test where the receiver doors were opened and data was collected in a continuous manner. The three points on each day were chosen for the steadiness of wind and continuity of other conditions in the minutes prior to taking the data.

TABLE 1. CONVECTIVE LOSS DATA

Data Point	Date	Time	Back Wall	Top Surface	Bottom Surface	Air Temperature
1	4 Oct.	1110	217.7	200.0	45.0	24.5
2	4 Oct.	1530	231.0	185.0	50.0	25.2
3	5 Oct.	1754	217.7	200.0	80.0	24.8
4	6 Oct.	1814	245.0	233.0	138.0	20.8
[illegible]	6 Oct.	1832	235.4	203.0	100.0	19.7
6	6 Oct.	1901	226.8	180.0	80.0	19.2
7	7 Oct.	1844	246.2	290.0	124.0	18.3
8	7 Oct.	1916	233.8	252.0	88.0	17.8
9	7 Oct.	1953	227.8	228.0	88.0	16.8

The results are shown in Table 2.

TABLE 2. CONVECTIVE LOSS RESULTS

Data Point	q_a	q_e	q_{cd}	q_c	T_w-T_∞	h
1	129.7	19.1	12.7	97.9	141.2	10.22
2	129.3	20.2	14.3	104.8	141.3	10.93
3	129.3	19.6	11.7	98.0	150.1	9.62
4	172.1	26.9	10.8	134.4	191.7	10.33
5	168.6	22.9	14.6	131.1	168.2	11.49
6	165.2	20.2	20.7	124.3	151.8	12.07
7	149.3	31.0	15.0	103.3	211.6	7.19
8	149.9	25.4	20.0	104.5	184.2	8.36
9	159.0	23.1	22.7	113.2	174.2	9.58

The power into the cavity is the mass flow rate multiplied by the specific heat and the temperature difference between the inlet and outlet of the receiver panel. The uncertainties in the absorbed power measurement is the largest source of error and ranges from 18 to 22%.

The radiation is determined from a zonal model of the receiver with four surfaces. The number of zones is limited by the available temperature measurements and the number of readily available radiation configuration factors. The computer program used to evaluate the radiation exchange solves the radiation exchange in simple enclosures. It contains a banded emissivity approximation, although this feature was not used here. Instead, measurements of the total emissivity of the various surfaces were used. The temperature of the receiver zone is the temperature of the midpoint of the heat exchanger panel. The temperature of the top and bottom inside surfaces was estimated from measurements taken by Pescatore[5] and Gaus[6]. The temperature of the aperture was assumed to be the outside air temperature. It was assumed that there was no radiation reflected back into the cavity. Uncertainties in the temperature were assumed to be ± 20 C on the heat exchanger and ± 50 C for the remainder of the cavity. The outside temperature was assumed to be known within ± 10 C. The emissivity was assumed to be known to within ± 0.1 ; this value is conservative due to the Hohlraum chamber effect of the cavity. The uncertainty in the radiation determination using these limits is ± 25%.

The conduction was estimated by examining the data listing before and after the tests when the doors were closed and the only losses were the conduction. The conduction was estimated by interpolating between the initial and final values. The uncertainty in the conduction is considered to be less than 6.0 kW or approximately ± 50%.

The convection is determined as described in equation 3. The values are similar to those predicted before the tests. The results show that the convection is not an excessive loss mechanism under the conditions shown. That is, the losses are within the bounds expected by pre-test predictions. There appears to be some bias in the results as evidenced by the similarity of the data in each of the last two sets of three and the fact that they differ from set to set. The cause of this bias is not clear at this time. It is possible, due to the small number of data points, that the source of the bias in this data set may never be found. The over-all uncertainty in the convective loss measurements ranged from ± [illegible] to ± [illegible]%. It is not unreasonable due to the complexity of the problem and the simplicity of the analysis to consider that the accuracy is within ± 50%.

The precise influence of wind is not clearly apparent in the data, again due to the small number of data points. The heat transfer coefficient is plotted versus the wind velocity in Figure 2, and there does appear to be some influence.

SUGGESTIONS FOR FUTURE TESTS

One of the most important results of the tests performed during the Fall Measurement Campaign was to learn what improvements should be made in the tests for the next campaign. The convection tests presented here need to be considered as relatively crude. They have provided an estimate of the convection from the receiver, but the uncertainties in those estimates are larger than desired. The following is a list of the limitations of these tests and suggestions for future tests.

1. The number of convection tests is very small. The limited amount of data makes it virtually impossible to evaluate the effect of wind on the convection. A large data set is necessary to ensure that the effects of the wind, from all quadrants and over a realistic range of velocities, may be observed. One method of ensuring a large data set is to routinely perform convection tests at the end of each operating day and in the morning before start-up as well as any day where the solar conditions are inadequate for normal operation. At other times, if unusual wind conditions prevailed that did not occur during other convection testing, operations could be stopped and convection tests could be performed. While it is realized that the purpose of the receiver is not to perform convection tests, the cost of an individual test is quite small, perhaps on the order of 100 kW thermal.

2. A larger number of data points would also provide more confidence in the results. The data presented appear to have some influence that is not correctly accounted for in the analyses. A larger number of data points might allow an understanding of the cause and provide a method for including a correction. A likely source of the error is the steady-state assumption. The conduction losses appear to be quite large shortly after shutdown, and gradually reduce through the night. If this variable conduction actually is the cause of the discrepancy, then appropriate tests could be devised to detect the source. For example, tests need to be performed with the receiver heating up and cooling down in order to bound the effects of unsteady behavior.

3. The availability of surface temperature measurements inside the receiver severely limits the accuracy and confidence in the radiation calculations. The uncertainties in the floor, ceiling, and side wall temperature measurements are too high simply because there is no redundancy check, and in some cases, no measurement at all. Installation of an array of thermocouples that would directly support the thermal

radiation program zonal model would greatly improve the accuracy and reliability of the radiation analysis.

4. The thermal radiation model used in these tests consists of only four zones, the heat exchanger, the top inside half and the bottom inside half of the receiver, and the receiver aperture. The radiation code used in the analysis is capable of solving for a much larger number of zones. However, the limited temperature data and the limited time and capability to determine radiation configuration factors prohibited a more detailed model. Development of a suitable model with an adequate number of zonal surfaces could take place prior to the next measurement campaign, and if properly done could guide the placement of the thermocouples in and around the receiver.

5. The ability to perform convection tests by operating the receiver with sodium from hot storage would increase both the accuracy of the convective loss measurements and the amount of time when convection testing could be performed. An effort should be made to determine if this is possible and then to do it often enough so that it becomes a routine operation.

6. Direct measurement of the emitted energy during the convection tests would probably provide a more accurate means of determining the radiation losses.

CONCLUSIONS

The conclusions of the convective loss studies during the Fall Measurement Campaign [illegible], may be summarized as follows:

1. The convection from the IEA SSPS central receiver system is of a reasonable size and is of similar magnitude to pre-test predictions. The importance of this finding is that the performance and economic predictions are reasonable and that no disastrously [illegible] will be found.

It was found that relatively accurate convection testing of actual receivers is possible given careful pre-test analysis and preparation. The pre-test preparation includes the steps utilized in these tests as well as those listed in the section on Suggestions for Future Tests.

3. The accuracy of the results presented herein is of the order of $\pm$ 50%. This level of uncertainty could be reduced by careful selection and calibration of the mass flow and temperature transducers, although the largest improvements are possible though proper preparation. It should be kept in mind, however, that accuracies beyond $\pm$ 25%or so are probably not possible.

ACKNOWLEDGMENTS

I would like to express my gratitude to the members of the ITET for their valuable assistance during the Fall Measurement Campaign.

REFERENCES

1. J. S. Kraabel, Receiver Efficiency Calculation Method, CRS Fall Equinox Measurement Campaign, 1982, IEA-SSPS Note R-77/82 JK 3222.

2. D. L. Siebers, Personal Communication, 1979.

3. R. K. McMordie, Convection Losses from a Cavity Receiver. Twentieth National Heat Transfer Conference, AIChE, Aug. 2-5, 1981.

4. T. van Steenberghe, Solar Absorptance Measurements on the First CRS Receiver Tubes, IEA-SSPS Note R-85/82 TVS 3220.

5. M. Pescatore, Personal Communication, 1982.

6. F. Gaus, Personal Communication, 1982.

7. J. S. Kraabel, An Experimental Investigation of the Natural Convection From a Side-facing Cubical Cavity, Thermal Engineering Joint Conference, ASME/JSME, Honolulu, Hawaii, Mar. 20-24, 1983.

APPENDIX 1 PERFORMANCE ESTIMATES FOR 6 SEPTEMBER 1982

The following data were obtained from the Data Acquisition System and were used to develop the strategy and pre-test accuracy estimates for the convection tests.

Incident Power:

$q_i = 2161kW$

$\delta q_i = \pm 15\%$ of q_i, or

$= \pm 324kW$

Absorbed power:

$q_a = 1863kW = \dot{m}c_p(T_o - T_i)$

$\Delta T = 530 - 270 = 260C$(assumed)

$c_p = 1.279kJ/kg - K \quad (T = 400C)$

$\dot{m} = \frac{q_a}{c_p(To - Ti)} = 5.60kg/s$

$\delta T = \pm 1.2\% of 400C = \pm 4.8C$

$\delta \dot{m} = \pm 2.6\% of 50m^3/hr = \pm 1.3m^3/hr$

$= \pm 0.304kg/s$

$$\delta q_a = q_a\left\{\left(\frac{\delta \dot{m}}{\dot{m}}\right)^2 + \left(\frac{\delta(T_o - T_i)}{T_o - T_i}\right)^2\right\}^{\frac{1}{2}}$$

$= \pm 113kW$

Reflected Radiant Power:

$\rho_{\text{effective}}$ = 0.03, somewhat less than the actual reflectivity due to the hohlraum

= chamber effect of the cavity receiver.

$\delta\rho_{\text{effective}} = \pm 0.25\rho_{\text{effective}}$(assumed)

$q_r = 0.903q_i$

$$\delta q_r = q_r\left\{\left(\frac{\delta q_i}{q_i}\right)^2 + \left(\frac{\delta \rho}{\rho}\right)^2\right\}^2$$

$= \pm 19kW$

Emitted Radiant Power:

$q_e = \epsilon_{effective} A_a (T_w^4 - T_\infty^4)$

$\delta q_e = \pm 0.33 q_e$(assumed)

$\epsilon_{effective} = 0.85$

A_a = Aperture area, $9.69 m^2$

T_w = $400C$, an average temperature

T_∞ = $20C$, outside temperature

Convective losses:

$Nu = 0.088 Gr^{\frac{1}{3}} \left(\frac{T_w}{T_\infty}\right)^{0.18}$

$h = \frac{Nuk}{L}$

$q_c = h A_c (T_w - T_\infty)$

L = height of floor to top of aperture, $3.63m$

A_c = effective inside area of receiver, $67.9m^2$

k = Thermal conductivity of air at T_∞

Nu = Nusselt number

Gr = Grashof number

$Gr = 2.74x10^{12}$

$Nu = 1430$

$h = 10.1 W/m^2C$

$q_c = 200.0 kW$

$\delta q_c = \pm 50\%\, q_c$(assumed)

Conduction: Assumed to be sufficiently small, so that uncertainties in q_{cd} are negligible.

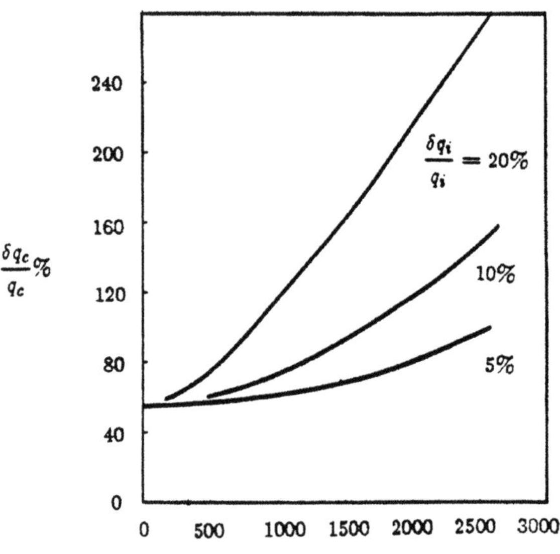

Figure 1. Convection measurement uncertainties as a function of incident power.

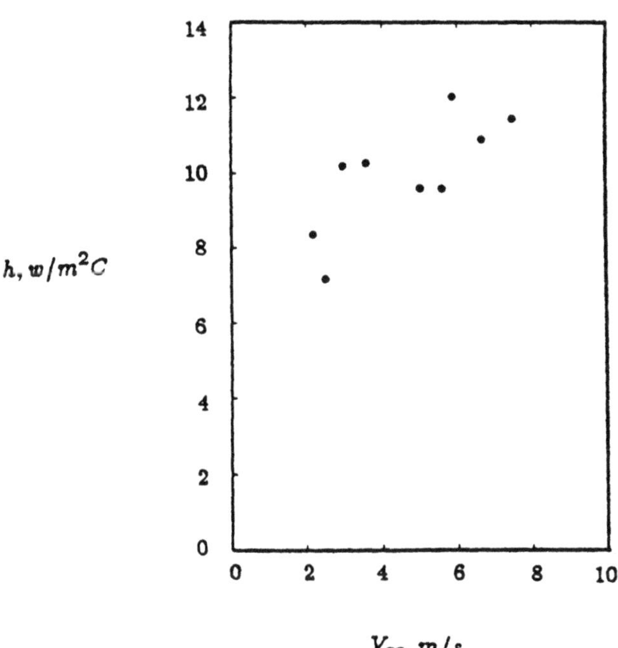

Figure 2. Heat Transfer Coefficient Versus Wind Velocity

APPENDIX D

CONVECTION LOSSES OF THE SULZER RECEIVER

By

H. Jacobs
IEA/SSPS Project - ITET

1. Introduction

Knowledge about the losses of the receiver is one of the most important factors needed in designing a CRS solar power plant.

In this report are published the results of convection losses measurement of the Sulzer receiver.

The lost energy was delivered by sodium from the storage tanks. All heliostats were in stow, so that the energy drop of the sodium is equal to the losses. To calculate and subdivide the numbers the computer code RADA was used. The equations are given in Reference 1.

2. Description of the Tests

The convection loss tests were performed in October 1982 and in March/April 1983.

During the test, sodium is passed through the receiver tubes and as the sodium cools down, temperature and flow rate are measured.

Temperature drop and flow rate are the input values for calculating the losses.

There were two different tests performed:

a) Normal Flow Test

Sodium from the cold storage was pumped in the normal way (from the bottom to the top) through the receiver back in the cold storage.

The accuracy of the results of this test are not very high (see Ref. 1). Better results can be expected by:

b) Reverse Flow Test

Sodium from the hot storage was passed through the receiver tubes from top to bottom. The temperature profile is more like the normal operation where sodium is heated as it flows bottom to top.

All the tests were performed with the following time schedule:

- The doors of the receiver are closed.
- The flowrate is kept constant for 15 minutes steady state. Flow with doors closed is necessary to calculate the conductive losses.
- The doors were opened and the flowrate increased.
- The flowrate was kept constant for at least 15 minutes. Steady state with doors open delivers the input values for calculating the:

    - convection losses plus
    - radiated losses

and the heat transfer coefficient.

3. Mode of Calculation and Results

The absorbed energy of the receiver is determined by:

$$P_{absorbed} = P_{into} - P_{refl} - P_{em} - P_{cond} - P_{convect} \quad (1)$$

During the convection losses tests the energy drop of the sodium has the same number as the losses. Because there is no radiation the reflected power becomes zero and, the equation for the losses becomes:

$$P_{loss} = P_{con} + P_{cond} + P_{em} \quad (2)$$

The calculation subdividing the losses is performed by the computer code RAD1.

The input values for this code are in Table 1.

Table 1

Input Numbers for RAD1

Nr	Date	Time h	T in deg	T out deg	flow m3/h	Twall deg	Ttor deg	Tbott deg	Tamb deg	Pin kW	Ped kW
10	29/3/83	18.00	399.02	207.81	4.94	303.42	283.00	45.00	13.8	301	21
11	7/4/83	16.00	442.19	253.52	6.48	347.86	300.00	75.00	20.0	381	25
12	8/4/83	15.53	399.61	245.75	7.06	322.68	275.00	62.00	23.3	343	26
13	12/4/83	9.18	241.06	216.50	20.42	228.78	180.00	75.00	13.5	166	16
14	14/4/83	10.01	232.13	214.06	26.60	223.10	195.00	45.00	11.5	160	15

The results for all tests which were performed are in Table 2.

Table 2

Results of the Convection Losses Tests

Nr	Date	Time h	Tmed deg	Power kW	Qrad kW	Qend kW	Qconv kW	Hconv W/m2/K	Wind m/s
1	4/10/82	11.10	193.2	129.7	19.1	12.7	97.9	10.11	5.0
2	4/10/82	15.30	205.8	139.3	20.2	14.3	104.8	10.93	5.8
3	5/10/82	17.54	192.9	129.3	19.6	11.7	98.0	9.62	3.5
4	6/10/82	18.14	224.2	172.1	26.9	10.8	134.4	10.33	5.5
5	6/10/82	18.32	215.7	168.6	22.9	14.6	131.1	11.49	6.6
6	6/10/82	19.01	207.6	165.2	20.2	20.7	124.3	12.07	7.5
7	7/10/82	18.44	227.9	149.3	31.0	15.0	103.3	7.19	2.2
8	7/10/82	19.16	216.0	149.9	25.4	20.0	104.5	8.66	2.5
9	7/10/82	19.53	211.0	159.0	23.1	22.7	113.2	9.58	2.9
10	29/03/83	18.00	289.6	300.7	39.6	21.2	239.9	16.45	2.5
11	7/04/83	16.00	327.9	380.8	51.5	24.5	304.8	18.83	4.7
12	8/04/83	15.53	299.4	342.5	42.8	22.6	277.8	19.13	4.1
13	12/04/83	9.18	215.3	166.0	20.6	15.6	129.8	12.18	7.5
14	14/04/83	10.01	211.6	159.6	20.2	15.2	124.2	11.85	6.5

Tests numbers 1 to 9 and 13 to 14 are normal flow tests. Numbers 10 to 12 are reverse flow tests using sodium from the hot storage.

To calculate the receiver efficiency it is necessary to calculate the losses under normal conditions with:

$$T_{inlet} = 270 \text{ degrees}$$

$$T_{outlet} = 520 \text{ degrees}$$

$$T_{ambient} = 35 \text{ degrees}$$

The medium temperature difference of the receiver then becomes:

$$T = \frac{270 + 520}{2} - 35 = 360K$$

It was not possible to perform tests with these temperatures. It is necessary to extrapolate the results from the tests that were run by the following. Fig.6.3-1 shows the total power and the convection losses versus medium temperature difference. A test of the measured values produced the following results:

$$P_{convect} = 1.62 \times T - 226 \text{ KW} \quad (1)$$

$$P_{loss} = 1.93 \times T - 253 \text{ KW} \quad (2)$$

Then the total losses under design conditions become:

$$P_{loss\ design} = 1.93 \times 360 - 253 = \underline{442 \text{ KW}}$$

This number should be constant for each load. It does not include any influence of the wind! Consequently, this number gives a range of the loss.

The total power does not include the absorption, which is described by the X-number. The average of X is determined in Reference 2 as:

$$X = 0.94 \text{ (average)}$$

When using the following equation (3) the efficiency of the Sulzer Receiver becomes nearly 80%.

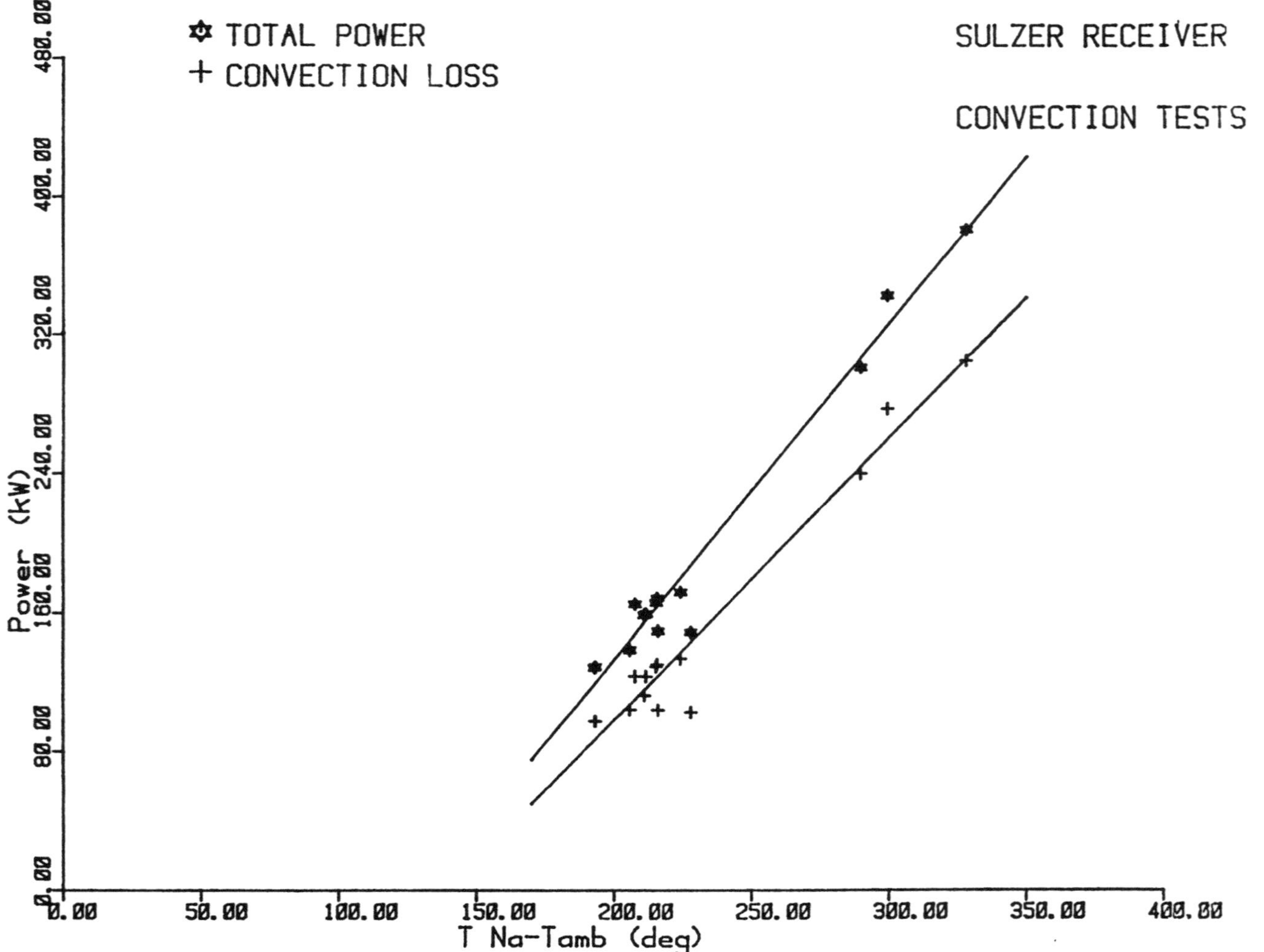

TOTAL POWER
+ CONVECTION LOSS
SULZER RECEIVER
CONVECTION TESTS
Power (kW)
0.00
80.00
160.00
240.00
320.00
400.00
480.00
T Na-Tamb (deg)
0.00
50.00
100.00
150.00
200.00
250.00
300.00
350.00
400.00

FIGURE 1

$$n = X \left(1 - \frac{P_{total\ loss}}{P_{absorbed} + P_{total\ loss}}\right) \qquad (3)$$

Figure 2 is a plot using this equation. The "+" are results of calculations in Reference 1. (Calculated points with error bands). Table 3 shows the efficiency for the different mass flows under design temperature conditions.

Mass Flow m^3/h	Absorbed Power	Efficiency %
5	380	43
10	761	59
15	1141	68
20	1522	73
25	1902	76
30	2282	79
35	2663	81

4. Reference List

1. John S. Kraabel
   Convection Testing of the Central Receiver System
   Fall, 1982, and
   Receiver Efficiency Fall Measurement Campaign, 1982

   Presentations of the CRS Workshop
   April 1983

2. Thierry van Steenberghe
   Solar Absorption Measurements on the First CRS Receiver Tubes

   R-85/86 29.10.1982

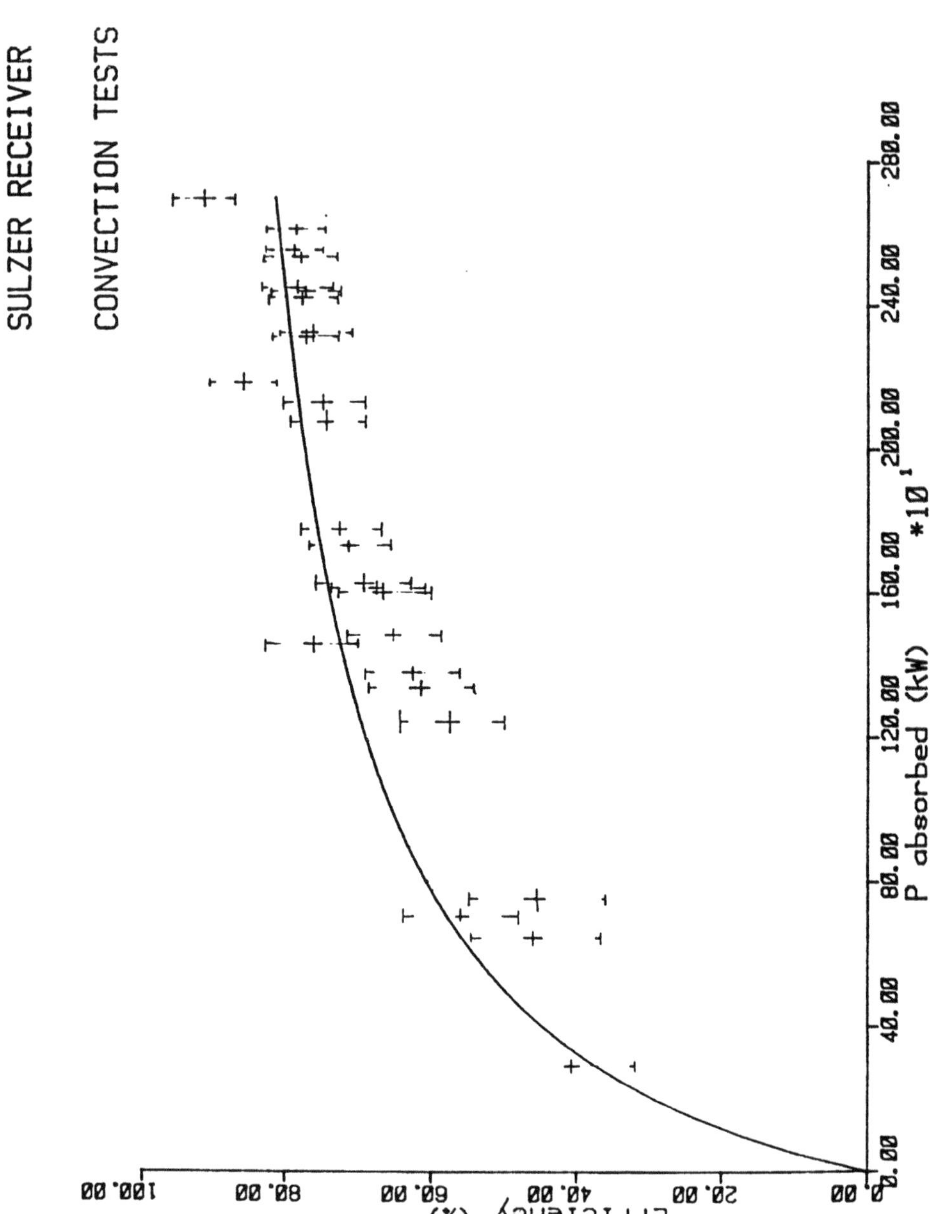
SULZER RECEIVER
CONVECTION TESTS
Efficiency (%)
P absorbed (kW) *10[1]
0.00
20.00
40.00
60.00
80.00
100.00
0.00
40.00
80.00
120.00
160.00
200.00
240.00
280.00

FIGURE 2

CRS PARASITIC CONSUMPTION: THE TRACE-HEATING

Antonio Cuadrado, Sevillana and Pierre Wattiez, ITET

1. INTRODUCTION

One of the first priorities when selecting the operation strategy for a solar power plant built to produce electrical energy is to reduce the plant internal electrical consumption, or plant "parasitics" to a minimum. This paper presents the results of a specific survey of the trace-heating electrical consumption during the first seven months of 1984, and the impact of modifications made to reduce its value. Trace-heating would have a significant effect in that it is the largest in-house load (68% of the total).

2. CRS TRACE-HEATING DESCRIPTION (Ref.1)

All of the sodium components and piping (except the receiver) are provided with an electrical trace-heating system (Fig.6.4-1) that preheats the complete HTS up to approximately 200°C to prevent the sodium from freezing. In addition to this preheating system, all the abovementioned elements are covered with mineral wool insulation.

The main characteristics of the trace-heating are:

- Connection : main diesel busbar
- Installed power : 135 kW
- Voltage : 380 V/220 V
- Average power (design point) : 12.8 kW

Fig.6.4-2 presents the total CRS in-house consumption and the impact of the trace-heating electrical consumption. Since the beginning of 1984, the total consumption has been reduced. This reduction is attributed first to the fact that the plant was operated during consecutive days ("T" in the figure), and second to several modifications made in February 1984 to the trace-heating system.

Trace-Heating Modifications and Results

The operational strategies used for some of the subsystems were changed, and in addition some parts of the trace-heating system were disconnected during this non-operation period; i.e., the purification system, plugging meter, and the upper trace-heating circuit of both tanks. Since then, they would be connected only when necessary. Also, to minimize the losses during weekends, the operational strategy was slightly modified: an amount of hot sodium in the hot tank is left in the hot tank and transferred to the cold tank when necessary to keep it above the temperature at which the trace-heating would be switched on. Therefore, there are two different periods to evaluate; the first one up to February 24, and the second one from May 21 to July 31.

Comparison of the two evaluation periods shows that a significant reduction has occurred in consumption as a result of the changes. The following table gives average values for a working day, Saturday, and Sunday before and after the modification.

	up to February 24			May 21 - July 31		
	Working Day	Saturday	Sunday	Working day	Saturday	Sunday
Trace-Heating (kWh)	1668	1706	1832	736	731	1320
Total CRS (kWh)	2420	2291	2467	1472	1302	1872

Trace-Heating Consumption Impact

Trace-heating consumption was suspected of being quite high; therefore at the end of September 1983, a computer was installed to specifically record the consumption of the SHTS trace-heating system. The data collected for ten months of operation are presented in Tab.6.4-1. Over that period, the trace-heating consumption represented 68% of the total CRS internal consumption without including the CRS plant parasitics shared in common with the DCS plant. The sources of electrical consumption were:

- Air conditioning in the control rooms
- Power to the main control building
- Power to the office building
- Power supply to Na electrical equipment, including electrical fans of the low and high voltage room, and the supply to the battery room

- Power in the heliostat field
- Lights in the main control building
  - " in the office building
  - " in the CRS and DCS building
  - " in the sanitary rooms
- Specific pump - water crude
  - drainage water
- Other items - warehouse
  - movable crane

The other 22% of the CRS internal consumption is divided between the three main CRS subsystems; the heliostat field, the sodium heat tansfer system, and the power conversion system for an average monthly electrical consumption of 20.5 MWh.

The trace-heating depends on the number of ambient temperature and solar hours.

Many other factors must also be considered i.e., the hours of operation of the receiver, the operation or nonoperation of the PCS, and the execution of some kind of test (when this test had some marked influence in the trace-heating consumption there is a notation on the graph). However, as shown in the following figures the main direct cause of the irregularities is the number of nonconsecutive days of operations.

1st day of operation	:	1080 kWh	1st day of operation	:	673 kWh
2nd " " "	:	593 "	2nd " " "	:	1234 "
3rd " " "	:	559 "	3rd " " "	:	1747 "
4th " " "	:	527 "	4th " " "	:	1907 "

The curve of Fig.6.4-3 represents the consumptions of the 1st, 2nd,... day of operation, and the curve of Fig.6.4-4 represents the 1st, 2nd,... day of nonoperation (only for the period from May 21).

When analyzing the curves, it can be observed that there is a discontinuity between the first day of nonoperation and the first day of operation. (The consumption on the first day of operation is more than on the first day of nonoperation). The explanation is as follows: readings of electronic energy consumed by the trace-heating are taken one or two hours before midnight, while operation usually starts around 8 a.m. This means that there is a difference of nine to ten hours between trace-heating consumption values and effective start of an operational day. If the points of the curves are taken ten hours later than their current position we get the continuous curve of Fig.C.4.5

The average consumption of the CRS plant (except trace-heating) is 570 kWh during a nonoperation day, and 750 kWh during an operation day (considering operation day to be when sodium is heated in the receiver, followed by PCS operation).

3. CONCLUSION

During the period covering the first seven months of 1984, the trace-heating consumption represented 60% of the total consumption for the CRS during an operation day and 72% during a nonoperation day.

After the modifications and changes in the operational strategy the internal consumption of the trace-heating was 41% on the fourth day of operation and 77% on the fourth day of nonoperation.

Just for comparison, the average consumptions of CRS and DCS common services are as follows:

	working day	weekend
CRS	750 kWh	570 kWh
DCS	501 kWh	332 kWh
Common	587 kWh	494 kWh

In summary one can state that trace-heating consumption is heavily dependent on the preceeding days' operational pattern (thermal memory). The curve obtained can be helpful in the estimation of long-term trace-heating consumption if the stastistical distribution of operational days is known. Based upon these data, careful management of operation can produce substantial savings in traceheating consumption.

4. REFERENCES

1. INTERATOM, CRS Solar Power Plant Description.

2. A. Cuadrado, "Trace-Heating Consumption", SSPS Internal Report R-36/84AC, 1984.

	COMMON PARA. (W)	COMMON PARA. (WH)	COMMON	TRACE-HEATING	% OF PARA. WH
OCTOBER 1983	84996.80	76654.1	8342.80	54968	72
NOVEMBER 1983	78520.30	70278.1	8242.20	50058	71
DECEMBER 1983	82827.85	73502.9	9324.95	50709	69
JANUARY 1984	84602.90	75839.4	8763.50	57553	76
FEBRUARY 1984	77207.90	68797.1	8410.80	47988	70
MARCH 1984	68221.65	59082.9	9138.75	37348	63
APRIL 1984	72637.50	64673.8	7963.75	46583	72
MAY 1984	67470.00	59289.3	8180.70	40013	67
JUNE 1984	54684.85	46079.3	8605.55	25637	56
JULY 1984	55417.30	44922.5	10494.80	23579	52
TOTAL FOR TEN MONTHS	726587.05	639109.3	85467.00	434416	68

Table 6.4.-1: CRS Parasitic Repartition

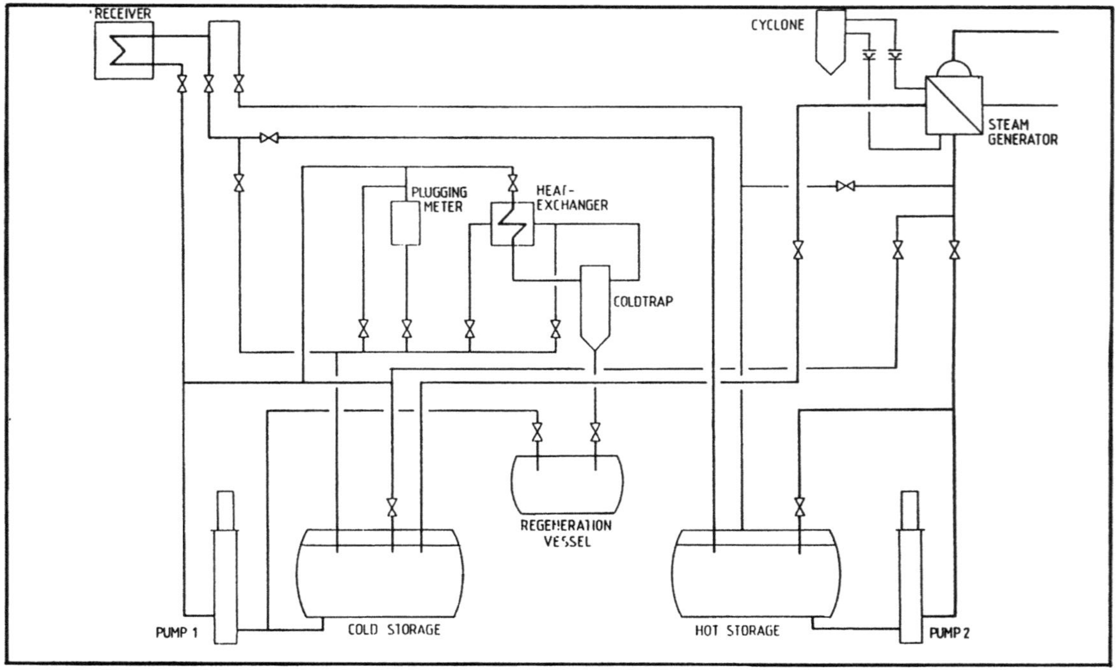

Fig. 6.4.-1: Sodium heat transfer and trace heating system diagram

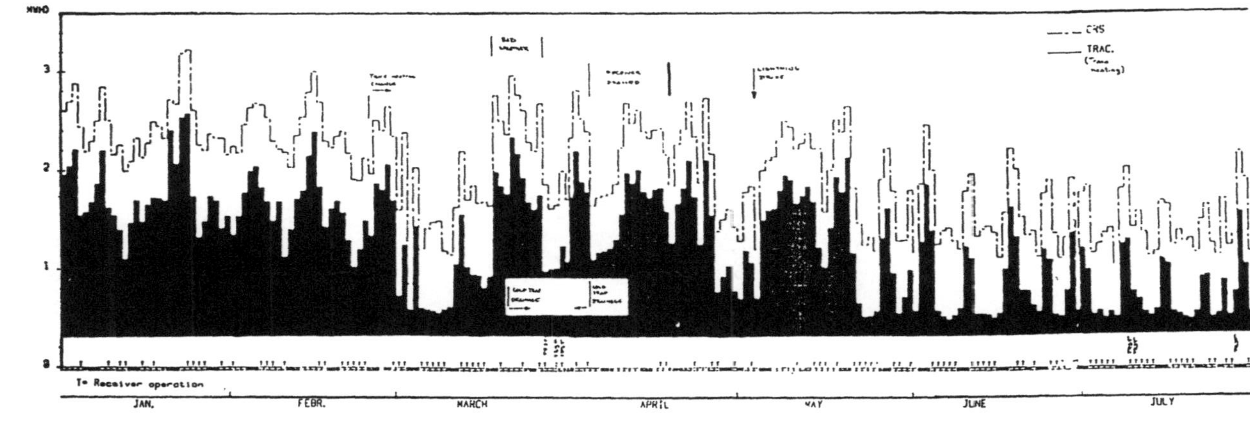

Fig. 6.4.-2: CRS Parasitic and Trace Heating Impact

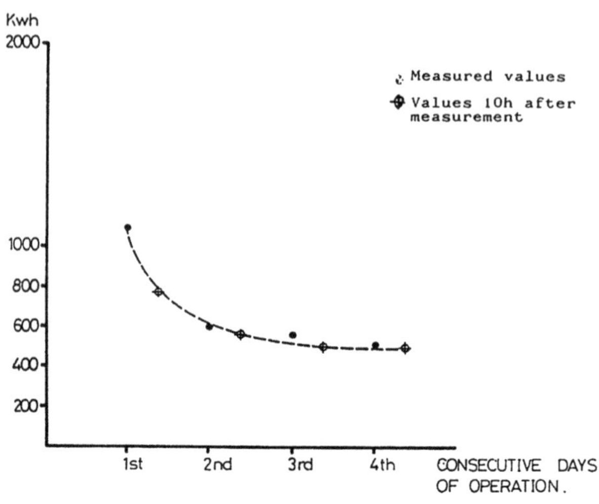

Fig. 6.4.-3: Variation of Trace Heating Consumption versus operating period

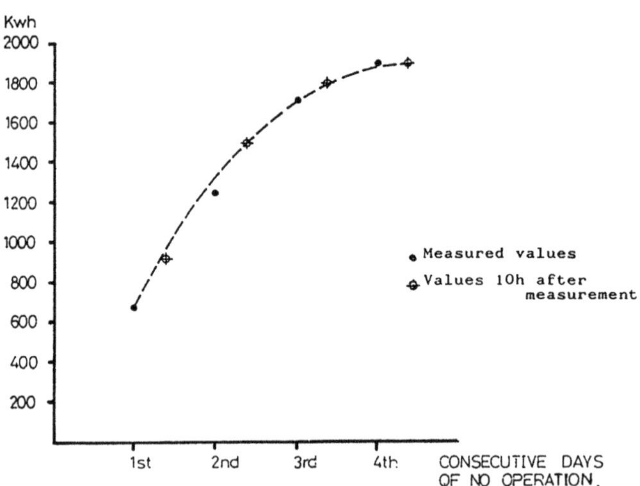

Fig. 6.4.-4: Variation of Trace Heating Consumption versus non-operating period

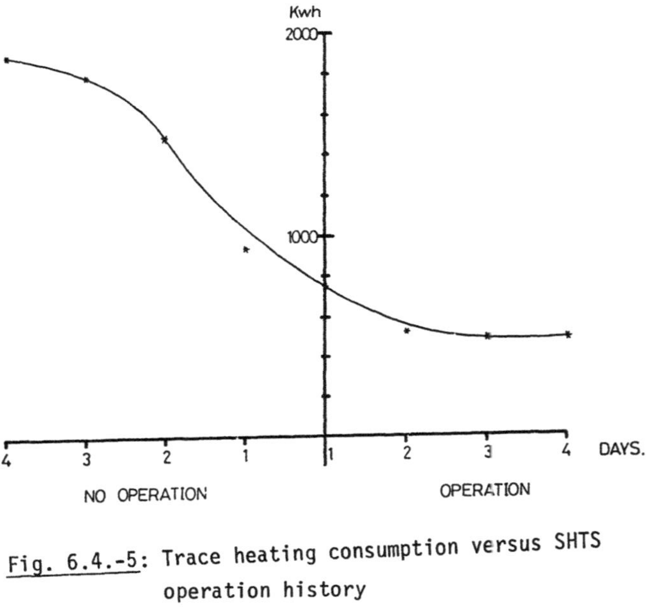

Fig. 6.4.-5: Trace heating consumption versus SHTS operation history

IMPLICATIONS FOR DESIGN AND OPERATION

C.S. Selvage, ITET

1. INTRODUCTION

Thermal inertia is present in all thermal systems and is usually a beneficial attribute because the systems tend to be continuous in operation and the inertia reduces or prevents thermal shock.

Solar thermal systems cannot be continuous operational systems because of the intermittent nature of the solar energy input. Designers of solar thermal systems have given a great deal of consideration to reducing the intermittency of the thermal system incorporating thermal energy storage and, in some cases, adding mass to the thermal elements to "smooth out" the thermal transients and thus reduce thermal shock. The addition of mass does have a negative aspect in that it adds thermal inertia to the total system.

Thermal inertia in a solar system causes an increase in start-up time of that system because the energy loss during the nonsolar hours must be made up when solar energy again becomes available. Some of the energy stored in the material mass of the system, the receiver, piping, pumps, etc., is lost to the environment during the nonsolar time period. This is evident and is quantified by a temperature change of the material in the system as a function of time. On the next attempt to start the solar thermal system, this energy loss (temperature drop) is made up by the solar collection/conversion system. Considering a day to day operation, the system always will start at a lower temperature than that required for operation. It takes time for the systems such as receivers, piping, pumps, steam generators and thermal-to-electric conversion equipment to reach rated temperature and the energy necessary to accomplish this is "thermal inertia".

During the early design phase of the SSPS project, a great deal of discussion of thermal inertia in the receiver occurred. There was much disagreement about the value of including thermal storage in the receiver and in the final design of the cavity receiver a rather large mass of ceramic material was included for thermal smoothing. INTERATOM, the receiver designer, describes the value of this ceramic wall and that description is repeated here as taken from Section 1.3.3 of the CRS Final Documentation.

2. CONSIDERATION OF THE BASIC RECEIVER DESIGN WITH RESPECT TO THERMODYNAMICS

In Chapter 1.3.1 possible receiver layouts with respect to basic requirements for the receiver tube bank, to the shape of the active heat transfer surface, and to different geometrical arrangements of aperture and active receiver surface have been discussed. In regard to these aspects a reference design of the receiver has been chosen. For a detailed design the following aspects had to be taken into account:

- short term thermal transients (minutes)
- long-term thermal transients (hours or overnight)
- thermal losses due to radiation, convection, and conduction
- temperature stresses
- minimum weight

The detailed thermodynamic calculation started with a consideration of the long-term behavior of the receiver during the night. Two important aspects have been considered:

- avoiding of sodium freezing inside the tubes
- preserving of the temperature profile in the tube bundle or - at least - providing the possibility to reprepare it

To realize this, there are three principle possibilities:

- to circulate the sodium (only the first aspect)
- to provide a heating system
- to provide an adequate storage capacity (inside the receiver)

After further layout considerations - if feasible - the provision of a receiver storage capacity was decided to be reasonable because of real advantages due to energy conversion, of operating conditions, and of a plain design.

A storage capacity could also simplify the sodium flow control with respect to short term transients. To reach this goal the storage system has to be arranged in close vicinity to the heat transfer surface and has to produce a similar temperature profile.

All these properties may, in principle, be attained by means of a ceramic layer, arranged behind the heat transfer surface. Ceramic materials have small thermal conductivity but large storage capacity.

Starting with this concept of a ceramic layer as a storage system, the following items have to be considered with the aid of a detailed thermodynamic calculation:

- steady-state temperature profile inside the cavity
- heat losses due to radiation, conduction, and convection as a function of time for different boundary conditions
- temperature profile as a function of time for different boundary conditions
- thickness of the ceramic and the insulation layer
- weight of the ceramic and the insulation layer

Calculations of all the above items are included in the final documentation justifying the value of this ceramic and showing how it can be beneficial rather than a problem.

In Spring 1983, Mr. F. Gaus from INTERATOM, as a member of ITET made detailed measurements of temperature variations within this ceramic wall. He reported on this work in the paper 3.5 of the April 1983 CRS Workshop. That paper is included here as Appendix 6.5-1. Shown are the heat-up and cool down characteristics of this ceramic showing that the wall can aid in at least one of the original objectives, which is to keep the over night temperature above the sodium freezing temperature. The very low conductivity of this ceramic (0.12 W/mk) inhibits this mass from increasing the thermal inertia by any great amount. However, the mass does absorb energy, and of course gives some of it up through the night, but this exchange is not a rapid action exchange so cannot help in a smoothing action on cloud passage. This is clearly shown in the paper by Mr. N. Gregory, 'Transient Response of the Sulzer Receiver', wherein he shows that the receiver outlet temperature drops from 525°C to 260°C in 5.4 minutes on removal of the energy input. The implication is that the ceramic thermal storage was not able to smooth out the temperature without loss of input energy. The Advanced Sodium Receiver (ASR) which has very little thermal mass responds in very much the same way as discussed by Mr. G.A. Magnani in the paper on ASR Temperature Regulation.

With storage-coupled systems such as the SSPS-CRS, thermal inertia becomes a very important concern because a large part of the total system mass must be heated at the start of operation, then all parts lose energy during the nonsolar hours and that loss must be replaced during the next solar operating period. A particular element for consideration is the thermal storage tank which is located between the receiver of the solar system and the steam generator. In the case of the SSPS-CRS sodium system that tank is made of stainless steel and has a large thermal capacity. This fact could be advantageous if the tank is hot and well-insulated, but if it is allowed to cool, it may take a long time to bring it (and the sodium) up to rated temperature. At this point, we must recognize that operating strategies must consider thermal inertia because it seriously influences how the system should be operated.

With this introduction, I must point out that Dr. J.G. Martín accepted the challenge to characterize some considerations of thermal inertia with a sodium-cooled, storage-coupled central receiver system and he subsequently developed the mathematical treatment of this situation which follows:

When one considers all the interfaces -- solar to thermal (the receiver), receiver to storage (the pipes and storage), and steam to water (the steam generator), it is apparent that careful consideration must be given to the inertia of the components or else the overall daily performance will be poor.

Consider the effect of different operating strategies on the time that it takes for the sodium in a steel tank to reach some operating temperature. The tank, made of steel that has a heat capacity C_p, has a mass M. Assume that, initially, there is a certain mass of sodium, m_o, in the tank. That sodium is at some temperature, T_o, which is lower than the rated temperature. For simplicity, we shall assume that heat transfer in the sodium and the tank is fast enough so that all the sodium in the tank can be characterized by some bulk temperature, T, which is the same as that of the tank.

At time $t = 0$, hot sodium at temperature T_{in}, flows into the tank at a mass flow rate Q_i. Sodium flows out of the tank at the rate Q_o. How long will it take for the tank and its contents to reach some rated temperature T_r? (Note: Assume that $T_{in} \gtrsim T_r \gtrsim T \gtrsim T_o$). To simplify this preliminary discussion, we shall assume that the heat capacities are not functions of temperature.

The rate of change of the total energy in the tank must equal the net rate at which energy is added to the tank.

The first rate has a contribution from the steel tank itself, equal to $MC_p(dT/dt)$, and one contribution from the sodium, $Cd(mT)/dt$, where m is the total mass of the sodium in the tank. In general, m changes with time; it is equal to m_o plus the time integral of the net flow Q into the tank (i.e., $Q = Q_i - Q_o$). Using the chain rule for differentiation, one may write

$$\text{rate of energy change} = (A + Q_m tC)(dT/dt) + CTQ \qquad (1)$$

where A is a constant, equal to $MC_p + m_oC$, and Q_m is the mean net flow into the tank up to time t.

The net rate at which energy is added to the tank is given by

$$\text{net rate in} = CT_iQ_i - CTQ_o - L' \qquad (2)$$

where L' is a general thermal loss term. On the basis of previous work by H. Jacobs (1), it is reasonable to state that

$$L' = LT \qquad (3)$$

where L is a constant.

Equating (1) and (2), using (3), and rearranging, one obtains

$$(A + Q_m tC)(dT/dt) = QC(T_i - T) - LT \qquad (4)$$

The term LT makes the time to reach T_r longer, but it is relatively small. For simplicity, we shall take L = 0 first, to emphasize the effect of operation modes on inertia. Then the integration of (4) is straightforward and yields

$$(T_i\ T_o)/(T_i - T) = \exp(\text{integral of } B(t') \text{ from } t'=0 \text{ to } t) \qquad (5)$$

where

$$B(t') = CQ_i(t')/(A + CQ_m t')$$

Substituting $T = T_r$, the operating temperature, and solving for t, gives the time that it takes for the system to reach T_r.

To perform the integrals in (5), we must know the time dependence of Q_i and Q_o. If these are constant, the integrations are trivial. For constant flows, consider these cases:

a) $Q_i = 0$
   In this case, $B(t') = 0$ and $T = T_o$ always until all the sodium flows out of the tank.

b) $Q_i = 0$

Here,

$$(T_i - T_o)/(T_i - T) = 1 + Q_i Ct/A$$

and solving for t when $T = T_r$

$$t = K(T_r = T_o)/(T_i - T_r)$$

where

$$K = A/Q_i C = (MC_p + m_o C)/Q_i C$$

c) $Q_i = 2Q_o$

Then,

$$(T_i - T)/(T_i - T) = (1 + Q_i(t/2)/A)^2$$

and solving for t when $T = T_r$

$$t = 2K((T_i - T_o)/(T_i - T)^{1/2} - 1)$$

Finally, if

d) $Q_i = Q_o$, i.e., when the mass of the sodium in the tank is constant, one has

$$(T_i - T)/(T_i - T) = \exp(-QCt/A)$$

or, for the time to reach T_r

$$t = K\ln((T_i - T_o)/(T_i - T))$$

All the times that have been derived are equal to some constant, K, times some factor. If the mass of the tank is 20500 Kg, the mass of the sodium in the tank at t_o is 87000 Kg (i.e., about 10 m^3), and the inflow rate is 26100 Kg/hr (i.e., about 30 m^3/hr), that constant is

$$K - 0.62 \text{ hours}$$

We have supposed that, in the ranges of temperatures of interest, C_p 590 J/Kg° and C 1290 J/Kg are representative values for the heat capacities of steel and sodium, respectively.

What will then be the time that it takes for a tank that is initially at T_o = 450°C to reach a temperature of 500°C if the inlet temperature is 520°C?

If there is no outflow, $t = 2.5$ K.
If the outflow is 1/2 the inflow, $t = 1.74$ K.

Finally, if the sodium mass in the tank is constant,
$t = 1.25$ K.

In other words, this simple model shows that the time to reach operating temperatures may be halved by bleeding the tank at the same rate at which it is being filled. Clearly, this does not cover all the possibilities, and there may be many reasons why, in some cases, this strategy of operation may not be desirable. However, a design and operating strategy that does not give due consideration to the implications of this simple example can lead to disappointments in terms of overall daily performance.

In the above example, we had taken $L = 0$. The inclusion of L is straightforward. The general solution is

$$(T_i - bT_o)/(T_i - bT) = \exp(\text{time integral of } B(t'))$$

where $B(t') = bCQ(t')/(A + CQ_m t)$

and $b = (CQ_i + L)/CQ_i$, a number close to 1.

For $Q_i = 0$, this equation reduces to the 'cool-down' expression,

$$T = T_o \exp(-Lt/A)$$

The assumption of constant flows is not realistic. Also, sodium heat capacity and specific volume are functions of temperature.

M. Sánchez and R. Carmona have written a simple program for the HP-85, which simulates the tank problem. The program estimates the tank and sodium temperatures and the volume of sodium in the tank, starting from specified initial conditions (T, m_o) and using as input the volume rates for the flow in and out, and the inlet temperature. It uses five minute intervals. It compares the estimated results with the 'actual' (DAS) results, plotting temperatures and volumes as functions of time, and calculating relative errors. An example for a run for 16/3/84 is attached.

This program assumes that the tank and the sodium are at the same temperature, but it does not take into account thermal losses and the temperature dependence of the sodium heat capacity and specific volume.

Some system analysis of all the above have shown the effect of the inertia and have provided some inclusions of how to avoid or minimize the effect of the thermal inertia. The attached charts show two days of CRS operation which were plotted with the objective of portraying thermal inertia. In the first chart for April 9, 1983, with the cavity receiver hot storage tank temperature at a little over 400°C in the morning, approximately 2.5 hours of operation with a receiver outlet temperature of over 500°C was required to heat the tank and allow an acceptable storage outlet (or steam generator inlet) temperature to exist. All of this energy was required to heat the tank. A similar situation is obvious for the December 23, 1983 operation with the ASR.

3. OPERATIONAL CONSIDERATIONS

Recently, there have been statistical analysis of the operational data of the SSPS-CRS in which some start-up times for the receivers and steam generators are presented. It seems from a brief review of these presentations that receiver start-up times are rather long and there has been some considerable discussion as to what creates this long delay. Some have argued the case of heliostat pointing problems and the fact that with the ASR there are three aiming points and a smaller target than with the cavity receiver. Also that the shield around the external receiver is more sensitive than with the cavity.

These are all probably true conditions. However, none of the reasons are what caused the analysis to indicate these large numbers or why both receiver systems have essentially the same numbers. The very simple reason is the requirements placed on the operation of this system. In the analysis, time for receiver start-up started when the receiver doors were opened and stopped when the outlet temperatures reached 500°C. No considerations of receiver coolant flow was used in the analysis. Further, the operational program was not considered and it seems to have more of an influence on these times of receiver response.

The method of starting the CRS creates this long--1.5 to 2 hours--start-up for the following reason. When opening the receiver doors sodium is flowing at a rather high flowrate, in the order of 20 m^3/hr, from the cold tank through the receiver and back to the cold tank. Heliostat energy is added to the receiver raising the outlet temperature of the sodium to a temperature near 300°C. Flow is adjusted to maintain a temperature near 300°C at the outlet of the receiver with this sodium flowing back to the cold tank. This condition is maintained as the bulk temperature of the cold tank is raised from the starting cold tank temperature, which is related to how long the system has been nonoperational down to the trace-heating temperature, until the bulk temperature is over 275°C. At that time, the flow is reduced to raise the receiver outlet temperature to 500°C which at this time takes only 15 to 20 minutes because all parts are warm and most or all of the heliostats are in track. Then the flow is switched so that the receiver outlet is directed to the hot tank. From this time, approximately 2 to 3 hours are required to obtain an acceptable temperature out of the hot tank so that acceptable steam is produced and the Spilling engine can be run. Much of these 2 to 3 hours spent heating the hot tank is used to heat the steam generator, so all is not lost; but again the analysis did not, or perhaps could not, consider this operation situation.

The message: One should be very careful in just looking at the statistics without looking and attempting to understand what is the cause and effect.

4. CONCLUSIONS

The thermal inertia is a problem that will not disappear. There are several methods that minimize it as a problem. One has been addressed mathematically and demonstrated experimentally. Another simple method to minimize thermal inertia as a problem is to provide a thermal-to-electrical conversion system (PCS) that can operate at the expected overnight cool-down temperature of the hot storage tanks. This would then allow an early start-up of the PCS as soon as the energy flow is directed to the hot storage tank.

5. REFERENCE

1. H. Jacobs, "Sodium Cycle, Storage, Trace-Heating", Proceedings of SSPS-CRS Midterm Workshop, Tabernas, Almería, Spain 1983, M. Becker, editor, DFVLR, Cologne, Germany (1983).

APPENDIX 1

S S P S

CENTRAL RECEIVER SYSTEM (CRS)

MIDTERM WORKSHOP

Tabernas, April 19 + 20, 1983

RECEIVER TEMPERATURE MEASUREMENTS

by

F. Gaus - I.T.E.T.
Interatom

CONTENTS:

1. NOMENCLATURE

A	- area, cross section, exchange area	(m^2)
C	- radiation constant	$W/m^2 \cdot K^4$
C_s	- black-body-radiation constant	$W/m^2 \cdot K^4$
M	- mass	Kg
$\dot{M}$	- mass flow	Kg/s
Q	- heat quantum	J = Ws
$\dot{Q}$	- heat flow	W = 1J/s
T	- temperature thermo dynamic	K
V	- volume	m^3
a	- temperature conductivity	m^2/s
c	- heat capacity	J/Kg·K
K	- thermal transmission coefficient	W/m^2K
$\dot{q}$	- surface power	W/m^2
s	- distance, depth of wall, thickness of a film	mm
t	- time	s
α	- heat transmission coefficient	W/m^2K
α	- absorbed term	
ε	- emission relation	
τ	- transmitted term	
ρ	- reflected term	
ϑ	- temperature	°C
λ	- thermal conductivity	W/m·K

Subscripts

A = ambient
Ce = ceramic wall
M = measurement point in the centre
R = receiver
S = measurement point on side
So = sodium
T = tube
Tb = tube bundle

a = beginning
b = bottom
e = evening
m = morning
t = top

Receiver data

h_{Tb}	= height of the tube bundle	3607 mm
r_{Tb}	= radius of the tube bundle (inside)	2250 mm
d_t	= tube diameter of the tube bundle	38 x 1,5 mm
h_{Ce}	= height of the ceramic wall	4000 mm
rce	= radius of the ceramic wall (inside)	2400 mm
s_{ce}	= depth of the ceramic wall	70 mm
ε_t	= emissivity of the tube	0,9
ε_{Ce}	= emissivity of the ceramic wall	0,75
λ_{Ce}	= thermal conductivity of the ceramic wall (for 600oC)	0,12 W/mK
ρ_{Ce}	= density of the ceramic	2500 kg/m3
ρ	= density of the tubes	7740 kg/m3
ρ_{So}	= density of the sodium at 500oC	832.3 kg/m3
$c_{p,Ce}$	= specific isobaric heat capacity	1005 J/kg·K

2. INTRODUCTION

This report deals with the temperature distribution in the ceramic wall of the SULZER receiver and with the cooling down behaviour of the receiver.

The report is divided into three main parts:

- temperature distribution in the ceramic wall
- heat conduction through the ceramic wall
- cooling down behaviour of the receiver.

Knowing the temperature distribution in the wall, one can determine:

- heat storage in the wall
- shock absorption
- temperature balance.

3. PROBLEM DEFINITION

The first examination of the CRS receiver ceramic wall was made with thermodynamic calculations by Merkel/ Grönefeld (Reference 1) and Weitzenkamp/Wild (Reference 2). These calculations were based on theoretical deliberations. Now the theoretical deliberations must be confirmed with practical, scientific data. For this reason, several thermocouples were installed in the ceramic wall to measure temperature at 22 measuring points. (Reference 3).

4. EXPERIMENTAL SET-UP

Thermocouples were used to measure temperature distribution in the ceramic wall, conduction of heat through the wall, and the cooling down behaviour of the receiver. All thermocouples were made of Ni-Cr/Ni and had a diameter of 3.2 mm. The thermocouples which were installed in pipes with a diameter of 21 mm, were pressed against the ceramic wall with a spring (Figure 1). The inside of the pipes was packed with insulation to prevent convention through the pipes. The thermocouples were connected with the Servogor recorders over a 60 degree cold junction to the Marshall Kiosk.

4.1 Temperature distribution in the ceramic wall

To measure temperature distribution in the ceramic wall, 2 vertical rows of thermocouples were installed into the ceramic wall from the back wall of the receiver (Figures 2 and 3). One row of thermocouples was installed at 11 degrees from the center vertical (center pattern), and another at 55 degrees (side pattern). The thermocouples were situated in this fashion because the construction of the receiver did not permit them to be placed in a better position on the mean perpendicular.

The position of the thermocouples rows had to allow for the different thermal heat expansions of the ceramic wall and the supporting surface wall of the receiver. Therefore, the thermocouples were placed where the supporting surface wall was sufficiently separated, due to its trapezoidal form, from the ceramic wall (i.e., 11 and 55 degrees from the mean perpendicular of the receiver). Severing of the thermocouples as a result of the different thermal heat expansions, thus was not a problem.

The distance between the thermocouples in the vertical direction was 258 mm; this distance equals the height of one tube bundle. The uppermost thermocouple was positioned at the uppermost tube bundle. The depth of each thermocouple in the ceramic wall was 52 mm.

4.2 Conduction of heat through the ceramic wall

To measure heat conduction through the ceramic wall, 2 sets of 3 thermocouples were placed in 2 horizontal positions: one at the center of the wall and one along the side of the wall (position I and position II, respectively, in Figures 1 and 2). For each position the thermocouples in the set were placed at different depths in the ceramic wall:

- For position I (center pattern)
  - (1) thermocouple 3 mm deep
  - (1) " 27 mm deep
  - (1) " 52 mm deep
- For position II (side pattern)
  - (1) thermocouple 1 mm deep
  - (1) thermocouple 27 mm deep
  - (1) thermocouple 53 mm deep

Both thermocouple positions were at the same height in the receiver at the centre of the heat focusing area. As with the vertical rows, these 2 sets of thermocouples were installed in the back wall where they were sufficiently separated from the ceramic wall.

4.3 Cooling down behaviour of the receiver

To measure cooling down temperatures, only 12 measuring points were chosen from the 22 measuring points available because of low (limited) recording capacity.

INTERPRETATION OF TEST RESULTS

4.1 Comparison of calculated and measured temperatures

Figure 4 graphically presents the temperatures of thermocouples M1, M2, M3, M5, M7, M9, and M10 over the 21-hour recording period. The data show a rise in the temperature of the ceramic wall up to 14:00 h, which was expectted. After 14:00 h, the temperatures of thermocouples M1, M2, M3, M5, and M7 continue to increase, although the sun radiation decreases.

Comparing the temperatures' graphs in Figure 5, one can see that the calculated and measured temperature trends, as well as the maximum calculated and measured temperatures in the ceramic wall, are different. Furthermore, the measured temperature in the ceramic wall is influenced by the sodium temperature in the tube bundle in the receiver.

Figure 5 and Figure 6 also indicate the nodal points selected for each temperature curve. For the calculated temperatures of the ceramic wall 18 nodal points were selected for the complete ceramic wall. For the measured temperatures of the ceramic wall, 30 nodal points were chosen. Nodal points from the lower third of the ceramic wall are not included. This selection was made because the same distribution of temperature exists at the east and west sides of the ceramic wall.

For the measured sodium temperatures in the receiver, only 8 nodal points were selected because there are only 8 measurement points on the east side of the receiver at each pass of the receiver tube bundle.

Figure 7 allows comparisons between 2 calculated and 3 measured sodium temperature curves. One calculated temperature curve was based on 18 nodal points, the other on 42 nodal points. The sodium temperature curves reflect data measured at noon, 14:00 h, and 18:00 h.

The differences between the calculated and measured temperatures of the sodium are less when the calculations are based on 42 nodal points than on 18 nodal points. However, the calculated sodium temperatures and the measured sodium temperatures are not that different. On the other hand, the calculated temperatures and the measured temperatures of the ceramic wall are quite different.

The sodium measurements had to be recorded on 2 separate days because of the recording capabilities. However, measurements taken later with selected thermocouples from both the center and side vertical rows showed the same trend.

Heat storage capacity and losses during the night in the ceramic wall

With the different temperatures from each thermocouple, estimates of the heat capacity of the ceramic wall can be calculated.

An average temperature can be calculated from the temperature of the top and bottom thermocouples (from the center and side).

The heat capacity in the evening can be calculated with Equation (1):

$$Q_e = c_{p,Ce} \cdot V_{Ce} \cdot \rho_{Ce} \left(\left(\frac{\left(\frac{t_{M1} + t_{M10}}{2}\right) + \left(\frac{t_{S1} + t_{S2}}{2}\right)}{2} \right) - t_A \right) \quad (1)$$

The heat capacity in the morning can be calculated with Equation (2):

$$Q_m = c_{p,Ce} \cdot V_{Ce} \cdot \rho_{Ce} \left(t_{M5} - t_A \right) \quad (2)$$

The heat losses of the ceramic wall during the 12-hour night-time period, i.e., the difference between Equation (1) and(2), are:

$$Q_{losses} = Q_e - Q_m \quad (3)$$

$$= 351{,}3 \text{ kWh} - 238{,}5 \text{ kWh}$$

$$Q_{Losses} = 112{,}8 \text{ kWh} \quad \text{(with Sodium circulation)}$$

The calculations do not take into account that the temperrature at the bottom of the ceramic wall is lower than the temperature measured at thermocouples M10 and S10 (there are no thermocouples at the bottom of the ceramic wall).

Thermal shock absorptance

Passing clouds can cause large temperature drops in the tube bundle. These drops can be evened out only by varying the sodium flowrate.

Generally, from the law of conservation of energy, the following equation is valid:

$$\rho + \alpha + \tau = 1 \tag{4}$$

When τ = 0, Equation (4) follows:

$$\rho + \alpha = 1 \tag{5}$$

In the radiation balance, the following is valid:

$$\dot{q} = \sigma \cdot T^4 \qquad \sigma = 5.67 \cdot 10^{-8}\ W/m^2 \cdot K^4 \tag{6}$$

Stefan-Boltzmann-equation

with $C_S = 5,67\ W/m^2 \cdot K^4$

$$\Rightarrow \dot{q}_S = C_S \cdot \left(\frac{T}{100}\right)^4 \tag{7}$$

If the heat transfer of the emission radiation can be examined without considering the wavelengths, then:

$$\dot{q} = \varepsilon \cdot \dot{q}_S = \varepsilon \cdot C_S \cdot \left(\frac{T}{100}\right)^4 \tag{8}$$

The exchanged radiation energy between two areas is the difference between the radiation that is absorbed by each area from the other area. The exchanged heat flux is the difference between the emission ratios of 2 areas, when these areas are in any position, as shown in Equation (9):

$$dQ_{1/2} = \varepsilon_1 \cdot \varepsilon_2 \cdot C_S \cdot \left(\left(\frac{T_1}{100}\right)^4 - \left(\frac{T_2}{100}\right)^4\right) \frac{1}{\pi} \cdot \frac{\cos\beta_1 \cdot \cos\beta_2}{S^2} \cdot dA_1 \cdot dA_2 \tag{9}$$

To simplify the radiation exchange between the tube bundle and the ceramic wall, two load areas, which are parallel to each other and larger in size than their distance from each other, were selected. If the reciprocal radiation is neglected, then:

$$\dot{Q}_{1/2} = C_{1/2} \cdot A \left(\left(\frac{T_1}{100} \right)^4 - \left(\frac{T_2}{100} \right)^4 \right)$$

with

$$C_{1/2} = \frac{C_S}{\frac{1}{\varepsilon_1} + \frac{1}{\varepsilon_2} - 1} \qquad (10)$$

In order to calculate the heat return radiation from the ceramic wall to the tube bundle, the ceramic wall was divided into 3 horizontal strips between the upper and lower measurement points and into 3 vertical sectors each covering 40o of the receiver arc, as shown in the diagramme. The ceramic wall temperature measurements used for the calculation were made at:

14:00 h on 05.08.82 (centre), and
14:00 h on 11.08.82 (on side)

A tube bundle temperature of 250 °C was chosen for the calculation.

The spot temperature readings from each sector have been averaged for use in the calculations.

Since two thermocouples were defective (M6, S8), the average temperature of the two adjacent spots has been used.

$\frac{M1+M2+M3}{3} + \frac{S1+S2+S3}{3}$	$\frac{M1+M2+M3}{3}$	$\frac{M1+M2+M3}{3} + \frac{S1+S2+S3}{3}$
$\frac{M4+2M5+2M7}{5} + \frac{S4+S5+S6+S7}{4}$	$\frac{M4+M5+M6+M7}{4}$ with $M6 = \frac{M5+M7}{2}$	$\frac{M4+2M5+2M7}{5} + \frac{S4+S5+S6+S7}{4}$
$\frac{M8+M9+M10}{3} + \frac{S8+S9+S10}{3}$ with $S8 = \frac{S7+S9}{2}$	$\frac{M8+M9+M10}{3}$	$\frac{M8+M9+M10}{3} + \frac{S8+S9+S10}{3}$ with $S8 = \frac{S7+S9}{2}$

According to the calculations, the exchanged radiation between the ceramic wall and the tube bundle is at a maximum in the sectors above the focus area.

In the sector below the focus area, the radiation exchange is very small; consequently, there is a very small shock absorbance in this area.

In the lower area, where no spot thermocouples are fitted, the ceramic wall temperatures are thought to be much lower than the tube temperatures. Therefore, there is no significant return radiation effect.

S^*	t	T_a	$\dot{q}$	$\dot{Q}$	M	t	T_a	$\dot{q}$	$\dot{Q}$	S	t	T_a	$\dot{q}$	$\dot{Q}$	
1	462				1	457				1	462				
2	450	728	8094	10,27	2	467	735	8525	10,82	2	450	728	8094	10,27	774 mm
3	428				3	463				3	428				
4	402				4	459				4	402				
5	356	653	4203	7,11	5	437	697	6332	10,72	5	356	653	4203	7,11	1032 mm
6	311				6	412	focus area			6	311				
7	276				7	386				7	276				
8	250				8	368				8	250				
9	223	565	1064	1,35	9	360	630	3249	4,13	9	223	565	1064	1,35	774 mm
10	207				10	326				10	207				
≙ 40°					≙ 40°					≙ 40°					

M + S = measurement points T_a = average temperature K

S^* = the same points as S $\dot{q}$ = surface power W/m^2

t = temperature °C $\dot{Q}$ = heat flow kW

6.5.-A14

Temperature balance effece

The tube bundles in the receiver are heated on only one side by solar insolation. Therefore, the surface of the tube bundle facing the sun has a much higher temperature than the side facing the back of the ceramic wall. Consequently, the ceramic wall should reflect as much heat as possible to the back of the tubes so that the difference in the wall temperature between the sun-side and the back-side of the tubes is minimal.

It is possible to calculate the radiation exchange between two areas with Equation (10), and as is already known, heat transfer occurs from a hot area to a cold area (second main theorem of the thermo-dynamic law). Figure 4 shows that the ceramic wall absorbs energy during receiver operation. This behaviour is evident from the measurements taken by the thermocouples in the centre and side rows. Thus the conclusion can be made that the ceramic wall absorbs energy mainly from the receiver tubes.

During receiver operation, the back-side of the tube bundle sends energy to the ceramic wall. No temperature balance between the sun-side and the back-side of the tubes can occur through the ceramic wall. To achieve a temperature balance, materials other than the ceramic wall must be chosen, materials which have a greater reflection capacity and less heat capacity.

5.2 Conduction of heat through the ceramic wall

Figure 8 shows the measurements taken in the center of the wall; Figure 9 shows the measurements taken at the side of the wall. The temperatures were measured at 3 different depths.

The side measurements in Figure 9 at 11:00 h show near-stationary heat conduction. This situation only occurs in the center of the receiver at 14:00 h and continues until evening (Figure 8). The side measurements do not exhibit this as clearly, but if the inaccuracy of the thermocouples is taken into account, this nearly stationary condition can also be applied to the side of the wall.

For stationary heat conduction, the following formula is valid:

$$Q = -\lambda \cdot A \cdot t \, \frac{d\vartheta}{dx} \tag{11}$$

$$\dot{Q} = \frac{Q}{t} \; ; \quad \dot{q} = \frac{\dot{Q}}{A}$$

$$\dot{q} = -\lambda \cdot \frac{d\vartheta}{dx} \tag{12}$$

$$\dot{q} \cdot \int_{x=0}^{x=s} dx = -\int_{\vartheta_1}^{\vartheta_2} \lambda(\vartheta) \cdot d\vartheta \qquad \text{with } \lambda = \text{const} \tag{13}$$

$$\dot{q} \cdot \int_{x=0}^{x=s} dx = -\lambda \int_{\vartheta_1}^{\vartheta_2} d\vartheta \tag{14}$$

$$\dot{q} \cdot s = -\lambda \cdot (\vartheta_1 - \vartheta_2) + C_{1/2} \qquad C_{1/2} = 0 \tag{15}$$

$$\dot{q} = -\frac{\lambda_{ce}}{s} \cdot (\vartheta_1 - \vartheta_2) \tag{16}$$

Time	Position	$\dot{q}$ [W/m2]
13:00	II	1,06
14:00	I	0,68
17:30	I	0,39
18:30	II	0,91

Pos. I (centre)

Date of measurement: 04.08.82

Pos II (side)

Date of measurement: 31.07.82

This calculation shows that the surface power is greater on the side of the ceramic wall than in the centre. This will need to be certified through further measurements.

5.3 Cooling down behaviour of the receiver

Pure sodium solidifies at approximately 100°C. In general, one has to maintain an operating temperature in the sodium system of at least 150°C. These temperatures can be maintained with the help of an electrical trace heating system that is applied to all components and pipes from the outside environment. In the case of the CRS receiver, it was not possible to install trace heaters in the receiver. Consequently, sodium is kept circulating during the night, so that the receiver tubes always maintain the minimum operating temperature. The possibility of maintaining operating temperature during the night with the help of the stored heat in the ceramic wall should be examined.

Figure 10 shows 2 different cooling down curves. The continuous lining in Figure 10 represents the cooling down behaviour with sodium circulation; the dotted curve shows the cooling down behaviour without sodium.

The cooling down behaviour indicates that during the first 6 hours of receiver operation, the sodium absorbs energy from the receiver; the cooling down is much faster with sodium than without sodium. After 6 hours, an equilibrium is reached, and the sodium gives energy to the receiver. The cooling down curve without sodium starts at 300 degrees for the measuring point S1. The cooling down after 6 hours is faster without sodium than with sodium circulation.

The area between the two curves represents the emission of heat from the sodium to the receiver. After 35 hours, the receiver reaches a temperature of 150 degrees. These 150 degrees have to be interpreted conservatively, for during the first hours the sodium draws energy from the receiver.

6 SUMMARY

The temperature measured by each thermocouple in the ceramic wall provides information about the heat capacity of the wall and the thermal shock absorptance and temperature balance effects that are dependent on that capacity.

The data shows that the temperature increase in the ceramic wall is similar to the increase in sodium temperature in the tube bundle.

The temperature ascent at the side of the wall is greater than in the center. With the transparent tube bundle, the maximum temperature was expected to be in the focus area of the tube bundle, but this was not confirmed.

Before a conclusion can be drawn all available measurement results must be evaluated. Only after interpretation of the data can a definite statement be made.

7. REFERENCES

Merkel/Gronefeld: Temperature distribution in the Receiver
CRS project report No. 65.422 04.10.78

Weitzenkamp/Wild: Receiver-redesign, thermodynamic design
report 62.00806.6 10.03.83

F. Gaus: Description of test equipment
Test programme, Doc. No. R-1/83FG. 25.02.83

CRS Final Documentation file 3

W. Tischler: Sodium properties ITB 76.128 Interatom
Sept. 76

VDI: Receiver redesign, thermodynamic design
Report 62.00806.6 10.03.83

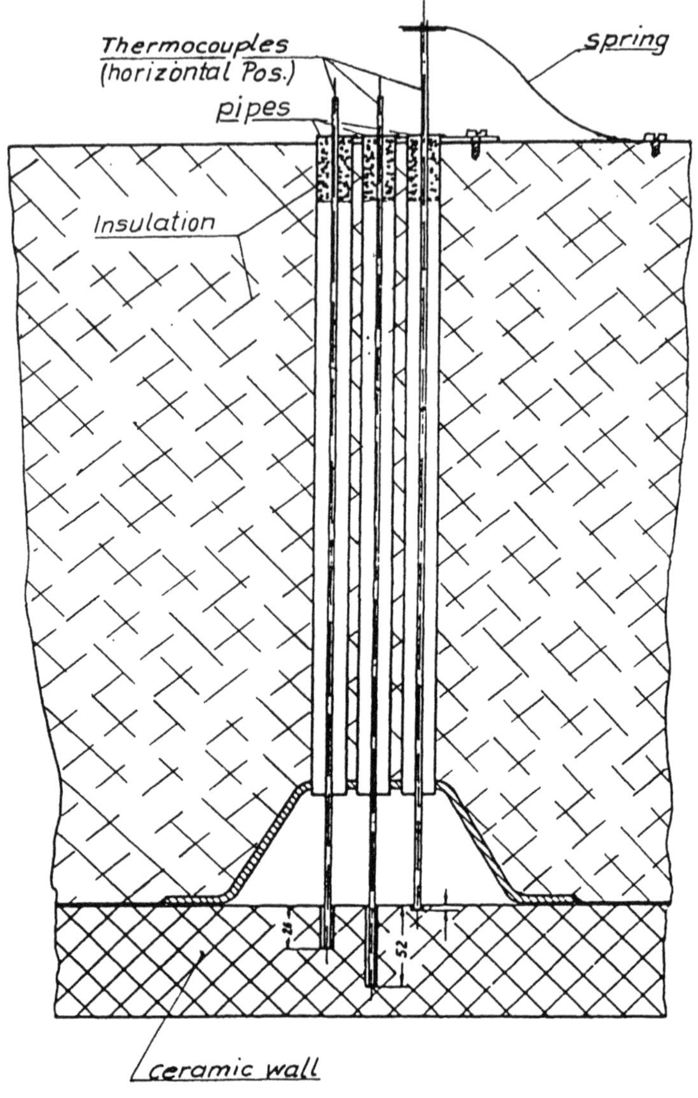

Figure 1
Sectional of Thermocouple Set-Up

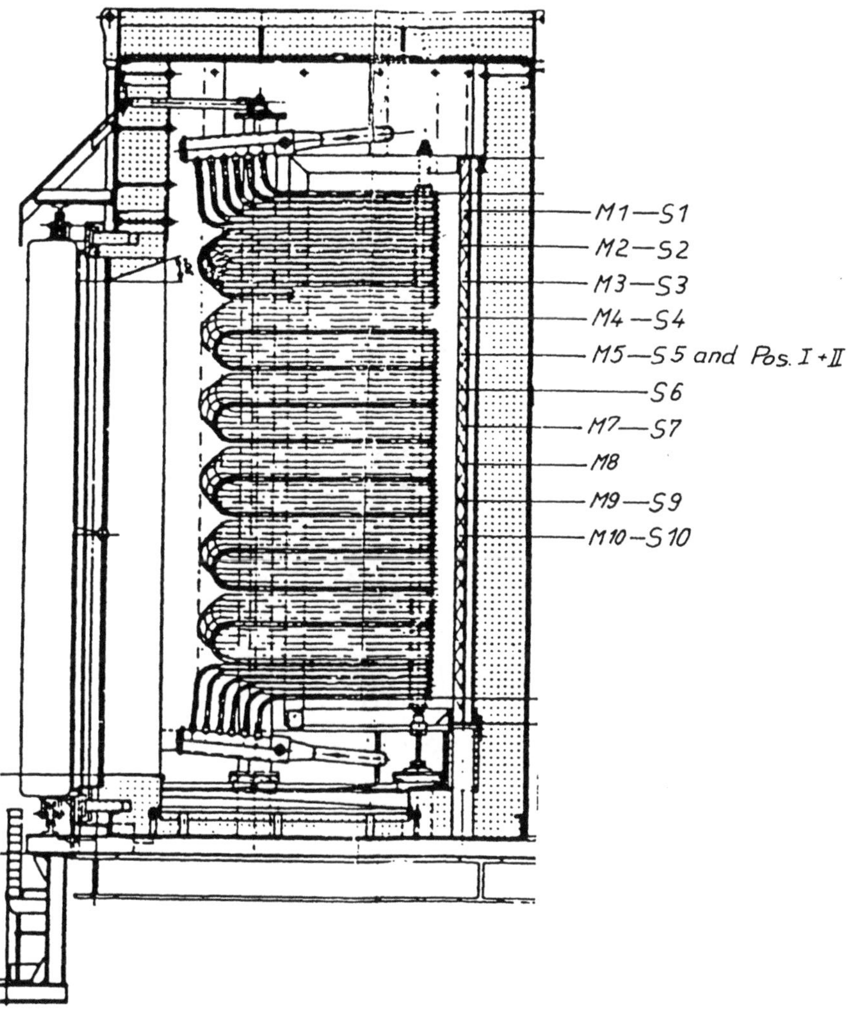

Figure 2
Position of Thermocouples

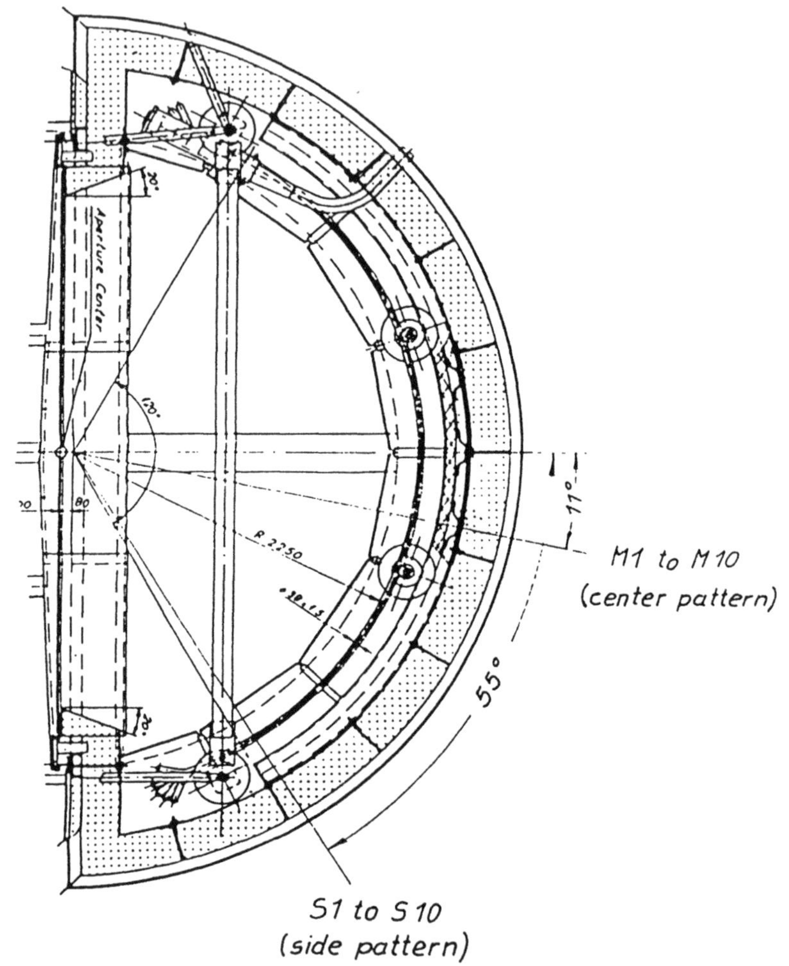

Figure 3
Position of Thermocouples
(Top View)

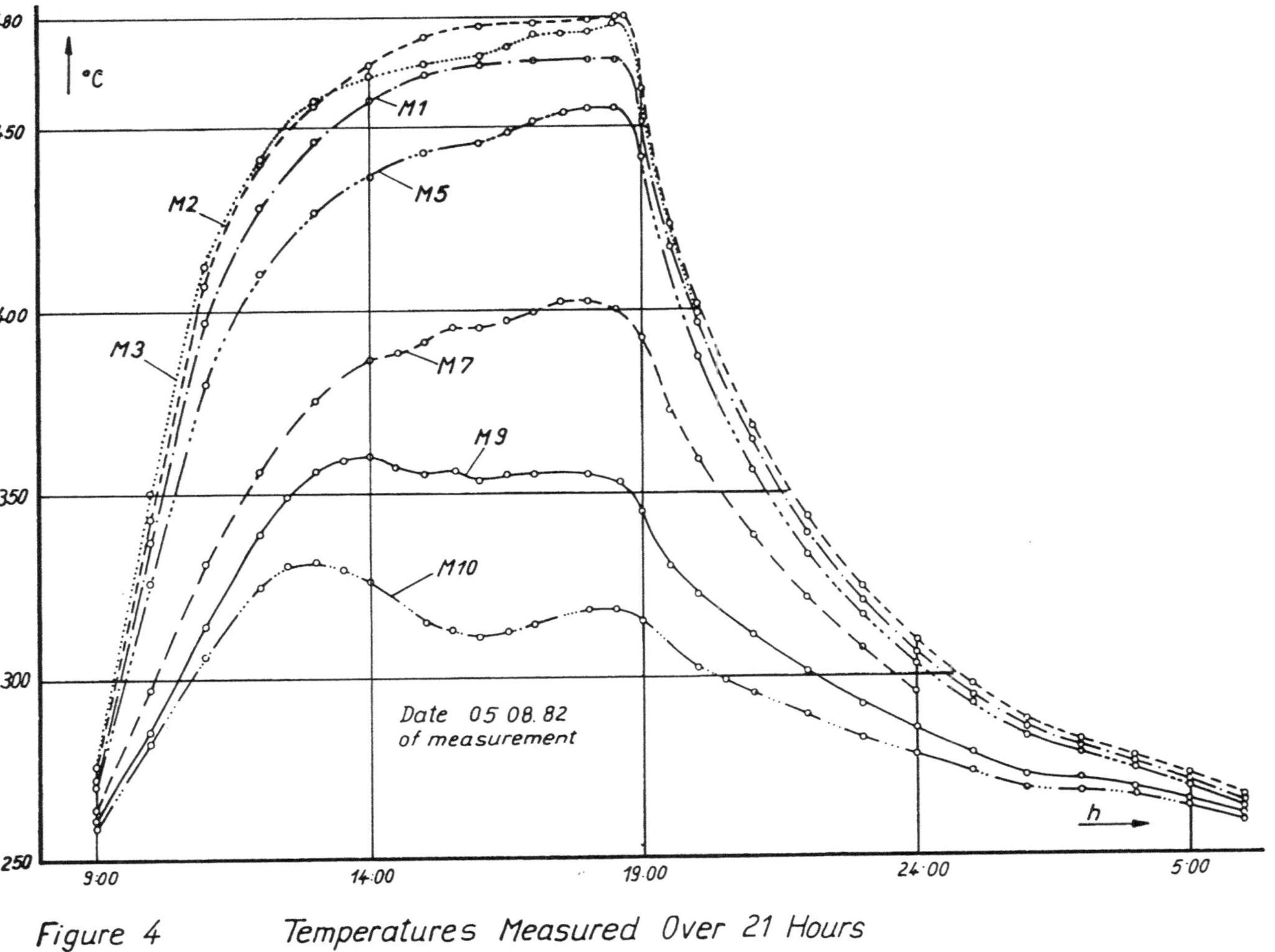

Figure 4 Temperatures Measured Over 21 Hours

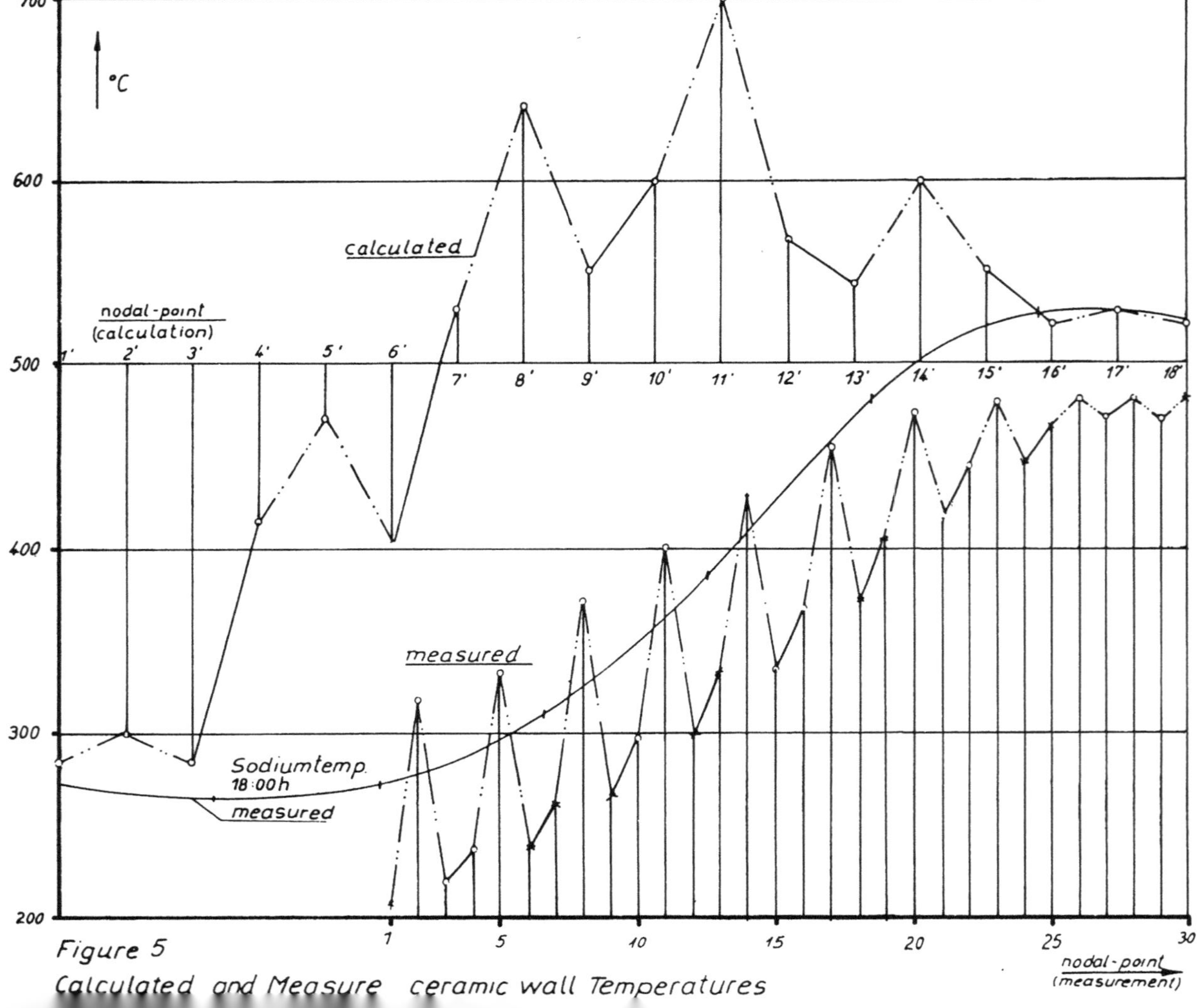

Figure 5
Calculated and Measure ceramic wall Temperatures

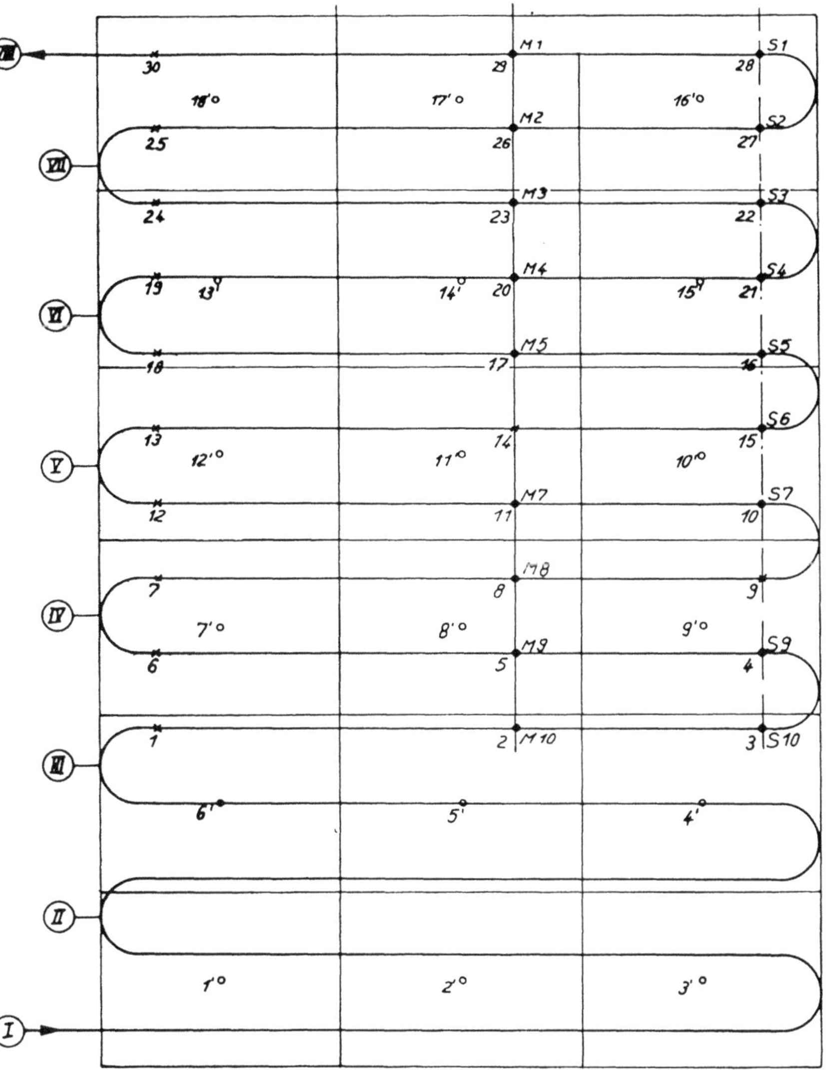

1' to 18' nodal-points for the calculation
1 to 30 nodal-points for the measurements; ceramic-w
I to VIII nodal-points for the Sodiumtemperatures

Figure 6
Nodal Point Systems

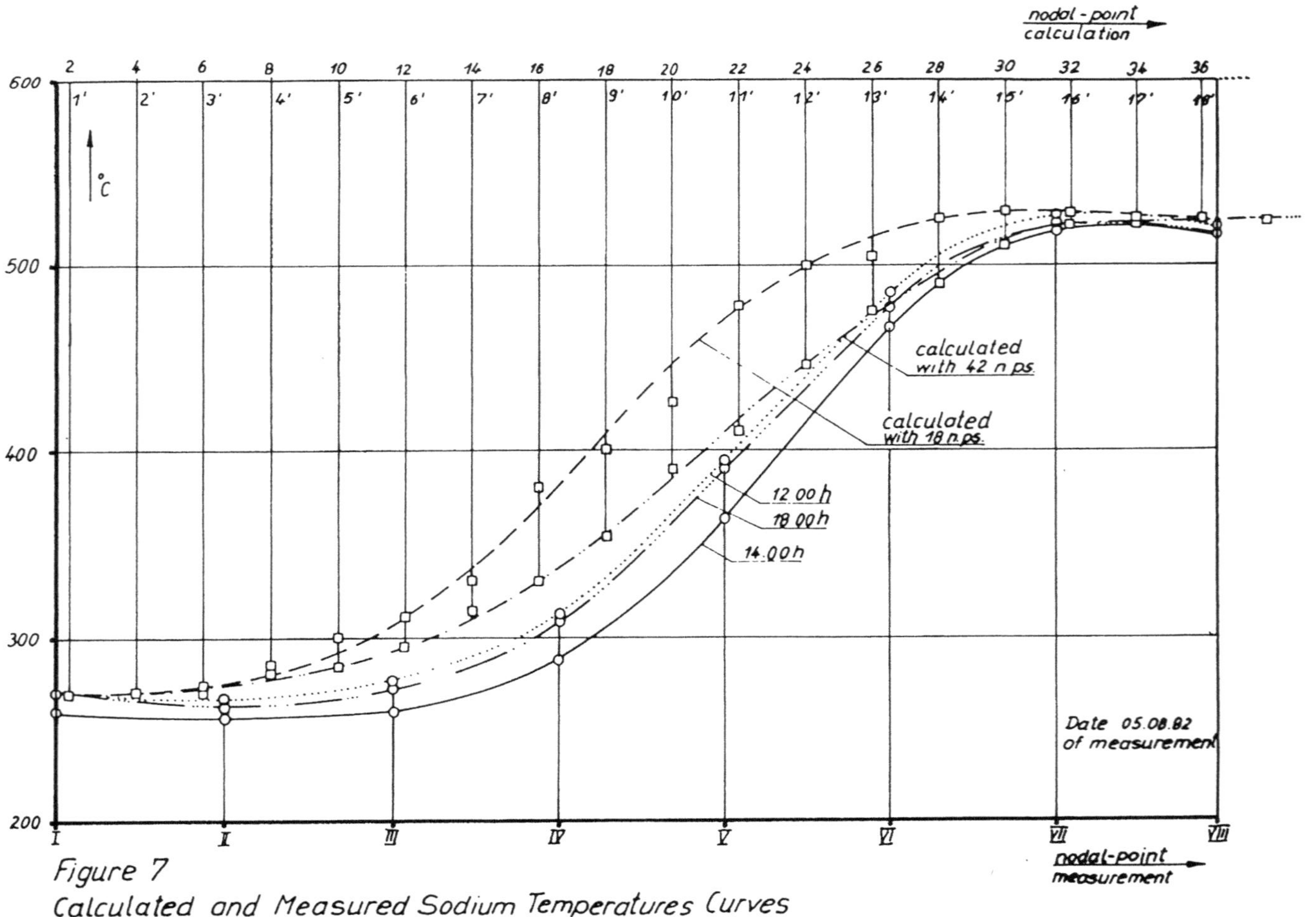

Figure 7
Calculated and Measured Sodium Temperatures Curves

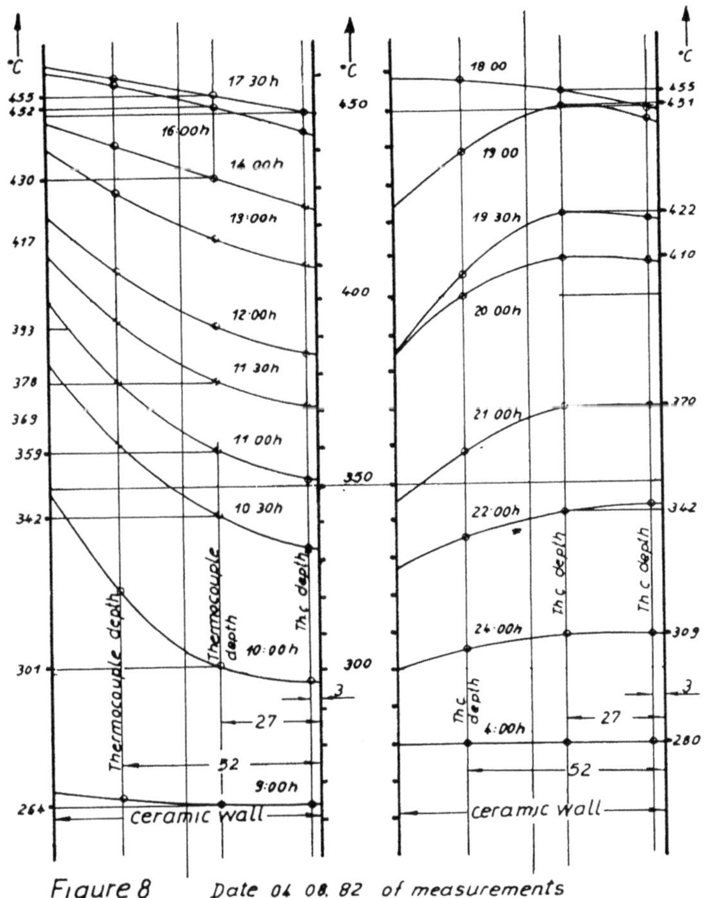

Figure 8 Date 04 08. 82 of measurements
Measurements taken in Position I (center)

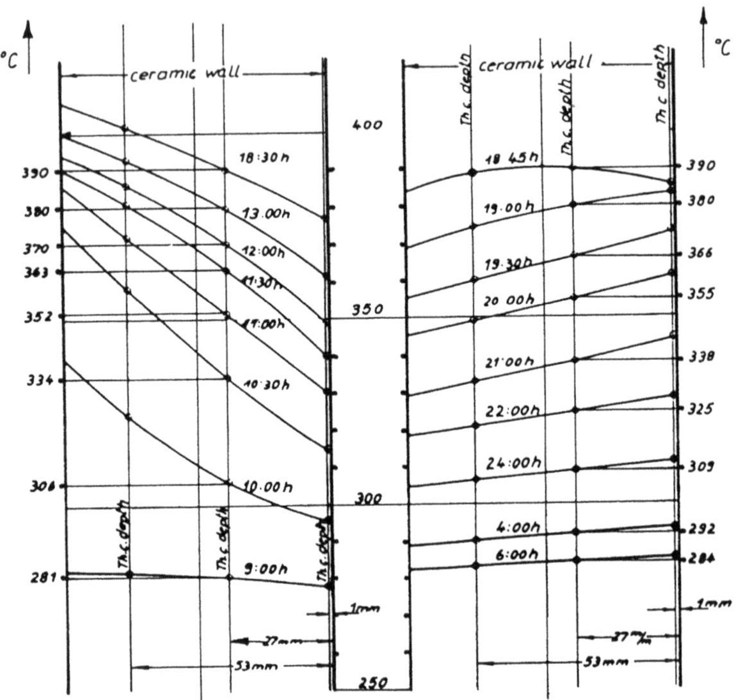

Figure 9 Date 31 07 82 of measurements
Measurements taken in Position II (side)

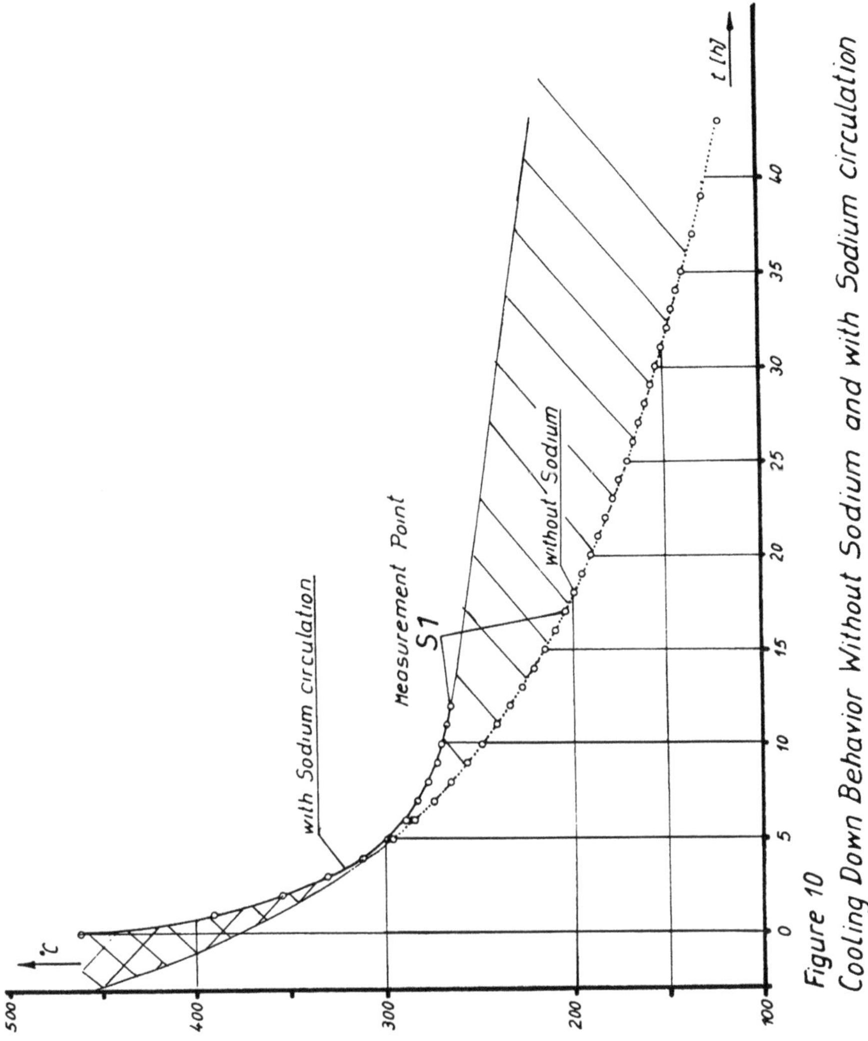

Figure 10
Cooling Down Behavior Without Sodium and with Sodium circulation

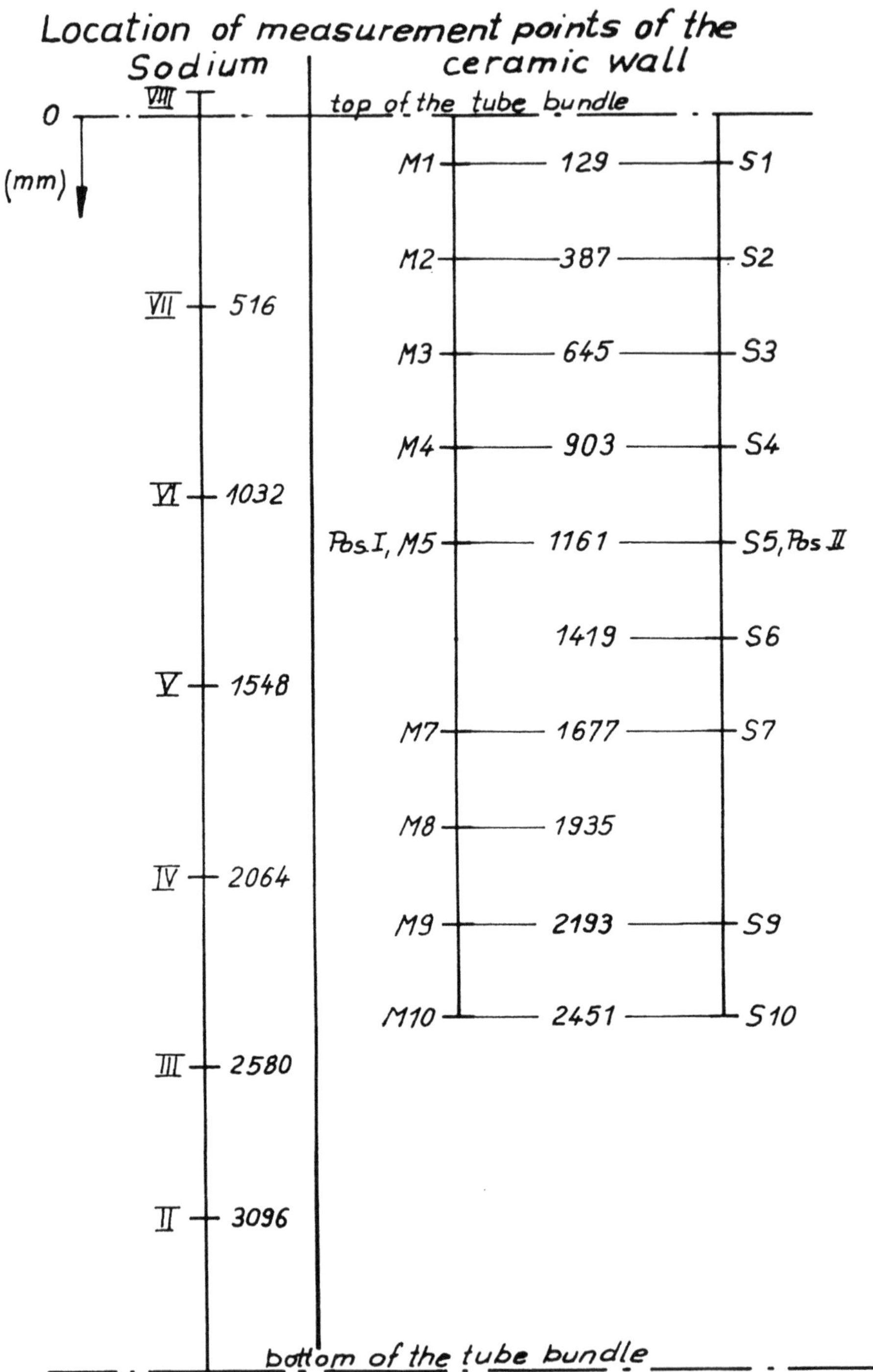

Location of measurement points of the
Sodium
ceramic wall
0
(mm)
top of the tube bundle
VIII
VII 516
VI 1032
V 1548
IV 2064
III 2580
II 3096
I
M1 129 S1
M2 387 S2
M3 645 S3
M4 903 S4
Pos.I, M5 1161 S5, Pos II
1419 S6
M7 1677 S7
M8 1935
M9 2193 S9
M10 2451 S10
bottom of the tube bundle

6.5.-A30

TANK LOADING PROGRAM

```
10 ! *************************
20 ! *         TANK 1          *
30 ! *************************
40 !
50 ! LECTURA DE LAS COND INIC.
60 E1=0 @ E2=0
70 DISP "        ESTADO INICIAL,D
   ATOS DAS"
80 DISP "  INTRODUCIR H0,T0,H4,
   T4[m,C]"
90 INPUT H0,T0,H4,T4
100 H=H0 @ GOSUB 750
110 V0=V
120 H=H4 @ GOSUB 750
130 V4=V
140 DISP "     INTERVALO DE TIEMP
    O (sg)"
150 INPUT I
160 PEN 1 @ PLOTTER IS 705
170 LIMIT 2,270,8,187 @ FRAME
180 LOCATE 20,140,10,90
190 SCALE 0,120,0,600
200 XAXIS 0,10
210 YAXIS 0,25
220 FOR Y=0 TO 600 STEP 50
230 MOVE -10,Y @ LABEL VAL$(Y)
240 NEXT Y
250 SCALE 0,120,0,70
260 PEN 1
270 YAXIS 0,5 @ CSIZE 2
280 FOR Y=0 TO 70 STEP 5
290 MOVE -4,Y @ LABEL VAL$(Y)&"
    "
300 NEXT Y
310 PRINT "---------------------
    ----------"
320 PRINT " VOL   VOL    E    TEMP
      TEMP   E "
330 PRINT " CAL   DAT          CAL
      DAT"
340 PRINT " m3     m3    %      C
       C     %"
350 PRINT "---------------------
    ----------"
360 FOR T9=1 TO 120
370 DISP "  ENTRADA TEMP Y FLUJO
    S (DAS) "
380 INPUT T3,H,T1,V1,V2
390 GOSUB 750
400 V3=V
410 PRINT USING 420 , V0,V4,E1,T
    0,T4,E2
420 IMAGE X,DD D,X,DD D,DDD D,X,
    DDD D,X,DDD D,DDD D
430 IF V0<0 OR V0>70 THEN 440 EL
    SE 470
440 BEEP 25,1000 @ BEEP 50,1000
450 PRINT "  VOLUMEN FUERA DE LI
    MITES "
460 BEEP 25,100 @ BEEP 60,1000
```

```
470 T=T1 @ GOSUB 730
480 R1=R
490 T=T0 @ GOSUB 730
500 R2=R
510 M0=V0*R2
520 M1=V1*I*R1/3600
530 M2=V2*I*R2/3600
540 M=M0+M1-M2
550 T=T0 @ GOSUB 770
560 C0=C
570 T=T1 @ GOSUB 770
580 C1=C
590 T=C1*M1*(T1+273.16)+C0*M0*(T
    0+273.16)-45.2*(T0-20)*I+618
    69000*(T0+273.16)
591 T=T/(61869000+C0*(M+M2))-273
    .16
600 GOSUB 730
610 V=M/R
620 E1=(V-V3)*100/V3
630 E2=(T-T3)*100/T3
640 SCALE 0,120,0,70
650 MOVE T9-1,V0 @ PEN 1 @ DRAW
    T9,V
660 MOVE T9-1,V4 @ PEN 3 @ DRAW
    T9,V3
670 SCALE 0,120,0,600
680 MOVE T9-1,T0 @ PEN 2 @ DRAW
    T9,T
690 MOVE T9-1,T4 @ PEN 4 @ DRAW
    T9,T3 @ PENUP
700 V0=V @ T0=T @ V4=V3 @ T4=T3
710 NEXT T9
720 END
730 R=950.1-.22976*T-.0000146*T^
    2
740 RETURN
750 V=-2.5863*H^3+12.723*H^2+7.9
    69*H-.15007
760 RETURN
770 C=4.1868*(343.24-.13868*(T+2
    73.16)+.00011044*(T+273.16)^
    2)
780 RETURN
```

VOL CAL m3	VOL DAT m3	E %	TEMP CAL C	TEMP DAT C	E %
14.2	14.2	0.0	372.8	372.8	0.0
15.0	15.2	-1.2	376.2	380.6	-1.2
15.2	15.4	-1.7	379.5	387.2	-2.0
15.3	15.7	-2.5	382.9	395.3	-3.1
15.3	15.7	-2.7	386.2	402.1	-4.0
15.4	15.7	-2.0	389.5	408.5	-4.7
15.3	15.7	-2.8	392 8	414.8	-5.3
14.6	15.4	-5.3	396.0	420.7	-5.9
13.4	14.5	-7.3	399.3	426.6	-6.4
12.1	13.0	-7.0	402.6	432.6	-6.9
10.9	11.9	-8.3	405.9	438.6	-7.4
9.8	10.8	-9.5	409.2	444.3	-7.9
8.7	9 7	-10.6	412.5	449.6	-8.2
7.6	8 9	-13.7	415.8	454.5	-8.5
7.0	7.8	-11.4	419.1	457.7	-8.4
6.8	7.7	-11.4	422.3	464.5	-9.1
6.4	7.3	-12.3	425.5	469.2	-9 3
5.9	6.9	-13.6	428 6	473.3	-9.4
6.1	6.7	-9.1	431.7	477.3	-9.6
6.9	7.1	-2.4	434.7	480.6	-9.6
8.1	8.0	.1	437.5	484.0	-9 6
9.2	9.1	1.9	440.2	487.0	-9.6
10.0	9.9	.7	442.2	489.2	-9.6
10.6	10.6	.3	443.9	491.1	-9.6
11.2	11.2	.2	445.6	492.5	-9.5
11.9	11.7	2.3	447.2	493.3	-9.3
12.3	12.3	-.1	448.4	494.3	-9.3
12.5	12.6	-.9	449.2	494.7	-9.2
12.6	12.6	.0	449.9	495.1	-9.1
12.8	12.8	.1	450.9	494.8	-8.9
13.5	13.3	1.5	452.3	495.4	-8.7
14.7	14.0	5.1	454.2	495.6	-8 3
16.5	15.7	4.9	456.4	496.5	-8.1
18.2	17.4	4.2	458.4	497.5	-7.9
19.9	19.3	3.3	460.3	498.7	-7.7
21 6	21.1	2.4	462.1	499.3	-7.4
23.3	23.0	1.4	463.8	499.8	-7.2
25.0	24.9	.5	465.5	500.5	-7.0
26.8	26.9	-.4	467.0	501.5	-6.9
28.4	28.6	-.5	468.5	501.7	-6.6
30.1	30.6	-1.6	469.9	501.7	-6.3
31.6	32.3	-2.2	471.2	501.8	-6.1
33.2	34.0	-2.5	472.5	502.7	-6 0
34.7	35.7	-2.9	473.7	503.7	-6 0
36.2	37.5	-3.4	474.9	503.4	-5.7
37.3	39.2	-4.7	476 0	504.6	-5.7
38.5	39.2	-1.8	477.1	504.6	-5.4
38.9	40.0	-2.8	478.1	504.2	-5.2
38.7	40.3	-4.1	479.1	504.6	-5.0
38 1	39.8	-4.2	480.1	505.7	-5.1
37.4	38.9	-3.7	481 0	505.5	-4.9
36 8	38.3	-3.9	481.9	506.0	-4.8
36.2	37.5	-3.3	482.8	506.7	-4.7
35.7	36.6	-2.4	483.7	506.8	-4.6
35.9	36.3	-1.1	484.6	507.3	-4 5
35.8	36.6	-2.3	485.4	507.7	-4.4
35.1	36.0	-2.7	486.1	508.0	-4.3
34.2	35.2	-2.8	486.9	508.5	-4.2
33.2	34.0	-2.4	487.6	509.0	-4.2
32.2	32.9	-2.0	488.2	509.1	-4.1
31.1	31.4	-.9	488 8	509.1	-4.0
30.0	30.3	-1.0	489.4	509.4	-3 9
28.8	28.8	-.1	490 0	509.7	-3.9
27.7	27.4	.8	490.6	510.2	-3.9
26.5	26.0	1.9	491.1	510.3	-3.8
25.4	24.9	1.8	491.7	510.4	-3.7
24.2	23.5	2.7	492.2	510.6	-3.6
23.0	22.2	3.7	492.7	510 7	-3 5

TIME, HOURS	MEASURED SODIUM TEMPERATURE, C	SODIUM LEVEL, m	INLET TEMPERATURE, C	INLET FLOW RATE, m^3/hr	OUTLET FLOW RATE, m^3/hr	INSOLATION, W/m^2
11: 0	372.81	0.85	521.41	19.54	7.04	768.69
11: 5	380.63	0.89	521.54	19.97	9.94	775.00
11:10	387.19	0.90	522.25	20.40	17.59	773.19
11:15	395.31	0.91	522.56	20.84	18.91	779.75
11:20	402.13	0.91	522.56	21.23	19.47	781.63
11:25	408.53	0.91	523.22	21.61	20.95	783.94
11:30	414.84	0.91	523.19	21.96	22.86	776.63
11:35	420.69	0.90	523.44	22.36	29.23	778.56
11:40	426.56	0.86	523.56	22.72	36.61	776.05
11:45	432.56	0.80	524.00	23.05	37.79	771.75
11:50	438.59	0.75	523.97	23.25	37.20	771.13
11:55	444.28	0.70	524.05	23.47	36.59	760.56
12: 0	449.56	0.65	524.34	23.66	36.04	753.44
12: 5	454.53	0.61	524.19	23.99	35.49	748.81
12:10	459.72	0.56	524.31	24.05	31 73	746.31
12:15	464.53	0.55	524.13	24.26	25.63	736.19
12:20	469.19	0.53	524.78	24.61	28.88	816.13
12:25	473.28	0.51	524.31	24.71	29.15	853.38
12:30	477.28	0.50	524.31	24.89	22.61	856.50
12:35	480.59	0.52	525.25	24.94	14.39	861.81
12:40	484.00	0.57	524.56	24.92	10.47	857.25
12:45	487.03	0.62	525.09	25.04	10.34	855.13
12:50	489.16	0.66	522.81	19.63	10.35	849.31
12:55	491.13	0.69	521.59	17.93	10.16	853.94
13: 0	492.50	0.72	522.66	17.95	9.80	854.56
13: 5	493.25	0.74	522.19	17.91	9.16	853.13
13:10	494.34	0.77	522.56	14.04	9.99	856.69
13:15	494.69	0.79	511.84	10.39	8.73	860.88
13:20	495.06	0.78	515.88	10.29	8.67	851.19
13:25	494.75	0.79	520.47	11.79	8.73	862.69
13:30	495.44	0.81	523.78	16.75	8.63	866.13
13:35	495.56	0.84	524.44	23.56	8.40	863.25
13:40	496.50	0.91	525.03	26.57	5.13	860.94
13:45	497.50	0.98	526.03	26.25	5.17	852.63
13:50	498.66	1.05	526.16	26.02	5.06	849.38
13:55	499.25	1.12	526.44	25.85	4.94	846.06
14: 0	499.75	1.19	526.81	25.77	4.86	846.06
14: 5	500.47	1.26	525.63	25.74	4.69	844.94
14:10	501.53	1.33	526.13	25.68	4.86	847.44
14:15	501.72	1.39	526.59	25.73	5.37	845.63
14:20	501.66	1.46	525.84	25.40	5.37	839.00
14:25	501.81	1.52	526.38	25.25	6.95	837.81
14:30	502.69	1.58	526.09	25.34	6.17	833.94
14:35	503.69	1.64	525.91	25.13	6.42	931.56
14:40	503.44	1.70	526.16	24.93	6.82	833.56
14:45	504.56	1.76	526.63	24.96	10.86	825.94
14:50	504.16	1.79	525.97	24.64	19.44	818.38
14:55	504.59	1.80	526.06	24.50	27.19	815.44
15: 0	505.72	1.78	525.88	24.25	31.23	816.44
15: 5	505.53	1.75	525.41	24.01	31.28	807.25
15:10	506.03	1.73	526.41	24.24	31.27	805.89
15:15	506.69	1.70	525.59	23.90	31.28	786.63
15:20	506.81	1.67	526.25	23.83	29.61	785.00
15:25	507.25	1.66	525.69	23.26	20.97	772.25
15:30	507.72	1.67	525.25	22.83	24.18	765.50
15:35	507.97	1.65	525.34	22.24	30.55	731.56
15:40	508.47	1.62	524.88	21.65	32.26	743.63
15:45	509.00	1.58	524.72	20.82	32.24	727.69
15:50	509.09	1.54	524.72	20.20	32.19	713.75
15:55	509.13	1.49	523.66	19.64	32.33	699.56
16: 0	509.44	1.45	524.47	19.22	32.88	692.06
16: 5	509.69	1.40	524.06	18.90	32.67	689.91
16:10	510.22	1.35	523.47	18.81	32.45	683.56
16:15	510.38	1.30	523.56	18.95	32.39	682.63
16:20	510.44	1.26	522.75	18.42	32.15	672.50
16:25	510.63	1.21	522.69	17.97	32.15	663.17
16:30	510.72	1.16	522.50	17.52	31.56	647.50
16:35	510.78	1.10	522.29	17.94	31.13	629.25
16:40	511.47	1.01	[illegible]	[illegible]	[illegible]	[illegible]

THE TANK. COMPARISON BETWEEN SIMULATION AND DAS

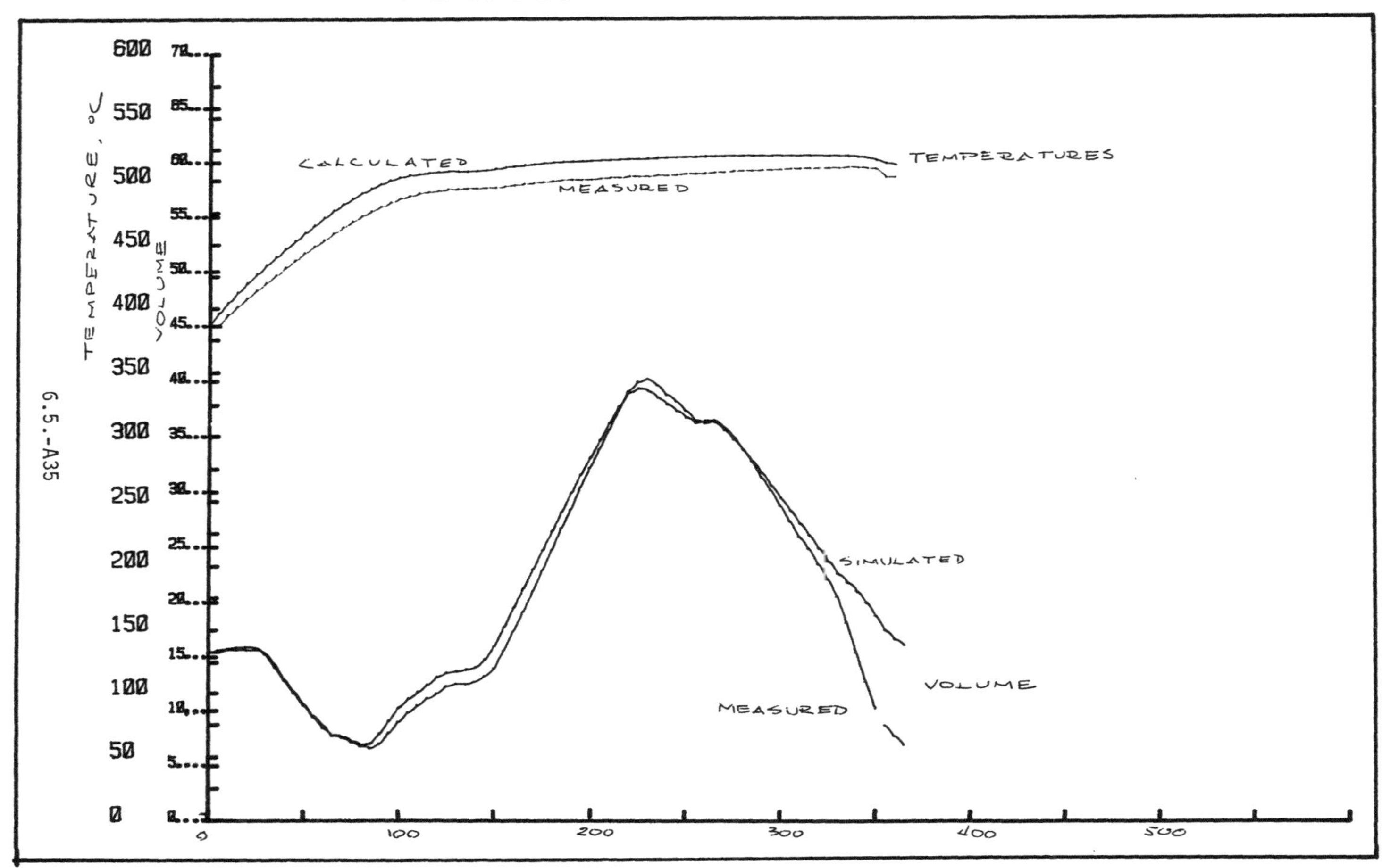

6.5.-A35

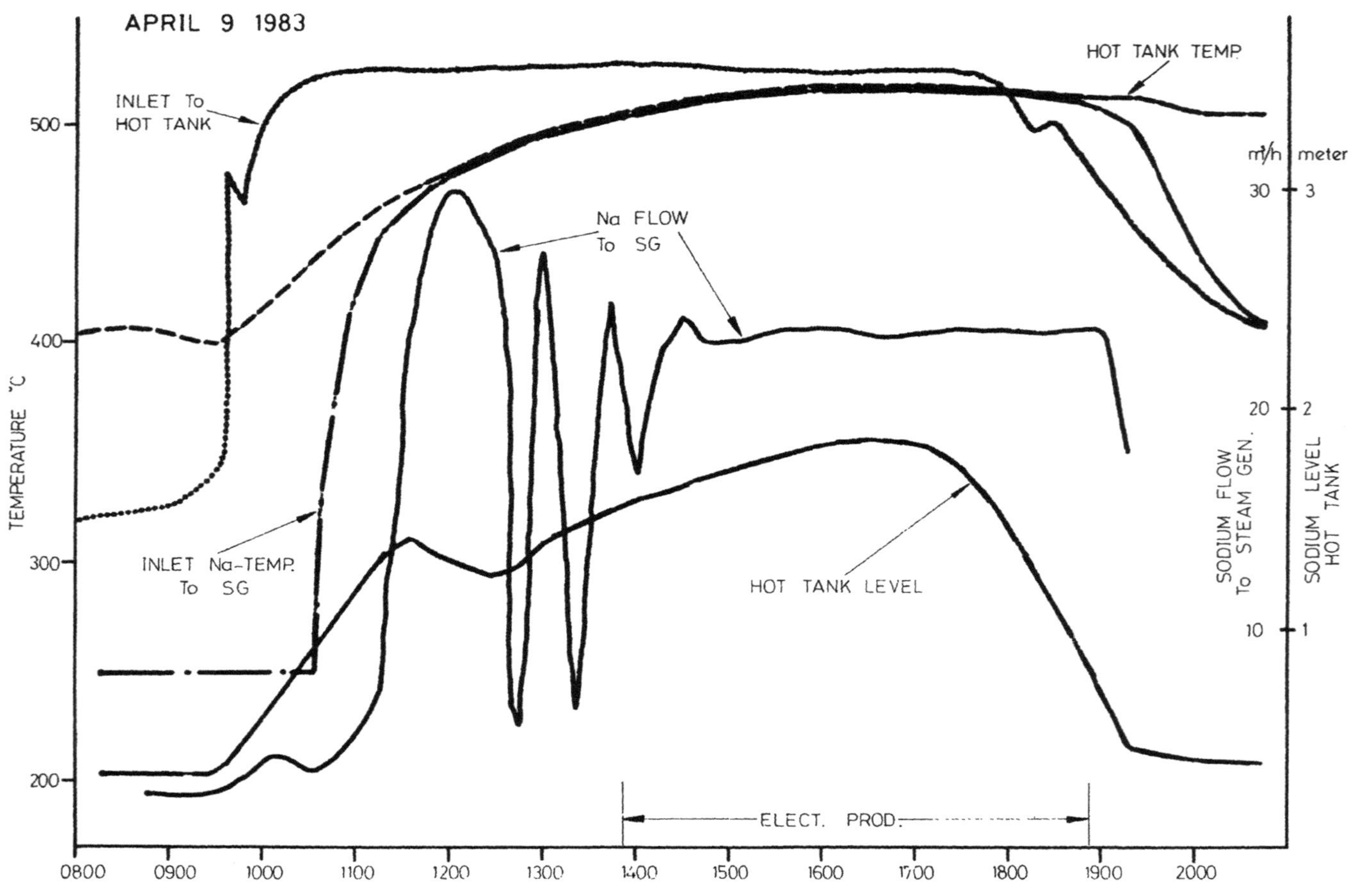

APRIL 9 1983
INLET To HOT TANK
HOT TANK TEMP.
Na FLOW To SG
INLET Na-TEMP. To SG
HOT TANK LEVEL
ELECT. PROD.
TEMPERATURE °C
500
400
300
200
m³/h meter
30 3
20 2
10 1
SODIUM FLOW To STEAM GEN.
SODIUM LEVEL HOT TANK
0800
0900
1000
1100
1200
1300
1400
1500
1600
1700
1800
1900
2000

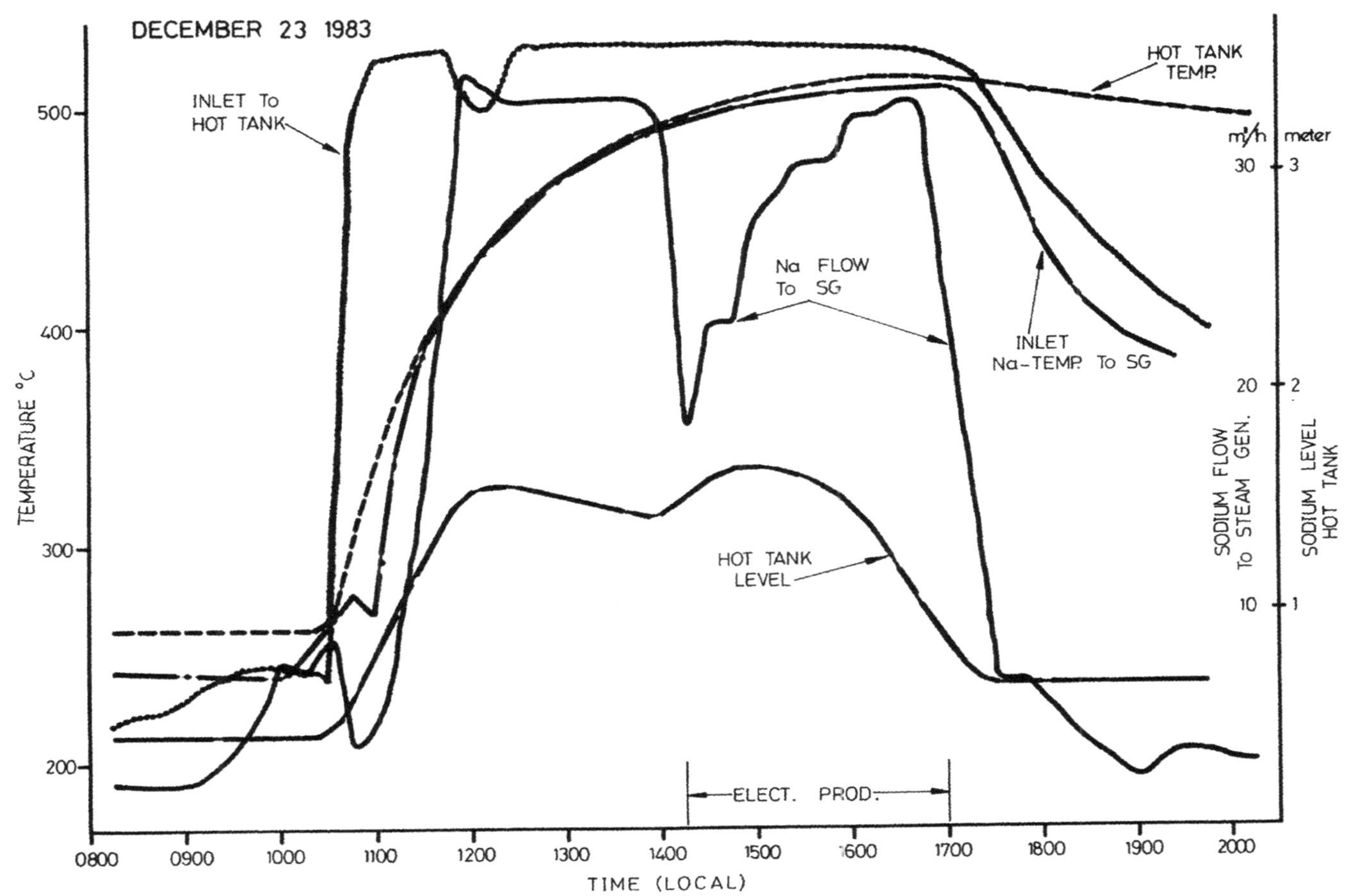

DECEMBER 23 1983
INLET To HOT TANK
HOT TANK TEMP.
Na FLOW To SG
INLET Na-TEMP. To SG
HOT TANK LEVEL
ELECT. PROD.
TEMPERATURE °C
500
400
300
200
0800
0900
1000
1100
1200
1300
1400
1500
1600
1700
1800
1900
2000
TIME (LOCAL)
m³/h
meter
30
20
10
3
2
1
SODIUM FLOW To STEAM GEN.
SODIUM LEVEL HOT TANK

DECEMBER 23 1983

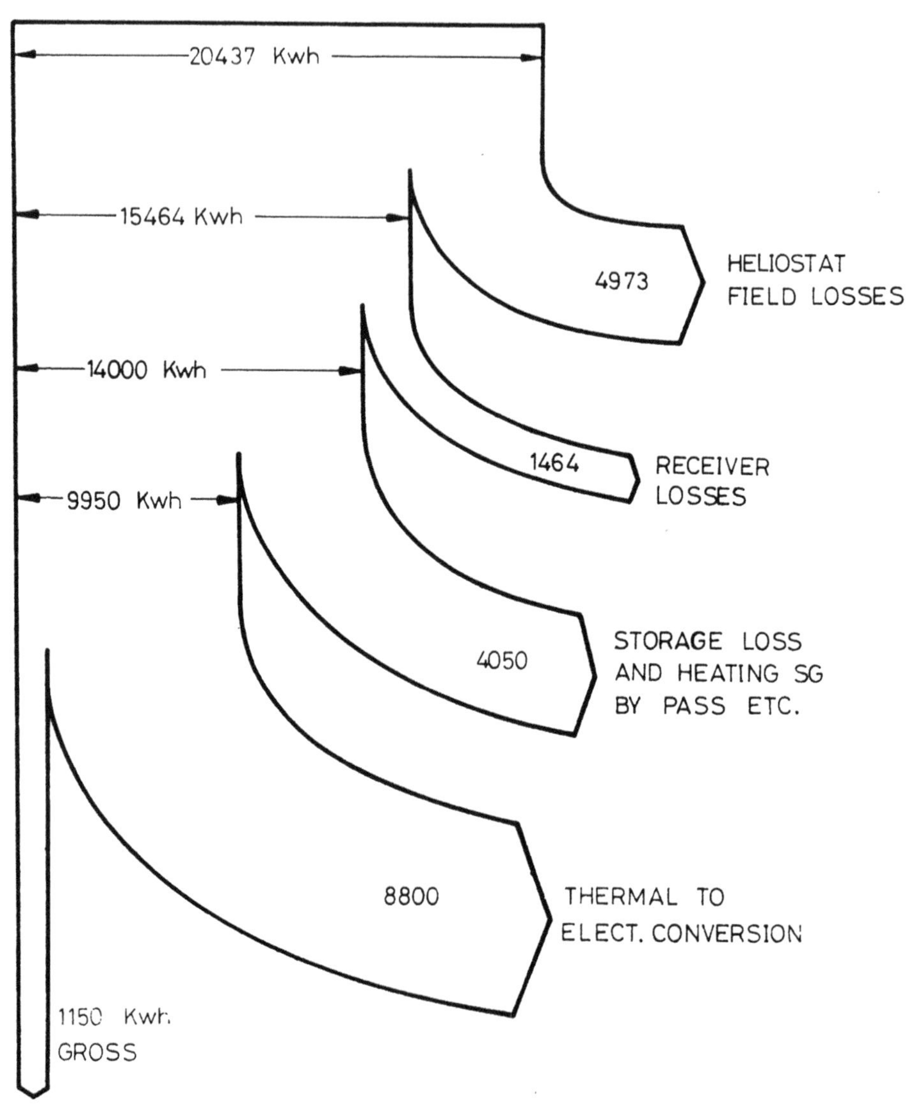

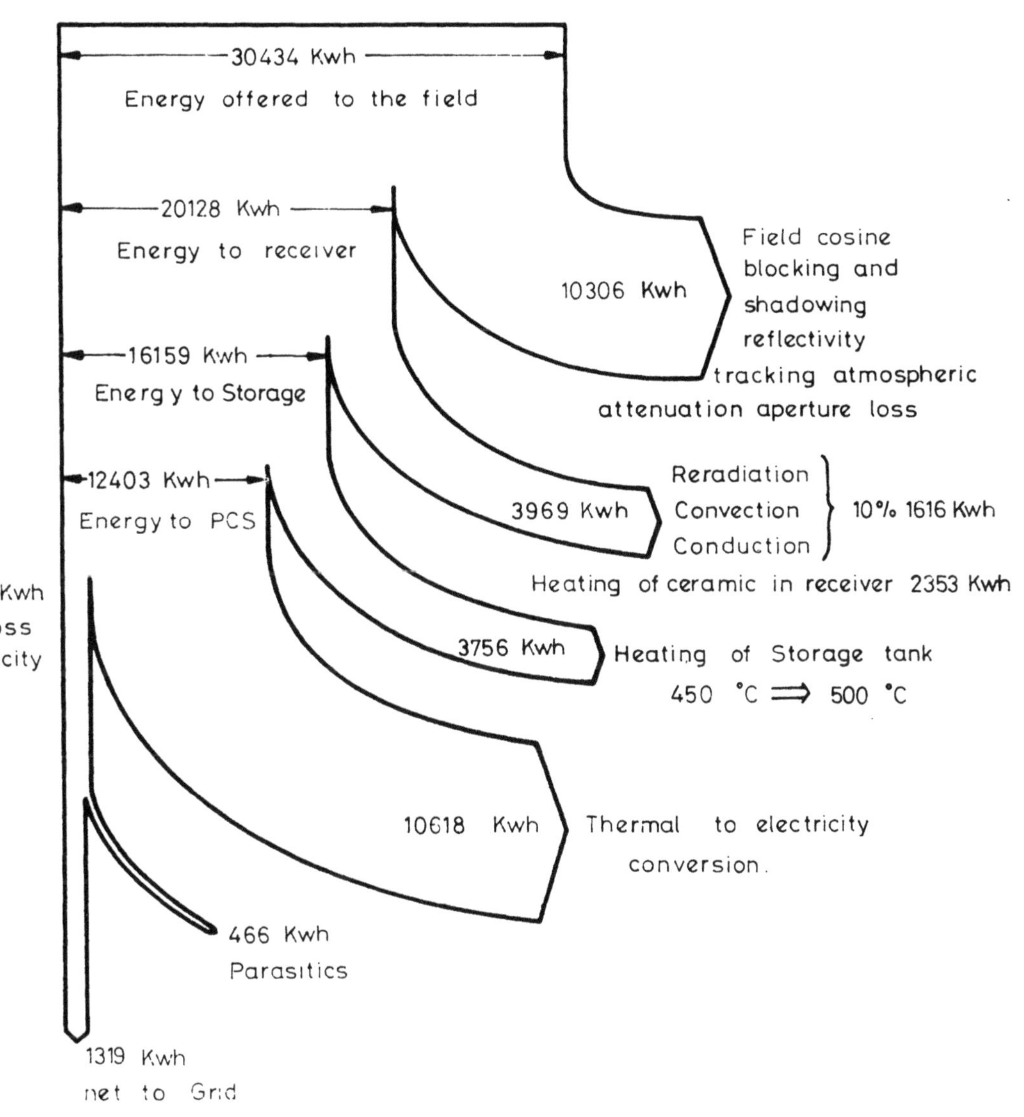

ENERGY BALANCE FOR CRS

April, 9, 1983

SYSTEMS ASPECTS/CONTROL

SYSTEMS ASPECTS/CONTROL

INTRODUCTION

The favorable heat transfer characteristics of sodium makesit possible to collect concentrated solar radiation at medium high temperatures in small active areas, and therefore, with relatively low losses. However, because of the intermittent nature of the solar source, the receiver and the sodium pump controls must be designed and operated so as to obtain adequate responses to insure receiver structural integrity. The two receivers installed and tested at the site represent two different coolant flow approaches, each of which pose different demands on the control system. This evaluation area reports on studies of this problem situation and in the final analysis, attempts to make clear the necessitity of system integration; that is, the need to design a coolant flow control system specifically for a receiver. This was not the completly true case at SSPS for either receiver. Even with the cavity installation the control was designed for an input power level much higher than was finally realized. The ASR, of course, was required to accommodate the coolant flow/pumping situation that was installed with minimum modification.

In a comprehensive review of the general problem of temperature control of central receivers, C. Maffezzoni, in the report TEMPERATURE REGULATION; CRS, discusses the process dynamics for different types of receivers, possible control approaches and practical control schemes for receivers cooled by sodium. This is a rather theoretical analysis which leads the reader through the theory and to the conclusions.

The conclusions are:

1 - Traditional temperature control concepts are unacceptable for fast responsive solar systems.
2 - Newer concepts are workable.

The cavity (Sulzer) receiver was designed with a heliostat field of 160 heliostats of 39.4 m2 each. As a result of insufficient funds, only 93 heliostats of 39.4 m2 each were installed,thus dramatically reducing the maximum possible energy input to the receiver. D.Weyers does not dwell on this fact in the report TEMPERATURE REGULATION; SULZER , which was the cavity receiver, but a reasonable description of the control system is provided along with some observations of the control system's performance.

The conclusions are:

1 - The traditional feedback control system, used with the cavity receiver worked rather well.

2 - The prime control variable was outlet temperature.

The external receiver designers were required to interface with the coolant pumping system and pump control system that was designed for the original receiver/heliostat field system. This receiver was also designed to be very responsive to solar energy input variations, which with the constraint of given pumping capabilities required the design and implementation of a very sophisticated control system. G. Magnani describes this control system in the report TEMPERATURE REGULATION; ASR.

The conclusions are;

1 - Feedforward control was essential for the fast response receiver, the ASR.
2 - The feedforward control system eliminated the concern for rapid increases in absorbed power.

Where previous control studies and designs approached the control problem in a responsive mode, M. Blanco and M. Sanchez address the control problem by controlling the energy input. Their efforts are described in the report TRACKING; CONTROL OF INCIDENT POWER AT THE RECEIVER. Following the development of a theoreti method a test of this approach was made at SSPS and that effort is included in the report.

The conclusions are;

1 - A computer program is developed to control the heliostat field in response to receiver energy needs.
2 - Experiments demonstrated that the system can be made to work.

TEMPERATURE CONTROL OF SOLAR RECEIVERS

Claudio Maffezzoni, ENEL

1. INTRODUCTION

The effective use of new energy processes essentially relies on the possibility of ensuring operation continuity to maximize the integral energy output. This is particularly true in the case of central receiver solar plants where a reasonable return of the capital investment requires daily production periods to be maximized, in spite of severe environmental disturbances like those due to clouds passing over the mirror field. Moreover, optimization of receiver efficiency may lead to the adoption of very high thermal fluxes on the receiver surface, which implies very high temperature derivatives and thermal stresses in connection with solar disturbances. These loads call for receiver temperature control systems with very good response times to keep the temperature variations within acceptable limits during fast cloud passages.

This paper demonstrates that traditional temperature control concepts are unacceptable in that they are generally not designed for very fast feedback control. The reason for this deficiency is essentially because receiver dynamics is affected by relevant transport delays. These delays prevent simple temperature control loops from the receiver outlet temperature to the coolant flow demand from being effectively closed.

A critical analysis of traditional control concepts is performed to point out that a new control concept can effectively be applied to high temperature receivers. The new concept proposed is based on the feedback of temperature measurements taken at intermediate receiver points. The concept is called Predictive Feedback Temperature Control (PFTC), as it incorporates a process behavior predictor fed by these temperature feedbacks. Since the prediction is based on a (possibly simplified) model of a suitable part of the process, the design of the PFTC requires an accurate analysis of the process dynamics, particularly the role of the non-minimum phase transfer functions characterizing the receiver input-output response.

Thus, the paper is organized as follows: Section 2 is devoted to the dynamics of sodium-cooled receivers and its relations to the achievable control performance; Section 3 considers the extension of dynamic analysis results to different kinds of receivers (e.g., gas and water cooled receivers); Section 4 illustrates the basic principles for the control of sodium receivers, comparing the possible different concepts; and Section 5 describes the practical realization of PFTC for sodium receivers. Finally, the extension of the control design principles to different types of receivers is discussed in the concluding remarks (Section 6).

2. RECEIVER DYNAMICS

Process dynamics of high-temperature solar receivers exhibits a number of peculiar properties which are very important to the control system design, and therefore have to be analyzed.

Two different basic tools are very useful in dynamic analysis:

1. Simulation, typically performed by the numerical integration of the nonlinear partial-differential equations describing the dynamic energy balance of any receiver tube.

2. Transfer function computation and frequency analysis, performed by linearizing the abovementioned partial-differential equations and applying Laplace-transformation.

As is well known, the second approach only takes into account the process behavior for small variations. On the other hand, it is especially suited to understanding the fundamental features of the process behavior under the effect of closed loop control; it examines amplitude and phase relations between any input variable and any output variable. The extrapolation of the process behavior to the case of large variations requires that nonlinearities be taken into account and precisely simulated.

Since the main concern here is receiver control, details on simulation are not discussed--the reader should refer to other papers (Ref.1-5).

Considering a typical situation where the receiver consists of a series of panels (each one consisting of parallel tubes), the receiver dynamics can be evaluated from the three relevant step-responses of a single panel:

(R.1) the response of the outlet temperature to a step of the coolant flowrate,
(R.2) the response of the outlet temperature to a step of the inlet temperature,
(R.3) the response of the outlet temperature to a step of the radiation intensity (or the thermal power on the panel).

For receivers using sodium as a coolant, the following physical properties have a major influence on process dynamics:

1. Since the fluid is a liquid, the working pressure is generally low, and therefore the tube walls may be very thin;

2. The heat exchange coefficient between the coolant and the tube wall is very high;

3. Since the sodium is a very good coolant (heat exchange and thermal capacity), high thermal fluxes are generally applied;

4. Since the heat losses in heat exchanging tubes are proportional to the fluid density and to the square of the fluid speed, the fluid speed must be limited (generally below 2 m/sec).

The dynamic counterpart of property (1) is that the energy storage within the metal is not as great as in water-cooled receivers and, together with property (2), that the time lag between radiation power and the power delivered to the fluid is quite small. Furthermore, property (2) implies that the receiver exchange surface is generally small so that the total energy storage of the receiver is limited and radiation power variations very quickly affect metal and fluid temperature (also in view of property (3)). A further consequence of property (2) is that tube metal and sodium are "thermally coherent"; that is, their temperatures, at a given section, move together without significant time lag between the two.

Finally, property (4) means that the fluid transport time in any receiver tube is not small, and therefore the receiver dynamics is dominated by the transport phenomena.

Assuming for the sake of simplicity that the radiation flux is uniform over a panel or at least over any panel tube, the above dynamic features can well be characterized on the step responses (R.1)-(R.3), whose typical shape is approximated in Fig.7.1-1a/c.

The approximations made in representing the step responses of Figure 1 significantly affect their behavior only in the neighborhood of the encircled points, where the neglected fast dynamics essentially results in a smoothing of the edges (dotted lines).

It is apparent from Fig.7.1-1that the panel temperature dynamics is essentially characterized by time τ' which can be approximately evaluated according to the formula (Ref.1,2):

$$\tau'^{*} = \tau\,(1+ \tau_m/\tau_f) \tag{1}$$

where $\tau = l/v$ (2)

$\tau_f = c\,\rho\,A/\pi\,\gamma_t d_i$ (3)

$\tau_m = c_m \rho_m A_m/\pi\,\gamma_t d_i$ (4)

with l = length of a panel tube
v = fluid velocity
c = specific heat capacity of the coolant
ρ = density of the coolant
A = internal cross section of a panel tube
γ_t = heat exchange coefficient between tube wall and coolant (including the equivalent of the tube wall resistance)
d_i = internal diameter of a panel tube
c_m = specific heat capacity of the tube metal
ρ_m = density of the tube metal
A_m = cross-section of the tube wall

Note that parameter τ', by definition, is the fluid transport time multiplied by the factor $(1 + \tau_m/\tau_f)$ which is greater than 1 and takes into account that the wall metal is almost thermally coherent with the adjacent fluid (see property (2) above and the subsequent remarks). This effect of lengthening the apparent transport time is a very peculiar phenomenon of heat exchanger known as "percolation" (Ref.6). It is applicable to those heat exchangers where transport phenomena dominate storage phenomena, that is where

$$\tau_m,\ \tau_f \ll \tau \tag{5}$$

Moreover, it is of greatest relevance when τ_m/τ_f is not small. In sodium receivers, the strong inequality (3) is generally verified due to the very high value of γ_t, and τ_m/τ_f is generally about 1.

For instance, for the Advanced Sodium Receiver installed in the Almería SSPS, the above parameters have the following values at the maximum load (for a single panel):

$$\tau \cong 2.5 s$$
$$\tau_f \cong .06 s$$
$$\tau_m \cong 0.1 s$$

It is finally worth noting that, when the load is varied, the parameter τ varies as the inverse of the load while the ratio τ_m/τ_f is nearly independent of the load. Thus, the responses of Fig.7.1-1 apply to any receiver load if the time scale is inversely proportional to the final flow rate.

In the case of cascaded panels, the temperature responses can easily be obtained via the superposition of the simple diagrams in Fig.7.1-1. The case of two identical panels in series is worth considering and is described in Fig.7.1-2, where both the receiver outlet temperature T_2 and the temperature T_1 at the outlet of the first panel are given. In the tutorial case of Fig.7.1-2, the assumption has been made that the two panels are geometrically identical but with different radiation power on them (the first one absorbs twice the power of the second one).

The transient responses can be easily explained, bearing in mind that the temperature variation at the outlet of the second panel is due to the superposition of the temperature variation occurring when its inlet temperature is constant and the temperature variation due to the inlet temperature (T_1). For instance, one can see that response (R.1) consists of two parts:

-a first part, or "prompt component" of the response (between t = 0 and t = τ') where the effect of the flow rate variation in the second panel takes place;

-a second part, or "delayed component" of the response (between τ' and 2 τ') where the effect of the variation of the first panel cutlet temperature takes place (delayed by the transport time τ').

Extending the reasoning to multi-panel receivers is straightforward, since the responses can be obtained by the repeated application of the above outlined superposition concept. As a result, in a real receiver the temperature variation of the fluid at a given section due to variations either of the coolant flow rate or of the radiation intensity turns out to consist of a "prompt component", due to the power incident on the upstream panel close to the considered point and of "delayed components" due to the power incident on the preceding panels which exhibit their effects with the corresponding total transport delay (multiple of the single panel delay τ').

They are actually the "delayed components" which make the temperature control problem very critical because the more the effects of the control variable on the output variable are delayed, the more difficult is the realization of a fast regulation of the output variable itself.

It is also quite obvious that receivers in which no definite splitting into panels is recognizable can be analyzed with the same line of reasoning. Those points where temperature sensors can be practically installed are defined as potential panel boundaries, assuming that the thermal power distribution is approximately uniform on the so-defined "virtual panels".

It can be shown (Ref.1,2) that for the above described behavior, there corresponds a very meaningful interpretation in terms of transfer-functions as defined by the block diagram of Fig.7.1-3 relative to the case of n panels in series. In the block diagram of Figure 3, the following notations hold:

- Δx is the variation of any variable x around its value at a given receiver steady-state
- β is the radiation intensity, i.e., the ratio between the current radiation power on the receiver and its nominal value
- w is the sodium mass flow rate
- T_o is the coolant temperature at the receiver inlet
- T_i, i =1,...,n, is the coolant temperature at the outlet of the ith panel (the node i being taken according to the fluid stream)
- DT_{pi}, i = 1,...,n, is the temperature increment which would take place on the ith panel if the corresponding panel inlet temperature is maintained constant (i.e., it is the temperature increment due to the thermal power absorbed by the ith panel)
- τ_m is the parameter given by (4) accounting for the time lag between radiation power and power entering the coolant stream

-F_i, i = 1,...,n, is the transfer function (with changed sign) between the coolant flowrate (in p.u. of the reference steady-state) and the temperature increment DT_{pi}, whose expression is given below

-D_i, i = 1,...,n, is the transfer function between inlet and outlet temperature for the ith panel (assuming $\Delta\beta$ = 0, Δw = 0), given below.

In the simple case in which the panels are identical and the flux distribution is uniform on each individual panel, the transfer function F_i is given by

$$F_i(s) = \lambda_i \frac{\bar{\beta} Q_t}{\bar{c}\,\bar{W}} \quad \frac{1 - e^{-B(s)}}{B(s)} \tag{6}$$

where $\bar{\beta}$, $\bar{W}$, and $\bar{c}$ are the values of β, W, and c at the reference steady-state, Q_t is the total nominal radiation power on the receiver, λ_i (with $\lambda_1 + \lambda_2 + \ldots + \lambda_n = 1$) is the fraction of the radiation power absorbed by the ith panel and

$$B(s) = s\tau \left(1 + \frac{\tau_m/\tau_f}{1 + s\tau_m}\right) \tag{7}$$

Furthermore, when the panels are geometrically equal, $D_1 = D_2 = \ldots D_n = D$ are given by:

$$D(s) = \exp\left(-B(s)\right) \tag{8}$$

When, as is generally the case of sodium-cooled receivers, inequality (5) applies, then we have

$$B(s) \simeq s\tau \;(1 + \tau_m/\tau_f) \triangleq s\tau' \tag{9}$$

and neglecting the fast dynamics of the lag $1/(1+s\tau_m)$ filtering the input $\Delta\beta$, the transfer function diagram of Fig.7.1-3 corresponds to the approximate response of Fig.7.1-1/2.

It is apparent from Fig.7.1-3 that the so-called "prompt" component is the response of the ith panel outlet temperature given by DT_{pi} and is computed by the transfer function F_i only. Moreover, in the considered case of geometrically equal panels and uniform radiation distribution on each panel, the functions F_i, i=1,...,n, coincide up to the factor gain λ_i, which accounts for the fraction of radiation power absorbed by the corresponding panel.

Some remarks are appropriate concerning the applicability of the above reasoning to real receivers.

Remark 1 (Uniformity of Flux on Each Panel): To evaluate temperature responses and transfer functions, it has been assumed that the power distribution on any panel i is uniform, and thus characterized by the only scale factor λ_i. However, similar results hold for typical flux densities occurring in real receiver panels like rather broad Gaussian or cosine density functions. The effect of the panel flux density is essentially that of changing the expression of the transfer functions F_i, which still remain equal up to the scale factor λ_i if the panel density functions are similar (i.e., the same up to a scale factor). Also, for the abovementioned Gaussian-like densities the resulting step responses (Fig.7.1-1) undergo minor changes, and the superposition of "prompt" and "delayed" components still applies. Transfer functions D_i are, in any case, not affected by the flux density but only by the panel geometry so that Equation (7) is of general applicability.

Remark 2 (Receivers Without Panel Splitting): It was noted that when the receiver is not split into a series of panels (as in the case of the SSPS receiver), "virtual" panels can be defined as tracts of parallel tubes where the two following conditions are verified:

1. The flux density of any "virtual" panel has a shape similar to that of a panel (e.g., tracts between two boundaries of the receiver surface).

2. The boundaries of any virtual panel are such that temperature sensors can be installed at the corresponding physical points.

In general the above two conditions are verified both in wall and cavity receivers or can easily be met during the mechanical design phase. Therefore, the concepts illustrated in the present paragraph are likely to apply to a wide range of receivers.

Remark 3 (Representation of Headers and Connection Pipes): So far, receivers have been considered where panels in series are connected without any intermediate element relevant to process dynamics. However, the typical way of constructing receivers split into panels is that shown in Fig.7.1-4 where an outlet header, a connection pipe, and an inlet header are used.

Headers and connection tubes have a non-negligible effect on the temperature dynamics which can be described by the block diagram of Fig.7.1-5, where $H_{oi}(s)$ is the transfer function of the outlet header of panel i, $C_i(s)$ is the transfer function of the connection pipe, and $H_{y(i+1)}(s)$ is the transfer function of the inlet header of panel (i+1). Of course, the connection elements have the basic effect of further delaying the "delayed" components of the response of the panel outlet temperature as the transfer function between ΔT_{oi} and $\Delta T_{o\,(i+1)}$ is given by $D_{i+1}(s)H_{y(i+1)}(s)C_i(s)H_{oi}(s)$ instead of simply $D_{i+1}(s)$. Headers generally behave as well as stirred tanks with significant storage and negligible transport time, while connection tubes behave like nonradiated tubes; thus we may assume:

$$H_{y\,i+1}(s) \cong \frac{1}{1+sT_y} \cdot \frac{1+sT_y'}{1+sT_y''} \tag{10}$$

$$H_{oi}(s) \cong \frac{1}{1+sT_o} \cdot \frac{1+sT_o'}{1+sT_o''} \tag{11}$$

$$C_i(s) = \exp(-B_c(s)) \tag{12}$$

where T_y, T_y', T_y'' and T_o, T_o', T_o'' are suitable time constants (commonly with $T_y, T_y' = T_o$, $T_y'' = T_o''$) and $B_c(s)$ is given by a formula similar to Equation (7) with suitable definitions of parameters τ, τ_m, and τ_f.

As a consequence, limiting the representation to one panel only, the diagram of Fig.7.1-3 has to be modified as is Fig.7.1-6, where it can be realized that the basic features of the block diagram of Figure 3 are maintained.

Remark 4 (Nonlinearities): Receiver processes are affected by the typical nonlinearities of heat exchangers which are not taken into account in the transfer function description. It may be useful to recall that the main nonlinearities in such processes are due to:

1. The fact that in the energy balance equation the energy transport term has the form $\partial wh/\partial x$, where x is the tube abscissa, w is the coolant mass flow rate, and h is the coolant enthalpy. The main consequence of this fact is that (see Equations (2) and (6)), in the linearized model, the parameter τ and the gain of transfer function (6) vary as $1/\bar{w}$, being $\bar{w}$ the mass-flow rate at the linearization point. This implies that receiver dynamics slows down at reduced load, while the input/output gain increases at low loads.

2. The heat transfer coefficient varies with the mass-flow rate; such a variation is of minor importance because it is rather limited for liquid metals (with respect to a gas or water) and it does essentially affect the transfer functions only through the ratio τ_m/τ_f in which the coefficient γ_t is cancelled.

It is worth noting that the first source of nonlinearity (point (1)) is very important for practical control design and can be easily taken into account by suitably modifying the control system based on the linearized receiver model.

3. PROCESS DYNAMICS FOR DIFFERENT TYPES OF RECEIVERS

The basic concepts dealt with in the preceding paragraph directly apply to receivers cooled by liquid metals or molten salt, that is when the coolant is a high density fluid not subject to phase transition. They could also be applied with minor changes to receivers using water as coolant but only in the case where neither boiling would take place nor supercritical high temperature (and low density) states would be encountered.

However, usual designs of water-cooled receivers are conceived so as to have the coolant in the three different conditions of liquid, gas-liquid mixture, and superheated steam. This fact generally defines three separated zones in the receiver, that is, the compressed liquid zone (CLZ) (or preheating zone), the evaporation zone (EVZ), and the superheating zone (SHZ).

The main peculiar aspects for such kind of receivers can be summarized as follows (see Ref.4,5):

1. To generate steam at sufficiently high enthalpy for mechanical power generation, water-cooled receivers are generally operated at high pressure, usually many tens of bars, so that receiver tubes must have much thicker walls than in sodium-cooled receivers.

2. Within the receiver, the three different zones CLZ, EVZ, and SHZ exhibit very different dynamic behaviors and moreover, they have strong mutual interactions, which mainly depend on which type of steam generator is adopted (once-through or with drum and recirculation) and on its operating criteria (e.g., constant or sliding pressure).

3. Heat exchange coefficients are very high only in zones of fully developed boiling, while they are lower in CLZ and very low in SHZ and in the deficient liquid region of EVZ; this implies that the average radiation flux on the receiver must be considerably lower than in sodium-cooled receivers and thus the total radiated surface considerably larger (at equal input power).

The analysis performed in paragraph (2) directly applies to CLZ where the heat transfer phenomena are quite similar to those in liquid metals. However, there are some important differences from sodium-cooled receivers which should be pointed out:

1. Due to the aspects (1) and (3), the total mass and the total coolant volume in the receiver is much larger with water than with sodium so that the energy storage is considerably larger in water receivers. Since the response time of heat exchanging systems is roughly proportional to the time constant

$$T_g = \frac{C_g}{wc} \tag{13}$$

   where C_g is the global heat capacity, w the fluid flow rate, and c the average specific heat capacity of the coolant, then water-cooled receivers are generally much slower than sodium-cooled receivers. This fact makes quite difficult and slow the stabilization of receiver temperatures in occasion of disturbances (e.g., cloud passages) and at plant start-up.

2. In SHZ the fluid has low density and high speed so that transport times are not so important, while the dynamics is dominated by the thermal inertia of the receiver metals which can be quite large in SHZ, where low fluxes and large exchanging surfaces must be used.

3. The three different zones mentioned above are not spatially fixed, since the beginning and the end of the evaporation (e.g., in a once-through boiler) takes the position implied by the energy balance and by the pressure value. Thus solar disturbances and/or pressure variations cause even larger motions of EVZ, which can be very difficult to control and give rise to large variations of the steam temperature in SHZ. Interaction among heat flux, temperature in CLZ, steam production, pressure, and steam temperature in SHZ can be very strong and

complex without a tight control of the position of EVZ (and consequently of the pressure). Moreover, since the heat flux is designed according to the nominal position of EVZ (where heat exchange coefficients are high), dynamic motions of EVZ should be carefully evaluated also during the receiver design.

The complexity associated with the (distributed) phenomena occurring in EVZ prevents from applying a simple transfer function approach to the analysis of water-cooled solar receivers, for which only detailed simulation (Ref.4) does represent a reliable approach. Transfer functions between input and output variable for control design can be derived by fitting the step responses of a detailed physical model (Ref.5). Some of the most unpleasant dynamical properties of such receivers can be significantly improved in the design phase by suitably shaping the flux map on the receiver and by reducing the total receiver length and mass (e.g., adopting a design with as many parallel tubes as possible).

The other case of a receiver cooled by gas has not been studied in detail by the author. However, some guidelines to understand their dynamics can be derived by the behavior of superheated steam zones in water-cooled receivers. The first remark is that the aspects (1) and (3) pointed out for the latter ones even more apply to gas-cooled receivers. This implies that the global dynamics is likely to be very slow; however, since the fluid speed is generally much higher, such processes will not be dominated by heat transport phenomena but by storage phenomena, that is by the storage of energy taking place in the relatively large mass of the receiver tubes. Since the heat exchange coefficients are rather low, it can be shown that the two characteristic time constants are approximately given by:

$$T_a = \frac{C_m}{wc} \tag{14}$$

$$T_b = \frac{C_m}{\gamma_t S_g} \tag{15}$$

where C_m is the total metal heat capacity, γ_t the average heat exchange coefficient, and S_g the global heat exchange surface. These two time constants do not affect the system dynamics as pure time delays (like in transport phenomena) but approximately as time lags (like in lumped storage systems). However, due to the large values generally taken on by T_a and T_b, the dynamics turns out to be quite slow and therefore fast temperature control is generally not simple to realize.

To conclude this paragraph, it may be observed that the variety of physical solutions and the different importance of the basic dynamic phenomena in the three considered classes of high temperature solar receivers used in CRS's, generally require a special dynamic analysis for any of those classes, for which the above remarks aim at giving only basic guidelines.

4. CONTROL PRINCIPLES FOR SODIUM RECEIVERS

4.1 Introduction

The illustration of control principles for sodium-cooled receivers (easily extendable to the case of liquid metals and molten salt) can conveniently be carried out with reference to the simplest receiver structure which still keeps the basic properties of real receivers. To this aim let us consider an ideal receiver consisting of two panels which could a priori be connected either in series or in parellel or fed by independent feed pumps (Fig.7.1-7).

It is worth to examine the advantages and disadvantages of the three flow control schemes. They can be summarized as follows.

Scheme PC: The coolant flow rates feeding the two panels cannot be arbitrary varied since only their sum is under control of the feed-pump. On the contrary, the ratio w_1/w_2 cannot be varied during operation since it only depends on the friction losses of the two panels fixed in the design phase. This type of flow control is obviously not acceptable because it is unable to meet the unavoidable variations of the thermal power absorbed by the panels and of their ratio.

Scheme IC: The drawbacks presented by scheme PC are naturally removed by adopting the scheme IC, where the independent control of w_1 and w_2 allows the actual power distribution between the panels to be met. The obvious disadvantage of this drastic solution is that two pumps and two separate flow control systems are required. This generally implies greater cost and reduced reliability. It is, however, feasible and satisfactory for control.

Scheme SC: It does not exhibit the basic disadvantage of scheme PC in that the possible variation of the power distribution between the two panels does not affect the required (common) flow through the panels. In fact, variations of the power distribution only affect the way in which the total temperature increase is split between the two panels; what can be considered of minor importance for receiver operation. However, there is a more subtle drawback in scheme SC due to dynamic reasons: since only the final temperature can be regulated by the unique flow rate, control should in principle rely on the outlet temperature measurement which in SC exhibits a more delayed response than in the other two cases (see Fig.7.1-2), essentially because of the "delayed" response component due to the first panel. Nevertheless, scheme SC appears to be attractive because of its simplicity and its conceptual feasibility.

The consideration of the three basic process schemes allows the following conclusion to be drawn: scheme SC would be preferable, or at least acceptable, if it is possible to devise a control system yielding approximately the same dynamic performance as scheme IC.

In solar receivers, control performance can be naturally evaluated by the resulting reduction of the effect of solar disturbances (e.g., cloud passing) on receiver outlet temperature and possibly on intermediate panel temperature. In most rigorous terms, the following performance index can be used:

$$J_p = \frac{(\Delta T/\Delta \beta)_{cl}}{(\Delta T/\Delta \beta)_{ol}} \qquad (16)$$

where $(\Delta T/\Delta \beta)_{cl}$ is the frequency response function between the solar disturbance $\Delta \beta$ and the relevant temperature when the control is applied, while $(\Delta T/\Delta \beta)_{ol}$ is the same function when no control is working (open loop). Actually J_p is a function of frequency and gives the rate of reduction of disturbance components for any frequency. Referring to Fig.7.1-7, the most significant temperature to be used in Equation (16) is the receiver outlet temperature T; however, also panel outlet temperatures T_1 and T_2 are often relevant because of possible local overheating.

Then we are going to evaluate performances of control concepts; first in linearized form and without any feedforward action based on possible direct measurements of the radiation intensity, since this can possibly be applied in addition to the basic control system assumed to be based on temperature feedbacks.

The two feasible process schemes IC and SC are now separately considered.

4.2 Temperature Control in Scheme IC

In this case the only applicable temperature control concept is that of providing two separate and identical control loops, where each feed pump controls the corresponding panel outlet temperature at the set-point value prescribed for the coolant temperature at the receiver outlet. One of these loops is described in Fig.7.1-8, where for the sake of simplicity, the pump flow control loop has been neglected because it is sufficiently fast.

The problem is to define the performance achievable for the scheme of Fig.7.1-8 Assuming that the temperature sensor is sufficiently fast, it is obvious that the achievable control performance is determined by the transfer functions between the control variable w_1 and the controlled variable T_1, and by the transfer function between the disturbances ß and T_1 itself.

In view of the block diagram of Fig.7.1-3, in this case the simple scheme of Fig.7.1-9 applies.

Denoting with R_1 the transfer function of the regulator, it is easy to show that the performance index (16) is given by:

$$J_p = \frac{1}{1+R_1F_1} \quad ; \tag{17}$$

thus the disturbance rejection capability is determined by the "return difference" $(1+R_1F_1)$: the larger the amplitude of the return difference over the frequency band of the disturbance, the higher the feedback disturbance compensation or rejection.

The achievable performance of the simple control loop of Fig.7.1-8 depends on how large the band Ω is on which the amplitude of R_1F_1 can be made much greater than 1, i.e., the maximum possible Ω for which we obtain

$$\left| R_1(j\omega)F_1(j\omega) \right| \gg 1 \quad , \; 0 \leqslant \omega \leqslant \Omega \tag{18}$$

It is well known that the maximum Ω primarily depends on the phase displacement introduced by $F_1(j\omega)$, because at those frequencies where $F_1(j\omega)$ exhibits large phase displacements increasing the loop gain via the regulation leads to poorly damped oscillations and to instability.

Moreover, phase lags in a process transfer function can conveniently be classified according to the feasibility of their compensation by the regulator:

(A) phase lags which can be compensated by derivative-like actions within the regulator

(B) phase lags which cannot be compensated in any possible way by the regulator

It is well known (Ref.7) that process transfer functions which do not incorporate phase lags of type (B) are the so-called "minimum phase" transfer functions (MPTF), the other ones are called non-MPTF. However, any non-MPTF P(s) can always be factored as:

$$P(s) = P_m(s)P_{nm}(s) \tag{19}$$

where $P_m(s)$ is a MPTF and $P_{nm}(s)$ is an all-pass function, i.e., a function which does not exhibit amplitude drop but only phase lags. It is just the phase lag of $P_{nm}(S)$, also called the non-minimum phase (NMP) of P(s), which cannot be compensated.

It can be shown that without any additional constraint on the regulator design but its feasibility, the achievable band Ω for the control loop cannot be larger than the value of ω for which the NMP of the process transfer function is greater than about 60°.

On the other hand, it can also be shown that the panel transfer function F_1 (Equation (6)) is a MPTF, so that no stability limitations are theoretically encountered in designing the simple control loop of Figure 8, whose theoretical performance (Equation (17)) can be made as large as necessary. This is because the single panel transfer function F_1 does not incorporate delayed components (Fig.7.1-2/3) which typically arise in cascading panels. Time-delay systems are actually classical non-minimum phase systems.

However, it should be mentioned that, in practice, there are a number of additional limitations (control effort, parasitic dynamics, complexity of the regulator, robustness of the control to parameter variations) that prevent the achievement of arbitrary large performance. It is nevertheless worth noting that limits to performance due to NMP are more severe as they are not removable in any way.

4.3 Temperature Control in Scheme SC

There are two basic alternative control structures for scheme SC, the one based on a single loop with the outlet temperature feedback only (OTF), the other exploiting the two available temperature measurements T_1 and T_2 (MTF). For the second one, further possibilities arise from the way in which temperature feedbacks are used.

The structure of OTF is trivial and formally identical to that of Fig.7.1-8, provided that temperature T_1 is replaced by T. There is, however, a substantial difference in the dynamic relations between the relevant variables since the transfer function scheme is that of Fig.7.1-10 where connection elements are neglected for the sake of simplicity.

In view of Equation (6), F_1 and F_2 differ only for the gain so that the scheme of Figure 10 can be drawn as in Fig.7.1-11 where

$$P(s) \triangleq F_1/\lambda_1 = F_2/\lambda_2 \tag{20}$$

$$D(s) \triangleq \exp(-B(s)) \tag{21}$$

It is easy to realize that the performance index J_p can still be expressed as

$$J_p = \frac{1}{1+R(s)F_t(s)} \tag{22}$$

where R(s) is the regulator transfer function and $F_t(s)$ is the total transfer function between $-\Delta w_1$ and ΔT, i.e.

$$F_t(s) = P(s)\,\lambda_1(D(s)+\rho) \tag{23}$$

with $\rho = \lambda_2/\lambda_1$.

It is worth noting that P(s) has the same form as $F_1(s)$ in scheme IC, so that what makes the difference in the process transfer function is the factor $(D(s)+\rho)$, where the second term accounts for the prompt variation and the first one for the delayed variation through the typical delay term $D(s) = \exp(-B(s)) \cong \exp(-s\tau')$. Furthermore it can be easily proven that taking the approximation $D(s) \cong \exp(-s\tau')$, the term $(\rho + D(s))$ is a MPTF if and only if $\rho \geqslant 1$, that is if the delayed component of the response is "smaller" than the prompt component. In other terms, if the power absorbed by the second panel is greater than the power absorbed by the first one ($\lambda_2 \geqslant \lambda_1$).

In the case of $\rho < 1$, $(\rho + \exp(-s\tau'))$ can be factored as:

$$\left(\rho + e^{-s\tau'}\right) = \left(1 + \rho e^{-s\tau'}\right)\left(\frac{\rho + e^{-s\tau'}}{1+\rho e^{-s\tau'}}\right) \tag{24}$$

where the first term is a MPTF and the second is an all-pass function (see Equation (19)).

It could be interesting to note that the NMP of Equation (24) attains 90° when

$$\omega\tau' = \arccos(-2\rho/(1+\rho^2)). \tag{25}$$

Since 90° of NMP essentially entails system instability, we can claim that the OTF control for a two panel receiver cannot have a disturbance rejection band larger than that given by Equation (25), if the radiation power on the outlet panel can, in any operating condition, be lower than the power on the inlet panel. If headers and connection tubes are considered, then the situation would become even worse. The effect of the delayed temperature response components due to the upstream panels is therefore the main reason which prevents single OTF control systems to be effectively applied to cascaded-panel receivers.

In most cases the more complex solution MTF (Multiple Temperature Feedback) must be applied.

When considering MTF designs, the non-trivial problem arises of how to use and combine temperature feedbacks.

The classical approach to the control of cascade-like processes is that of using any intermediate variable feedback to minimize the effect of the external disturbance on the process section whose output is the considered variable. Since the control input is the coolant mass flow rate only, this concept leads to the multi-loop system (MLS) of Fig.7.1-12.

The design of the two-loops system of Fig.7.1-12 is usually carried out in two steps:

1. First, the feedback compensation (FC) using temperature T_1 is synthesized so as to reduce the effect of the disturbance on temperature T_1.

2. Second, the final temperature regulator (FTR) is synthesized so as to minimize the residual effect of the disturbance on the final temperature T_l assuming that the FC is already present.

It is not difficult to realize that the MLS concept does not work for the considered solar receiver. To show this fact, it is useful to redraw the control scheme on the process transfer-function diagram of Fig.7.1-11. This is made in Fig.7.1-13 where ΔT_0 is neglected as irrevelant to the control design.

At step (1), attenuation of disturbance effects on ΔT_1 requires that C(s) be chosen so as to make the return difference $(1+\lambda_1 P(s)C(s))$ sufficiently large over a wide frequency band. This can be done in the same way as for the control of scheme IC, since the process transfer function $\lambda_1 P(s)$ is an NMP function. Then, at step (2), the modified process transfer-function $\Delta T/\Delta X$ on which the synthesis of R(s) must be based becomes:

$$\frac{\Delta T}{\Delta X} = \frac{-\lambda_1 P(s)}{1+\lambda_1 P(s)C(s)} \cdot (\rho + D(s)) \tag{26}$$

where it appears that, whatever C(s) is chosen, the considered transfer-function will incorporate the possibly non-minimum phase function $(\rho + D(s))$. This fact implies that the receiver outlet temperature control loop cannot realize a tight temperature control over a sufficiently wide band. Actually, the principal cause of performance limitation in the simple OTF has not been removed by the addition of the inner temperature loop. On the other hand, increasing the tightness of the inner temperature loop practically implies that the coolant mass flow rate is controlled so as to regulate temperature T_1, while the final temperature T follows.

This solution is clearly not acceptable because, due to the usual variations of the ratio $\lambda_2/\lambda_1 \triangleq \rho$, keeping temperature T_1 at a constant value does not guarantee that also the final temperature T is automatically controlled. We can therefore conclude that MLS is not a good solution for receiver control, in that it does not tackle the most critical problem, which is that of reducing or removing NMP within the principal control loop.

To overcome drawbacks of traditional multi-loop systems unable to modify the phase characteristics of multi-panel receivers, a different concept has to be adopted for MTF design, principally aiming at removing NMP due to the cascade process structure.

The concept, called predictive feedback temperature control (PFTC) is easily illustrated with reference to Fig.7.1-10.

Observe that, if the temperature variation ΔT_1 is "filtered" through a transfer function given by

$$H_2(s) = 1 - D_2(s) \tag{27}$$

and added to the temperature variation ΔT, we obtain a combined temperature signal $\Delta\hat{T}$, given by

$$\Delta\hat{T} = \Delta T + (1 - D_2)\Delta T_1 \quad (28)$$

in which the delayed component due to $D_2 \Delta T_1$ is dropped out. In fact, it results

$$\Delta\hat{T} = \Delta T_1 + \Delta DT_{p2} = (F_1 + F_2)\left(-\frac{\Delta w_1}{\overline{w}_1} + \frac{\Delta \beta/\overline{\beta}}{1 + s\tau_m}\right) + D_1 \Delta T_o \quad (29)$$

where it appears that the effect of Δw_1 on $\Delta\hat{T}$ is through the "prompt" components only, that is $F_1 + F_2$. Moreover, it should be stressed that the transfer function $\Delta\hat{T}/\Delta w_1$, is given by

$$\Delta\hat{T}/\Delta w_1 = -(F_1(s) + F_2(s))/\overline{w}_1 = -P(s)/\overline{w}_1 \quad (30)$$

which coincides, apart from the gain, with the transfer function of a single panel, which has been shown to be MPTF.

Then, if the temperature control problem could be stated as that of regulating $\Delta\hat{T}$ instead of ΔT, it would be possible to design a temperature control loop with the same performance as that of scheme IC, while using only one control pump. Therefore, it is worth analyzing the performance of the control system of Fig.7.1-14, where the temperature combination $\Delta\hat{T}$ is the regulated variable.

First, we observe that at the steady state, $\Delta\hat{T}$ coincides with ΔT since $(1-D_2(s))$ has zero gain at the steady state. Thus, as for the static performance, regulating $\Delta\hat{T}$ is the same as regulating ΔT. Moreover, since $\Delta\hat{T}/\Delta w_1$ is given by Equation (30), the transfer function R(s) can be chosen so as to yield the same return difference as that of scheme IC. That is:

$$R(s)P(s) = R_1 F_1 \quad (31)$$

To evaluate performance, it is convenient to compute the Equation (16), both for the outlet temperature ΔT and for the intermediate temperature ΔT_1 one obtains:

$$\frac{(\Delta T/\Delta\beta)_{cl}}{(\Delta T/\Delta\beta)_{ol}} = \frac{1}{1+R(s)P(s)} \quad (32)$$

$$\frac{(\Delta T_1/\Delta\beta)_{cl}}{(\Delta T_1/\Delta\beta)_{ol}} = \frac{1}{1+R(s)P(s)} \quad (33)$$

Therefore, in view of Equation (31), the same disturbance rejection ratio is obtained for both panel temperatures as in the case of scheme IC, that is both panel temperatures can be controlled with the same dynamic accuracy as in the case of independent flow control of each panel. The drawback of the cascade structure of the receiver with the consequent NMP effects is fully overcome by the predictive feedback temperature control system, whose name has been chosen to mean that the feedback of the intermediate panel temperatures must incorporate a predictor (i.e., $D_2(s)$) of the effect of the temperature variations on the subsequent panel.

The concept of PFTC can also be given the following more physical interpretation: to emulate performance of independent control (IC) of panel coolant flow, the temperature control of the different panels should be made independent of each other, in spite of the temperature interaction due to the cascade structure of the receiver. To achieve this goal, it can be observed that the independent contributions to the receiver outlet temperature are ΔT_1 and ΔDT_{p2} (or, in the general case of Fig.7.1-3, ΔT_1, ΔDT_{p2}, $\Delta DT_{p3}, \ldots,$ ΔDT_{pn}), as there is not any effect of ΔT_1 on ΔDT_{p2}. The regulator R(s) of Fig.7.1-14 will in fact command flow rate variations proportional to the sum of ΔT_1 and ΔDT_{p2}, which are proportional to the power absorbed by panels 1 and 2, respectively, through the same filtering function P(s). Thus, the same flow rate will be obtained as though the receiver consisted of a single panel absorbing the whole thermal power, independently of the way in which the radiation power is actually split between the real receiver panels. Therefore, the control operates as though the quantity to be regulated were $\Delta T_1 + \Delta DT_{p2}$, instead of ΔT: the main advantage is that $\Delta T_1 + \Delta DT_{p2}$ reveals power flow unbalances as they affect the different panels without any influence of the actual panel connection and of the consequent delayed temperature components.

The ability of PFTC to promptly detect (and compensate) possible unbalances between radiation power and coolant flow is of primary importance for temperature control performance. This aspect is given in particular evidence by representing the process behavior as in Fig.7.1-15, where ΔT_0 is neglected as irrevelant to the analysis.

It is apparent from Fig.7.1-15 that for any power distribution (λ_1, λ_2), the temperature increment on each panel does not change if at any time the following balance equation is verified:

$$\frac{\Delta W_1}{\bar{W}_1} = \frac{\Delta \beta / \bar{\beta}}{1 + s\tau_m} \tag{34}$$

Therefore, the difference signal Δ p can be assumed as power unbalance index for the receiver. The power unbalance affects "in parallel" the temperature increments DT_{p1} and DT_{p2} through the delay-free transfer function P(s), weighted with the percentage of the radiation power absorbed by the corresponding panel. In fact, we might estimate the power unbalance by looking at anyone of the variable DT_{p1} or DT_{p2}, if parameters λ_1 and λ_2 and function P(s) were exactly known. Actually, in view of Equations (6) and (20), it can be observed that the gain of P(s) given by $\bar{\beta}Q_t/\overline{wc}$ coincides with the total temperature difference on the receiver at the steady-state, which is a priori known as a design parameter. Also the dynamics of P(s) is subject to little uncertainty because it only depends on B(s), computable from the receiver panel geometry and the coolant flowrate. Only the power distribution parameters λ_1 and λ_2 are quite uncertain and variable with time (sun position). This means that no reliable temperature control can be done by looking only at a single panel. However, if the sum of DT_{p1} and DT_{p2} is used as power unbalance detector as is done in PFTC, this major uncertainty is removed; in fact, by definition $\lambda_1 + \lambda_2 = 1$ and the considered sum is totally independent of the actual power distribution.

Therefore we can conclude that:

- PFTC is able to yield very fast temperature control because its control loop does not include the delay-like effects due to the cascading of panels.

- PFTC is a robust control scheme, as its control loop is not essentially affected by possible variations of the power distribution among the panels.

- Through PFTC the temperature control for the cascaded scheme SC is essentially equivalent to that of scheme IC, though only one control pump is necessary.

Finally, it is quite obvious that the control concept discussed above can easily be extended to the case of many cascaded panels.

5. PRACTICAL CONTROL SCHEMES FOR SODIUM RECEIVERS

Practical realizations of PFTC concept can easily be designed, bearing in mind that the principal feedback signal must be the sum of the outlet temperature T_{o1} of the first panel and of the temperature increments DT_{pi}, i = 2, ..., n due to the power absorbed by the subsequent n-1 panels (Fig.7.1-3). In the general case, where also headers and connection pipes are present, the variations of the temperature increments DT_{pi} can be calculated as follows (Fig.7.1-6):

$$\Delta DT_{pi} = \Delta T_{oi} - D_i \Delta T_{yi} \tag{35}$$

That is the difference between the panel outlet temperature ΔT_{oi} and the expected effect of the panel inlet temperature ΔT_{yi} on the outlet temperature itself (computed by the predictor D_i). Since D_i is the transfer function between panel inlet and outlet temperature variations, it can be simply modelled by a model of the panel (or an equivalent tube) without any absorbed power: this is actually the nonlinear version of the predictor D_i. Furthermore, if account is taken of the main process nonlinearity (Paragraph 2, Remark 4), also the regulator R(s) of Fig.7.1-14 must be given an adaptive form, letting the regulator parameters be varied according to the measured coolant mass flowrate.

A possible scheme using PFTC is therefore that of Fig.7.1-16, where M_i represents the nonlinear model of panel i with no radiation (i.e., the dynamic relation between inlet temperature and outlet temperature with no radiation) and w_d is the demand of coolant flowrate w to the flowrate loop.

It is worth noticing that both models M_i and the regulator R are affected by the measurement of w, to take into account the dependence of M_i by the coolant flowrate and the adaptation of the regulator R with the receiver load.

Obvious simplifications of the scheme of Fig.7.1-16 are applicable in the following practical cases:

(1) When all the panels are geometrically equal, all the models M_i, i=1,...,n, coincide and the signal $\hat{T}$ can be simply computed as in Fig.7.1-17, where M denotes the common model of any panel.

(2) When the receiver does not have headers and connection pipes, that is when the receiver consists of many parallel tubes with intermediate temperature measurement points and no panel splitting, the signal $\hat{T}$ can be computed as in Fig.7.1-18 where M_i models the part of parallel tubes between the temperature measurements.

Of course, if the temperature measurements are located at points equispaced along the receiver, also in this case only one "panel model" M is used, by feeding it with the sum of $T_{o1}, T_{o2}, \ldots T_{o\ n-1}$.

The complete control scheme also can incorporate additional control actions to improve actual performance:

- An outer slow control loop allowing accurate trimming of receiver outlet temperature at the steady-state, to compensate possibly for thermal losses from headers and connection tubes which make the sum of temperature increments different from the total temperature increase across the receiver.

- A feedforward action based on an estimation of the incoming solar radiation to promptly (though roughly) compensate for fast cloud passages.

Since PFTC allows tight control of coolant temperatures at the points where temperature feedbacks are applied, practical PFTC implementations should also be based on the following remarks. The most critical temperatures are generally immediately downstream of receiver sections with high flux densities; therefore, temperature sensors should be placed at those points if practically possible. Placing the temperature sensors at uniformly equispaced points (or at the borders of identical panels) only one tube model is required for the PFTC realization: this greatly increases control simplicity.

6. CONCLUSIONS

As appears from the two preceding paragraphs, the concept of the Predictive Feedback Temperature Control can be applied as it is to any receiver structure using a nonboiling liquid as a coolant, i.e., liquid sodium or molten salt. PFTC may also be applied to any case of one-phase coolant though real benefits have not been analyzed for gas-cooled receivers, where storage phenomena are more important than transport delays, as briefly discussed in Paragraph 3.

The most complex case appears to be that of water-cooled receivers because of the presence of the coolant phase-transition where temperature measurements are not of any significance, while release of energy in that zone is very important and requires accurate control. Without the use of special nonconventional sensors able to yield reliable enthalpy estimates, the PFTC concept can only be used to stabilize the compressed water zone and possibly to obtain a global control of the evaporation zone. However, prediction of the evaporation zone behavior requires quite complex models so that fitting PFTC concept to water-cooled receivers is felt to require further investigation, though approximate PFTC realizations were proved to have acceptable performance in real cases (i.e. the Eurelios Project(Ref. 5)).

7. REFERENCES

1. C. Maffezzoni, G. Magnani, and S. Quatela, "Integrated Process and Controller Design of High Temperature Solar Receivers", Proceedings of the 21st IEEE Control and Decision Conference (CDC), December 1982.

2. C. Maffezzoni, "The Concept of the Integrated Design of Process, Control, and Operation in Solar Central Receiver Plants", 2nd International Workshop on the Design, Construction, and Operation of Solar Central Receiver Projects, Varese, Italy, June 1984.

3. C. Maffezzoni, G. Magnani, and S. Quatela, "Process and Controller Design of High Temperature Solar Receivers", to appear in IEEE Trans. on Automatic Control, March 1985.

4. C. Maffezzoni and F. Parigi, "Dynamic Analysis and Control of a Solar Power Plant, Part I: Dynamic Analysis and Operation Criteria", Solar Energy Vol. 28, No. 2, pp. 105-116, 1982.

5. C. Maffezzoni and F. Parigi, "Dynamic Analysis and Control of a Solar Power Plant, Part II: Control System Design and Simulation", Solar Energy Vol. 28, No. 2, pp. 117-128, 1982.

6. L.A. Gould,"Chemical Process Control: Theory and Applications", Reading, MA: Addison-Wesley, April 1978.

7. I.M. Horowitz, "Synthesis of Feedback Systems", Academic Press, 1963.

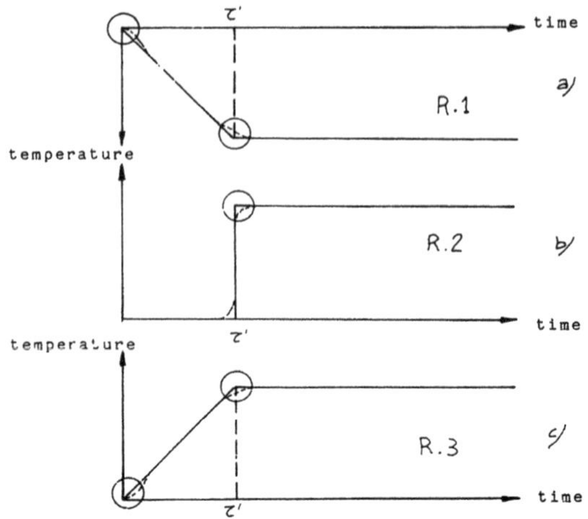

Fig. 7.1.-1: Panel Outlet Temperature Response to Steps of Variables

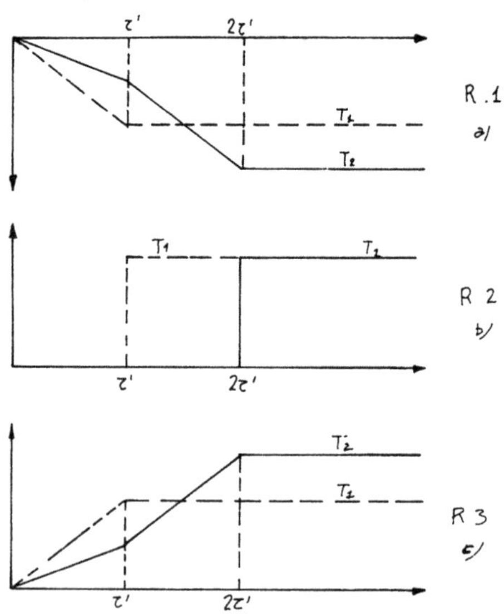

Fig. 7.1.-2: Two-panel Receiver Responses

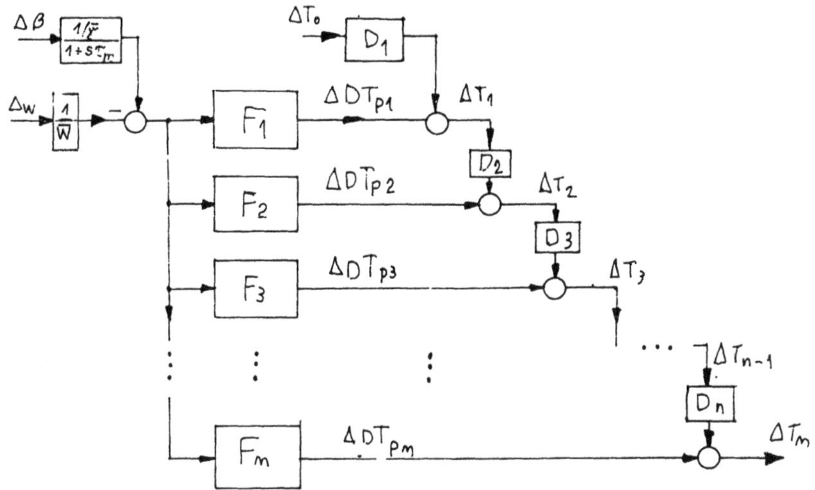

Fig. 7.1.-3: Transfer-function Block Diagram of a Wall-panel Receiver

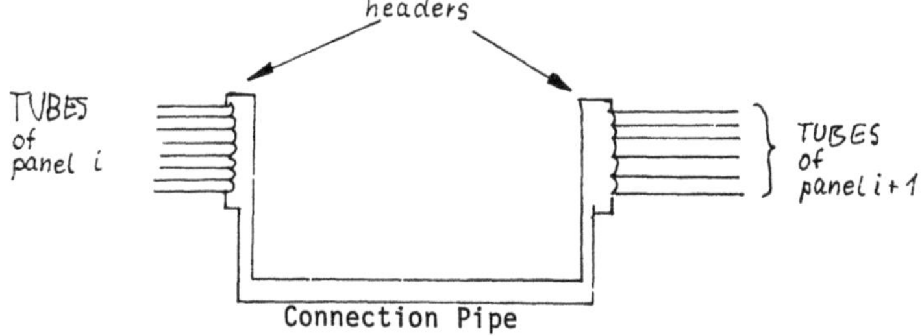

Fig. 7.1.-4: Typical Series Connection of Panels

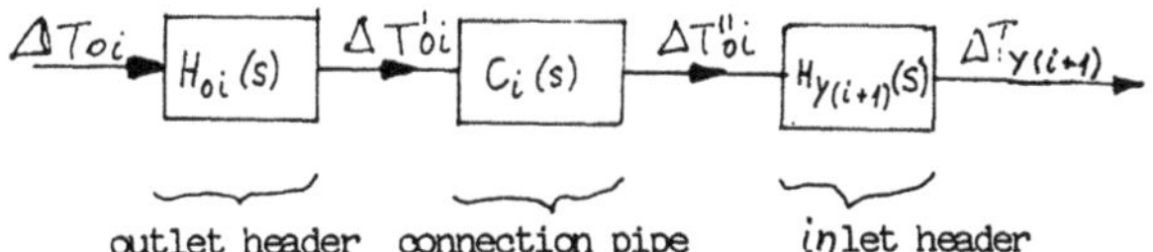

Fig. 7.1.-5: Transfer-function of the Connection Elements between Two Panels

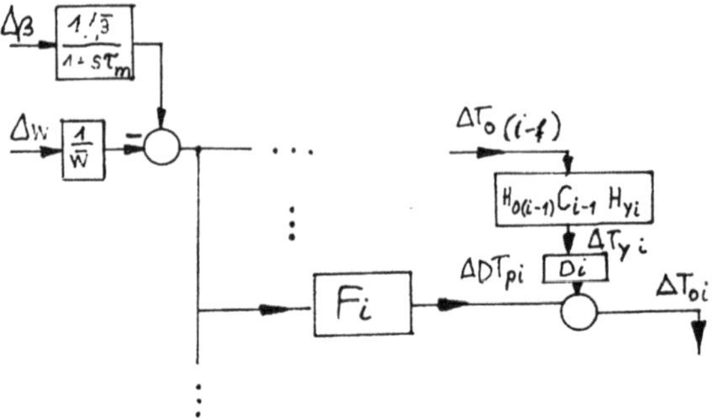

Fig. 7.1.-6: Block Diagram of Headers and Connection Tubes

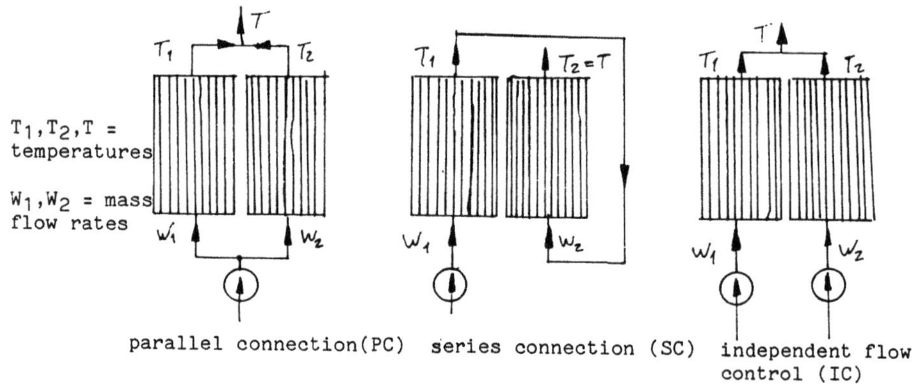

Fig. 7.1.-7: Possible Flow Control of a Two-panel Receiver

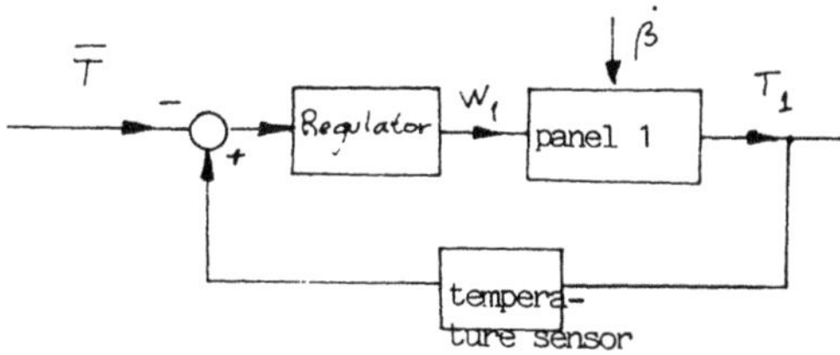

Fig. 7.1.-8: Panel Temperature Control Loop for IC

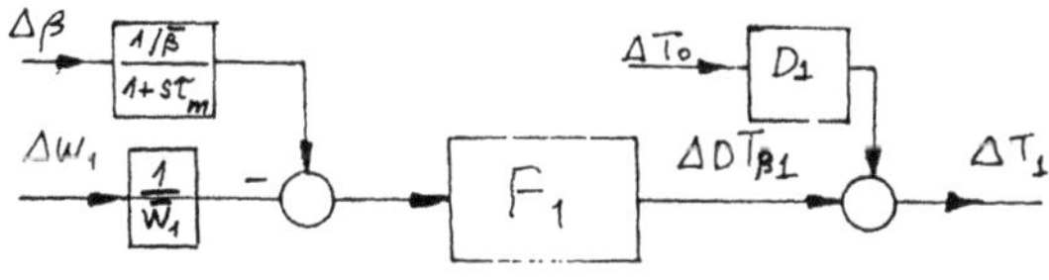

Fig. 7.1.-9: Transfer Function Scheme for One Panel

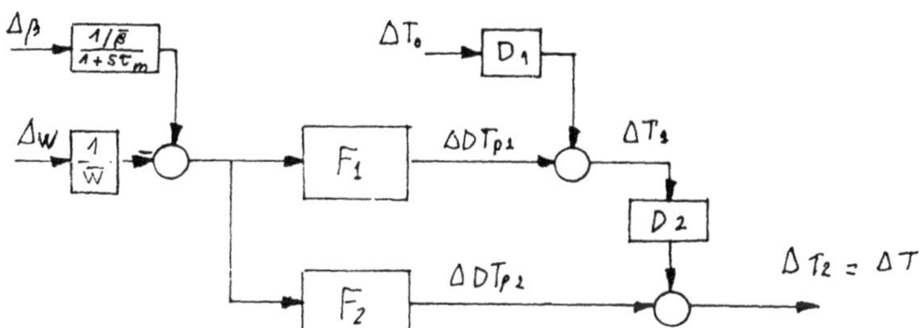

Fig. 7.1.-10: Transfer Function Scheme for SC

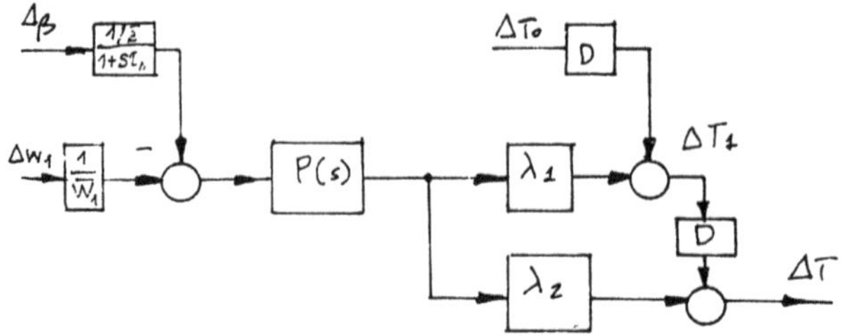

Fig. 7.1.-11: Rearrangement of Scheme for SC

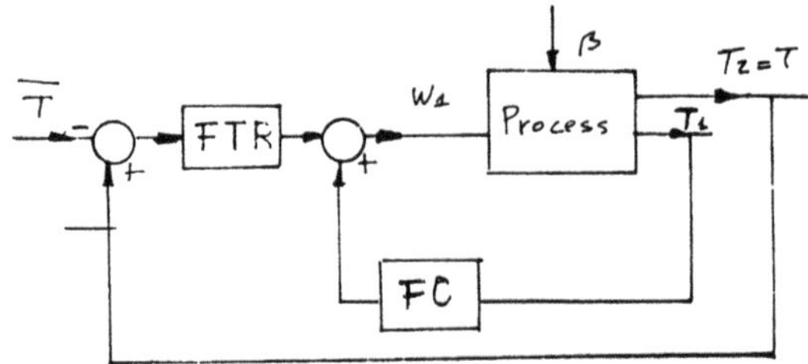

Fig. 7.1.-12: Two-loop Control for SC Scheme

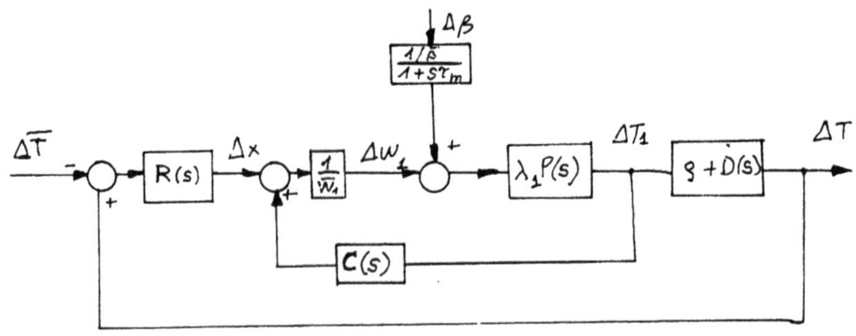

Fig. 7.1.-13: Two-loop Temperature Control System for SC

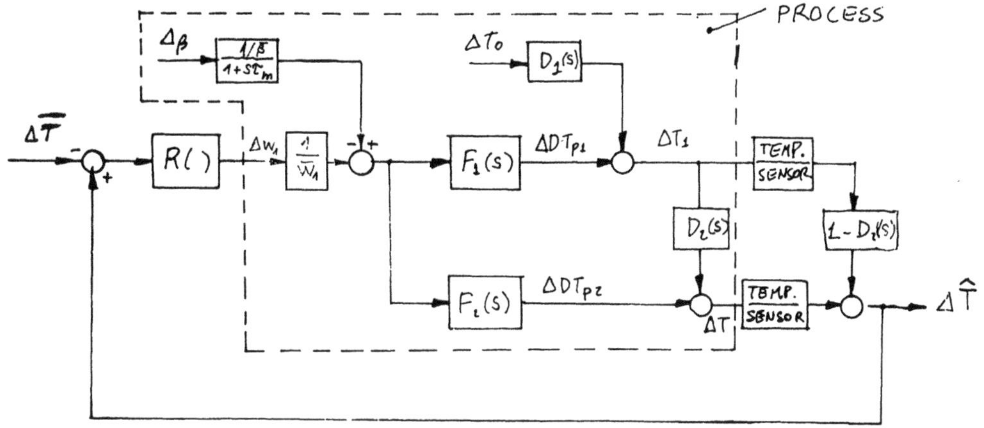

Fig. 7.1.-14: PFTC for Scheme SC

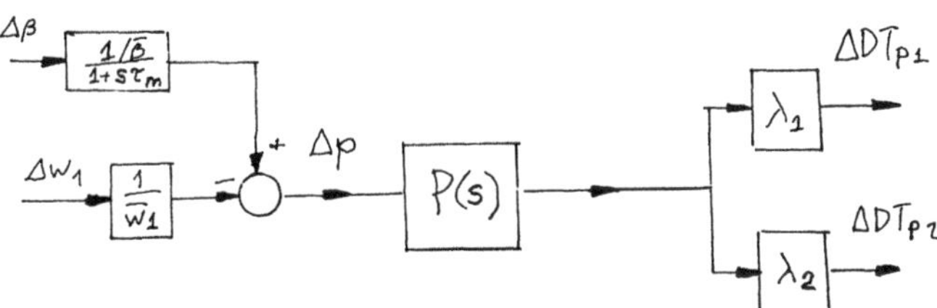

Fig. 7.1.-15: Effects of Thermal Unbalances

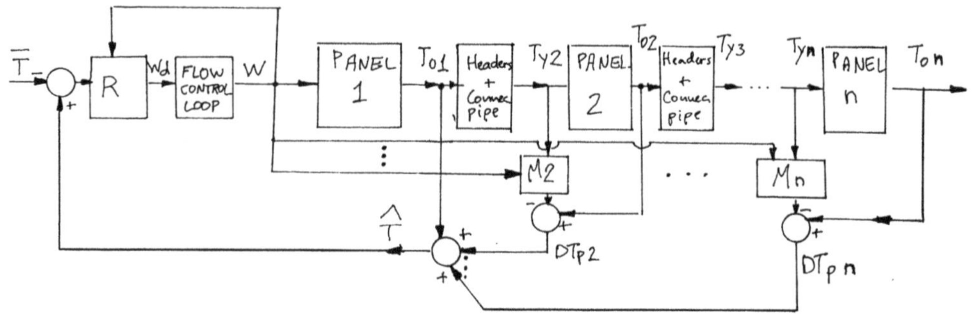

Fig. 7.1.-16: PFTC Realization Scheme

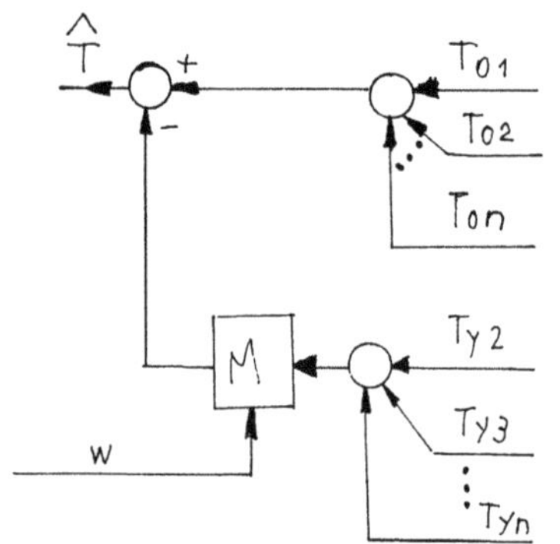

Fig. 7.1.-17: PFTC in the Case of Identical Panels

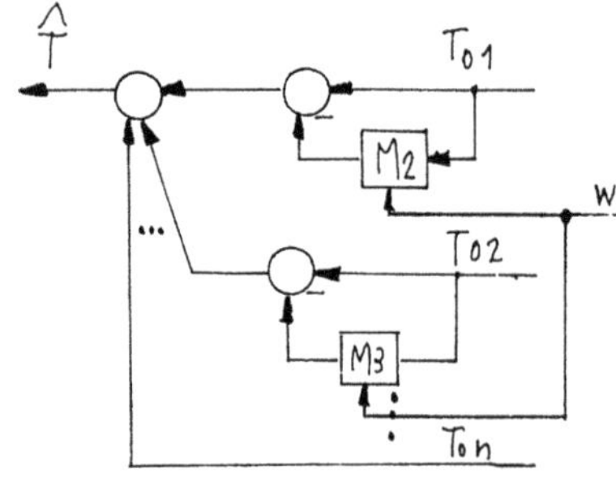

Fig. 7.1.-18: PFTC in the Case of no Panel Splitting

SULZER FEEDBACK CONTROL CONCEPT

H.D. Weyers, INTERATOM

1. INTRODUCTION

In 1979 INTERATOM was assigned the task of designing a feedback control circuit for the sodium-cooled solar receiver. The design parameters were:

- outlet temperature constant at 530°C
- load changes from 0-100% in 15 sec (cloud passage)
- protection of tube bundle in case of overheating

The resulting feedback control system for the receiver was simulated on a computer to study its operation before it was actually installed and tested.

2. DESCRIPTION OF THE DESIGN

The feedback control circuit for the receiver maintains the outlet temperature of the receiver at a constant value. In the event that normal operation is disturbed, the feedback control circuit counteracts the disturbance variables and protects the receiver. There are several input variables for the control circuit:

- Outlet temperature LK01CT25, which constitutes the control variable.

- Bundle temperatures LK01CT06-CT08, which are the auxiliary command variables. In case of local superheating at the tube bundle, these variables will increase the flowrate independently of the outlet temperature.

- Flow measurement LK01CF01, which serves as the control variable of the subordinated feedback control circuit to ensure the minimum flowrate of 10%.

- Sun presence sensor LK01CU01, which serves as the disturbance variable to detect and control rapid changes of the insolation in the field.

- Sodium pump LK01AP01, which is a manipulated variable controlled by means of the feedback control circuit.

Fig.7.2-1 presents a detailed block diagram of the receiver feedback control. The location of the thermocouples that are used to control the tube bundle is shown in Fig.7.2-2.

3. MODIFICATIONS DURING CONSTRUCTION

Since this control circuit was simulated before testing, changes required during the test phase were minimal and limited to:

- temperatures of tube bundles from LK01CT06, LK01CT07, and LK01CT08 to LK01CT07, LK01CT08 and LK01CT09 to produce better indications and quicker responses.

- outlet temperature from LK01CT25 to LK01CT12 (which is also one of the temperatures of the outlet header) to provide quicker response to the control circuit.

- sun presence sensor hardware configuration to simplify the adjustment of influence.

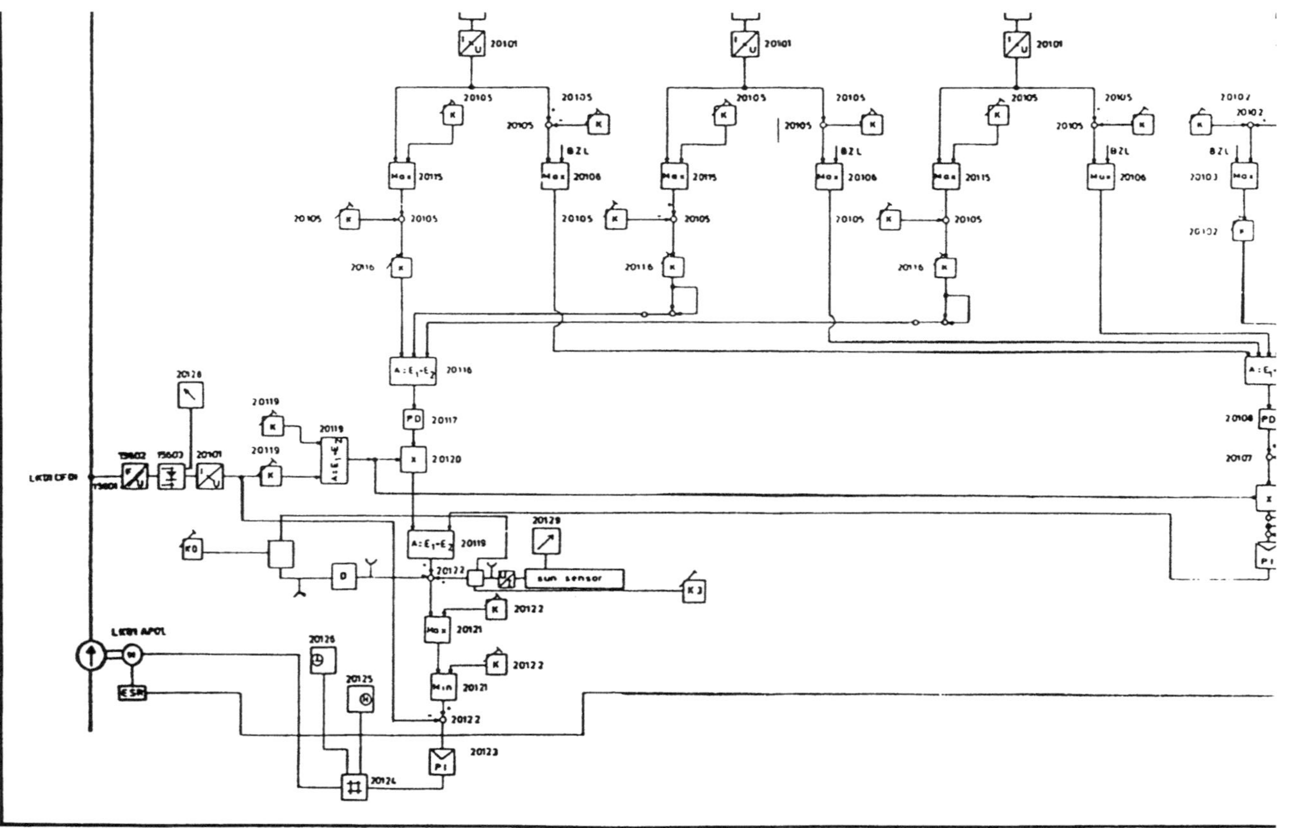

Fig. 7.2.-1: Receiver Feedback Control

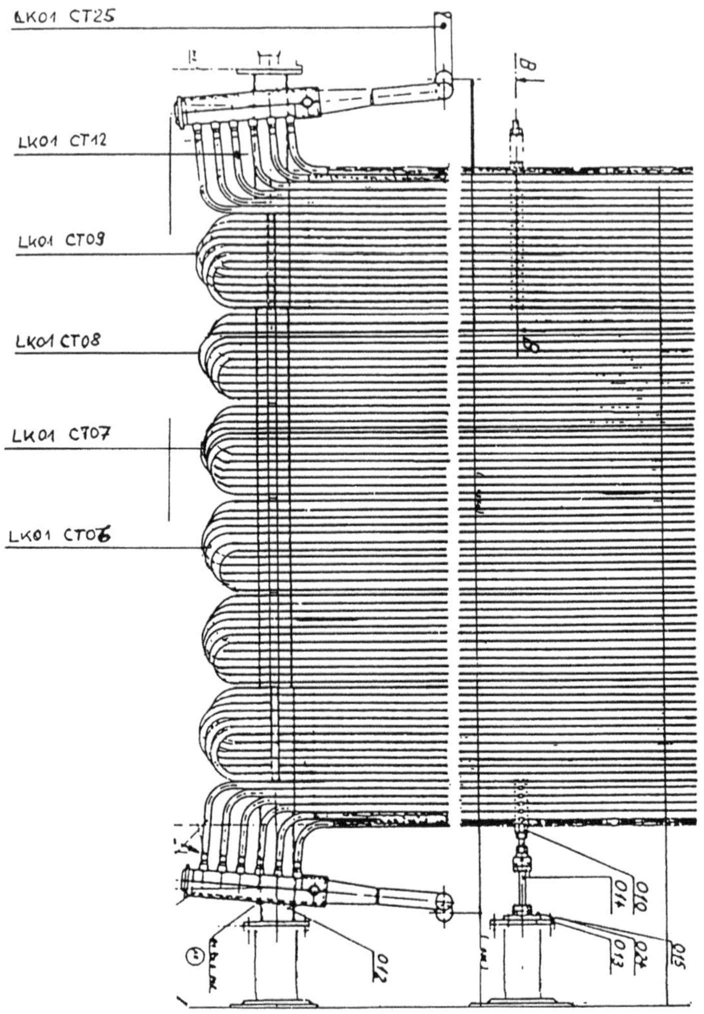

Fig. 7.2.-2: Location of Thermocouples

THE ALMERIA ADVANCED SODIUM RECEIVER: DYNAMIC ANALYSIS AND CONTROL

G.A. Magnani, (°)

1. INTRODUCTION

The Advanced Sodium Receiver (ASR) of the IEA/SSPS Almería plant is an uncommon heat exchanger because of its high thermal fluxes (138 W/cm^2 peak at the design point) and because the absorbed power is not controllable and is variable from its maximum value to zero and viceversa in a few seconds.

In the lifetime analysis of the receiver about 6500 full range transients per year have been considered besides daily morning start-up and evening shutdown. It is clear that, in addition to the steady-state temperature profiles, the dynamic behavior of receiver temperatures plays a crucial role in the useful life of the receiver; therefore transients are an important consideration in the structural design of the receiver.

It appears that both an accurate dynamic simulation and an effective control system are important. The simulation objectives are to identify the most critical transients and evaluate the thermal stresses in the most stressed sections during the receiver design phase. Effective control is required to limit receiver overheating and temperature gradients as much as possible when transients occur in the temperature, such as in the case of a cloud passing over the heliostat field. It should also be emphasized that an adequate knowledge of process dynamics is required in any involved control system design.

This paper deals first with ASR distributed parameter dynamic analysis and simulation, and then it describes the ASR control system based on the

(°) At present, the author of this document is with SdI - Studio di Informatica S.p.A., Milan, Italy.

The work here described was carried out at ENEL - Automation and Computing Research Center, Cologno Monzese (Milan), Italy by S. Quatela and G.A. Magnani.

SdI designed the ASR direct digital controller.

new concept of Predictive Feedback Temperature Control (PFTC), together with the additional practical expedients necessary to cope with full load variations due to cloud passages. Finally the direct digital controller is briefly described, and some experimental results are shown to validate modelling and to evaluate control system performances.

2. ASR MODELLING

2.1 Receiver Structure

A sketch of the ASR is shown in Fig.7.3-1. The absorber area consists of five equal panels, connected in series, and assembled side by side. Each panel is composed of 39 tangent tubes connected to an inlet and an outlet header; connection tubes link the headers of cascaded panels.

The sodium is fed from the cold storage tank (270°C) by a centrifugal pump and flows upwards in all the panels: first in the two lateral ones, then in the two intermediate, and finally through the central one. It leaves the receiver at a temperature of 530°C and flows into hot storage.

The cold and hot storage allow the receiver to operate independently of the electric power generation loop operation. Further details on the ASR structure can be found in Ref.1 and 2 and a detailed description of the IEA/SSPS plant is in Ref.3.

2.2 Basic Modelling Assumptions

The receiver modelling has been approached with the following physical assumptions:

1. The thermal power supplied to the external wall of the receiver tubes is essentially an exogenous variable, so that the thermal power per unit length $q(x,t)$ absorbed by a heating tube at time t and abscissa x is expressed as:

$$q(x,t) = \beta(t)\, Q_t\, \phi(x) - q_l(x,t) \tag{1}$$

   where $\beta(t)$, $0 \leq \beta(t) \leq 1$, represents the radiation intensity on the receiver at time t; Q_t is the total nominal power absorbed by the tube, $\phi(.)$ is the normalized thermal power map on the tube (i.e. with $\int_0^l \phi(x)dx = 1$, l = tube length) and $q_l(x,t)$ accounts for tube convection and reradiation losses.

Since the thermal power map can vary slowly with time, $\emptyset(.)$ is assumed to be fixed during any operating transient. The loss term $q_l(x,t)$ depends on the tube wall temperature according to the formula (Ref.4):

$$q_l(x,t) = h_c\, d_e \pi (T_w(x,t) - T_a) + k_r\, d_e (T_w^4(x,t) - T_a^4)$$

where T_w is the absolute external wall temperature, T_a the absolute ambient temperature, h_c the convection coefficient, k_r the tube radiation constant, and d_e the external tube diameter. Variations of thermal losses can be relevant only for large temperature variations and are neglected for linearized modelling.

2. The fluid (liquid) compressibility is neglected, so that the fluid density is independent of the pressure. Only the dependence of density on temperature is taken into account for large variations.

3. The heat exchange coefficients are computed by Seban-Shimazaki correlation (Ref.5):

$$Nu = 5. + 0.025\, (Re\ Pr)^{0.8} \qquad (2)$$

where Nu, Re, and Pr are the Nusselt, Reynolds, and Prandtl numbers respectivdey. This correlation accounts for the variation of the coefficient with the mass flow rate, neglected for linearized modelling.

4. The distributed nature of the process is fundamental in the dynamical analysis and is taken into account in the energy conservation equations of heating and connection tubes.

5. The bulk temperature of the fluid in a given section of a tube is considered uniform. Moreover, in view of the dominant effect of the heat transport of the fluid flow, heat diffusion within the fluid is neglected.

6. The sodium is "well-mixed" within inlet and outlet headers.

7. Any set of identical heating tubes connected in parallel is represented by an equivalent fictitious tube, assuming the process variables of each tube to have the same value.

2.3 Reference Model Equations

The non-linear reference model needed for full scale transient simulation is obtained by applying the basic assumptions stated in Section 2.2, together with the following equations for the various receiver components (see Appendix 1 for the list of the symbols).

Heating Tube

- Energy conservation for the tube metal.

$$c_m \rho_m A_m \frac{\partial T_m(x,t)}{\partial t} = q(x,t) - \gamma_t \pi d_i \left(T_m(x,t) - T(x,t)\right) \tag{3}$$

- Energy conservation for the fluid.

$$c \rho A \frac{\partial T(x,t)}{\partial t} + cw \frac{\partial T(x,t)}{\partial x} = \gamma_t \pi d_i \left(T_m(x,t) - T(x,t)\right) \tag{4}$$

Connection Tube

Same as (3) and (4) with $q(x,t) = 0$.

Header

- Energy conservation for the header metal.

$$M_m c_m \frac{dT_{mc}(t)}{dt} = -\gamma_h S_i \left(T_{mc}(t) - T_{oc}(t)\right) \tag{5}$$

- Energy conservation for the fluid.

$$c \rho V_c \frac{dT_{co}(t)}{dt} = \gamma_h S_i \left(T_{mc}(t) - T_{oc}(t)\right) + w\left(h_{ic}(t) - h_{oc}(t)\right) \tag{6}$$

Temperature Sensor

$$\frac{dT_e(t)}{dt} = \frac{1}{\tau(w)} \left(T(t) - T_e(t)\right) \tag{7}$$

where the sensor time constant $\tau(w)$ depends on the fluid flow rate w through the heat exchange coefficient.

Feedpump and Feedpump Motor

$$\delta p = k_{p1} w^2/\rho + k_{p2} \rho (\Omega/\Omega_o)^2 \tag{8}$$

$$J\dot{\Omega} = \Omega (k_{m1}V - \Omega)/k_{m2} - w \delta p/\eta \rho \Omega \tag{9}$$

Nonlinear simulation of the receiver process has been performed by a modular simulation program, which carries out the numerical integration of the system of partial and ordinary differential equations obtained by putting together (according to the receiver configuration) Equations (3) to (9) relative to the different components.

To keep the distributed transport phenomena incorporated in Equations (3) and (4), those partial differential equations have been integrated by using a finite difference method (Ref.6) with a convenient number of spatial nodes: about 300 spatial nodes have been adopted for the whole receiver, where the fluid crossing time is about 35 seconds at the maximum load, to get a suitable representation of heat transport along the tubes and of its associated delays in temperature behavior.

2.4 Linearized Model

A linearized model in the comlex variable domain suitable for control system design has been developed, and in view of the relevance of the heat transport phenomena, it has been given the form of transcendental transfer functions for heating and connection tubes. The model has been developed from Equations (3) to (9) by linearization, Laplace transformation, and analytical integration of the resulting ordinary differential equations with respect to spatial abscissa x.

In particular, assuming zero initial conditions for the variational equations, the following Laplace Transform relation holds for a generic heating tube or a set of parallel tubes (see Appendix 1 for the meaning of the symbols):

$$\Delta T(1,s) = e^{-B(s)}\left(\Delta T(0,s)+\Gamma(s)\ \frac{Q_t ß}{c\overline{W}}\left(\frac{\Delta ß(s)}{1+s\tau_m} - \frac{\Delta w(s)}{\overline{W}}\right)\right) \tag{10}$$

where

$$B(s) = s\tau\left(1 + \frac{\tau_m/\tau_f}{1+s\tau_m}\right) \tag{11}$$

$$\Gamma(s) = \ell\int_0^1 e^{B(s)v}\ \emptyset(vl)dv \tag{12}$$

Note that Equation (10) incorporates the basic assumptions stated in Section 2.2, and, in particular, the approximations specified at points (1) and (3).

From these equations we obtain the transfer function between inlet and outlet temperatures of a heating tube

$$G_{tr}(s) = \frac{\Delta T(1,s)}{\Delta T(o,s)} = e^{-B(s)}$$

Because of the high heat exchange coefficient between sodium and the tube metal and the very thin tube wall, $\tau \gg \tau_m, \tau_f$.

Therefore, we approximately have (Ref.7)

$$B(s) = S\tau(1+ \tau_m/ \tau_f) = S\tau_d \qquad (15)$$

where $\tau_d = \tau(1+ \tau_m/ \tau_f)$.

Thus, the inlet and outlet temperatures are essentially related by a pure delay τd much greater than the crossing time.

Analog equations also hold for G_{t1}, the transfer function between inlet and outlet temperature of a connection tube, even in different parameters.

From Equations (10) to (12) we also obtain the transfer function between outlet temperature and sodium flowrate

$$G_W = \frac{\Delta T(1,s)}{\Delta w(s)} = -e^{-B(s)} \cdot \Gamma(s) \cdot \frac{Q_t \bar{ß}}{c\bar{w}^2} \qquad (16)$$

and the transfer function between outlet temperature and sun intensity

$$G_q = \frac{\Delta T(1,s)}{\Delta ß(s)} = e^{-B(s)} \cdot \Gamma(s) \cdot \frac{1}{1+s\,\tau_m} \quad \frac{Q_t \bar{ß}}{C\,\bar{w}} \qquad (17)$$

Note that since τ_m is quite small, G_w and G_q apart from the sign, have essentially th same dynamic characteristics.

To understand these characteristics we consider a tube irradiated with uniform thermal power along its abscissa x (Ø(.) = const).

In this case $$G_w(s) = -\frac{Q_t \bar{ß}}{c\bar{w}^2} \frac{1-e^{-B(s)}}{B(s)}$$

Once in the time domain, the response of outlet temperature to the unit negative step of the sodium flow rate is depicted in Fig.7.3-2.

This response looks like a limited ramp τd seconds long and without delays: as a function of frequency $G_W(j\omega)$ appears in fact as a minimum phase function.

It is worth noting that this is a very important property for the control point of view; in fact (see Equation (8)) the achievable performance of a simple control loop is essentially determined by the non-minimum phase incorporated in the process transfer function.

Luckily this property essentially also holds true for the actual thermal power distribution Ø(.) on the ASR panels (Fig.7.3-3), which is a Gaussian-like even function of the tube abscissa. This point is accurately discussed in Ref.9, where it is shown that symmetrical function Ø(.) likely implies G_W as a minimum phase function.

The responses of the ASR panel outlet temperatures to a step variation in the sodium flow rate are shown in Fig.7.3-4. The contribution of a single panel is shown in the first part (about 10 sec) of the responses. It appears quite prompt, ramp-like shaped with a small initial delay due to a portion of the tube not irradiated before the outlet header. The second part of the transient, which is slower and delayed, is due to the propagation of the temperature variation of each panel upstream from the one considered.

Because of this second part of the transient, the panel i outlet temperature (i = 2,5) versus sodium flow rate is a non-minimum phase function.

The transfer function between inlet and outlet temperatures of each header can be obtained from Equations (5) and (6):

$$G_{TC}(s) = \frac{\Delta T_{oc}}{\Delta T_{ic}} = \frac{1 + s\,\tau_{mc}}{1 + \dfrac{\tau_{mc}\tau_{fc} + \tau_{mc}\tau_{c} + \tau_{fc}\tau_{c}}{\tau_{fc}}\,s + \tau_{m}\tau_{c}\,s^{2}} \tag{18}$$

In view of the forward propagation of the thermal power it is possible to build the overall ASR linear model, cascading the above defined transfer functions according to the fluid path in the receiver (Fig.7.3-5). Using this scheme, a computer program has been written which computes the overall transfer functions and plot the Bode and Nyquist diagrams needed for control design.

A block diagram of the overall linear model of the ASR is depicted in Fig.7.3-6 as an aid in understanding the new control scheme concept.

As shown,

- All panels are identical.
- The effects of sodium properties variation with temperature on Gw(.) are negligible.
- The Ø(.) of each panel are similar.

Given these conditions,

$$G'_{wi} = \lambda_i \frac{Q_T \bar{\beta}}{c\, \bar{w}^2} \cdot e^{-B_i(s)} \Gamma_i(s) . G_{TCi} \cong \lambda_i G_w G_{TC} \qquad (19)$$

Thus, G'_{wi} differs only on the low frequency gain, which depends on the fraction λ_i of the total power Q_T absorbed by panel i.

$$G'_{tk} = G_{tlk} \cdot G_{tck}$$

where G'_{tk} is the transfer function between the temperature at the outlet of panel k-1 and that at the inlet of the radiated tubes of panel k and

$$G''_{tk} = G_{trk} \cdot G_{tck}$$

where G''_{tk} is the transfer function between the panel k radiated tubes inlet temperature and the panel k outlet temperature.

In view of the above hypothesis,

$$G_{tck} = G_{tcj} \, , \quad \forall k, j$$

$$G_{trk} = G_{trj} \, , \quad \forall k, j$$

and hence $G''_{tk} = G''_{tj} = G''_t \, , \; \forall k, j$.

3. CONTROL SYSTEM DESIGN

3.1 Introduction

In principle, ASR temperature control may be performed by varying the sodium flow rate according to the variation of the incoming thermal power

(Equation (10)). Unfortunately there are many conditions under which the apparently simple single input (sodium flow rate) single output (outlet temperature) control problem becomes quite complex:

- The disturbance (entering power variations) can be very large and fast (design condition: full load variation within eleven seconds), are not predictable and not measureable with sufficient accuracy for a satisfactory compensation.

- The ASR has to be automatically operated not only over an unusually wide normal operational range (10% to 100% load), but also during transients in which a dense cloud temporarily covers the whole heliostat field (U-type transients) so that first the receiver cools down and then returns to the normal temperature profile.

- The outlet temperature versus sodium flow rate transfer function is affected by a large amount of non-minimum phase (Ref.9) essentially due to variations in the outlet temperature of the upstream panels, which propagate toward the receiver outlet at approximately the speed of the coolant; thus the variations in the receiver outlet temperature are considerably delayed.

The design of the control system has been carried out according to the following steps:

- linear feedback around an equilibrium point
- extension to the full normal operation load range with inclusion of the feedforward action
- U-type transients control

3.2 Linear Feedback Synthesis.

The first point is related to sodium flow rate regulation. Indeed this is not particularly meaningful in that with a standard PI regulator a closed loop response can be obtained that is fast enough to consider sodium flow rate regulation and receiver temperature regulation as decoupled. Consequently, the temperature controller can be designed separately, considering sodium flow rate as the control variable.

In Fig.7.3-7 the Bode diagrams of the receiver outlet temperature versus sodium flow rate transfer function (100% load) are shown. It can be seen that it is not possible to get a closed loop regulation band width that is consistently greater than about 0.03 rad/sec with reasonable phase and gain margins. In the same way it can be seen that feeding back intermediate panel temperatures does not significantly improve regulation performances. Only the first panel temperature response has good dynamic characteristics but unfortunately it is not very meaningful in that the fraction of the total power absorbed by the first panel is quite small (few percents).

As a consequence of these points, it was necessary to use a new control scheme called PFTC (Predictive Feedback Temperature Control (Ref.9)). Its basic idea is to feedback a signal ($\hat{T}$) that uses a predicting action (model of a radiated tube) to control the harmful effects due to temperature propagation toward the downstream panels.

The feedback signal $\hat{T}$ is obtained according to

$$\Delta\hat{T} = \sum_{K=1}^{5} \Delta T_{OK} - \sum_{i=2}^{5} \Delta T_{iK} G''_{T} \tag{22}$$

where

$$\Delta T_{OK} = (G'_{WK} + \Delta T_{iK} G''_{T})\,\Delta w, \quad \Delta T_{i1} = \emptyset$$

The above equation gives

$$\Delta\hat{T} = \Delta w \sum_{K=1}^{5} G_{WK} = -\frac{Q_T \bar{\beta}}{c\,\bar{W}^2}\, e^{-B(s)}\,\Gamma(s)\, G_{TC}\,\Delta w \tag{23}$$

Equation (23) points out two significant properties of $\Delta\hat{T}$:

1. Its dynamics look like the outlet temperature of a single panel which absorbs the total receiver power, without being affected by non-minimum phase due to propagation delay. As shown in Fig.7.3-8, a regulation loop which uses $\hat{T}$ as feedback signal can have the same band width as the temperature regulation loop of a single receiver panel.

2. It is independent of the actual power distribution λ between the panels; it depends only on the total ASR absorbed power Q_T.

To obtain $\Delta\hat{T}$ from Equation (22) requires that G''_t be known. According to Equation (21) G''_t is the transfer function between radiated tube inlet temperature and panel outlet temperature.

An ideal realization of G''_t could be obtained cascading the model of a radiated pipe with no thermal power and that of a header (Ref.9); however the following simple rational function has been used:

$$\hat{G}''_t = \frac{1}{1 + sT_1} \cdot \frac{1 - sT_2}{1 + sT_2} \tag{24}$$

where T_1 and T_2 have been chosen by fitting the Bade diagrams of G''_t in a suitable frequency band.

One important point which still has to be considered is related to PFTC's capability to compensate for the effects of the incident power variations on each panel's outlet temperature.

If $\hat{G}''_t = G''_t$, with reference to Fig.7.3-6 and taking into account Equation (23), then

$$\frac{\Delta T_{oj}(s)}{\Delta ß(s)} = \tilde{G}_{qj} \, (1+G_W G_{tc} G_{pid})^{-1} \qquad j = 1,...,5$$

where

$$\tilde{G}_{Qj} = -\frac{1}{1+s\tau_m} G_w \left[\lambda_j + \sum_{K=1}^{j} \lambda_K G_T''^{\,j-K} \cdot \prod_{i=K+1}^{j} G'_{Tj}\right]$$

is actually the open loop transfer function between $\Delta ß(s)$ and $\Delta T_{oj}(s)$.

Thus, variation of the panel outlet temperatures due to solar disturbances are divided by the return difference $(1+G_W G_{tc} G_{pid})$ which has a band as wide as the return difference of a single panel temperature regulation loop.

In the actual implementation the following points have been considered:

- An outer slower loop with an integral action regulator has been added to ensure that the ASR outlet temperature will be equal to its set point at steady state.

- Instead of using the panel outlet temperatures which are downstream in a non-radiated portion of the tube, thermocouples welded on the tubes near the radiated zone have been used to obtain $\hat{T}$. This does not influence the above considerations but T_1 and T_2 of Equation (24) must be modified.

- The inner loop PID regulator parameters have been tuned on the basis of the Bode diagrams of the $\Delta\hat{T}/\Delta W$ transfer function (Fig.7.3-9). It is worth noticing that the D action has little effect and has been chosen so as to compensate the sodium flow rate loop and thermocouple lags. A cut-off frequency of about 0.35 rad/sec has been obtained. During the field testing of the regulator, the simulation values of the parameters were maintained.

3.3 Non-Linear Scheme and Feedforward Action

Process parameters have a variation of the order of ten based on the sodium flow rate within the normal operation load range (10% to 100%); therefore it is necessary to vary the PFTC parameters with load. Keeping in mind that $G_W(o)$ is nearly inversely proportional to the sodium flow rate and dynamic characteristics of Gw(s) (see τ_d Equation (15)) and directly proportional to the sodium flow rate, obvious laws of parameter variation as a function of sodium flow rate have been established.

The modified PFTC is shown in Fig.7.3-10. This regulation system has been tested by the non-linear simulator in which the regulator functions have been included.

The simulation results demonstrated that, in the case of large solar disturbances, the PFTC guarantees satisfactory temperature control without any feedforward disturbance compensation except for the case of a sudden increase of the solar intensity ß(t) from the minimum to the maximum value. This is explained as follows. At the minimum load, the process is very slow so that the response of the closed loop regulation is slow, especially if compared with local temperature gradients in those panel sections where the ratio ß(t)/W*(t) (where $W^* = w/W_n$) quickly becomes greater than unity. In this case where ß(t) rapidly increases from the minimum load, a feedforward disturbance compensation is very helpful. The compensation consists of incorporating in the coolant flow demand a feedforward signal proportional to a suitable estimate $\hat{ß}(t)$ of the solar radiation intensity ß(t).

As a matter of fact, quickly increasing w(t) not only directly reduces the unbalance between ß(t) and w(t), but also increases proportionally the speed of response of the PFTC.

Unfortunately, a precise and reliable estimate of ß(t) is difficult with the available instrumentation, as is the case at the Almería plant (Ref.7.3-10) However analytical simulation has verified that the PFTC, with the aid of a rough feedforward signal (50% error on ß(t)), can allow the ASR to safely operate in the worst foreseen operating conditions.

For safety's sake, the feedforward sodium flow rate demand W_f, which according to Equation (10) should be given by $W_f = W_N\hat{\beta}$, has been expressed as

$$W_f = K W_N \hat{\beta}$$

where $K = 1 \qquad \dot{\hat{\beta}} \leq 0$

$K = K_1 > 1 \qquad \dot{\hat{\beta}} > 0$

K_1 can be adjusted by the plant operator to take into account the cosine effect on $\hat{\beta}$.

3.4 U-Type Transients

"U-type" are those typical transients which give rise when a dense cloud passes over the heliostat field. In this case, the radiated power focused on the receiver becomes negligible when the cloud entirely covers the heliostat field, and then it goes up again when the cloud passes.

In the first part of such transients, the receiver cools down rapidly (Fig.7.3-11/13). Once the sodium flow rate reaches its minimum value no control exists. In the second part, the temperatures go up again in time as the power increases.

These transients are among the worst for the receiver, especially due to the high gradients present at the start of the second part. In fact, when the reflected power increases rapidly the receiver temperatures increase with high starting gradients, which essentially depend on the difference between the actual temperature distribution and the nominal temperature profile, on the rate of increase of power, and on the tube metal and sodium mass (Appendix 2). This means that they are independent of any control action except rapid control of the heliostats aiming, which is not possible. Thus, in order to operate the receiver on cloudy days a structural design constraint was established for the ASR.

-type transients also present unique problems for the control design:

- A loss of control capability occurs when, due to the lack of power, the sodium flow rate reaches the minimum allowable value (minimum saturation condition). As usual in control systems, this condition calls for actions to avoid regulator wind-up (i.e. delay in going out of minimum saturation at the next power increase).

- Poor control occurs when power increases and the receiver temperatures are lower than the nominal profile (set point). The feedback control does not fully operate until the set point has been reached, and even then the feedback control counteracts the proper feedforward action.

These two critical points have been overcome in a simple but effective way:

- When the sodium flow reaches its minimum rate, the temperature in the inner loop set point is automatically equal to the actual T signal (the outer regulator "follows" T). Thus, during rapid changes, the ASR temperatures are regulated to their actual value if the absorbed power increases. At low frequency the outer regulator will change the T set point so as to have T_0 equal to its reference value. This is why the transients in Fig.7.3-11/13 show high flow rate values (the maximum allowable by the pump), a reversal in temperature trend, and a long settling time. However, the sodium flow rate value could be easily limited to the design value (as an example) so as to considerably reduce the settling time. Nevertheless, during the first tests, the decision was made to not limit the sodium flow in order to reduce the mean temperature gradients during receiver heating.

Actually the adopted control algorithm behaves as a "bang" control which asks for the maximum sodium flow rate in the case of rapid temperature increases when the receiver is cooling down. This strategy seems correct in that absorbed power excursions are generally from zero to a maximum value, which, even if it is function of absolute time, does not deviate far from the design value during most of the operation.

4. THE DIRECT DIGITAL CONTROLLER

Even if the ASR control scheme appears complex and unusual, it can be performed by either analog or digital electronics. In view of its greater flexibility and lower cost, the digital solution has been chosen. Apart from temperature control in each operative condition, the microcomputer performs acquisition and presentation of many physical and computed variables, provides receiver protection, and provides an easy and extended man-machine interface.

The regulation algorithm has been programmed using an ad hoc pseudo-language that allows on-site modification of the regulation scheme.

Regulator and predictor parameters and alarm thresholds can be easily modified from the system's keyboard, as well as the relationship between the analog input and data base, the output pages, etc.

The hardware configuration can be summarized as follows:

- Single board computer Z80 cpu based, with floating point processor, 32K byte EPROM and 32K byte RAM

- 54 galvanically isolated analog input scanned each 250 ms, 8 optoisolated digital input, 8 optoisolated digital output

- Analog back-up for the sodium flow rate set point analog output

5. EXPERIMENTAL RESULTS

For completeness, a few experimental results are shown both to validate ASR modelling and to evaluate control system performances. Fig.7.3-14 shows a comparison between the simulated and experimental ASR temperature responses that occur as a result of step changes in the incident power and sodium flow rate. Note that the actual power distribution between the panels in the field tests is unknown, which could partially explain the differences between simulation and experimental results. Detailed analysis of the ASR transient response and a discussion of modelling validation can be found in Ref.10.

In Fig.7.3-16/17 two closed loop transients are shown:

- T.1, where an abrupt deviation and a subsequent refocusing of 25% of the heliostats were provoked, causing a corresponding variation in the ASR absorbed power. No feedforward disturbance compensation is applied in this case.

- T.2, where "spontaneous" transients caused by cloud passages were recorded, and the normal incident power estimate was used as feedforward compensation.

While test T.1 was artificially provoked to verify the capability of the PFTC to withstand large perturbations, test T.2 is simply a sample of the continuous recording stored on the supervision system during a "normal" operation day with cloudy sky (January 12, 1984). In particular, we note that the typical V-shaped variations of solar radiation could have been very severe for the receiver without automatic control, as they would have caused a temperature variation of about 100°C. Moreover, from Fig.7.3-17 it appears that the feedforward signal used for the disturbance compensation is rather poor because its variations are very different from those of the coolant flow rate actually needed to control the receiver temperatures.

6. CONCLUSIONS

This accurate distributed parameter mathematical modelling of the ASR has led to a good understanding of the temperature behavior related to both the sodium flowrate and absorbed power variations. Based upon this modelling, the new temperature control system concept (PFTC), has been synthesized. This system, which combines predictive and feedback action, appears particularly fit for monophase fluid cooled solar receivers in view of its capability to compensate for the non-minimum phase of temperature responses and its insensitiveness to receiver incident power map variations.

The simulation of cloud passage transients has shown a relative weakness in the feedback control action where the absorbed power increases rapidly from the minimum load (10%). However, a very rough (50% error) feedforward action allows the controlled system to function well even in these worst cases.

Simulation and the following stress analysis have shown that the most damaging transients to the ASR are due to rapid increases in the absorbed power following a time of heliostat field shadowing. In this case the receiver temperatures are lower than the nominal profile and high temperature gradients arise even with "perfect" control. The control algorithm has been improved to minimize these gradients, and the structural design has been modified so that the ASR can withstand them.

In this way the ASR can be operated automatically in any weather conditions, as the one-year normal operation has demonstrated.

7. REFERENCES

1. V. Bedogni, A. De Benedetti, A. Di Meglio, and C. Sala, "The Almería Advanced Sodium Receiver (ASR): Basic Design and Operating Conditions", ISES Solar World Congress, Perth (Aus), August 1983.

2. V. Bedogni,and A. De Benedetti, "Almería Advanced Sodium Receiver (ASR): Stress Analysis Considerations", ISES Solar World Congress, Perth (Aus), August 1983.

3. W. Grasse, "IEA - Small Solar Power Systems Project", International Symposium, Cologne (Ger).

4. R.W. Hallet, Jr. and R.L. Gervais, "Central Receiver Solar Thermal Power System, Phase 1", McDonnell Douglas Astr. Co., NTIS rep. n. SAN/1108, Vol. 4, Huntington Beach California, October 1977.

5. W. McAdams, "Heat Transmission", Int. Stud. ed. McGraw Hill, 1954.

6. G.D. Smith, "Numerical Solution of Partial Differential Equation Finite Difference Methods", Oxford University Press, 1978.

7. L.A. Gould, "Chemical Process Control: Theory and Applications", Reading, Mass.: Addison-Wesley, 1969.

8. I.M. Horowitz, "Synthesis of Feedback Systems", Academic Press, London, 1963.

9. C. Maffezoni, G.A. Magnani, and S. Quatela, "Process and Control Design of High Temperature Solar Receivers: An Integrated Approach", to appear in IEEE-AC Transactions on Automatic Control.

10. R. Carmona and J.G. Martín, "The SSPS Advanced Sodium Receiver: Transient Response", IEA/SSPS Deliverable Review, Tabernas (Almería), Spain, October 1984.

NOMENCLATURE

$q(x,t)$	= Thermal power per unit length absorbed by a heating tube.
$\beta(t)$	= Solar radiation intensity at time t.
Q_t	= Total nominal power absorbed by a tube.
$\emptyset(.)$	= Normalized thermal power map on a tube (i.e. with $_0{}^1\ \emptyset(x)dx = 1$).
l	= Tube length.
w	= Coolant mass flowrate.
c_m	= Specific heat of the tube metal.
ρm	= Density of the tube metal.
A_m	= $(d_e{}^2 - d_i{}^2)/4$.
$T_m(x,t)$	= Mean tube metal temperature at abscissa x and time t.
γt	= Heat exchange coefficient for a tube.
d_i	= Inner diameter.
d_e	= Outer diameter.
$T(x,t)$	= Temperature of the fluid in the tube at abscissa x and time t.
c	= Specific heat of the fluid.
A	= $d_i{}^2/4$.
ρ	= Density of the fluid.
M_m	= Total metal mass of a header.
γh	= Heat exchange coefficient for a header.
S_i	= Heat exchange surface for a header.
T_{mc}	= Mean temperature of the header metal.
T_{ic}, T_{oc}	= Fluid temperature at the header inlet and outlet.
V_c	= Inner volume of a header.
h_{ic}, h_{oc}	= Fluid enthalpy at the inlet and outlet of a header, respectively.
$T(t)$	= Fluid temperature at the sensor location, at time t.
$T_e(t)$	= Measurement of T(t).
δp	= Pump head.
Ω	= Motor and pump angular speed.
k_{p1}, k_{p2}, Ω_o	= Characteristic parameters of the pump.
km_1, k_{m2}, η	= Characteristic parameters of the motor.
$s = \alpha + j\omega$	= Complex variable.
$\Delta T(x,s)$	= Laplace transform of the fluid temperature variation at tube abscissa x.
$\Delta\beta(s)$	= Laplace transform of the variation of $\beta(t)$.
$\Delta w(s)$	= Laplace transform of the variation of $w(t)$.
$\bar{w}$, $\bar{\beta}$	= Values of the fluid mass flow rate and of the solar intensity at the linearization equilibrium point, respectively.

= ρ Al/w

= $c \rho A/\pi \gamma_t d_i$

= $c_m A_m \rho_m/\pi \gamma_t d_i$

= Number of parallel tubes of a panel.

= Total length of the receiver along the fluid stream.

= Number of panels in series.

= Thickness of a tube panel.

= Indexes identifying a panel (among 5). As subscripts, they are used to refer any quantity to panel i,k.

= Total thermal power absorbed by the receiver in nominal condition.

= Total fluid mass flow rate in nominal condition.

= e/d_e.

= Fluid temperature at the outlet of panel i.

= Fluid temperature at the inlet of panel i.

= Laplace transform of the variation of $T_{oi}(t)$ around a given equilibrium point.

= Q_{ti}/Q_T.

= $|\lambda_1, \lambda_2 \cdots, \lambda_m|'$.

As a simple example to illustrate a topic generally neglected consider the following:

Neglecting the tube metal dynamics for the sake of simplicity, from Equations (3) and (4), it can be obtained:

$$C_pA \frac{T(x,t)}{t} + \frac{Cw \quad T(x,t)}{x} = ß(t)Q_tØ(x) \tag{A.1}$$

and at steady state

$$CW_N \left.\frac{T(x,t)}{x}\right|_N = Q_tØ(x) \quad , \quad \text{where } ß_N = 1 \tag{A.2}$$

Replacing (A.2) into (A.1), we obtain

$$\frac{T(x,t)}{t} = \frac{1}{A} \left(ß(t)W_N \left.\frac{T(x,t)}{x}\right|_N - W(t) \frac{T(x,t)}{x} \right)$$

which, considering the case $W(t) = W_N$ ("perfect" regulation), gives

$$\frac{T(x,t)}{t} = \frac{W_N}{A} \left(ß(t) \left.\frac{T(x,t)}{x}\right|_N - \frac{T(x,t)}{x} \right)$$

this equation furnishes an indication of the uncontrollable temperature gradients which arise when the power increase and the receiver temperatures are lower than the nominal profile.

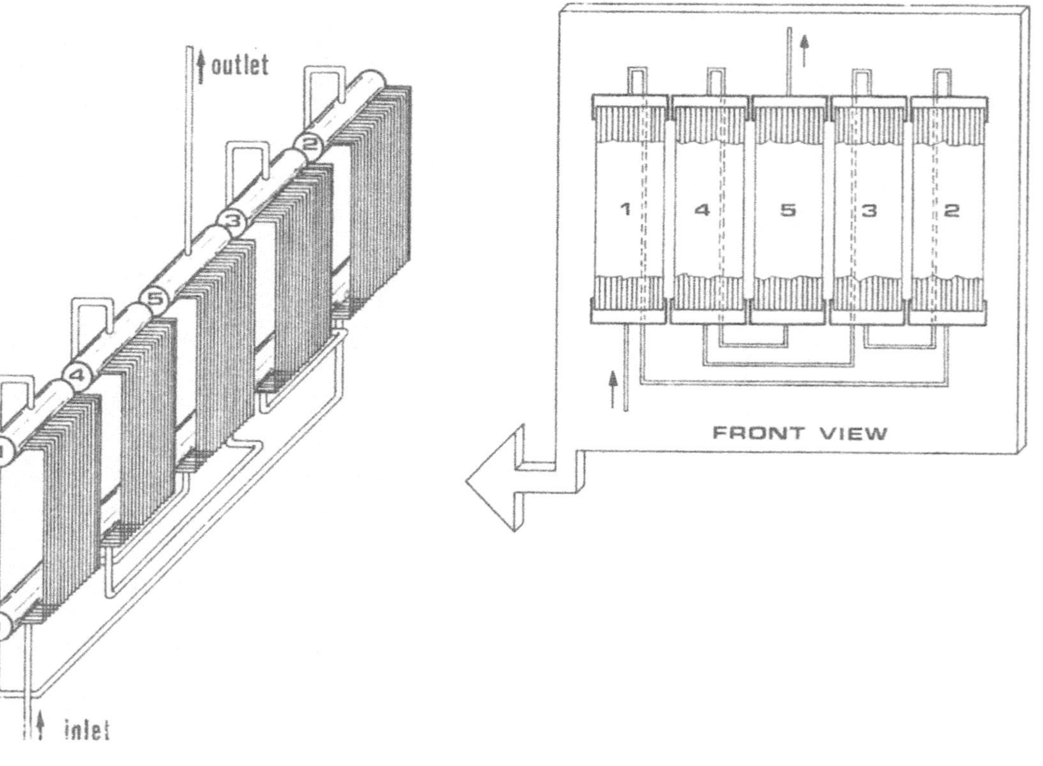

Fig. 7.3.-1: Sketch of the ASR

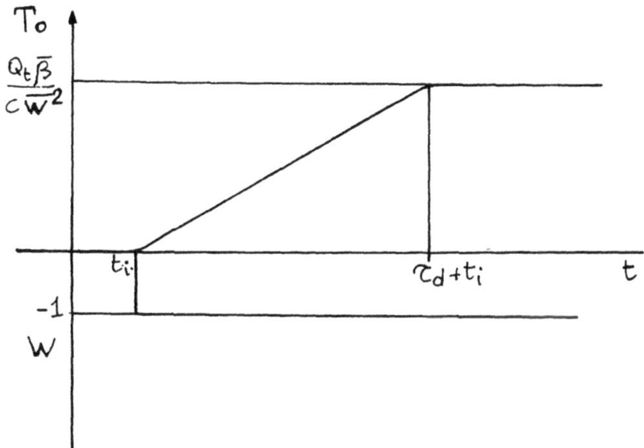

Fig. 7.3.-2: Tube Outlet Temperature Response to a Unit Negative Step in Sodium Flow Rate in Case of Uniform Power Distribution

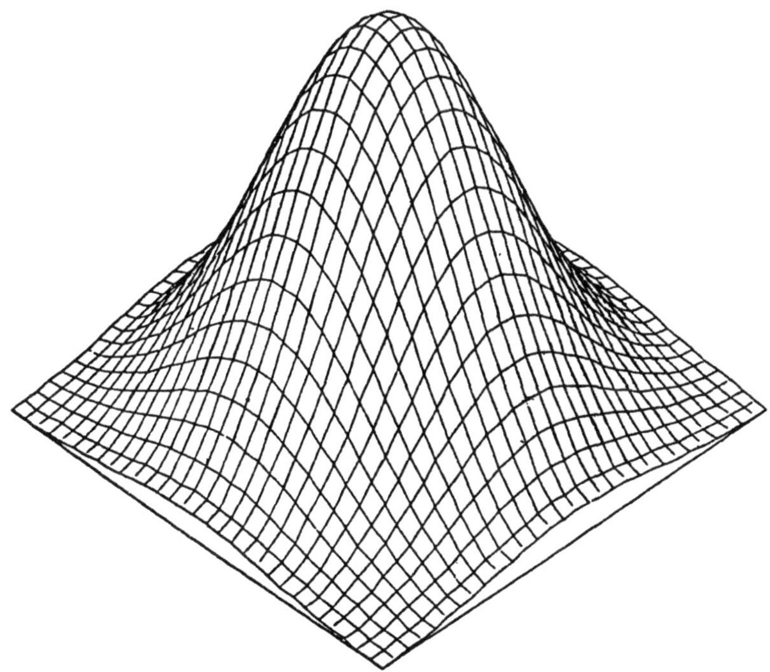

Fig. 7.3.-3: A Typical Flux Map

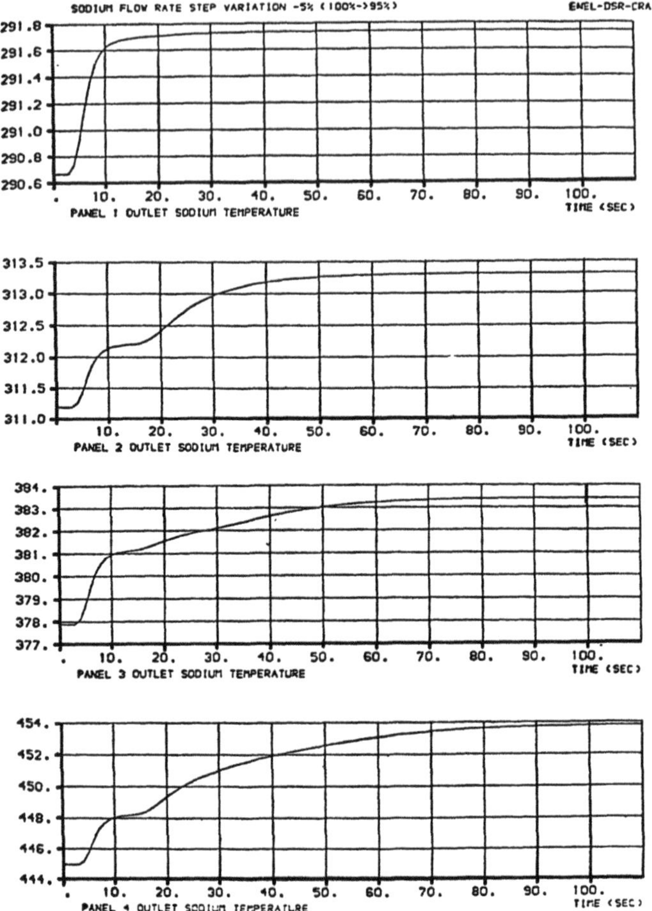

Fig. 7.3.-4: Panel Outlet Sodium Temperature Responses to a Step Variation on Sodium Flowrate Demand

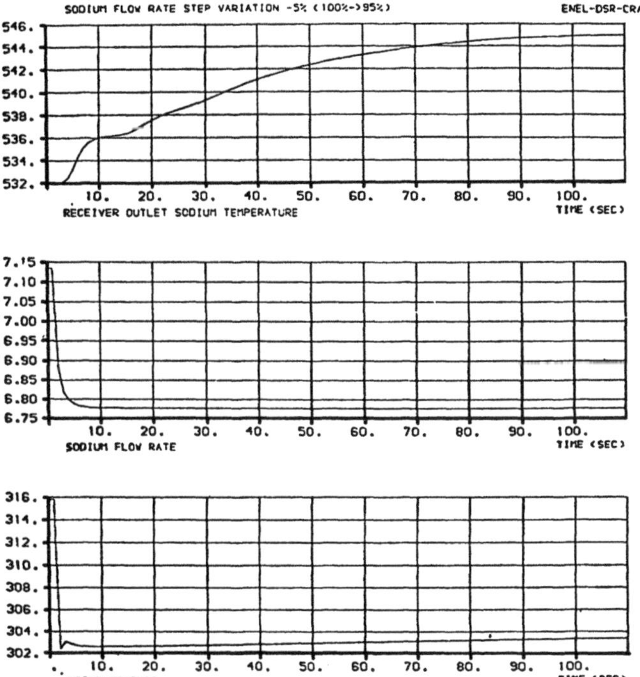

Fig. 7.3.-4b: (see text)

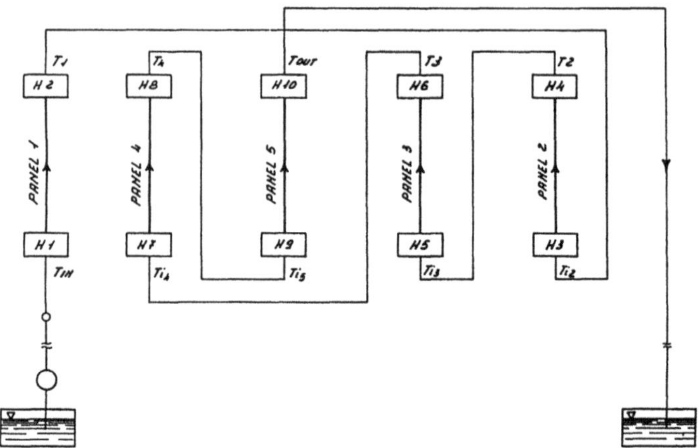

Fig. 7.3.-5: ASR Pattern

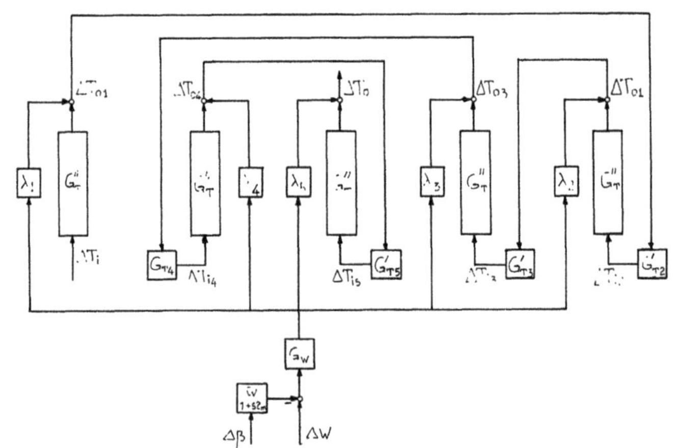

Fig. 7.3.-6: Block Diagram of the ASR Overall Model

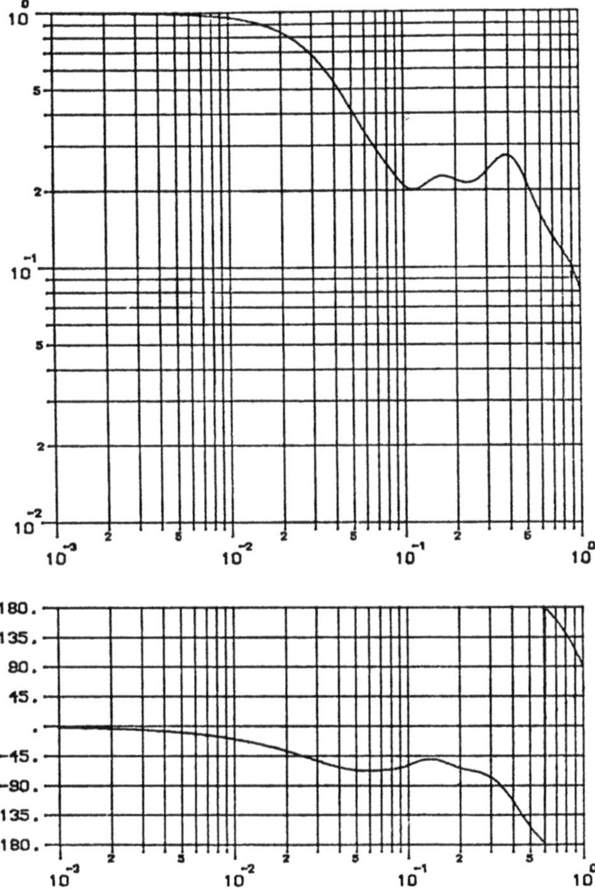

Fig. 7.3.-7: Plots of the ASR Outlet Temperature versus Sodium Flowrate Transfer Function(100% load)

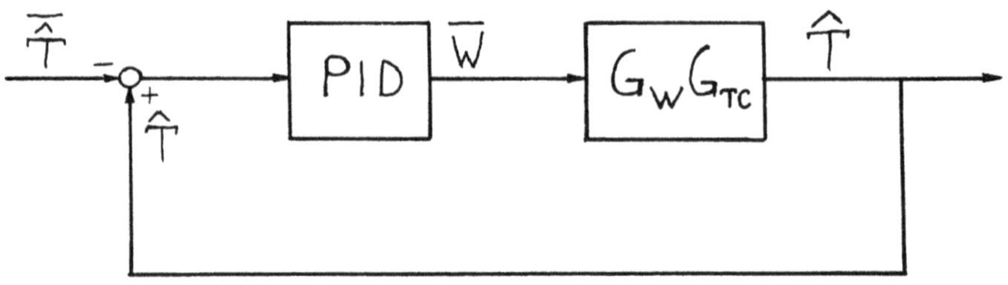

Fig. 7.3.-8: PFTC Loop

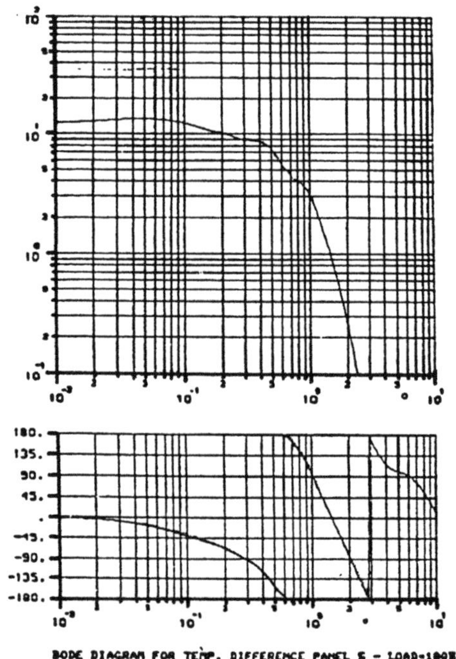

Fig. 7.3.-9: Plots of the Transfer Function of T versus Sodium Flowrate

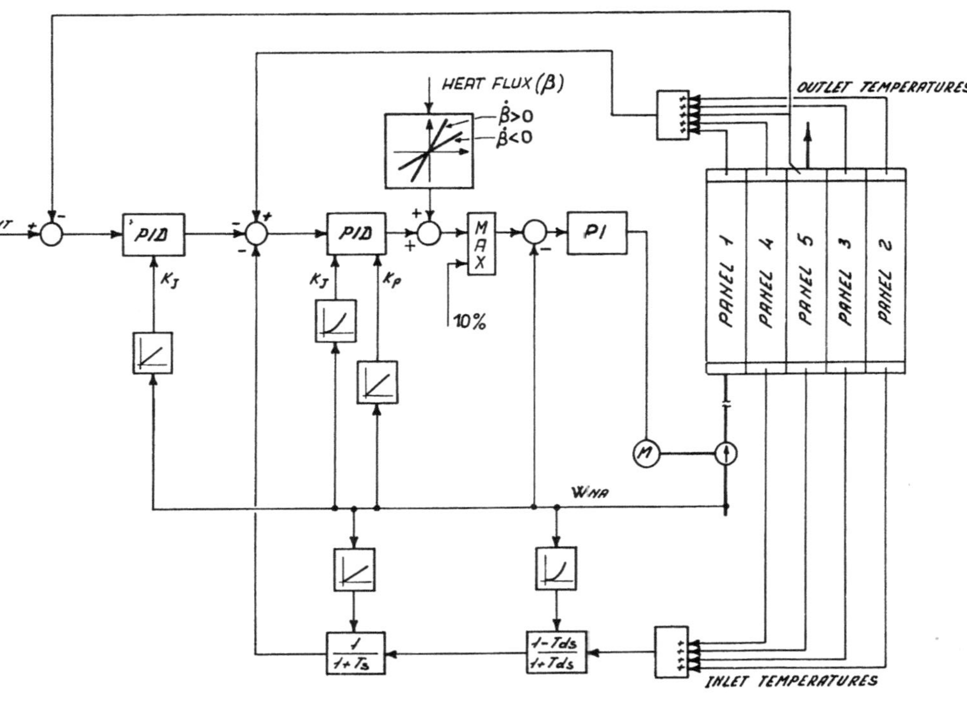

Fig. 7.3.-10: Functional Block Diagram of the ASR Predictive Feedback Temperature Control

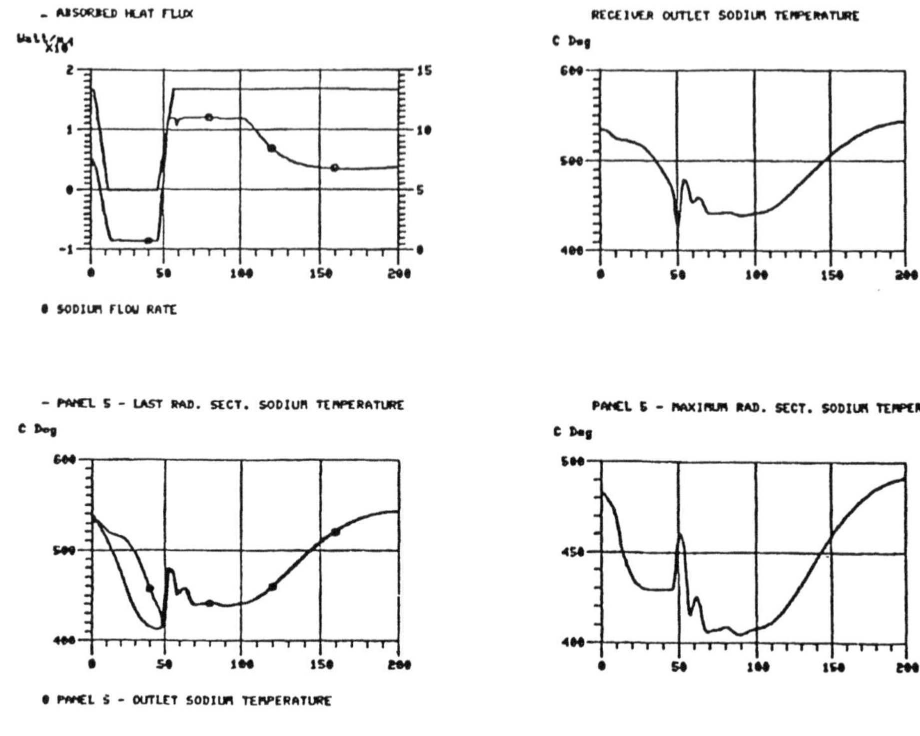

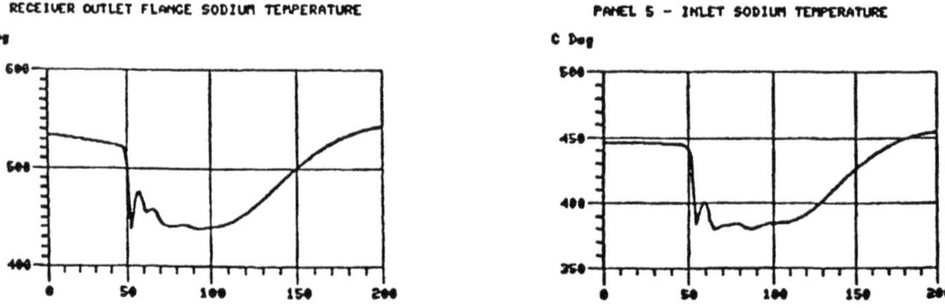

U-type TRANSIENT - Max. Sodium Flow Rate 10.9 Kg/s ENEL-CRA 19-NOV-82

Fig. 7.3.-11: U-Type Transients

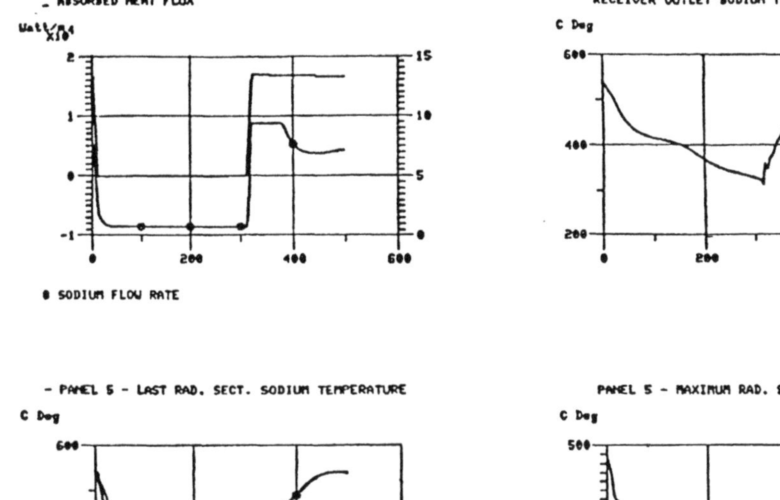

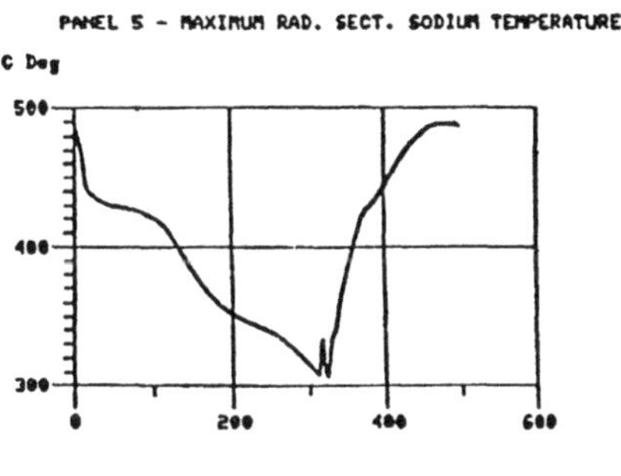

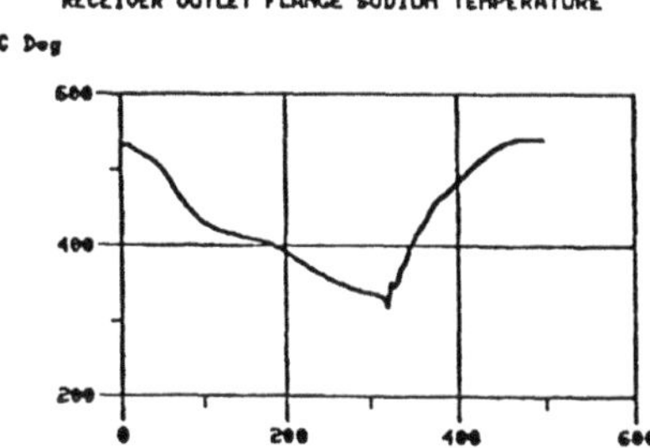

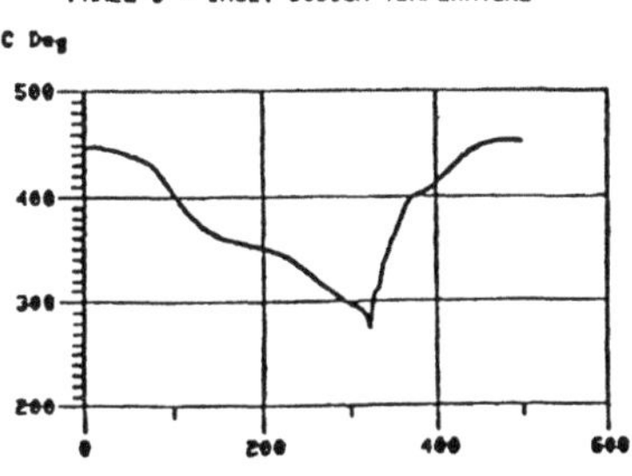

Fig. 7.3.-12: U-Type Transients

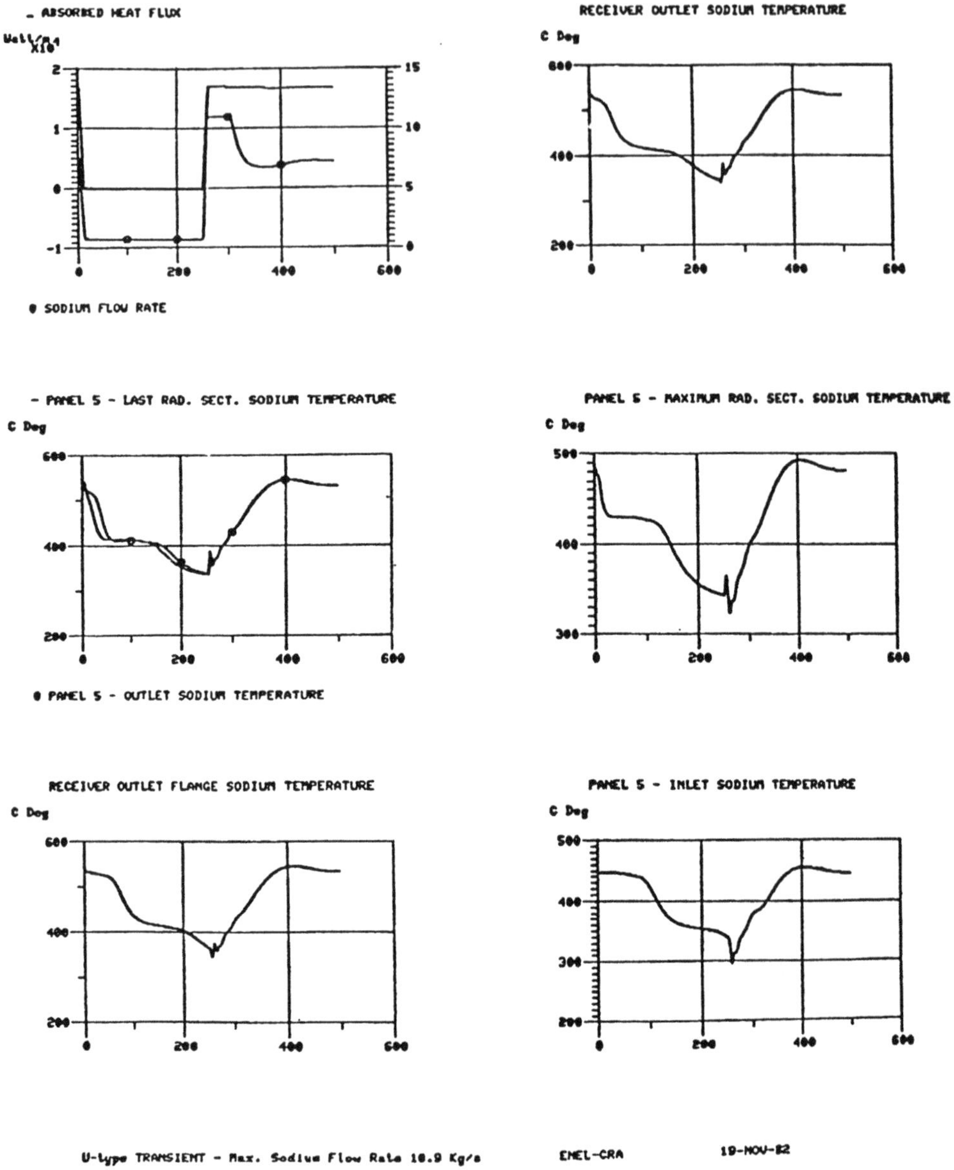

– ABSORBED HEAT FLUX

Watt/m² X10⁴

2 1 0 -1 | 15 10 5 0

0 200 400 600

● SODIUM FLOW RATE

RECEIVER OUTLET SODIUM TEMPERATURE

C Deg

600 400 200

0 200 400 600

– PANEL 5 – LAST RAD. SECT. SODIUM TEMPERATURE

C Deg

600 400 200

0 200 400 600

● PANEL 5 – OUTLET SODIUM TEMPERATURE

PANEL 5 – MAXIMUM RAD. SECT. SODIUM TEMPERATURE

C Deg

500 400 300

0 200 400 600

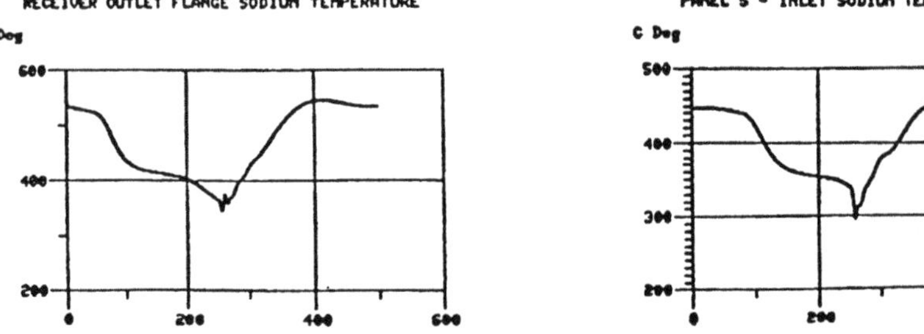

U-type TRANSIENT – Max. Sodium Flow Rate 10.9 Kg/s   ENEL-CRA   19-NOV-82

Fig. 7.3.-13: U-Type Transients

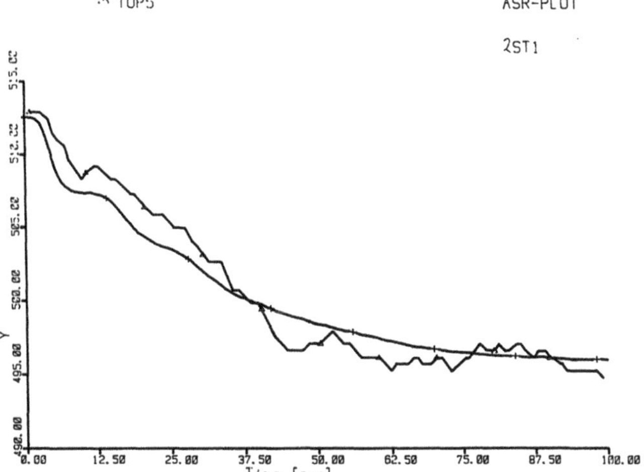

Fig. 7.3.-14: Response to a Down Step of 165 kW in the Flux,at High Flow (8.06 kg/sec)

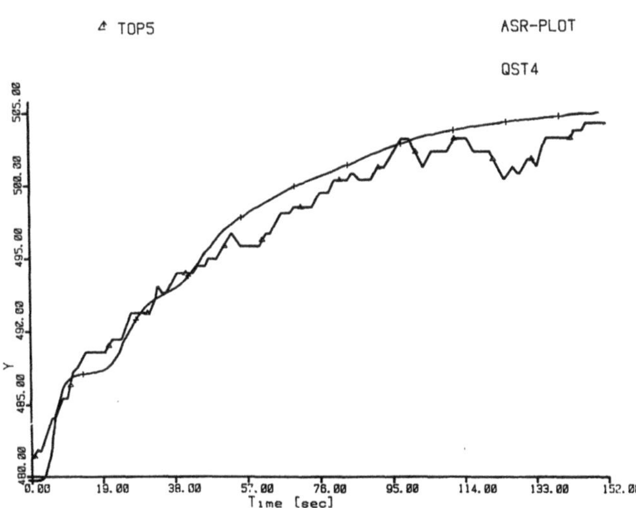

Fig. 7.3.-15: Response to an Upward Step of 165 kW in the Flux, at Medium Flow

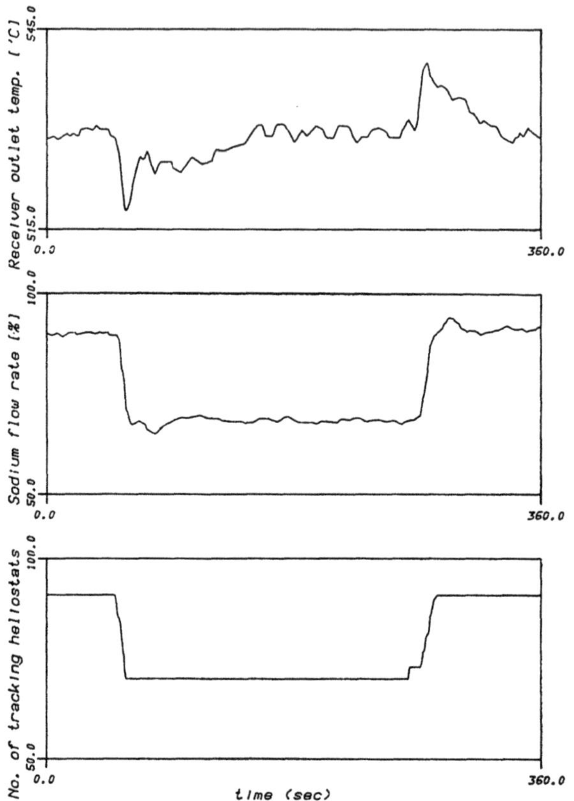

Fig. 7.3.-16: Field Test T.1; Variation of the Number of Tracking Heliostats

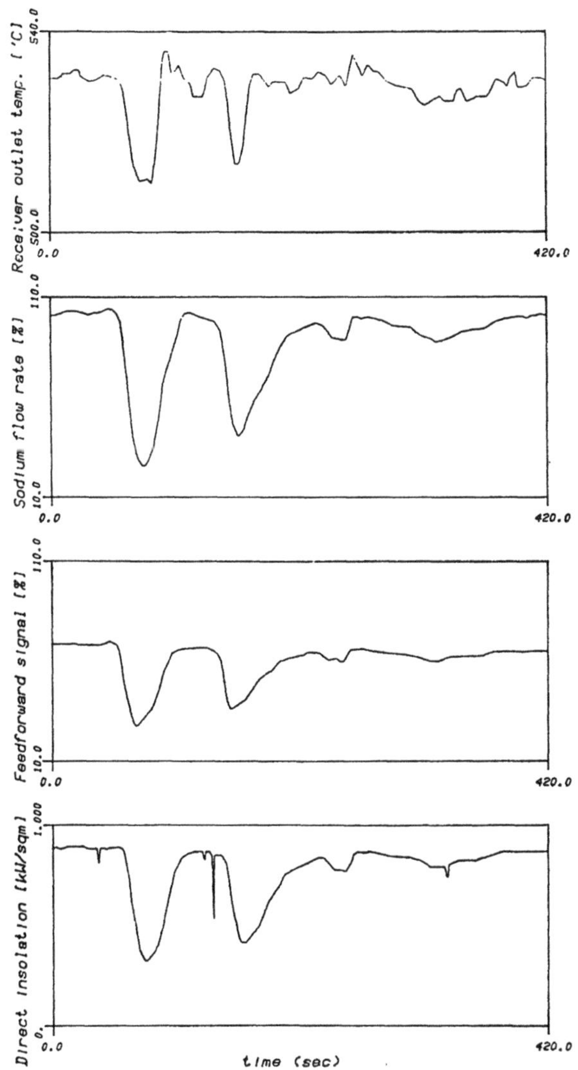

Fig. 7.3.-17: Field Test T.2; Cloud Passages

CONTROL OF INCIDENT POWER AT THE RECEIVER

Manuel Sánchez and Manuel Blanco, ITET

1. CONTENT

This paper describes a computer program which has been developed to control the power sent to a receiver. The program, 'CONTROL', takes heliostats in and out of track in 'real time' to achieve this goal.

2. OBJECTIVES

Control of the power makes it possible to achieve the following objectives:

- to raise the level of automation of the CRS, making possible the performance of routine operations such as receiver start-up, without manual intervention.
- to shape the power density profile, making possible the application of CRS technology to chemical and metallurgical processes where energy input must be carefully controlled.
- to enhance the potential of the CRS as an evaluation tool.

3. CONTROL VARIABLE

To control the power, one must know the useful power which each heliostat can send to the receiver at any given instant. The useful power depends on the effective heliostat surface and the reflectivity distribution, which must be estimated.

Effective Heliostat Surface

The program GEOFAC, also presented in this meeting, calculates the cosine of the incidence angle and the blocking and shadowing correction for each heliostat. From the superposition of the two corrections and the angle, it estimates the effective heliostat surface in real time.

Reflectivity Distribution

Based on studies and measurements of the reflectance variation in the heliostat field, another program has been written to characterize the reflectance over the heliostat field in terms of zones of homogeneous reflectance, which change as a function of time during each washing period.

Based on this characterization of the heliostats, a program has been prepared which identifies the heliostats which must be put in track to achieve a desired power level ('set-point') at any solar time, given the solar irradiance.

The preparation of this program was based on the following concepts:

° Heliostat Status
Each heliostat is classified according to the power sent to the receiver. The heliostat may be

1.1 - off line
1.2 - on standby
1.3 - tracking
1.4 - necessarily in tracking

(A heliostat may be 'necessarily in tracking' because of limitations imposed by thermal expansion in the receiver.)

° Heliostat Motion
This refers to the program output option which identifies the change, if any, of a heliostat from 'tracking' to 'standby' or viceversa. This change is decided in the calculated program before finishing the main step.

4. CALCULATION PROGRAM

'CONTROL' has been implemented on the VAX 11/730. The general block diagram for the program is shown in Fig.7.4-1. The open loop for calculation/control follows directly after the first initialization block (INIT). The real time for this loop is about three minutes in the VAX 11/730.

The decision to move the heliostats is based on the criteria that the number of heliostats whose status is going to be changed be as small as possible. To ensure this, the heliostats in the 'tracking' and 'standby' sets are arranged in the order of decreasing useful power. The program will choose alternately heliostats with high or low useful energy, as a function of the difference between the set-point and the power sent to the receiver at the earlier 'instant', until the error which is made when 'moving a heliostat' is larger than the error made if that heliostat is not 'moved'.

5. RESULTS AND CONCLUSIONS

The power sent to the receiver was calculated for October 10, 1984 from the five-minute irradiance DAS readings from 11:00:00 until 14:30:00.

The results of this simulation are listed inTab.7.4-1. A 'result' is recorded whenever the difference between the set point and the calculated power sent to the receiver is greater than 5 kW or whenever a new 'heliostat move' is required, in which case the difference is computed after each 'move'.

Fig.7.4-2 shows the irradiance, the set point, and the power sent calculated at the time that moves are made (the moves are those listed in Tab.7.4-1). The relative errors between the set point and the estimated power sent to the receiver are less than 1% except in two instances when the set point was less than the power contributed by the set of heliostats, which must necessarily be in track.

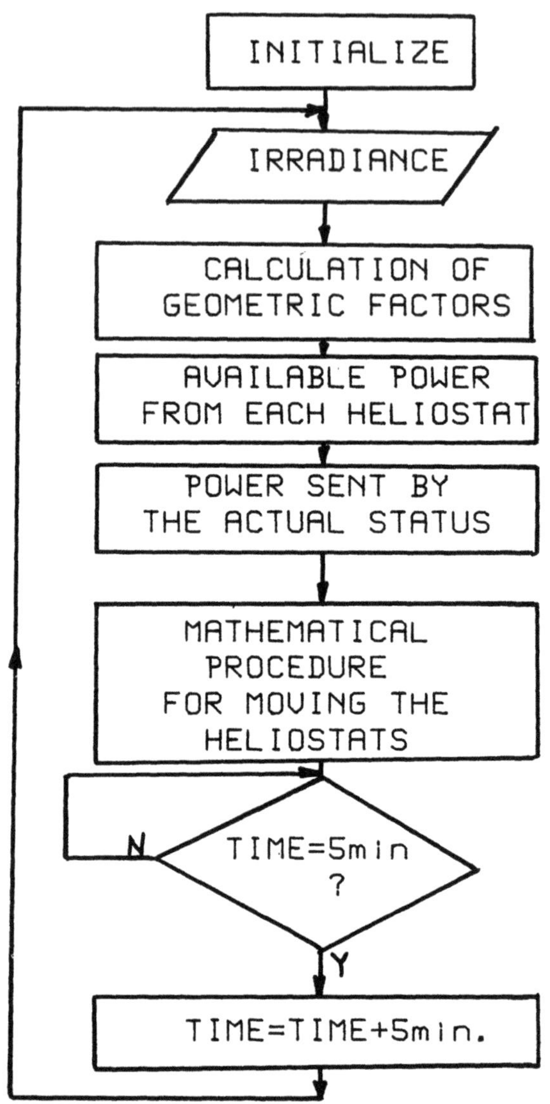

Fig. 7.4.-1: Block Diagram of "CONTROL" program

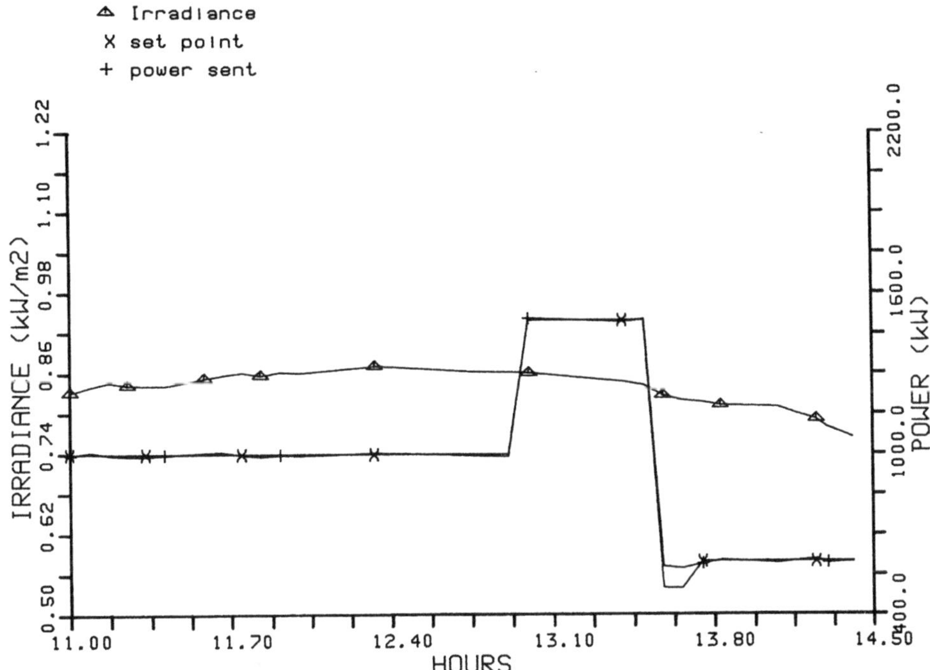

Fig. 7.4.-2: Irradiance and Control of Set Points and Power Sent
Simulation Date : 84/10/10

SOLAR TIME HH MM SS	IRRADIANCE (kW/m2)	SET POINT (kW)	DIFFERENCE (kW)
11 00 00 TO STANDBY 45. 35. 49. 43. 47. 53. 33. 75. 55. 81. 51. 89. 37	0 83175	1000 00	3 40
11 05 00 TO STANDBY 69. 65. 79. 85	0 83994	1000 00	-8 50
11 10 00 TO STANDBY 36. 41. 91. 87	0 84719	1000 00	4 06
11 15 00	0 84250	1000 00	8 57
11 20 00	0 84194	1000 00	7 99
11 35 00	0 85306	1000 00	-5. 97
11 40 00 TO STANDBY 111	0 85831	1000 00	-7 92
11 45 00 TO STANDBY 106	0 86137	1000 00	4 08
11 50 00 TO STANDBY 112	0 85725	1000 00	8 70
11 55 00 TO STANDBY 100	0 86269	1000 00	2 11
12 00 00	0 86069	1000 00	5 89
12 20 00	0 87100	1000 00	-5 72
12 45 00	0 96306	1000 00	5 74
12 55 00	0 86206	1000. 00	8 91
13 00 00 TO TRACKING 35 59 75. 43. 79. 37. 36. 81. 65. 87. 83. 49. 85 53 89. 112. 51. 91. 55. 41. 106. 100. 45. 47	0 86156	1500 00	-6 23

SOLAR TIME HH MM SS	IRRADIANCE (kW/m2)	SET POINT (kW)	DIFFERENCE (kW)
13 25 00 TO TRACKING 44	0 84781	1500 00	6 74
13 30 00 TO TRACKING 37	0 84212	1500 00	-0 40
13 35 00 TO STANDBY 44. 112. 102. 67 73. 35. 56. 65. 71. 41. 106. 97. 37 69. 100. 108. 43. 77. 101 75. 79. 103. 81. 49. 107. 87 83. 51. 99. 85. 53. 109. 45. 39. 93. 91. 47. 105. 55 95. 57. 59. 111	0 82694	500 00	-79 59
13 40 00	0 81981	500 00	-72 95
13 45 00 TO TRACKING 54	0 81725	600 00	9 11
13 50 00 TO TRACKING 111	0 81256	600 00	-3 77
14 05 00	0 81006	600 00	7 89
14 10 00 TO TRACKING 59	0 79937	600 00	1 50
14 15 00 TO TRACKING 57	0 79162	600 00	-7 29
14 20 00	0 77931	600 00	5 66
14 25 00 TO TRACKING 90	0 76437	600 00	0 96

Table 7.4.-1: Results of Calculations for the Power sent to the Receiver on Oct.10,1984(see text)

POTENTIAL FOR IMPROVEMENTS

POTENTIAL FOR IMPROVEMENTS

INTRODUCTION

This evaluation area (section) addresses one of the original objectives of the SSPS project -- the possibilities of this technology for the future. Certainly all of the evaluation reported on in this volume contributes to developing possibilities for improvement of the Central Receiver System. Each of the shortcomings of the SSPS system led to improvement suggestions so, as these were identified, modifications were indicated, and even small disappointments have led to suggestions for improvements.

An example of an effort in one area which led to improvement ideas resulted from a major calibration effort performed by A. Brinner from the DFVLR, Stuttgart, and led Brinner and W. Schiel to prepare the report, IMPROVEMENTS IN MEASUREMENT EQUIPMENT. In preparing this report they drew on experience from the SSPS - CRS to make concrete suggestions regarding measurements in future experimental power plants.

Their conclusions are:

1 - Measurement sensors should be calibrated frequently.
2 - More measurement points are needed.
3 - Control and measurement systems should be separated.

Reviewing the performance data from all of the evaluations reported in this volume, as well as some SSPS/ITET internal reports, J. Martin and R. Carmona suggest improvements which are practical with the SSPS - CRS, in POTENTIAL FOR IMPROVEMENTS. In this report they make estimates of the effect these suggested improvements could have on the overall efficiency and the reliability of the plant, presenting these effects in a graphical form. Further, they make a distinction between changes on the present plant at a reasonable cost and modifications which might affect the design of future and larger plants.

Their conclusions are:

1 - Small efficiency improvements are possible by frequent heliostat washing, heliostat tracking and checks of beam quality.
2 - Major improvements are possible by changing the PCS and redesigning the heliostat field layout - perhaps to 20%
3 - Plant loads (parasitics) must be reduced.
4 - Sodium is not a good storage medium.

In an attempt to bring "lessons learned" into this evaluation effort, and to make use of these experiences, C. Selvage comments on the positive aspects of the various solar thermal electric approaches, used to date by the community, and develops some possibilities for future design considerations -- specifically, the experiences with sodium as a heat transfer fluid and its limitations as an energy storage medium led to the suggestion of a sodium/ salt system.

The conclusions are:

1 - Thermal storage is a very important charateristic of solar thermal systems. This fact is often overlooked.
2 - Sodium is an excellent heat transfer fluid and has demonstrated its usefulness as a receiver coolant.
3 - Molten salt is a much better thermal storage medium than sodium.

IMPROVEMENTS IN MEASUREMENT EQUIPMENT

Andreas Brinner, DFVLR, ITP, Stuttgart

1. INTRODUCTION

Operational experience with the SSPS Central Receiver System has demonstrated that there are several aspects of the measurement system that need improvement. These are:

1) Periodic calibration of the measurement equipment to guarantee the availability and accuracy of the Data Acquisition System.

2) Additional measurement equipment to separate the measurement sensors from the control sensors and installation of spare sensors at all important measurement points.

3) Separation of the Plant Control System and the Data Acquisition System in terms of computers and data evaluation programs.

The modifications to the existing measurement system that are suggested have arisen because of errors in the layout or operating strategy of the CRS or new developments in measurement techniques.

2. PERIODIC CALIBRATION

Because the measuring sensors are the most sensitive parts of a data acquisition system, their behavior as a function of temperature, aging, and mechanical and electrical stresses must be well known. Thus, they have to be calibrated and checked from time to time. The following are the most important sources of error that can lead to the destruction or inaccuracy of a sensor:

1) Improper installation of a sensor can destroy or change the active sensor head or its connections.

2) Mechanical stress during mounting often can change the physical properties of the sensor.

3) Mechanical vibrations such as those caused by motors can slowly change the properties of some sensors; for example, the magnetic properties of a permanent magnetic flowmeter.

4) The measurement behavior of some sensors changes during its operational life because of changes produced in the material properties by high temperatures or temperature shocks. One example is a thermocouple which changes because of "aging".

5) Operation of a sensor above its normal operating range often leads to the destruction of the sensor or its signal conversion cards.

6) A continuous change in the cabling, signal-transmitting, or signal-conversion because of corrosion or constant overheating can produce a slow change in the sensor's output. This effect will not be immediately obvious because it occurs slowly and continuously.

7) Repeated breakdowns of the power supply or excess voltage can strongly influence the life span of sensors and signal conversion cards.

In addition to these sources of error there are others which depend on special physical properties of the sensor used. However, the net result of these malfunctions is an inaccuracy in the measured values, and therefore, an inexact data evaluation. For this reason, periodic calibration of the measurement equipment is absolutely necessary.

A time-table for calibration of the measurement equipment should be developed that contains as a minimum the following items:

1) All sensors which have to be calibrated.

2) For each sensor the time interval between two calibrations.

3) An exact description of how to do the calibration.

4) A list of all necessary funds for every calibration.

Exact records must be maintained on the calibration performed. Only when the calibration is performed properly and periodically, a guarantee is given that measurements recorded are accurate

3. SEPARATION OF MEASUREMENT AND CONTROL SENSORS

Besides calibrating the measuring sensors, it is important to separate the measurement equipment used for normal operations from the measurement equipment used specifically for data evaluation.

The most important reason for this separation is that the demands on a control sensor are different from those on a measuring sensor. A control sensor must have a long lifetime, must be easy to handle and robust, resistant against mechanical, thermal and electrical stresses, and should be low-priced. On the other hand, it is inaccurate ($\pm$3%) and has a long response time. At many parts of the plant even threshold value controllers are sufficient for the operation.

Conversely, a measuring sensor is sensitive to all stresses, can be difficult to handle, and is usually expensive. However, it is more accurate, has a lower response time, and a better signal-to-noise ratio. Thus, control sensors are not sufficient for making the exact measurements necessary for many data evaluations. Consequently, two independent measuring systems each with its own sensors, are necessary. This arrangement offers the additional advantage that the normal operation of the plant will not be disturbed by failure, exchange, or calibration of one or more sensors.

Experience over the last three years with the Central Receiver System has shown that the present operational and data acquisition equipment (thermocouples, level meters, flowmeters, pressure indicators, etc.) is sufficient. However, in the future plant control and measurement systems at solar plants should be completely separated and the following additional measurement equipment added:

1) An Automatic Control System for single heliostat alignment.
   This system allows periodic measurement of the deviations of the beam of each heliostat from the correct position without disturbing the normal operation. Such deviations result from high wind velocities, growing backlash, slackening of the mirrors, and sag of the foundation.

2) A flux measuring system like the HERMES developed by DFVLR-Stuttgart to measure optical power from the mirror field into the cavity and spillage losses. With this system all solar-specific efficiencies and losses can be measured.

3) A sunshape device to determine the exact stair-step by hand of the circumsolar factor. Without this measurement the stair-step error lies in the range of $\pm15\%$.

For the most important measurements (like the sodium flow or the receiver temperatures) it should be normal procedure to install two independent sensors for the same measurement in order to guarantee safe operation.

4. SEPARATION OF CONTROL AND DATA ACQUISITION SYSTEMS

To completely separate the control and measuring systems requires that each system have its own computer, neither system will be disturbed by the other.

To perform the measurements and store the resulting data, the Data Acquisition System should be equipped with a small front-end-computer like a DEC POP 11/23 with own programming. A dialog program is proposed that allows the evaluator to change the conversion tables of the sensors, select sensors for special measurements, connect additional sensors, vary the scanning rate of each sensor, control the sensor- and conversion card functions, and activate spare sensors in the event of failure without disturbing the normal measurements. All data should be stored on magnetic data tapes but with a data format that can be read and evaluated by all of the participants.

The Data Acquisition System should permit a direct data evaluation. This would require the installation of data processing programs to evaluate the measurement results.

The most important of these is a program for the evaluation and association of all daily, monthly, and yearly data both in its raw and edited forms.
One part of the program should allow to associate the data by polynomials (for example, 1. to 5. order) and another part should perform statistic evaluations of the data . Both parts offer the possibility of detecting statistical relations in the data itself and between groups of data.

The scaling of the plots and the arrangement of the lists must be programmable.

For the detection of program mistakes and their elimination, all program sources must be accessible, and it should be possible to incorporate modifications in the programs.

To concentrate on the data evaluation the handling of the programs should be as simple as possible; therefore, a programming mask is proposed.

CALIBRATION OF RELEVANT MEASURING SENSORS AT THE 500 KW_e CENTRAL RECEIVER SYSTEM AT ALMERIA, SPAIN

Andreas Brinner, DFVLR-Stuttgart, Germany

1. INTRODUCTION

To operate and evaluate the 500 kW_e central receiver system it is necessary to measure temperatures and flowrates of the sodium at different locations, particularly at the heat transfer system.

In the whole system thermocouples (mainly Ni-NiCv sensors) are provided for temperature measurements. One characteristic of thermocouples is that they change their output (voltage) after some time of operation; this is called aging. Thus, occasional removal of the sensors is necessary to check how their characteristics have changed. With the data of the examinations of the thermocouples, the conversion factors (voltage - output (mV) into temperature (°C)) must be corrected.

Also the flowrate is measured at different points of the sodium circuit. However, most of the points are only important for the operation and control of a power station. Thus, they may have a maximum deviation of ± 3% relative. Only the values at one point are used for the determination of the heat flows, losses, and efficiencies of the solar heat transfer system. The relability of the statements about the efficiencies depends decidedly on the accuracy of the measurements at the measuring point LK01CF01.

Therefore, this flowmeter was tested and exchanged by a new calibrated one to confirm the deviations which could be found since its installation.

The examination of the thermocouples and the exchange of the flowmeter including the evaluation of all test results were done by DFVLR-Stuttgart in 1984. All necessary instruments were provided by DFVLR.

2. CALIBRATION METHOD AND ARRANGEMENT FOR THERMOCOUPLES

Normally, a thermocouple is calibrated by measuring the output (mV) at some exactly known temperatures, converting the outputs into temperatures, and comparing these results with those from a reference. A calibration oven is normally used to produce different temperatures. The thermocouple-outputs are measured by a digital voltmeter.

The measurement arrangement is presented in Figure 1. To hold the temperature of the furnace chamber on a constant level, a copper compensation body with two holes for thermocouples is used.

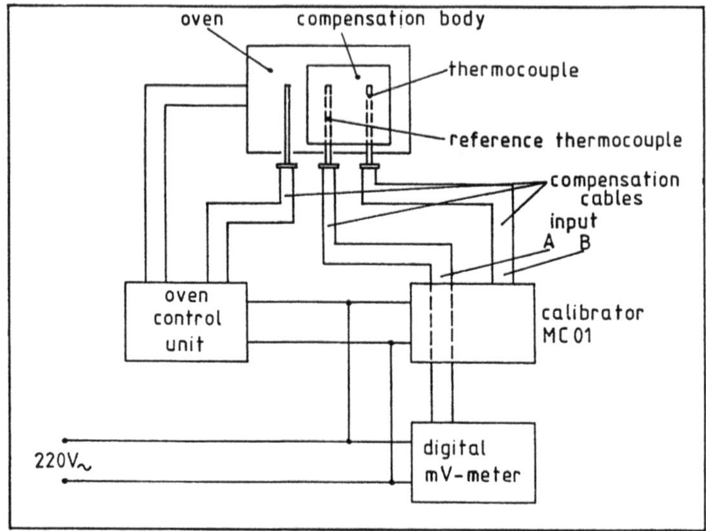

Figure 1: Block Diagram of the Measuring Arrangement

The thermocouples are placed in the holes of the compensation body and each thermocouple is connected with the input of the calibrator.

The calibrator is able to measure two thermocouples at the same time using two different inputs. On a LED-indicator the output (mV) of the thermocouples or of the temperature of the thermocouples in °C or °F can be displayed.

For compensating the influence of the ambient air temperature the calibrator is equipped with an internal 0°C, 20°C, or 50°C cold junction device.

3. CALIBRATION PROCEDURE FOR THERMOCOUPLES

First, every sensor had to be removed and exchanged against a duplicate of the same in order not to disturb the normal operation of the CRS. Then the sensor was put into one hole of the compensation body. Then the oven was switched on to heat up to a temperature of 450°C. The temperature was controlled by the reference thermocouple (Degussa) and the calibrator.

Close to the nominal point of the temperature the mV-output and also the temperature of the thermocouple and the reference type were measured by steps of 1 K. After the measurement was taken the compensation body in the furnace chamber was cooled down by nearly 100 K. This procedure was repeated until the compensation body was cooled down to finally 100 °C.

Four temperature ranges were selected:
1. 450 - 400°C
2. 350 - 300°C
3. 250 - 200°C
4. 150 - 100°C,

where the following data were measured:
1. Output (mV) of the reference type
2. Temperature of the reference type
3. Output (mV) of the thermocouple
4. Temperature of the thermocouple.

The following list of ten thermocouples was proposed by the ITET for calibration. Additionally, two thermocouples installed in the former Sulzer-receiver were measured.

Number of sensor	Object
LK01 CT01	Sodium temperature cold storage
LK01 CT01R	Inlet temperature receiver
LK01 CT36	Outlet temperature receiver
LK02 CT08	Inlet temperature hot storage
LK02 CT01	Sodium temperature hot storage
LK02 CT03	Inlet temperature steam generator
LK02 CT05	Outlet temperature steam generator
LA01 CT01	Inlet temperature heat exchangers
LA01 CT04	Inlet temperature water
LB01 CT01	Inlet temperature steam
Sulzer-Receiver	
LK01 CT03	Inlet temperature receiver
LK01 CT25/26	Outlet temperature receiver

Table 1: List of Recalibrated Thermocouples

4. RESULTS OF THE EXAMINATION OF THERMOCOUPLES

The relation between the measuring points of a thermocouple is described by a polynominal of second order. This acceptance is allowed for temperatures under 600°C. Above this temperature polynominals up to the seventh order have to be taken. A polynominal of second order is described by the following equation:

$$y = a + bx + cx^2 \qquad (1)$$

y = temperature (°C)
x = output thermocouple (mV)
a = coefficient (°C)
b,c = coefficient (°C/mV)

With the data of the measurements the fitting curves have been calculated for all thermocouples using the least squares method. These fitting curves, presenting the temperature as a function of the output voltage, have been plotted.

For comparison of 0°C- and the 20°C- reference lines of a standard NiCr-Ni-sensor are plotted, too. For an example of the plotted fitting curves, see Appendix 1.

Interpretating the plots of all thermocouples it has been confirmed, however, that the deviations between the linearized reference lines and the fitting curves of the thermocouples are as large as the temperature differences, which have to be measured for example between the thermocouples LK01 CT36 and LK02 CT08.

The deviations are in most cases nonlinear. This results probably from the installation of unaged thermocouples. These thermocouples age in operation otherwise than those aged under laboratory conditions.

Therefore, a polynominal of second order, as described at the beginning of this chapter, is proposed for the conversion to obtain higher accuracy. The complete set of the coefficients a, b, and c of the polynom are listed in Table 2.

Sensor	Coefficient a (°C)	Coefficient b (°C/mV)	Coefficient c (°C/mV)
LK01 CT01	- 5.47	25.54	- 0.0534
LK01 CT01R	+ 0.02	25.80	- 0.0537
LK01 CT36	- 0.34	25.41	- 0.0485
LK02 CT08	- 7.21	25.96	- 0.0788
LK02 CT01	- 0.80	25.63	- 0.0642
LK02 CT03	- 11.93	26.25	- 0.0816
LK02 CT05	- 11.68	25.90	- 0.0705
LA01 CT01	- 8.04	17.96	- 0.00889
LA01 CT04	+ 8.39	18.00	- 0.00869
LB01 CT01	+ 7.16	17.97	- 0.00616

Table 2: Coefficients a, b, c for polynominal fitting of the thermocouple mV-output according to Equation (1)

With the proposed polynominal and its coefficients a conversion can be programmed, which reproduces the behavior of the thermocouples. As a result of testing the thermocouples it has been shown that the deviations of most sensors are within the guaranteed tolerance of the manufacturer and the DIN-order (± 0.6 - 1% relative or ± 2 K absolute). However, because the temperature differences to be measured are in the range of the tolerance, a correction of the method of conversion is absolutely necessary.

5. CHECKING OF CABLING

5.1 Checking Procedure

Like the calibration of the thermocouples, it is also important to check the cabling of the sensors. Therefore, the cabling from the connection of the sensors up to the output in °C at the monitor of the data acquisition system was checked.

A voltage was injected at the connections of a thermocouple. The indicatedtemperature of the calibrator was compared with the temperature displayed at the monitor of the data acquisition system (DAS). In this way it was possible to check the whole signal line once at all.

5.2 Results of Testing the Cabling

Cables of the Sensor	Calibrator-display (°C)	Monitor-display in the control room (°C)	Mean Deviation (K)
LK01 CT01	250.0	251.67	+ 1.45
		251.23	
LK01 CT01R	251.67	252.18	+ 0.76
		249.83	
	530.0	530.68	
		531.67	
LK02 CT08	300.0	300.49	
	450.0	451.0	
	500.0	501.6	+ 1.02
	550.0	551.0	
LK02 CT01	460.0	460.49	+ 0.34
		459.82	
LK02 CT03	300.0	300.0	0.0
LK02 CT05	300.0	300.0	0.0
LA01 CT01	50.0	48.4	
	100.0	97.65	
	150.0	148.01	- 1.32
	200.0	249.57	
		250.0	
LA01 CT04	300.0	300.0	0.0
LB01 CT01	250.0	248.47	
	300.0	298.97	
		298.1	
	350.0	348.6	
	400.0	399.0	± 1.36
	450.0	449.0	
	500.0	500.54	
		501.0	
	550.0	553.0	

Table 3: Deviation in Temperature Display Between Calibrator and DAS Monitor

The monitor displays exactly the correct values of the sensors LK02 CT03, LK02 CT05, and LA01 CT04. It is, however, not possible to correct the displays of the sensors LK01 CT36 and LB01 CT01, because at these sensors there is a deviaton range around the correct value and not a constant but changing deviation above or below the correct value. The rest of the sensors can be corrected with the stated deviations.

The introduction of an individual reduction factor (Table 5) is proposed for each thermocouple.

6. PRINCIPLE FUNCTIONING OF A PERMANENT MAGNETIC FLOWMETER

Th principle function of the sodium flowmeters provided for CRS is based on Faraday's law of induction:

$$U = \frac{B \cdot V \cdot K_1 \cdot K_2}{d} \qquad (2)$$

where B = magnetic flux density
V = volumetric flow
d = inside diameter of pipe
K_1 = correction factor for the geometrical data
K_2 = correction factor for the magnetic field

According to the law,a voltage is induced into an electrical conductor moved through a magnetic field at a right angle. A flowmeter operated according to this law consists of two parts:

- a permanent magnet attached on the pipe, producing parallel magnetic field lines across the flow
- two electrodes fixed at a right angle to the magnetic field by which the induced voltage can be measured

The induced voltage is exactly proportional to the flow.

7. COMPARISON OF THE ORIGINAL AND THE NEW CALIBRATED FLOWMETER

To check the original flowmeter, that means to determine the single deviations and the total deviations in comparison with the new calibrated one, five process variables were recorded to check that at every experiment the measuring conditions were constant.

These process variables are:

Reference	Description	Units
1. LK01 CF01	Receiver flowrate	m^3/hr
2. LK01 CE01	Cold pump speed	RPM
3. LK01 CP01	Cover gas pressure	bars
4. LK01 CT01	Cold storage outlet temperature	°C
5. LK01 CT01R	Receiver inlet temperature	°C

Table 4: List of All Recorded Process Variables

At the first test of the flowmeter it was stated that the magnet had been shifted in z-direction (= flow direction) and that the magnet was displaced with respect to the electrodes.

Furthermore, it has been supposed that the magnet had changed its characteristics during its running time. For this reason, the magnet was sent to INTERATOM for recalibration and readjustment of its geometrical data. At the examination of the flowmeter it was stated that the magnet had slid down the pipe and that the magnet was displaced with respect to the electrodes.

8. ESTIMATE OF ERRORS DUE TO REVERSE FLOW OPERATION

A method to determine receiver losses is to perform experiments at low temperature levels with small sodium flowrates in reverse flow direction.

The most important result learned was that the flowmeters are not damaged by these experiments. There is only a deviation during the experiment, which at once disappears at the operation in right flow direction. There are two problems during the operation in reverse flow direction:

1. The voltage will be negative. According to the manual of flowmeters, however, the connections and the transmission cards have to be exchanged in order to convert only positive signals. Therefore, no exact statement about the displayed flowrate and about any possible deviation can be made.

2. The second source of error is the geometrical adjustment. For the operation in both directions the x- and y-adjustments are the same ones, but the adjustment in z-direction must be changed.

9. RESULTS OF THE EXAMINATION OF THE FLOWMETER LK01 CF01

The examination of a flowmeter at measuring point LK01 CF01 showed the following sources of error:

1. Over temperature shocks, mechanical stress during the mounting, and mechanical vibrations during the operation have a noticeable influence on the magnet. They cause a changing of the magnetic field. This causes a new adjustment of the mechanical mounting of the magnet on the pipe.

This changing of mounting causes a changing of the output voltage of the electrodes. The new output voltage comes to:

$$E = 0.24998 \ \frac{mV}{m^3/hr} .$$

The deviation in comparison with the original value comes to:

$$E_d = 0.096\% \text{ per } m^3/hr.$$

2. Because of an insufficient fixing of the magnet on the pipe there results another deviation than expected after the recalibration of the magnet.

Flowrate (m^3/hr)	Deviation (m^3/hr)	(%)
10	- 0.016	- 0.16
20	- 0.112	- 0.56
30	- 0.21	- 0.69

Table 5: Deviation in Z-Direction Converted to Even Flowrate Figures

This deviation can be explained by the magnet slipping down to the position z = -1.6 mm.

3. A displacement between the electrodes and the magnetic field together with a deviation in x- and y-direction causes a reduction of the output voltage.

Flowrate (m^3/hr)	Deviation (m^3/hr)	(%)
10	- 0.13	- 1.3
20	- 1.06	- 5.3
30	- 1.99	- 6.6

Table 6: Deviation Because of a Displacement, Converted to Even Flowrate Figures

4. It has been stated that the deviation does not become zero at a flowrate of 0 m^3/hr. The best possible explanation is the effect of an internal forced heat convection of the sodium.

   The deviation caused by this effect lies in the range of ± 0.5 m^3/hr at 0 m^3/hr rated flowrate.

5. The cabling of the flowmeter showed an average deviation of ± 0.13% over the whole measuring range.

The proving of the conversion card set of the original flowmeter had as a result a mismatching between the output voltage and the current signal:

Flowrate (m^3/hr)	Deviation (m^3/hr)	(%)
10	- 0.133	- 1.332
20	- 0.533	- 2.664
30	- 1.2	- 3.996

Table 7: Mismatching Between the Output Voltage and the Current Signal

Of all mentioned deviations only the influence of the cabling (deviation = 0.13%) can be ignored.

As a summary it can be stated about the inaccuracy of the flowmeter LK01 CF01:

In the range around a flowrate of 0 m^3/hr the accuracy is good, but in this range a deviation of 0.5 m^3/m must be subtracted from the displayed flowrate. At higher flowrates the examination had the following data as a result, which is obtained by adding all deviations:

Flowrate (m^3/hr)	Deviation (less displayed) (%)
10	+ 2.79
20	+ 8.52
30	+ 11.29

Table 8: Deviation of the Flowrate of Flowmeter LK01 CF01

These deviations are much higher than the expected inaccuracy of ± 2.5% over the whole measuring range.

10. REFERENCES

1. SSPS Technical Report Nº 3/81. Belgonucleaire, Belgium, 1981.

2. M. Becker, H. Ellgering, and D. Stahl, "SSPS CRS Construction Report SR2", DFVLR, Germany, 1983.

3. M. Becker, SSPS Technical Report Nº 4/83, DFVLR, Germany, 1983

4. SSPS CRS Final Documentation, Vol. 27. INTERATOM, Germany, 1980.

5. SSPS CRS Final Documentation, Vol. 33. INTERATOM, Germany, 1980.

6. Degussa, "Meßtechnik-Mantelthermoelemente", Degussa, Germany, 1980.

7. Degussa, "Compensating Cables", Degussa, Germany, 1981.

8. Degussa, "Meßtechnik-Thermoelemente. Internationale Grundwertreihen", Degussa, Germany, 1981.

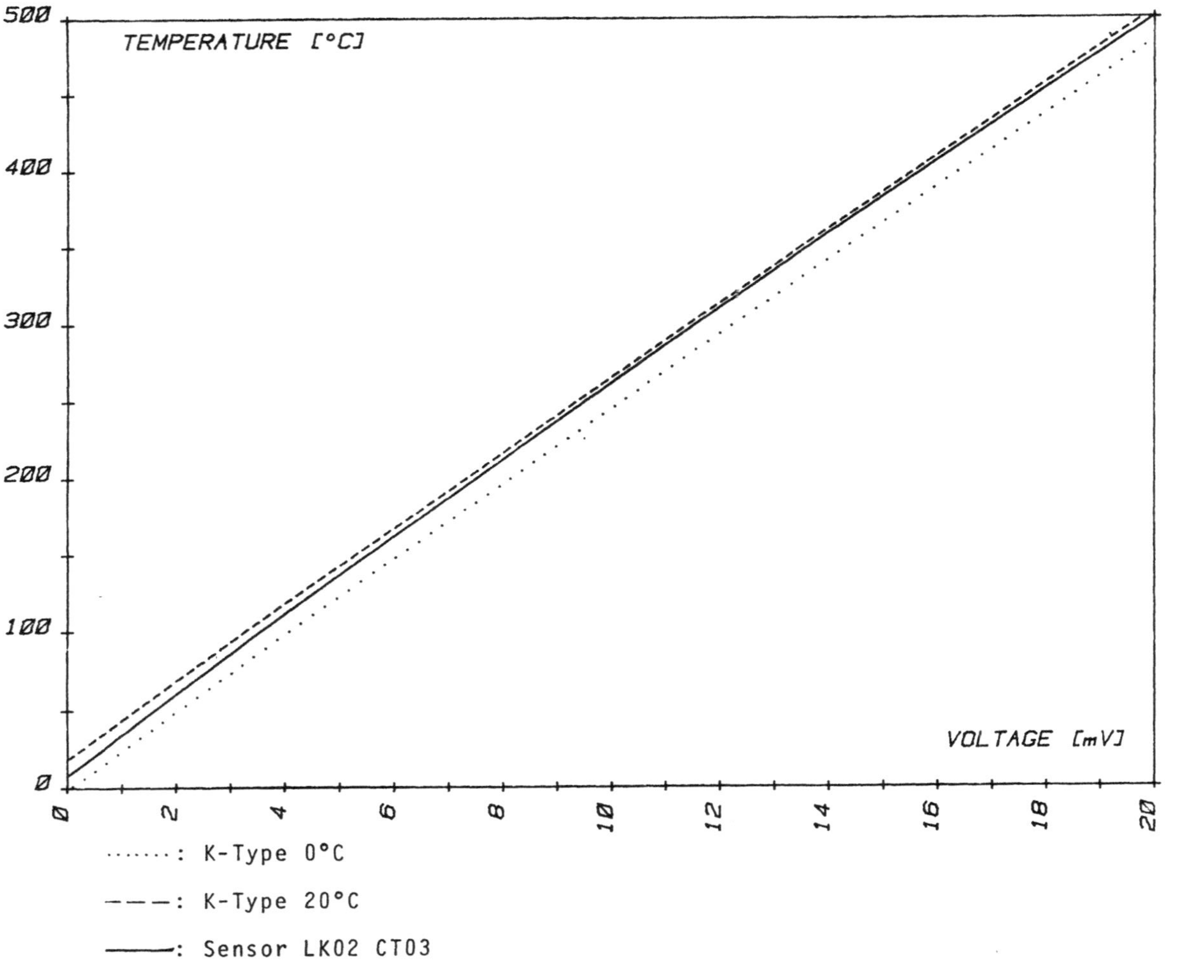

TEMPERATURE [°C]
VOLTAGE [mV]
0
100
200
300
400
500
0
2
4
6
8
10
12
14
16
18
20
: K-Type 0°C
: K-Type 20°C
: Sensor LK02 CT03

Calibrator: DEGUSSA MC01
Reference: DEGUSSA type E

Test Data of the Sensor: LK02 CT03

Reference V/mV	Reference T/°C	LK02 CT03 V/mV	LK02 CT03 T/°C
17.39	440.0	17.39	440.0
17.36	439.0	17.39	439.0
17.31	438.0	17.31	438.0
17.27	437.0	17.27	437.0
11.97	311.0	12.00	311.4
11.94	310.0	11.96	310.5
11.90	309.0	11.92	309.5
11.85	308.0	11.88	308.6
11.82	307.0	11.84	307.6
11.77	306.0	11.80	306.7
11.73	305.0	11.76	305.7
8.73	231.0	8.77	231.9
8.69	230.0	8.72	230.8
8.65	229.0	8.68	229.7
8.61	228.0	8.64	228.6
8.57	227.0	8.61	227.6
8.53	226.0	8.57	226.6
4.95	136.0	4.97	136.6
4.91	135.0	4.94	135.7
4.87	134.0	4.90	134.8
4.83	133.0	4.87	134.0
4.78	132.0	4.83	133.0
4.74	131.0	4.78	132.0

POTENTIAL FOR IMPROVEMENTS

José Martín and Ricardo Carmona, ITET

SUMMARY

There are lessons to be drawn from the SSPS experience that may help evaluate the potential of central receiver technologies. Our experience confirms that sodium-cooled receivers can collect solar energy efficiently and reliably; it also confirms the need for large scale, not only on the basis of manpower considerations but also because of the relative impact of plant loads and the increased efficiency of larger subcomponents.

Discussions on possible improvements should start with an assessment of what the performance of the CRS has been. 'Efficiencies' evaluated at some instant are usually misleading. Operating days can be found when noon efficiencies have been about 13%, even when the field was soiled and the PCS was operating at half of its rated capacity. With a clean field, and operating at rated high power, the efficiency could be as high as 15%.

A different picture appears when the estimates are made on the basis of a whole operational day. On a day which was deemed 'good', the overall efficiency has been estimated as 8.1% with 25350 kWh potential energy offered to the heliostat field and 2065 kWh generated. These numbers are disappointing, particularly with the CRS in house plant load at about 1472 kWh. Optimistic extrapolations are possible: the average PCS efficiency for that date was only 17%. This weighed heavily on the overall generation and could certainly be improved.

Unfortunately, few days were 'good'. To discuss possible improvements, it is helpful to differentiate between those which raise the average efficiency in a good day, and those which raise the plant availability.

With a modest effort on the present CRS only minor efficiency improvements are possible. A higher average field reflectance may be achieved by either increased washings or antisoiling coatings, and spillage may be reduced through corrections in tracking and better image quality. Major efficiency improvements require a new PCS and, to a lesser extent, a redesigned field layout. If all of these changes are implemented, the overall daily gross efficiency can be raised to 18%, even for a plant of the size of the CRS.

Availability improvements are harder to quantify. A better field/receiver match and, of course, a suitable PCS would raise availability. Other measures are software modifications to add flexibility to normal operation, start-up and filling-up sequences, lightning protection, better heliostat images and tracking accuracy, and an improved signal for the feedforward control action for the receiver. Some of these changes can be implemented in the present CRS.

Operation of the receiver at higher than rated temperature will reduce or eliminate the losses which arise now because hot sodium must be cooled wastefully. However, future designs must take thermal inertia effects into account, so that plants can be operated effectively within the limits intended in the design.

It has been proven that control technology can ensure safe transient operation even for high flux (i.e., very efficient) receivers.

Dramatic increases in the receiver efficiency are not to be expected. A plant which is about an order of magnitude higher than the CRS can operate at an overall gross efficiency of 20% on a good day. Extrapolation to even larger plants is not wise.

Energy storage in sodium is not an elegant or economical solution. Also the relative impact of plant loads must be lowered substantially. These disadvantages must be weighted against those of alternative receiver technologies. Finally, a major effort needs to be made in the matching of subcomponents and in improving system reliability. Control technology can make a sound system work, but it cannot make it sound.

1. INTRODUCTION

There are several problems with any discussion on the potential for improvements on central receiver systems. The open-endedness of the subject is not the least of those problems.

The discussion on the potential will start with an assessment of what the present CRS does. The plant will be represented in terms of blocks, chosen in such a way that actual measurements or accurate estimates can be made of how much energy goes into and out of each block.

The reader may use these blocks to estimate the effect of changes. If the reflectance of the field is 0.85 on any one day, he may speculate on how much better (or worse) the average reflectance, and the plant performance, may be.

In this paper, we shall propose possible changes of those numbers -- with investments which may be modest or not. We shall also propose different numbers for larger plants.

The blocks are interesting, but insufficient. They may fail to indicate where major problems lie. Reliability and subcomponent matching are major problems of the CRS, and they offer the most 'potential for improvement'. A discussion of thermal inertia illustrates to the need to match subsystems. Spillage losses do also.

Sodium-cooled receivers are highly efficient and controls can ensure their safe transient operation. Other solar central technologies, such as direct absorption receivers, are needed for high-temperature applications which may have a large impact on energy supply. Reliability and component matching are problems which must be emphasized in the future development of any of these technologies.

2. PRESENT CRS PERFORMANCE

A reasonable prerequisite to any discussion on the potential for CRS performance improvements is an assessment of the present plant performance. This assessment may be made at some instant or design point, or over a long-time period--a plant lifetime, a year, a 'normal' or 'good' day.

That noon or design loss stair-steps may be misleading is no news to anybody, although there is some value in such descriptions -- we shall attempt one.

Since the ASR has not been operated for a whole year, we may also choose a 'good' day as the evaluation period.

2.1 Noon Performance

The 'loss tree' for operation in a short period about noon on June 7 will be chosen as an example. The period sampled is between 12:10 and 12:30 solar noon, when the mean irradiance was 925 W/m^2, and 86 heliostats were in track.

The blocks in the energy diagram in Fig.8.2-1a are 'loss blocks' connected at nodes where energy can be measured or estimated. The numbers have been estimated for June 7.

The first three blocks refer to the heliostat field: the first accounts for cosine losses, shadows, and blocking. Using a program developed by the ITET (Ref.1), it has been computed that 0.116 of the power incident on the 86 heliostats on track was lost at noon because of these geometrical effects. The second block accounts for reflectivity losses. From SSPS/ITET reflectivity data and models (Ref.2, 3), we know that the mean field reflectivity at that time was 0.83. The third block is for spillage and attenuation losses. These have been adapted from results calculated from a HELIOS run (Ref.4). These represent 4% of the total power sent by the field to the receiver.

The losses in the receiver, which is the next block, are calculated from the data on the power absorbed in the sodium, from the DAS. The efficiency of the receiver is estimated as 0.89 (Ref.5); the total collected power is 1975 kW.

The losses in the pipe between the receiver and the tank have been estimated (Ref.6) to be about 20 kW. At noon on June 7, 1580 kW were sent to the PCS--the rest was being stored in the tank. The losses in the tank are about 30 kW (Ref.7). The total losses on the pipe and tank are about 3%. The reason why only 1580 kW are sent to the PCS on the day chosen is that on this day the PCS is being run at 50% of its nominal rating.

This introduces some ambiguity in the estimate of efficiency. The PCS efficiency at 330 kW is 20.8%. Using this number to complete the calculation, one obtains an overall efficiency of 13% for this period. On the other hand, if 'efficiency' is interpreted as the ratio between the gross electrical power output and the solar power offered to the field during this period without making any allowance for the extra power being stored, the non-efficiency is only 11%. The noon losses for June 7 are illustrated, in slightly different form (the 'Christmas tree presentation') in Fig.8.2-1b, prepared by P. Wattiez.

It may be a worthwhile exercise to discuss the potential for improvements in each of these blocks -- notably in the PCS. However, noon efficiencies do not give a complete picture of what one needs to know about a plant performance.

2.2 Performance During a Good Day

Consider the plant performance on a 'good' day: February 21, 1984(Fig.8.2-2a). The first heliostats were put in track at 09:15, and 82 heliostats were already on track at 09:35, when the insolation level was 525 W/m^2. The losses associated with the first block (shadow, blocking, and cosine) have increased in comparison with the noon losses, so that the geometrical factor has gone down to 0.781. For this day, the mean field reflectance was 0.853, and the spillage losses accounted for 8% of all the power sent by the field. The day average efficiency for the receiver is 86.8%. Almost all of the 13500 kWh are sent to the hot tank, because electricity has been generated the previous day and the cold tank was relatively close to its nominal temperature in the morning.

From the 12270 kWh sent from the tank to the PCS, approximately 2000 kWh were spent in start-up (Ref.7); 2065 kWh of electrical power were generated, for an overall daily efficiency of 16.8%.

The average efficiency for the CRS for the whole day is 8.1%. The 'tree' presentation of the losses for February 21 is given inFig.8.2-2b.

Fig.8.2-3 gives the daily data for April 9, 1983 with the Sulzer cavity receiver. These data are reported in Ref.8.

2.3 Why Are Good Days So Scarce?

What is a useful definition of a 'good' day for evaluation?

A good day for evaluation should be

(1) reasonably clear,
(2) one where the heliostat field is sent to track early and a relatively high fraction of the heliostats (say 90%) are in track continuously whenever the irradiance is above some minimum value (say above 400 W/m^2),
(3) one when electricity is generated without failures in the PCS, and finally
(4) one when the DAS operates without failures

There is one major problem with these criteria for choosing a 'good' day -- the data base. The number of such days for the CRS in 1984 is one (or zero, depending on what one means by reasonably clear). That day was February 21.

The effort to find the good day is not fruitless because it helps to identify major areas where improvements are needed:

(1) subcomponents are not sufficiently reliable,
(2) the PCS is not suitable for its purpose, and
(3) the heliostat field is not matched to the receiver.

Improvements on these areas may not affect the loss stairstep for any one normal day (or the noon efficiency) but will certainly help to have more 'good' days.

In the next section we shall analyze in more detail some of these areas. (External factors, such as grid failures and nonoperation of the heliostat field during weekends, also affect the plant unavailability. These factors will not be discussed here.)

Many design and operation changes can be suggested to improve the performance of a CRS plant. These may range from a minor suggestion on increasing heliostat washing frequency to a complete reorientaton of concept and scale.

We are going to analyze the effect of possible improvements by dividing their scope into two major categories: 1) changes which can be made realistically, with a small or modest investment, in the present CRS system, and 2) changes which are not limited by these immediate concerns.

3. MODIFICATIONS TO THE PRESENT CRS

The changes which could be made to the present CRS can be further subdivided into changes to improve the overall daily plant efficiency and changes to make the plant more reliable. Clearly, there is an element of arbitrariness in this division--if one were to consider changes to improve actual efficiency averaged over the year, for example, the reliability-related improvements would not be separable. The distinction is made simply to facilitate the discussion.

3.1 Changes to Improve Daily Thermal Performance

Increase of Power into the Receiver

The amount of power reflected by the present field of 93 heliostats is a strong function of time, and the (noon) peak power incident on the receiver is about 27 MW. The receiver itself could collect some more power (possibly about 20% more), if the heliostat field were larger.

The extra heliostats would not only allow a higher peak power, but they would make it possible to flatten the daily power curve.

As part of the GAST project, 30 new heliostats are being added to the CRS field. They are being installed here to test the heliostats, using the receiver as a cooled target. At the moment, it is not intended that their controls will be integrated with those of the rest of the field.

This integration is not a difficult or costly task. It will make the facility more flexible, and will increase the operation time. We suggest that it be carried out.

Incidentally, a program being developed at the site (Ref.9) aims at shaping the flux power level at the receiver. The combination of the added heliostats and this program will make it possible to improve the system performance and the system research potential.

Receiver

The efficiency of the present receiver is about 87%, averaged over the whole day. It is hard to conceive practical design changes which may raise this figure dramatically.

The concept behind the ASR was that the efficiency could be increased in a high flux receiver because the losses would be smaller in the smaller active area. The actual receiver performance seems to confirm the validity of this concept.

One possible area for marginal improvement is in the absorptance. The receiver may in fact be soiled: on tests reported in Ref.10, dusting Pyromark coated tubes lowered the average absorptance to 0.88. In the same test, when tap water was allowed to run down the vertically held tubes to simulate rain effects, and later allowed to dry, the absorptivity measurements ranged from 0.9 to 0.95, showing an almost complete recovery to the pretest values. Of course, washing can only be effective if the coating adheres well to the tubes. Still, it is possible that simple washing, or washing followed by repainting, may increase the receiver efficiency. Unfortunately, no tests of the absorptance of the receiver surface have been conducted on site. Absorptivity measurement equipment is now available at the site, on loan from Sandia. The measurements must be performed on cold tubes, and this means that the receiver will have to be emptied. Because of the ordering of evaluation priorities, a decision was made earlier to postpone these measurements.

Receiver Lateral Protection

Spillage and tracking errors heat the uncooled panels which surround the receiver and force the operator to put heliostats in standby to avoid overheating. This problem is likely to get worse with the heliostat additions, or if the receiver is operated at a higher power level. It is therefore suggested that the problem be evaluated so as to permit operation with higher incident fluxes at the lateral panels. It may be desirable to replace the lateral protection.

Because the day chosen for the daily characterization was a winter day, spillage was not a major problem and therefore the effect of improved lateral protection would not show on the energy budget for that date. However, for a given summer day (say June 7), the additional power that could be sent to the receiver if all the field could be placed on track can be roughly estimated as 1500 kWh.

Sodium Pump

The pump was originally designed to operate at flow rates higher than 3 m^3/hr. After the advanced sodium receiver was installed, the pump had to be modified so as to speed its transient response (Ref.11). With this change, there are stability problems at low flows and therefore the minimum rate has been raised to 6 m^3/hr. This high minimum imposes limitations on operation: at a given inlet temperature, the receiver cannot operate at insolation levels as low as it ordinarily would. It is suggested that the pump controls be modified so as to reduce the minimum flow rate.

Receiver Controls

The feedforward action of the CRS control system depends on the solar heat flux data: this action limits temperature peaks when the flux rises sharply. It is more important at low flow rates, when response times are longer. Unfortunately feedforward is imprecise, because the heat flux is not estimated accurately. There are ten photovoltaic sensors, mounted on heliostats in the field. The average of the output from these sensors is the signal used: there are several problems with this arrangement. The cells are mounted horizontally; the gain is not constant, but depends on sun angle. They measure global insolation: a cloud which causes a sharp drop in the flux at the receiver, may not induce as sharp a drop in the global insolation. Another difficulty occurs because of shading of the sensors, particularly in winter (Ref.12).

(Feedforward depends on the insolation and its time derivative. Because of the imprecise estimation of the solar flux, the derivative action is the relevant one: its main purpose is to protect the receiver from overheating when the insolation rises.)

It is desirable to modify the way in which the average flux is calculated and to improve the feedforward action of the control system.

This modification would not affect the overall efficiency in a clear day, but could make an important difference in days when there are passing clouds.

High Temperature Operation

Although the loss stair step is guiding this discussion, the 1984 CRS Program of Work (POW) has emphasized the concept of thermal inertia. In fact, the POW refers to a 'paradigm shift' when adressing the issue of inertial effects such as warm-up and start-up losses. This concept has been advanced and discussedin considerable detail (Ref.13).

An outstanding example of a 'thermal inertia effect' arises because of the need to reach, at the hot tank, the temperature required for PCS operation as quickly as possible. This time is shortened by taking hot sodium out of the tank and sending it to the steam generator -- to cool it down so that it can be sent to the cold tank. The amount of energy thrown away by this procedure has been as high as 4 MWh in some days; 1 or 2 MWh is a more characteristic number.

The time needed to heat up the tank has another undesirable consequence: in some days the operating temperature cannot be reached at all. The collected energy is kept in the hot tank until the next day when it is dispersed again.

A small program has been written to estimate hot tank temperatures and sodium volume as a function of sodium outlet temperature and flow rate. This program simulates the actual tank performance rather closely.

The results of calculations for sodium volume and temperature from actual initial and operating conditions and receiver output for a specific day are plotted in theFig.8.2-4. The volume in the tank at the start of normal receiver operation is 15 m^3: the temperature is 380°C. To hasten the heating-up, sodium is sent to the steam generator at the same time as sodium is fed to the tank. It takes about one hour and forty minutes for the sodium to be heated to 500°C, the normal operating temperature for the PCS when the receiver outlet temperature is 530°C. Obviously, this time can be reduced if the receiver outlet temperature is higher. The results of assuming that the receiver outlet temperature is raised to 560°C, assuming no change in the rate at which the operator allows sodium to flow out of the hot tank, and that the reciever collected power remains unchanged. The second assumption implies that the flowrate into the tank is reduced.

The result of the higher outlet temperature is that there is less hot sodium at any moment in the tank (although the stored energy is the same), but the nominal temperature is reached about 35 minutes earlier.

The second figure represents a fictitious situation when no sodium is allowed out of the tank -- i.e., no energy is dissipated. The initial conditions in the tank are that the temperature is 400°C and the volume is 7 m^3 (the minimum). Runs at identical receiver power at 530 and 560°C outlet temperature are simulated. At the lower temperature, the tank reaches 500°C in three hours, at the higher temperature, almost one and one-half hours earlier.

The assumption of 'constant power' is misleading because the losses increase with receiver and tank temperature. From simulation and the results of an experiment when the receiver has been operated at 560°C, it can be stated that the differences are small. The efficiency drops from 90.2 to 89.8% when going from 530 to 560°C.

An alternative or complementary solution to the problem of the hot tank is to connect the hot tank directly to the cold tank. If there is a need to bleed the hot tank, sodium can be sent to the cold tank without losing the energy. If there is no operation during one or two days, for example, both tanks will have cooled down. Sending relatively hotter sodium from the hot to the cold tank makes it possible to heat both tanks more quickly. There is a limit to how effective this technique may be: this limit depends on the relative sizes of both tanks, and the sodium inventory. This limit would be reached only in exceptional days. To take advantage of this alternative, one must add a pipe with two valves.

PCS

The present CRS overall performance is heavily penalized by the poor characteristics of the power conversion system (PCS). The appropriate modifications in the sodium heat transfer system (receiver, pipes, storage, etc.) will depend on what changes are to be made in the PCS, if any.

The present engine is not a successful or well-integrated subsystem. It has many reliability problems. A modification of an original four-cylinder design, the five-cylinder engine needs steam at 500°C so that steam heating its fourth cylinder is still at 400°C; however, nowhere in the engine is thermal energy converted above 400°C. Therefore, in effect, one needs 500°C to operate a 400°C convertor. A separate heat exchanger lowers the steam temperature.

The high start-up losses and the mediocre steady-state performance mean that the overall PCS efficiency during one day is about 16.8%.

It may be possible to modify the engine 'back' to the 'normal' Spilling engine design, so that it operates at lower pressure and temperature (32 K and 380°C). This modification will have to be made by the manufacturer. Then it would not be necessary to reach 500°C in the tank - in fact, the PCS could not use sodium at this high temperature. If the modifications were successful, the worst aspects of the present engine would be eliminated.

It is possible to feed the steam generated in the CRS to the DCS/PCS. In fact, a proposal to do this was made in 1982. The modifications involve the installation of new steam pipes, and of a water injector to reduce the steam temperature. The present CRS/PCS would simply be eliminated.

A major advantage of this approach would be the savings in workers' time. At present, the start-up of the two PCS place heavy demands on personnel. So does the maintenance of the Spilling engine.

The modification also eliminates the CRS start-up losses. The DCS turbine would operate for many hours, synchronized to the grid, at rated power. Finally, the overall conversion efficiency for the plant would rise, not only because of the elimination of the CRS/PCS start-up losses, but because the DCS/PCS turbine has in fact a higher efficiency than the Spilling engine, and it would be even higher if it were to operate at its rated power for a longer time.

Another major advantage is the possibility to operate the receiver at a lower temperature, adding flexibility to the operation of the CRS.

Of course, the idea of heating sodium to 530°C and then cooling it down to 300°C to generate electricity is apalling. It is bad enough to cool it down to 400°C, as it is does at present. The fact that this solution may make sense for this plant at all is an indication of the unsuitability of the present engine.

(It is instructive to keep in mind that the 10 MW turbine at Barstow operates at 35% efficiency. We shall return to this when discussing the CRS potential without limitations.)

3.1 Changes to Improve Reliability

In the discussion on why are good days so scarce (Section 3.3), we stated that

(1) subcomponents could be made more reliable,
(2) the PCS was not suitable, and
(3) the heliostat field is not matched to the receiver.

Operation may be affected or interrupted by multiple component failures: HAC and DAS blocks, electronic circuit, valve leak, etc. A reliability study, based on a fault tree analysis such as the one being carried out for the DCS, may help to identify those areas where redundancy in design or extra care in maintenance will repercute in longer normal operation.

On the item of the PCS, further references to its penalizing effect on performance is a sport very akin to beating dead horses. A discussion on the heliostat field is more profitable.

Suitability of the Heliostat Field

The heliostat field is relatively reliable (barring lightning strikes and transformer short circuits). However, the field is not tracking properly early in the morning and late in the afternoon. It is certainly not tracking well enough for the small ASR.

Spillage is a problem, particularly in the summer. Because of spillage, the walls which surround the receiver overheat and activate the receiver alarms which are set at 300°C. To avoid this type of 'receiver trip', the operation team sends some heliostats at the edges of the field out of track. Because less power reaches the receiver, the nominal receiver outlet temperature cannot be reached quickly. The time required to reach the operating temperature is made even longer because of the relatively high minimum flow in the receiver -- 6 m^3/sec. The operation implications reverberate the whole day.

So, evidently, at present, either the field or the software is not suited to the ASR.

Software Modifications: Increased Flexibility

The present software is rather inflexible: it requires a long time to perform some tasks. To change the aiming point, for example, the computer must be off line and the target coordinates must be changed in each heliostat, one by one. If it is noted that the aim of one heliostat needs to be changed, the whole field must be put on stow, first, and then the computer must be taken off line. Only then can the trouble heliostat be reaimed. Similar cumbersome procedures must be followed if one desires to change the flux distribution on the receiver.

The software may, and should, be made more flexible to allow file modifications while the field is operating.

Filling the receiver is one application which may benefit from software modifications. If these make it possible to set the aiming points for the heliostats independently of each other while on line, one may design a flux map on the receiver so as to minimize gradients and stresses. If the change in the aiming point has to be made while the 'HAC' is out of line--and the field in stow--this operation may take a long time indeed.

Software Modification: Start-up Sequence

At present, in order to place the heliostats in track, the operator must give a long and cumbersome sequence manually. For the whole field, the operation may take more than forty minutes.

The start-up procedure may be completely automated. In fact, the field control (HAC) is programmed to do this, according to the instructions contained in the so-called 'COMMAND FILE'. Unfortunately, the instructions can be followed automatically if and only if all the heliostats are operational. If any one is out of order, the procedure does not work.

It is suggested that the file be ammended to allow for automatic start-up even when some heliostats are out of order.

Lightning Protection

The field has been struck twice by lightning in two years. The strikes did much damage and forced the plant out of operation.

Several experts met on Thursday, July 19, 1984, to study ways to protect the field. A report from the meeting summarizing the expert's recommendations, is being prepared by the Plant Operating Authority. Their main suggestions were to improve the protection of the electronic units and to improve the grounding (earthing) of the lightning rod.

It is suggested that lightning protection should be improved, taking into account these recommendations.

Recovery - Avoidance of Receiver Overheating

After power failures in the grid, the heliostats 'lose' their orientation. When power returns, the computer must give each heliostat, in turn, its old position. It takes about two minutes to give instructions to our small heliostat field. During this time, the flux image moves away from its intended focus - the outer walls heat up, and the receiver is operated under conditions for which it was not designed; namely, with the flux concentrated in the outer panels. This is not acceptable, and the consequences would clearly be more dramatic in a larger field.

Separate feeding of the electronic units with a small uninterrupted power supply (UPS) would solve this deficiency.

At present, all heliostats must be stowed every time the grid fails. This means a loss of 20 to 50 minutes of operation for every failure. With the small UPS, the heliostats would not have to be stowed, and there would be no loss of operation.

Tracking Accuracy

It has been noted that either the heliostats or the software which controls the field are not tracking accurately enough to permit operation with the whole field early in the morning and late in the afternoon. The reasons for this difficulty and the possible remedies are not clear. A report of work on progress is being presented at this review (Ref.14).

4. PERFORMANCE OF AN IMPROVED PLANT DURING A GOOD DAY

What would a 500 kW_e demonstration plant be like if it was designed now, taking advantage of what we have learned?

Heliostat Field

It has been estimated that the power sent to the receiver from a heliostat field with the same area as the present field could be 8% more on an annual basis for a redesigned plant (Ref.15).

To maintain high reflectivity, the environment at the site is not particularly benign. If a successful anti-soiling coating or washing procedure can be found, it will be particularly appreciated here. Preliminary results are promising (Ref.16). To simulate the conditions prevalent in less dusty sites, it may be appropriate to wash the mirrors more often. The clean average value of the reflectivity is 91%. Doubling the washing frequency, one may use 88% for the representative field reflectivity.

Spillage may be reduced to 4% by correcting tracking errors and improving image focussing. The receiver efficiency may be improved by 2% by optimizing the operating strategies and preparing as good an absorbing surface coating as possible.

An 'adequate' power conversion system operating at the steam conditions which can be attained at the CRS may have an efficiency of 30%.

Taking all these improvements into consideration, the overall efficiency, defined as the ratio between gross electrical power and the potential power offered to the field, on the same day chosen earlier, can be raised from 8.5 to 17.8%. The improved performance is illustrated in Fig.8.2-5a/b.

(We have made no reference to the plant load. The SSPS-CRS is a small plant: the plant load is disproportionally large compared with the plant output. Trace-heating alone consumed on the average 36 kWh(e) per day in the period between May 21 and July 31 (Ref.17). The total CRS parasitic consumption for the CRS for the same period was 1472 kWh per day. This is close to the gross generation of a good day - and certainly higher than the average. Clearly, a plant of the scale of the SSPS-CRS cannot be expected to generate much net electricity, and careful attention should be paid to minimize parasitics in the design of larger plants.)

5. LONG RANGE POTENTIAL

The importance of scale on the long range feasibility of solar thermal generation in CRS has been the object of considerable attention. Optimization models prepared in different countries have identified that the 'optimum scale' is somewhere between 75 and 100 MW_e (Ref.18). Manpower considerations by themselves may establish the need for large plants.

Large plants also lessen the impact of plant load on net power output and help justify the investment in equipment with high performance. We have already noted the high efficiency of the Barstow turbine, for example. Further, careful design may take advantage of pressure heads to drive flows, and make it possible to lower or eliminate trace heating. They may also justify the complexity of configurations which combine the best features of different technologies -- for example, those which use sodium as a heat transfer medium and salt as a storage medium.

It is not wise to try to extrapolate performance data from a system to another which is more than two orders of magnitude higher. For example, using the efficiency reported for the Barstow turbine in the calculation of an improved version of the small plant, the computed efficiency would be 20% -- the same as the noon efficiency predicted from careful studies by the GAST project for a 20 MW_e plant with an air-cooled 800°C receiver (Ref.19). Such extrapolation is not warranted.

The concept behind the billboard receiver was to increase the efficiency by diminishing the losses. This was accomplished by operating at high flux levels and therefore making it possible to reduce the active absorber area. As added bonus, the receiver has a very short warm start-up time. The controls system has succeeded in protecting the receiver from overheatings during transients.

Interestingly, problems have arisen because of the components between which the receiver is the interface: the heliostat field and the rest of the sodium loop. The heliostats (or the software that controls them) have not tracked accurately enough, and/or their images have not been good enough. At the receiver, the lateral wall protection has been the weak link.

The related problem of heliostat errors and receiver protection may be alleviated by emphasizing the same concept on which the billboard is based, i.e., high flux density. A secondary concentrator scheme (Ref.20,

21) may help alleviate concerns about focusing and tracking errors, and help to smooth the flux profile. The reflecting surfaces may act as protection and possibly may act as preheating zones.

This note should not end without emphasizing the importance of subcomponent matching in any optimization effort. The difficulties encountered with the CRS-PCS should serve as a lesson: one is often forced to contemplate design or operation changes so as to accomodate the requirements of a PCS which is not particularly suitable. A wiser approach may be to contemplate those changes by using the sodium temperature at the outlet from the hot tank as a parameter. Carried one step further, this may help give insight into the potential for central receivers.

It is certainly not clear that electricity generation is the most advantageous application of this technology. For example, copious amounts of electricity are needed to produce calcium carbide, an endothermic process which seems eminently suited to be assisted by a central solar receiver (Ref.22). Other processes, such as steam reforming of methane, the cracking of hydrocarbons, coal and shale gasification, are also promising processes (Ref.23, 24).

The high conductivity of sodium makes it possible to have an efficient high flux receiver at medium high temperature ranges (550°C). Further, the properties of sodium are well known and the receiver performance can be predicted accurately (Ref.25). Sodium does not fare as well as an energy storage medium. High fluxes are also possible with other concepts, notably with the solid particle curtain (Ref.26, 27); at a recent review, it has been reported that bauxite pellets have been effectively heated to temperatures up to 1000°C without breaking or agglomeration (Ref.28). It has been suggested that direct absorption receivers may be used to process carbon dioxide atmospheres (Ref.22). We do not know what the efficiency of a particle curtain receiver will be -- but it is not likely that it will be higher than the one for the ASR, which on the other hand has limited applications for high temperature processes. These two approaches may well be suited to different applications and must be compared with other competing technologies (air-cooled receivers, etc.). The problems of reliability may well give one of these technologies an edge.

We have seen that digital controls can make a sound system work. However, that technology cannot make the system sound. We are reminded here of a statement from the members of the faculty at the University of Salamanca, in answer to a complaint that the University did not do much to improve princely intellectual prowess: 'Lo que natura non da, Salamanca non presta'. Control technology cannot make a CRS plant work economically and realiably unless that plant is designed with due care to subsystem matching in realistic conditions, and designed and maintained to be reliable.

6. ACKNOWLEDGMENTS

The authors want to thank C.S. Selvage for his insights on subcomponent matching, realistic conversion efficiency, and thermal inertia. P. Wattiez drew the loss trees. Some of the improvements suggested here to improve plant reliability were suggested by the POA staff. The authors are grateful for this help.

7. REFERENCES

1. M. Blanco, "Calculation of Field Geometrical Efficiency", paper presented at this review (1984).

2. P. Wattiez, "Reflectivity Degradation", op cit.

3. M. Sánchez, "Soiling Effects: Method for Estimating the Field Reflectivity Distribution", op cit.

4. F. Biggs and C.N. Vittoe, "The HELIOS Model for the Optical Behavior of Reflecting Solarconcentrators", SAND6-0347.

5. H. Jacobs and C.S. Selvage, "Results of the Performance of the Sulzer Cavity Receiver and the Franco-Tosi External Receiver", paper presented at this review.

6. M. Andersson and H. Jacobs, "Losses of Piping and Tank", op cit.

7. H. Jacobs and R. Carmona, "PCS", op cit.

8. F. Martinez and R. Carmona, "Statistics, Performances, Stair-Steps", SSPS CRS Midterm Workshop, SSPS Technical Report 4/83 (1983).

9. M. Blanco and M. Sánchez, "Control of Power Sent to the Receiver by the Heliostat Field", paper presented at this review.

10. Pyromark dusting.

11. R. Carmona and J. Martín, "Transient Response: ASR", paper presented at this review.

12. R. Carmona and J. Martín, "The SSPS ASR Transient Response", Second International Workshop on the Design, Construction, and Operation of Solar Central Receiver Projects, Varese, Italy, June 4-8, 1984.

13. C.S. Selvage, "Remarks on Receiver Losses", paper presented at this review.

14. M. Blanco and R. Carmona, "Tracking: Simulation of Errors", op cit.

15. F. Ramos and L. Crespo, "CRS Collector System Optimization Review by Using the ASPOC Program", op cit.

16. I. Susemihl, "Soiling Effects: Coating and Investigation on Glass and Mirror Samples", op cit.

17. A. Cuadrado and P. Wattiez, "Trace-heating Consumption", op cit.

18. J. Gretz, C.S. Selvage, and A. Skinrood, "Report on Lessons Learned", paper read at the IEA Workshop on Large Thermal Arrays, San Diego, California (1984).

19. "Programa Tecnologico GAST", Asinel-INTERATOM (1983).

20. P. Kesselring, "The Secondary Mirror Concept", International Seminar on Solar Fuels and Chemicals, Stuttgart (1983).

21. J. Martín, "A Secondary Sunlight Concentrator", Proceedings of International Solar Forum, Berlin, September 11-15, 1984.

22. J. Andújar and J. Martín, "Closing the Carbon Dioxide Cycle: Solar Manufacture of Acetylene", op cit.

23. R.W. Cawling, J.D. Fish, L.G. Radosevich, and J. Vitko Jr., "Solar Central Receiver Fuels and Chemicals Project Status Report; October 1980 - June 1981", SAND81-8232, Sandia National Laboratories, August 1981.

24. M. Becker, "The Future of Central Receiver Systems", paper presented at this review.

25. A. De Benedeti and M. Blanco, "ASR Performance: Comparison with Simulation", op cit.

26. J. Martín and J. Vitko Jr., "ASCUAS: A Solar Central Receiver Utilizing A Solid Thermal Carrier", SAND82-8203, Sandia National Laboratories, January 1982.

27. C. Royere, "High-Temperature Large Scale ....", International Seminar on Solar Fuels and Chemicals, Stuttgart (1983).

28. P.K. Falcone, "Technical Review of the Solid Particle Receiver Program", SAND84-8229, Sandia National Laboratories, January 1984.

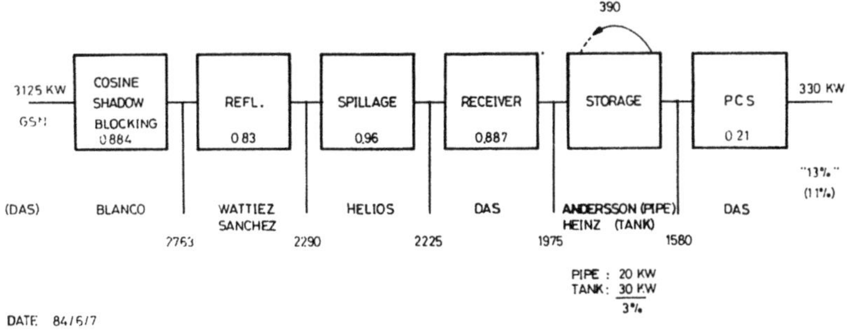

Fig. 8.2.-1a: CRS Noon Performance with the High Flux Billboard Receiver
The Efficiencies

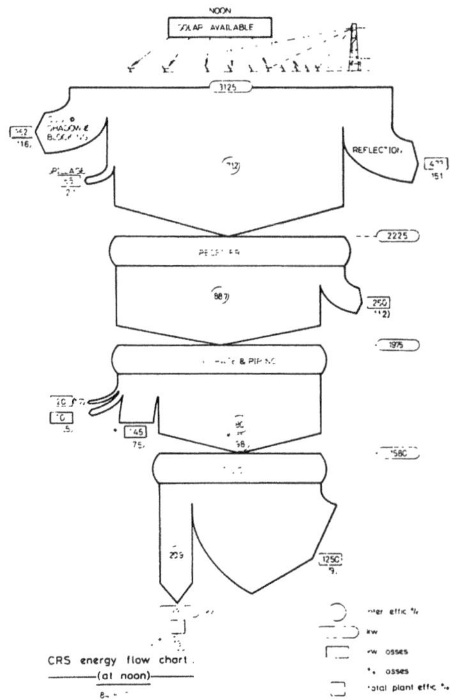

Fig. 8.2.-1b: The Losses

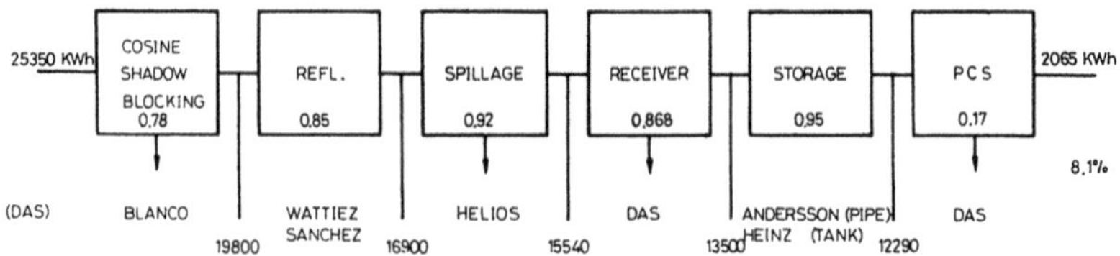

DATE · 84/2/21

Fig. 8.2.-5a: Improved Daily Performances
The Efficiencies

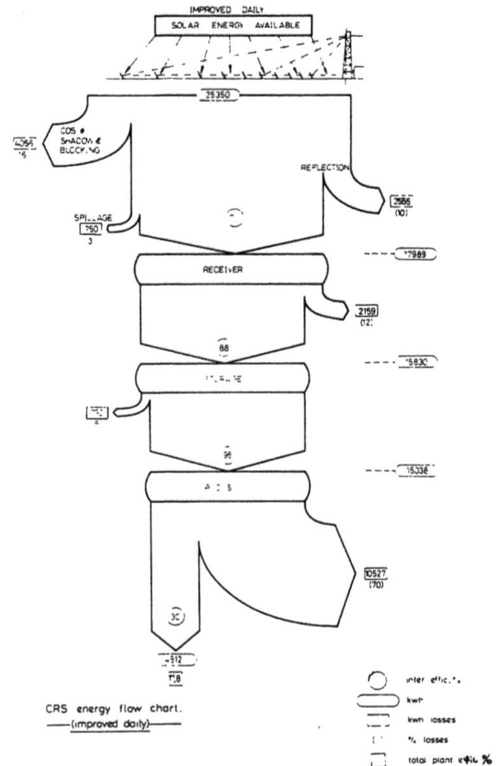

Fig. 8.2.-5b: The Losses

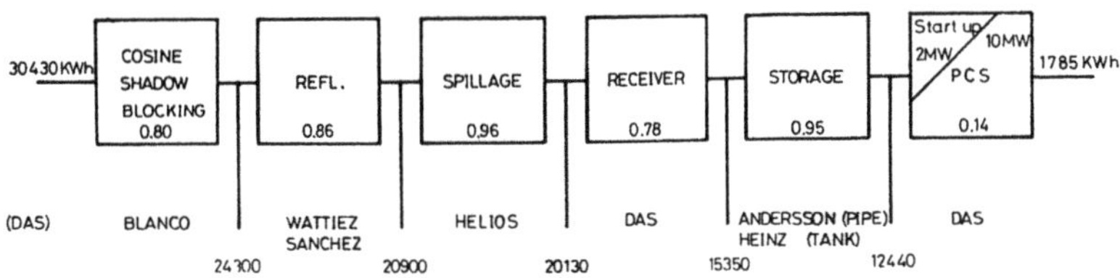

Fig. 8.2.-3: CRS Performances During a Good Day, with the Shallow Cavity Receiver

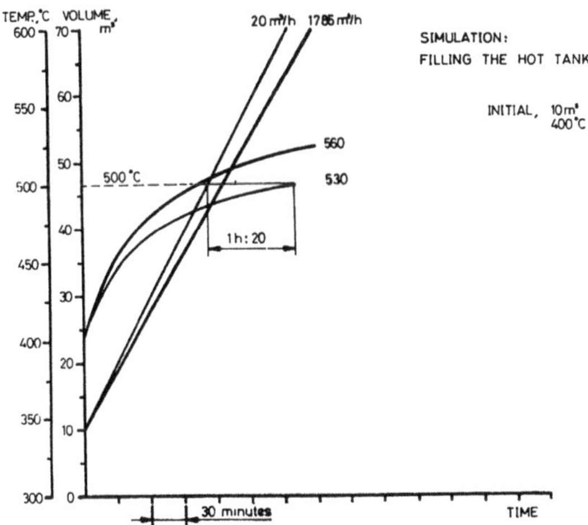

Fig. 8.2.-4: Effect of Raising the Outlet Temperature on the Time Required to Reach Operating Conditions at the Hot Tank

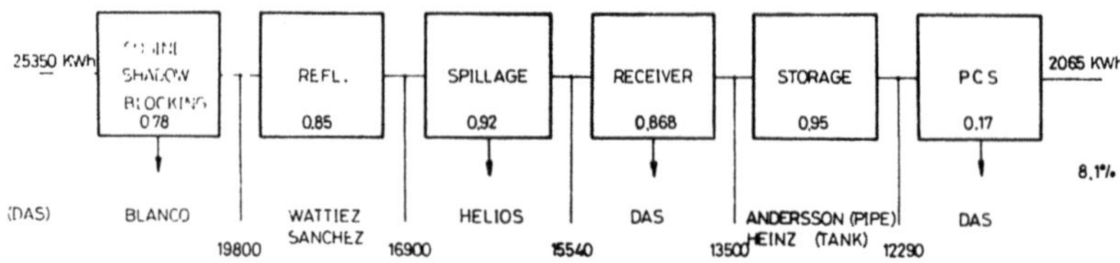

DATE : 84/2/21

Fig. 8.2.-2a: CRS Performance During a Good Day, with the High Flux Billboard Receiver

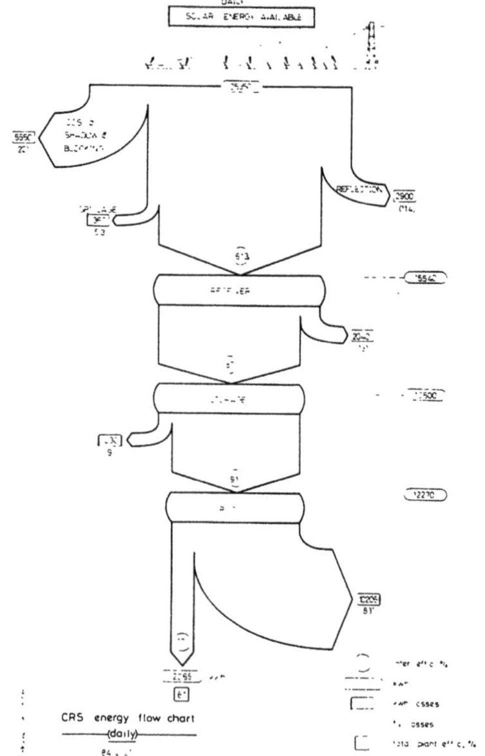

Fig. 8.2.-2b: The Losses

SYSTEMS CONSIDERATIONS

Clifford S. Selvage, ITET

1. INTRODUCTION

Present day solar electric systems are limited to solar photovoltaic systems and solar thermal electric systems. Present indications are that photovoltaic systems are very expensive, and although those costs are decreasing, it will be many years before they are economically feasible. Solar thermal systems seem closer to achieving an acceptable pay back period; however no commercial venture has materialized. Solar thermal has one real advantage over photovoltaic systems in that it contains an achievable, practicable method of energy storage. In this paper, this system advantage is addressed with some suggestions for the improvements that might help reduce the barrier to industrial application.

2. SOLAR ELECTRIC SYSTEMS

Producing electricity from the sun's radiant energy can be accomplished by direct conversion using photovoltaic materials or indirect methods in which the radiant energy is first converted to thermal energy and then converted to electricity. Photovoltaics do not seem to be cost-effective at this time because of the cost of producing the necessary quality of silicon material. Certainly this cost is decreasing as improved methods of purifying silicon are developed and as quantity demands encourage investment in silicon production facilities.

Some rather large photovoltaic systems are in commercial operation. ARCO has placed in service both the 1 MW_e tracking photovoltaic system at Hisperia, California and the 6 MW_e improved tracking photovoltaic system at Carrisa Plains, California. Both were designed, manufactured, and installed as an ARCO investment, with the electricity produced being sold to the electric utility of the specific service area. The energy tax credits laws of California provided the stimulus for this commercialization and as ARCO is a large California company, they are independently able to benefit from this effort. A similar attempt to define and build a solar thermal central receiver plant given the same stimulus has not been successful. So, although much large engineering design and analysis has been accomplished, no solar thermal central receiver plant has been funded by industry for construction in California or anywhere else.

Thermal systems seem to have a particular advantage over photovoltaic systems: the ability to store energy for delivery when there is demand for that energy. With a solar thermal system, it is possible to have a storage system that could allow production of electricity all day and night. Such an approach is possible but seems impractical at this time with present thermal storage systems.

The storage system at SSPS uses the liquid sodium heat transfer fluid that flows through the receiver. Sodium is an excellent heat transfer fluid because of its very high conductivity, and therefore a very poor thermal storage medium because of its low heat capacity. (High conductivity materials have a poor heat capacity; conversely materials with a high heat capacity have a poor conductivity.) So it seems some compromising would be productive in attempting to design a cost-effective central receiver thermal electric system.

The proposed 30 MW_e Carrisa Plains central receiver system was to be sodium-cooled and was to use the receiver coolant as the thermal storage medium just as the SSPS project was arranged. A basic piping design diagram is shown in Fig.8.3-1. The system was to have 70 minutes of thermal storage at full rating, with 400,000 gallons, or 1,514,000 liters of sodium stored in a tank 12 meters in diameter and 16.5 meters high -- and all that is accomplished is storage of approximatley 102 MWh of thermal energy. The cost of this amount of sodium is over one million dollars. Actually the proposed system stored this energy in 1400 cubic meters with a 243 K temperature drop. Sodium has a capacity of approximately 1.27 KJ/KgK or about 73 kWh/m^3. If a 30% conversion efficiency is assumed, that amount of storage seems correct.

In the early consideration of this system, the prime contractor for the design had blocked out the sodium flow as a function of time as shown in Fig.8.3-2. The hot tank levels as a function of time are shown at the top of this figure and the cold tank levels are at the bottom. It is interesting to note that the tank temperatures for the start of the day (925°F or 496°C) are higher than those used at the SSPS. However, the overnight cool-down temperatures (566°C - 496°C = 70°C) are in the same realm as the SSPS. This returns to the subject of thermal inertia, which is presented earlier in these papers. Note that at 16:00, where 370,000 gallons of sodium has a temperature of 566°C, the system is turned off. This is the 1400 m^3 of storage representing about 102 MWh of thermal energy stored.

A molten salt such as 60% $NaNO_3$, 40% KNO_3 which has a specific heat of approximately 1.5 KJ/KgK and a density of 1741 Kg/m^3 at 540°C could store the same 102 MWh in a volume of 575 m^2. Using molten salt, the tank volume at Carrisa Plains (1514 m^3) could store approximately 273 MWh of thermal energy, which would allow the plant to operate for nearly three hours, compared to 70 minutes with sodium. Certainly these figures are approximate because the specific heat of this salt is not that certain and the density of the salt at 530°C was used rather than average density over the temperature limits in the thermal cycle. However, the approximations are close.

As stated earlier, sodium had a high conductivity: in the range of 65 W/mK° compared to 0.398 W/mK for salt at about 540°C. As a receiver coolant, sodium is far superior, in that it allows a higher flux on the receiver, and therefore a smaller receiver and a higher efficiency of the receiver circuit as discussed in the previous paper by H. Jacobs and C.S. Selvage.

3. CONCLUSION

A great deal has been learned about the use of sodium in the central receiver thermal electric system at SSPS. The benefits of using sodium as a receiver coolant have included lower losses, faster response, and lighter weight. At the same time problems have been experienced in using sodium as the storage medium. Another concern has been the possibility of a water/sodium reaction in a sodium-powered steam generator; the possibility of this type of violent reaction would not exist if a salt storage system and salt-powered steam generator were used. All of this leads to the suggestion of a sodium-cooled receiver circuit, a salt storage system, and a salt-powered steam generator in the thermal-to-electric conversion cycle. The additional potential improvement in overall efficiency probably in the range of 10% by use of salt-powered steam reheat should not be overlooked.

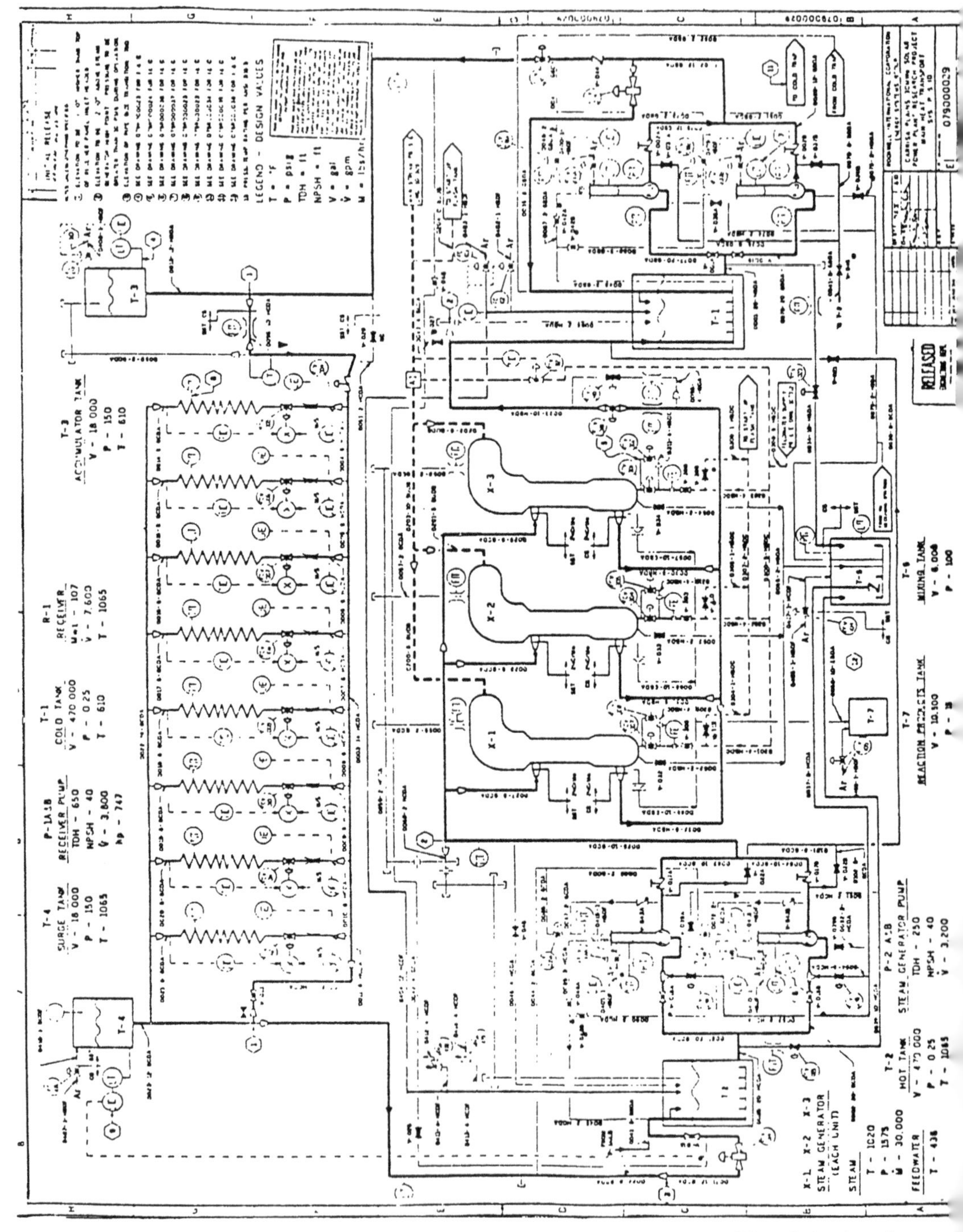

Fig. 8.3.-1a: (see t

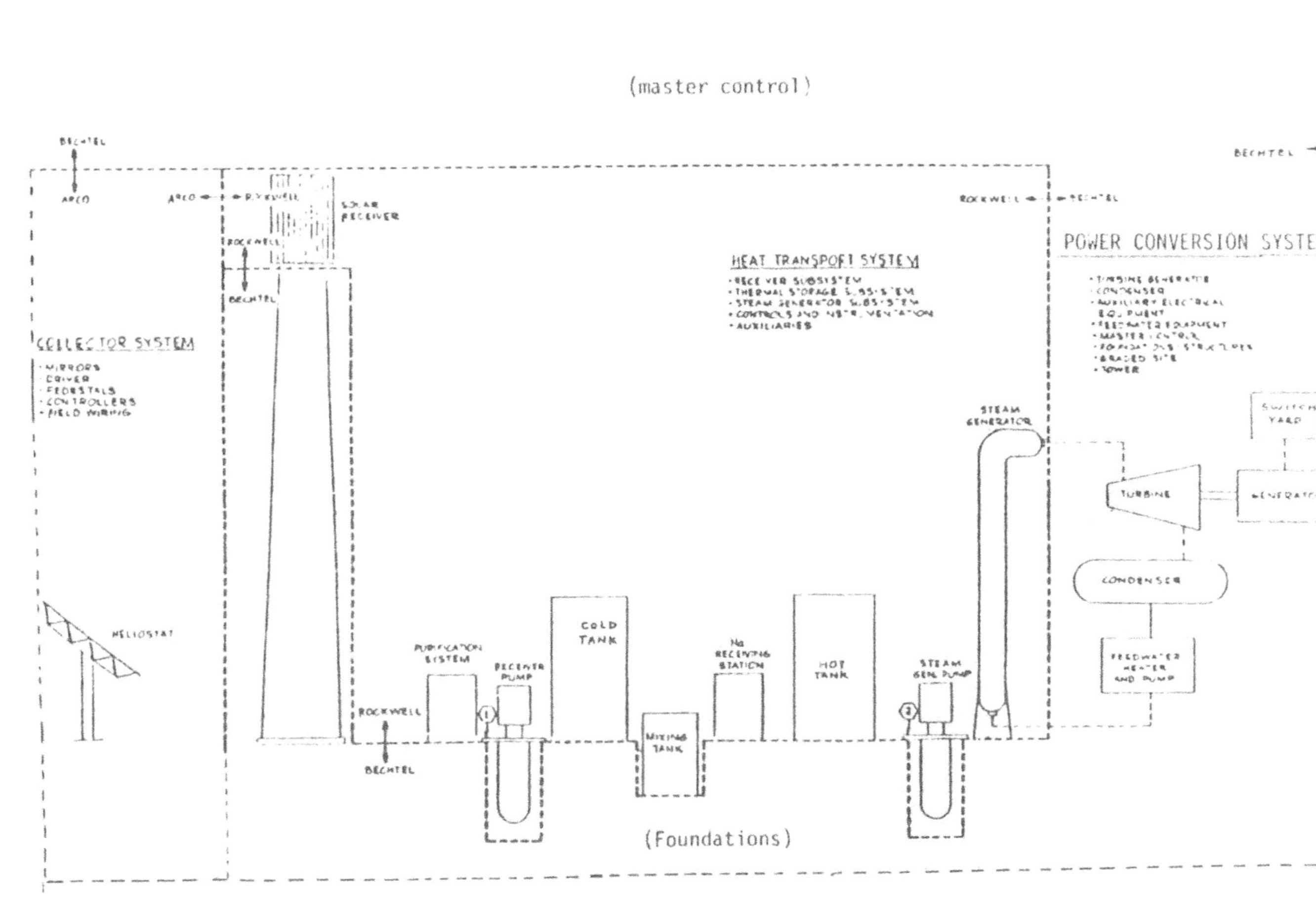

Fig. 8.3.-1b: Systems Identification

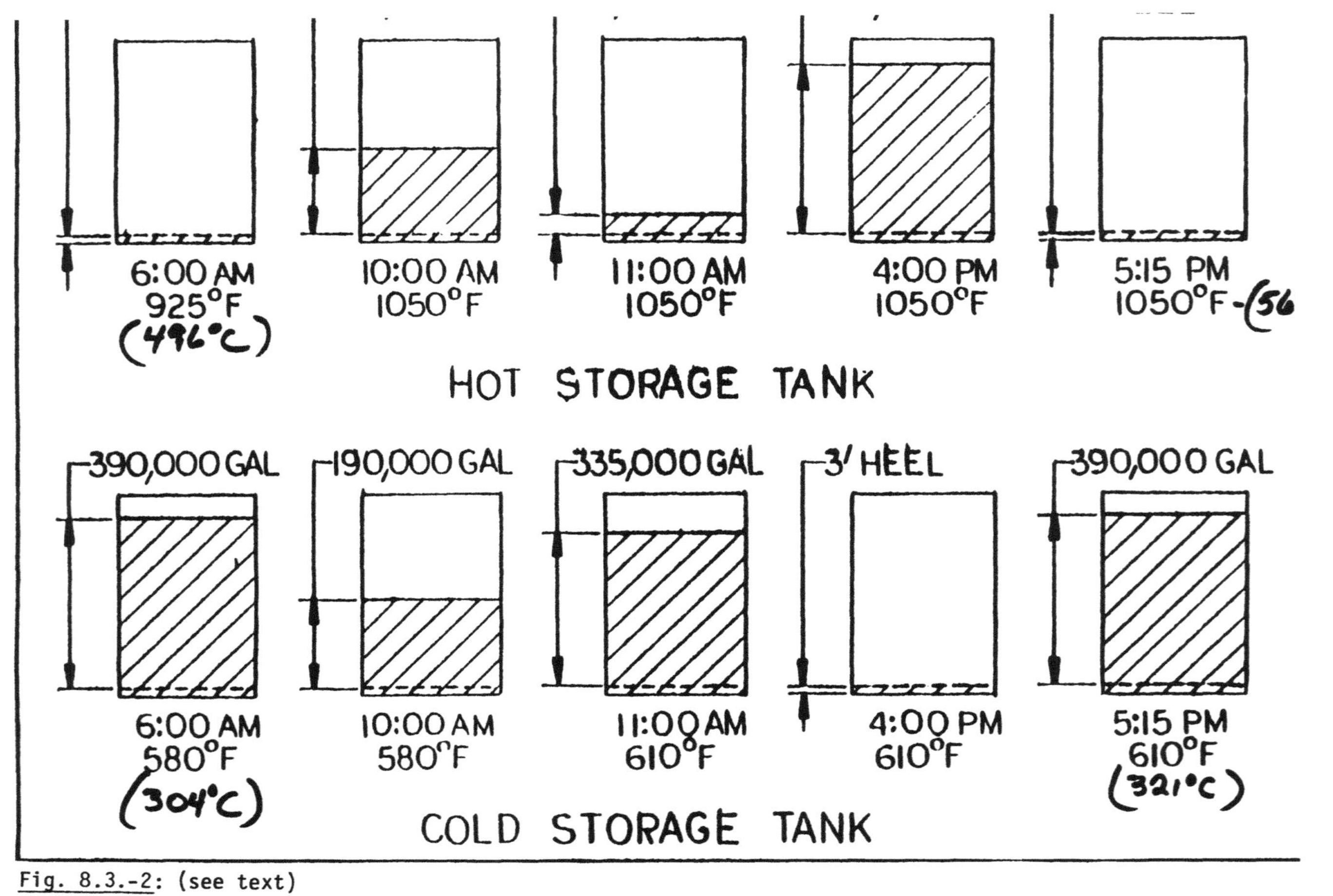

Fig. 8.3.-2: (see text)

APPENDICES

SSPS - CRS BIBLIOGRAPHY

CRS-SSPS SEMIANNUAL REPORTS

Report Nº	Title	Author	Date
SR II	CRS - Construction Report	M. Becker et al	Mar. 1983
SR IV	CRS - First Period of Operation	W. Bucher	May 1984
SR V	CRS-ASR Construction Experience Report	J. Hansen	Sep. 1984

CRS-SSPS TECHNICAL REPORTS

Report Nº	Title	Date
1/79	Heliostat Field and Data Acquisition Subsystem for CRS (by Martin-Marietta)	Dec. 1979
2/79	CRS-Heliostat Field, Interface Control and Data Acquisition System (by McDonnell Douglas)	Dec. 1979
2/80	Analysis of Special Hydraulic Effects in the SHTS Piping System (by Belgonucleaire)	Nov. 1980
3/80	Redesign of the CRS - Almería Receiver Aperture and Comparison of INTERATOM and MMC Heliostat Field Performance Calculations (by INTERATOM)	Nov. 1980
3/81	CRS Instrumentation Review (by Belgonucleaire)	Jun. 1981
5/81	Device for the Measurement of Heat Flux Distributions (HFD) Near the Receiver Aperture Plane of the Almería CRS Solar Power Station (by DFVLR)	Nov. 1981
1/82	SSPS Workshop on Functional and Performance Characteristics of Solar Thermal Pilot Plants, Part II, Results of the Tower Facilities Session (by M. Becker, DFVLR)	Jul. 1982
2/82	Concentrated Solar Flux Measurements at the IEA SSPS Solar Central Receiver Power Plant, Tabernas, Almería (Spain) (by G. von Tobel, C. Schelders, M. Real, EIR)	Apr. 1982
3/82	Effect of Sunshape on Flux Distribution and Intercept Factor of the Solar Tower Power Plant at Almería (by G. Lemperle, DFVLR)	Sep. 1982
3/83	The Advanced Sodium Receiver (ASR) - Topic Reports (by AGIP Nucleare and Franco Tosi)	May 1983
4/83	CRS-Midterm Workshop (edited by M. Becker, DFVLR)	Jun. 1983
5/83	Investigations and Findings Concerning the Sodium Tank Leakages (edited by W. Bucher, DFVLR)	Jul. 1983
7/83	Thermal Losses of the Sodium Storage Vessels of the Central Receiver System (by H. Jacobs, SSPS-ITET)	Nov. 1983
1/84	Executive Summary - IEA/SSPS - CRS Workshop (April 1983) (by C.S. Selvage, SSPS-ITET)	Mar. 1984

CRS/DCS TECHNICAL REPORTS (common)

Report Nº	Title	Date
1/81	Tabernas Meteo Data Analysis Based on Evaluated Data Prepared by the SSPS O.A. (by Belgonucleaire)	Jun. 1981
4/81	International Energy Agency Small Solar Power Systems (SSPS) Project Review (January 1981) (by A.F. Baker, SANDIA)	Jul. 1981
6/81	Determination of the Spectral Reflectivity and the Bidirectional Reflectance Characteristics of Some White Surfaces (by DFVLR)	Dec. 1981
2/83	FH-PTL Wedel Reflectometer, Type 02-1 No. 3 Final Report and Report on the Test Program (by G.Lensch, K. Brudi, P. Lippert,Fachhochschule Wedel)	Mar. 1983

CRS-SSPS INTERNAL REPORTS

Subject: Heliostat Field

Report Nº	Title	Author	Date
R-1/81	Test Report: Heliostat Reflectance Spot Check	R.P.Stromberg	24.7.81
R-9/81	Technical Changes Incorporated in the MMC Heliostat Controllers (HAC) after Acceptance	W. Grasse	26.8.81
R-17/81	Heliostat Reflectance Estimate	R.P.Stromberg	24.9.81
R-27/81	Heliostat Status Report for Month of October	E. Madigan	17.11.81
R-43/82	Belgian Heliostat Evaluation since September 1981 until March 1982	T. van Steen-berghe/ P. Wattiez	11.5.82
R-76/82	Reflectivity Measurements over the Heliostat Field: 15.5.82-3.8.82	P. Wattiez	13.9.82
R-86/82	F.M.C. (2.9-14.10.82) Heliostat Field Reflectivity Behavior	P. Wattiez	8.11.82
R-94/82	Heliostat Field Reflectivity	LR/MS	13.12.82
R-26/83	Heliostat Field Investigation Program	P. Wattiez	16.5.83
R-27/83	Investigation of Water Contained in Heliostat Modules	P. Wattiez	13.6.83
R-29/83	Wind Studies in Respect of Mirror Field Stowage	M. Loosme	8.8.83
R-6/84	Mirror Corrosion and Heliostat Condition	M. Sánchez/ J. Ramos	30.3.84
R-12/84	Simulation of Tracking Errors	M. Blanco	5.4.84
R-14/84	Status of Heliostat Field Alignment	J. Ramos	24.4.84
R-24/84	Heliostat Flux Distribution, Weekly Report (week nº 25)	A. Cuadrado	29.6.84
R-25/84	Heliostat Flux Distribution, Weekly Report (week nº 26)	A. Cuadrado	3.7.84
R-27/84	Heliostat Flux Distribution, Weekly Report (week nº 28)	A. Cuadrado	24.7.84
R-28/84	Heliostat Flux Distribution, Weekly Report (week nº 29)	A. Cuadrado	30.7.84

CRS-SSPS INTERNAL REPORTS (continued)

Subject: Heliostat Field (continued)

Report Nº	Title	Author	Date
R-29/84	A Mathematical Model for Estimating Average Heliostat Field Reflectivity	M. Sánchez/ P. Wattiez	10.5.84
R-30/84	Heliostat Field Reflectivity	M. Sánchez/ L. Ruiz	17.7.84
R-31/84	Heliostat Flux Distribution, Weekly Report (week nº 30)	A. Cuadrado	6.8.844
R-32/84	Reflectivity Measurement and Meteo Data Collection from Feb.20 to Apr. 10, 1984	H. Winter	3.5.84
R-35/84	Heliostat Flux Distribution, Weekly Report (week nº 32)	A. Cuadrado	21.8.84
R-41/84	Multiple Seasonal Model for Predicting the Average Reflectivity (Computer Program Application to SSPS-CRS Heliostat Field)	M. Sánchez	10.5.84
R-48/84	On Site Calculations with the HELIOS Code	H. Jacobs	25.10.84

Subject: Receiver

Report Nº	Title	Author	Date
R-22/81	Inputs for the Design of the Advanced Sodium Receiver	INTERATOM	16.10.81
R-56/82	CRS Receiver Reradiation and its Measurement Calculation Note	T. van Steenberghe	28.6.82
R-58/82	CRS Receiver - Reradiated Flux Distribution (RFD) Measurement System	T. van Steenberghe	7.782
R-70/82	Heat Flux Distribution Measurement System: Special Versions of the Software for the Kendall Radiometer	T. van Steenberghe	2.9.82
R-77/82	Receiver Efficiency Calculation Method CRS Fall Equinox Measurement Campaign 1982	J. Kraabel	16.9.82
R-78/82	Almería Steam Generator. Part Load Characteristic with Reduced Steam Pressure	Sulzer AG	24.9.82
R-85/82	Solar Absorptance Measurements on the First CRS Receiver Tubes	T. van Steenberghe	29.10.82
R-1/83	Measurement and Evaluation of Temperature Inside the CRS Receiver	F. Gaus	25.2.83
R-12/83	Test Program for Measurement and Evaluation of the Receiver Tube Bundle Movements with Displacement Transducers	F. Gaus	9.3.83
R-23/83	Solar Hemispherical Absorptance of the CRS First Receiver Ceramic Wall	T. van Steen-	4.8.83
R-24/83	ITET Participation in the ASR-Functional Test	H. Jacobs/ M. Pescatore	10.8.83
R-28/83	Convection Losses of the Sulzer Receiver	H. Jacobs	24.8.83

CRS-SSPS INTERNAL REPORTS (continued)

Subject: Receiver (continued)

Report Nº	Title	Author	Date
R-33/83	Measurement of the Distance Between Receiver Tubes and a Reference Line, and Control of the Central Support in the Sulzer Receiver	F. Gaus	7.9.83
R-37/83	ASR Incident on September 7, 1983	A.DeBenedetti A. di Meglio J. Hansen	19.9.83
R-11/84	Steady-State and Transient Conditions in a Tube with Incident Thermal Flux and Inlet Condition Assigned: Program ASRTTHB (ASR Tube Thermal Balance)	A.DeBenedetti	12.4.84
R-17/84	A Comparison of the SULZER Cavity Receiver and the FRANCO-TOSI External Receiver	M. Pescatore	22.5.84
R-19/84	Receiver Vent Valves Control System Modification	A. Cuadrado	30.5.84
R-21/84	Preliminary Results on the Performance of the SULZER Cavity Receiver and the FRANCO-TOSI External Receiver	C.S. Selvage/ H. Jacobs	26.6.84
R-33/84	Heliostat Field - Receiver Overall Simulation	A.DeBenedetti	6.8.84
R-34/84	ASR Temperature Increase at Central Panel Outlet	A.DeBenedetti J.G. Martín	6.8.84
R-37/84	ASR Ceramic Border Temperatures Versus Incident Radiation	A. Cuadrado J. Ramos	28.8.84
R-38/84	THERESA: A Thermal Analysis Code for Billboard Receivers	A.DeBenedetti	17.8.84
R-46/84	Sulzer Receiver Transient Response Data During Period March-April 1983	N. Gregory	11.10.84
R-47/84	Available Point and Class Summary Data: Sulzer Receiver	N. Gregory	18.10.84

Subject: Tank

Report Nº	Title	Author	Date
R-5/82	Determination of Sodium Tank's Volume as a Function of Level	C. Gómez	4.2.82
R-26/82	Failure and Repair of the Cold Storage Vessel LK01 BB01 - Preliminary Incident Report	D. Stahl/ INTERATOM	26.3.83
R- 68/82	Sodium Leakage at the SSPS-CRS Cold Storage Tank	W. Bucher	11.8.82
R-84/82	Thermal Shock as a Possible Cause of Sodium Leaks	J.G. Martín	21.10.82
R-4/83	Thermal Losses of the Sodium Storage Vessels	H. Jacobs	9.2.83

CRS-SSPS INTERNAL REPORTS (continued)

Subject: Power Conversion System

Report Nº	Title	Author	Date
R-32/82	Test Program for Improving the Load Change Behavior in the Steam Generator Casing	H. Jacobs	5.4.82
R-44/82	CRS PCS Gross Efficiency from November 11, 1981 to December 12, 1981	C. Gómez/ R. Carmona	14.5.82
R-63/82	CRS Steam Motor Alternatives-Discussions held at SSPS, Tabernas 15.-16.7.1982	J. Hansen	22.7.82
R-66/82	Steam Generator Load Change Behavior Improvement	H. Jacobs	29.7.82
R-73/82	Spilling Steam Motor 10 Day Test	J. Hansen/ F. Martinez	23.8.82
R-79/82	Supplement to the Spilling Steam Motor Test Report Nº R-73/82	J. Hansen	27.9.82
R-10/83	Review of Spilling Performance Status Report at Final Review on CRS Repair, 24.3.83	J. Hansen	22.3.83

Subject: Miscellaneous

Report Nº	Title	Author	Date
R-3/81	Information Concerning the Status of CRS Performance Determination	M. Becker	4.8.81
R-11/81	Receiver Platform Incident due to Power Failure	W. Grasse	1.9.81
R-11a/81	Final Incident Report on Receiver Platform Incident	E. Madigan/ F. Martinez/ D. Stahl	10.12.81
R-23/81	Reporting System	W. Grasse/ P. Wattiez/ A. Baker	29.10.81
R-30/81	Input for SSPS Status Report given at the CRS-Department of Energy Annual Meeting (October 13-16, 1981)	E. Madigan	17.11.81
R-21/82	Revision of CRS-DAS Calculation	C. Gómez	2.3.82
R-48/82	Recommendations for CRS/HAC after visit of Ed Madigan	E. Madigan	1.6.82
R-49/82	Program of the CRS Measurement Campaign around Fall Equinox - Status June 1982	M. Becker	22.6.82
R-83/82	CRS Daily Summary, Description, and Comments	P. Wattiez	27.10.82
R-2/83	Comparison of the SSPS-CRS Heat Flux Measurements and Corresponding Theoretical Predictions	Dr. Kiera/ INTERATOM	25.1.83
R-5/83	SSPS Stage 3 as Promoting Program of the Solar Energy Diffusion	P. Wattiez	3.2.83
R-6/83	CRS Fall Equinox Measurement Campaign	T. van Steenberghe	16.1.83
R-13/83	HFD Repair Report	J. Ramos	24.3.83

Subject: Miscellaneous (continued)

Report Nº	Title	Author	Date
R-17/83	Status Report of CRS / SHTS	F. Ruiz	19.5.83
R-18/83	SOLTES Modelling of the SSPS Plants at Sandia National Laboratories Mission Report	T. von Steenberghe	17.5.83
R-31/83	Status of CRS Before Start-Up ASR Operation	P. Laguía	22.8.83
R-36/83	CRS Plant Operation and its Documentation with Data Tape from November 16, 1981 to August 31, 1983	M. Pescatore	5.9.83
R-3/84	Modifications to the Cold Sodium Pump Control	J. Ramos	20.1.84
R-16/84	Water Physical Property Package	A.DeBenedetti	3.5.84
R-36/84	Trace Heating Consumption	A. Cuadrado	27.8.84
R-42/84	IEA/SSPS Calibration Report- Calibration of Relevant Measuring Sensors	A. Brinner	27.9.84
R-44/84	An Operational Usage Factor for the CRS Plant	N. Gregory	2.10.84
R-45/84	Available Data and Approximation of the Average CRS Plant Start-up Time	N. Gregory/ H. Jacobs	2.10.84
R-51/84	Progress on Lightning Protection Improvements for CRS Heliostat Field (2	J. Ramos	8.10.84

Subject: CRS-DCS Common

Report Nº	Title	Author	Date
R-2/81	Measurement Device Test for Solar Insolation	G. von Tobel	28.7.81
R-5/81	Plant Optimization Phase (POP) Test Requirements	B. Wilson/ R.P.Stromberg	10.8.81
R-5a/81	Idem.	A. Baker/ W. Wilson/ R.P.Stromberg	28.9.81
R-5b/81	Idem.		Oct.81
R-5c/81	POP Test Requirements DCS System Tests (new Tests Nº 34-37) DCS Updated POP Test Listings CRS Updated POP Test Listings	P. Wattiez	5.11.81
R-6/81	Cost of Electricity	A. Rieger	17.8.81
R-10/81	Cooling Towers for Future Solar Power Plant Projects	A. Rieger	27.8.81
R-12/81	Report of Damages caused by the 2nd Flooding of SSPS Plant, Tabernas, Almería	P. Heintzelmann	1.9.81
R-14/81	Mirror Reflectance Study, SSPS Projects, Tabernas, Spain	R.P.Stromberg	15.9.81
R-16/81	Recommendations for Improvement of the 25 kW Grid Behavior	A. Rieger	23.9.81
R-18/81	Reflectivity Measurement Corrections	R.P.Stromberg	30.9.81
R-19/81	Optimization of the Time Period Between Mirror Washing	A. Rieger	2.10.81
R-21/81	Comparison of Solar Radiation Measurement Devices	G. von Tobel	15.10.81
R-3/82	Measurement Program of Diffuse Component on Clear Skies	A. Sevilla	29.1.82

CRS-SSPS INTERNAL REPORTS (continued)

Subject: CRS-DCS Common (continued)

Report №	Title	Author	Date
R-23/82	Solar Heating for the SSPS Office Buildings	A. Sevilla	15.3.82
R-27/82	Comparitive Tests on Two Portable Reflectometers D&S-15R + FH-PTL-02.1	P. Wattiez	30.3.82
R-28/82	Reflectivity Measurement Procedure	P. Wattiez	31.3.82
R-41/82	A General Purpose Polynomial Regression Program	T. van Steenberghe	11.5.82
R-45/82	Plant Safety Meeting	M. Loosme	29.4.82
R-46/82	Results of Reflectivity Variation Over the SSPS Mirror Fields	P. Wattiez	27.5.82
R-52/84	Insolation at the SSPS Site in Tabernas (1981 and 1982 till May)	P. Wattiez	17.6.82
R-60/82	Lightning Protection of the SSPS Plant	M. Loosme	13.7.82
R-65/82	Reflectance Properties of Mirror Modules from the IEA Project	R.B. Petit/ A.R. Mahoney/ SNLA	26.7.82
R-71/82	Cleaning Cost of the SSPS Fields	P. Wattiez	10.9.82
R-72/82	SSPS Field Cleaning Frequency	P. Wattiez	11.8.82
R-75/82	Intermediate Report on the Photovoltaic Experiment at the IEA/SSPS Site in Tabernas (Spain) August 1982	P. Toggweiler	8.9.82
R-11/83	Computer Plotting as Evaluation Tool	M. Andersson	9.3.83
R-20/83	US Trip of Dr. Kalt March 1983 Travel Report	A. Kalt	Apr. 1983
R-22/83	1983 Solar Ephemeridis for the Plataforma Solar, Tabernas (Almería)	T. van Steenberghe	8.7.83
R-25/83	CRS and DCS Common Parasitic Sources	P. Wattiez	23.2.83
R-30/83	Passive Solar Retrofit of the SSPS Building Office	A. Sevilla	26.8.83
R-34/83	Energy Calculations of an Office Building Using the DOE-2.1a	L. Chien-Kuo You/ J.G. Martín	23.9.83
R-41/83	Monthly Direct Insolation Data	L. Castillo	7.10.83
R-5/84	Earthing System Inspection	B. Calatrava	26.3.84
R-8/84	Report of Work Performed at the Small Solar Power Systems (SSPS) Site 1.7-31.8.83	W. Banhardt/ S. Lauties	23.3.84
R-10/84	Safety Meeting 2/84	B. Calatrava	8.4.84
R-18/84	Safety Meeting	B. Calatrava	22.5.84
R-20/84	Safety Meeting	B. Calatrava	6.6.84
R-40/84	Wind Conditions on the SSPS Site	L. Castillo	27.9.84
R-50/84	The IEA/SSPS Office Building Cooling and Heating Loads: Effect of Retrofit Options	L. Chien Kuo You/ J. Martín	12.7.84

SOLAR TERMINOLOGY

Absorber	The blackened surface of a collector which absorbs solar radiation and converts it to heat energy: a flat black paint is a good absorber (Argue).
Absorptance	The ratio of absorbed to incident solar radiation (Kreith).
Absorptivity	The ratio of absorbed radiation by a surface to the total incident radiated energy on that surface.
Active Solar System	A system in which a transfer fluid (liquid or air) is circulated through a solar collector where the collected energy is converted, or transferred to energy in the medium.
Aiming Point	The focalization point of a heliostat on a receiver.
Air Mass	The ratio of the actual distance traversed through the Earth's atmosphere by the direct solar beam to the depth of the Earth's atmosphere, normal to the surface (McVeigh).
Ambient Temperature	The surrounding air temperature.
Antireflection Coating	The application of a thin film of dielectric material to a surface to reduce its reflection and to increase its transmission of light or other electromagnetic radiation (Lapedes).
Aperture Area	The net area of the collector that intercepts radiation (Kreith).
Array	An assembly of a number of collector elements, or panels, into the solar collector for a solar energy system.

Attenuation	The reduction of radiation flux over a given path length, due to absorption and scattering (McVeigh).
Auxiliary Energy	In solar energy technology, the energy supplied to the heat or cooling load from other than the solar source, usually from a conventional heating or cooling system. Excluded are operating energy, and energy which may be supplemented in nature but does not have the auxiliary system as in origin.
Auxiliary Energy Subsystem	In solar energy technology the Auxiliary Energy System is the conventional heating and/or cooling equipment used as supplemental or backup to the solar system.
Azimuth	The angle between the south-north line at a given location and the projection of the earth-sun line in the horizontal plane (Kreith).
Beam Radiation	Radiated energy received directly, not from scattering or reflecting sources.
Black Body	A term denoting an ideal body which would absorb all and reflect none of the radiation falling upon it (McVeigh).
Bypass Loop	A piping arrangement which bypasses or circles the flow of a heat absorbing medium around rather than through a piece of mechanical equipment (Burt).
Carnot Cycle	A hypothetical cycle consisting of four reversible processes in succession: an isothermal expansion and heat addition, an isentropic expansion, and isothermal compression and heat rejection process, and an isentropic compression (Lapedes).

Carnot Efficiency	The efficiency of a Carnot engine receiving heat at a temperature absolute T_1 and giving it up at a lower temperature T_2 (Lapedes).
Collector	A device which absorbs solar radiation and converts it to heat energy (Argue).
Collector Area	The area of a collector which traps the sun (Argue).
Collector Efficiency	The ratio of th energy collected by a solar collector to the radiant energy incident on the collector (Williams).
Collected Solar Energy	The thermal energy added to the heat transfer fluid by the solar collector.
Collector Subsystem	The assembly of components that absorbs incident solar energy and transfers the absorbed thermal energy to a heat transfer fluid.
Concentrating Collector	A collector which uses reflective devices or optical lens arrangements to concentrate the sun's rays onto a small collector/absorber area (Montgomery).
Concentration Ratio	Ratio of radiant energy intensity at the hot spot of a focusing collector to the intensity of unconcentrated direct sunshine at the collector site (Williams).
Control System or Subsystem	The assembly of electric, pneumatic, or hydraulic, sensing, and actuating devices used to control the operating equipment in a system.
Conversion Efficiency	Ratio of thermal energy output to solar energy incident on the collector array.
Cooling Tower	A heat exchanger that transfers waste heat to outside ambient air.

Design Point	The specific moment chosen to define for performance characteristics of a system.
Diffuse Radiation	Solar radiation which is scattered by air molecules, dust, or water droplets and incapable of being focused.
Direct Radiation	Solar radiation that comes directly from the sun.
Drain Down	An arrangement of sensors, valves, and actuators to automatically drain the solar collectors and collector piping.
Effective Heat Transfer Coefficient	The heat transfer coefficient, per unit plate area of a collector, which is a measure of the total heat losses per unit area from all sides, top, back, and edges.
Emissivity	The ratio of the radiation emitted by a surface to the radiation emitted by a perfect blackbody radiator at the same temperature (Lapedes).
Emittance	The power radiated per unit area of a radiating surface (Lapedes).
Energy Gain	The thermal energy gained by the collector transfer fluid. The thermal energy output of the collector.
Flow Rate	Velocity at which a fluid travels, usually through an opening or duct (Burt).
Focusing Collector	A concentrating type of collector using parabolic mirrors or optical lenses to focus the energy from a large area onto a small absorbing area.
Fossil Fuel	Petroleum, coal, and natural gas derived fuels.

Gross Area	The total frontal area of a collector, including framing and structural supports (Kreith).
Header	Or: manifold. The pipe running at either end of a solar collector which distributes the heat transfer fluid to, or collects it from the collector (Argue).
Heat Capacity	In a general sense, refers to the ability of material to store heat (Argue).
Heat Engine	A thermodynamic system which undergoes a cyclic process during which a positive amount of work is done by the system; some heat flows into the system and a smaller amount flows out in each cycle (Lapedes).
Heat Exchanger	A device used to transfer energy from one heat transfer fluid to another while maintaining physical segregation of the fluids. Normally used in systems to provide an interface between two different heat transfer fluids.
Heat Gain	An increase in the amount of heat contained in a space, resulting from direct solar radiation and the heat given off by people, lights, equipment, machinery, and other sources (Dresser).
Heat Loss	A decrease in the amount of heat contained in a space, resulting from heat flow through walls, windows, roofs, and other building envelope components (Dresser).
Heat Storage	An insulated container in which energy collected by a solar system can be held for use when the sun is not shining or during extremely cold weather; the heat may be stored in a variety of media including water, rock-beds, paraffin wax, or eutectic salts (Argue).

Heat Transfer Fluid	The fluid circulated through a heat source (solar collector) or heat exchanger that transports the thermal energy by virtue of its temperature.
Heliostat	An electro-optical-mechanical device that orients a mirror so that sunlight is reflected from the mirror in a fixed specific direction, regardless of the sun's position in the sky (Williams).
Honeycomb	Used to suppress free-convection heat transfer across the air gap between a collector plate and its glass cover and to reduce radiation losses from the collector (Kreith).
Hot Spot	The location on a focusing collector at which the concentrated sunlight is focused and the highest temperatures are produced (Williams).
Incident Angle	The angle between the sun's rays and a line normal to the irradiated surface (Kreith).
Incident Solar Energy	The amount of solar energy irradiating a surface taking into acount the angle of incidence. The effective area receiving energy is the product of the area of the surface times the cosine of the angle of incidence.
Insolation	The total amount of solar radiation striking a collector area.
Instantaneous Efficiency	The efficiency of a solar collector at one operating point, $(T_i-T_a)/I$, under steady state conditions (see Operating Point).
Instantaneous Efficiency Curve	A plot of solar collector efficiency against operating point, $(T_i-T_a)/I$ (see Operating Point).

Insulation	Used to minimize heat loss from the nonilluminated side of the absorber plate (Kreith).
Irradiance	The amount of solar radiant energy falling on a surface per unit area and per unit of time (Kreith).
Linear Concentrator	A solar concentrator which focuses sunlight along a line (Williams).
Load	That to which energy is supplied, such as internal electrical consumption of each subsystem. The system load is the total solar and auxiliary energy required to satisfy the required functioning.
Long Term Storage	The heat storage capacity of a solar heating system that works on a long term or annual cycle (Argue).
Manifold	The piping that distributes the transport fluid to and from the individual panels of a collector array.
Mass Flow	The mass of a fluid in motion which crosses a given area in a unit time (Lapedes).
Microclimate	Highly focalized weather features which may differ from long-term regional values due to the interaction of the local surface with the atmosphere.
Nocturnal Radiation	The loss of thermal energy by the solar collector to the night sky.
Operating Energy	The amount of energy system has a dynamic operating range due to changes in level of insolation (I), fluid input temperature (T), and outside ambient temperature (T_a). The operating point is defined as: $$\frac{(T_i - T_a)}{I} \; \frac{(°C \times hr. \times m^2)}{BTU}$$

Operational Collector Efficiency	Ratio of collected solar energy to incident solar energy <u>only during the time the collector fluid is being circulated with the intention of delivering solar-source energy to the system</u>; in other words, when the collector is in track position.
Parallel	Two or more equivalent systems working side by side.
Peak Watt	Unit used for the performance rating of solar electric power systems; a system rated at one peak watt will deliver one watt at the specified working voltage under peak solar irradiation (Kreith).
Process Heat	Heat produced in the form of hot water, steam, etc, for industrial process applications.
Pyranometer	An instrument used for measuring global radiation (McVeigh).
Pyrheliometer	An instrument used to measure the direct irradiance of the sun along a surface perpendicular to the solar beam; diffuse radiation is excluded from the measurement (McVeigh).
Radiant Flux Density	The amount of radiant power per unit area that flows across or onto a surface (Lapedes).
Radiation	When unqualified, usually refers to electromagnetic radiation (Lapedes).
Radiometer	An instrument used to measure radiant energy.

Rankine Cycle	A closed heat engine cycle using various components, including a working fluid pumped under pressure to a boiler where heat is added; an expander (turbine) where work is generated; and a condenser used to reject lowgrade heat to the environment (Kreith).
Reflectance	The ratio of radiation reflected from a surface to that incident on the surface.
Reflected Radiation	Solar radiation which strikes an exposed surface after being reflected from the grounds, trees, buildings, snow, etc.
Rejected Energy	Energy intentionally rejected, dissipated, or dumped from the solar system.
Reflecting Surfaces	Usually highly polished metals or metal coatings on suitable substrates.
Reflectivity	The property of reflecting radiation possessed by all materials to varying extents (Kreith).
Reradiation	Radiation resulting from the emission of previously absorbed radiation (Webster).
Scattering	Interaction of radiation with matter where the direction is changed but the total energy and wavelength remain unaltered (McVeigh).
Selective Surface	A surface which has a high absorptivity for incident solar radiation but also has a low emissivity in the infra-red region (McVeigh).
Sensible Heat	Heat stored in a medium (such as water, bricks, or another substance) in which there is a temperature rise (Argue).
Series	Two or more systems connected in an additive manner (Argue).

Short Term Storage	The heat storage of a solar heating system that is able to supply heat for only a few days (Argue).
Solar Constant	The amount of solar radiation which is received immediately external to the Earth's atmosphere and incident upon a surface normal to the radiation taken at mean Earth-sun distance; it is not a true constant as it varies, mainly due to sun-spot activity (McVeigh).
Solar Energy	Generally describes those renewable energy sources which directly or indirectly are powered by the sun; they include: direct solar radiation, wind, falling water, biomass, and waves; solar energy is also used to specifically describe solar radiation (Argue).
Solar Engine	An engine which converts thermal energy from the sun into electrical, mechanical, or refrigeration energy (Lapedes).
Solar Multiple	The ratio of the thermal energy, actually collected by the heliostat/collector field of a solar plant, divided by the amount of thermal energy necessary to achieve the rated output of that plant under design point conditions.
Solar Power Farm	An installation for generating electricity on a large scale using solar energy, consisting of an array of solar collectors, steam or gas turbines, and electrical generators (Crowther).
Solar Radiation	The electromagnetic radiation emitted by the sun (Argue).
Solar Tower	A tall tower, positioned to collect reflected direct solar radiation from an array of heliostats; the top of the tower contains the heat exchange chamber and the hot working fluid is used in a conventional electrical generating system at ground level (McVeigh).
Specific Heat	The quantity of heat required to raise a unit mass of homogeneous material one degree in temperature in a specified way (Lapedes).

Specular Reflection	Mirror-like reflection from a surface (Williams).
Stirling Cycle	A regenerative thermodynamic power cycle using two isothermal and two constant volume phases (Lapedes).
Stirling Engine	An engine in which work is performed by the expansion of a gas at high temperature; heat for expansion is supplied through the wall of the piston cylinder (Lapedes).
Storage Efficiency	Measure of effectiveness of transfer of energy through the storage subsystem taking into account system losses.
Storage Subsystem	The assembly of components used to store solar-source energy for use during periods of low insolation.
Stratification	A phenomenon that causes a distinct thermal gradient in a heat transfer fluid, in contrast to a thermally homogeneous fluid. Results in the layering of the heat transfer fluid, with each layer at a different temperature. In solar energy systems, stratification can occur in liquid storage tanks or rack beds, and may even occur in pipes and ducts. The temperature gradient or layering may occur in a horizontal, vertical, or radial direction.
Sun Tracking	The ability to follow apparent motion of the sun across the sky (Crowther).
Thermal Mass	The heat storage capacity of a structure provided by large quantities of heavy material (Argue).
Tracking Collector	A solar collector that moves to point in the direction of the sun.
Total Radiation	The total of diffuse and beam radiation (Kreith).

Transmittance — The ratio of radiant energy transmitted through a transparent surface to energy incident on it (Dresser).

Trough Concentrator — Single-curvature (or cylindrical) concentrator characterized by one plane of symmetry (Kreith).

ABBREVIATIONS

ACU — ACUREX Corporation, manufacturer of the one-axis tracking collector

ASR — Advanced sodium receiver, type billboard - external

CRS — central receiver system

DCS — distributed collector system

DFVLR — Deutsche Forschungs-und Versuchsanstalt für Luft -und Raumfahrt e.V.

HFS — heliostat field subsystem

HTS — heat transfer system

IEA — International Energy Agency

ITET — International Test and Evaluation Team

kWh — kilowatt hours, a measure of electrical energy. The product of kilowatts of electrical power applied to a load times the hours it is applied. One kWh is equivalent to 3,413 BTU of heat energy.

MAN — Maschinenfabrik Augsburg - Nürnberg Aktiengesellschaft

OA — Operating Agent of the project, entrusted to the DFVLR

PCS — power conversion system

POA — Plant Operation Authority, entrusted to Sevillana

SHTS — sodium heat transfer system

SSPS — Small Solar Power Systems

CONVERSION TABLES

The multiple systems of units and measures used in the basic biological, physical, and engineering sciences is often confusing to those with inter-disciplinary interests. The following tables have been selected from the Arnold Engineering Development Center Document AEDC TDR 62-2 as most pertinent to such interests. Tables of electrical and thermodynamic units not covered in this extract may be found in the AEDC document.

The reader may convert from the measure of quantities in the units listed on the left to those across the top of the page by multiplying by the given factor. The small number to the upper right of each factor is the power of ten by which the factor is to be multiplied. For example, 8.68977^{-2} is equivalent to 0.0868977. Underscores indicate exact values.

Slight numerical differences will be noticed between the values herein and those in most currently existing texts because of the redefinition of the foot in 1960.

Tables

1. Length
2. Area
3. Volume
4. Mass
5. Time
6. Angle
7. Velocity
8. Force
9. Pressure
10. Energy
11. Power
12. Mass Flow Rate
13. Pumping Speed
14. Temperature

TABLE 1

Length (ℓ, d)

	Å	cm	ft	in.	km	m	μ	mils	miles (US)	mm	miles (Nautical)	rods	yd
Angstrom Units	1.00000	1.00000^{-8}	3.28084^{-10}	3.93701^{-9}	1.00000^{-13}	1.00000^{-10}	1.00000^{-4}	3.93701^{-6}	6.21371^{-14}	1.00000^{-7}	5.39957^{-14}	1.98839^{-11}	1.09361^{-10}
Centimeters	1.00000^{8}	1.00000	3.28084^{-2}	3.93701^{-1}	1.00000^{-5}	1.00000^{-2}	1.00000^{4}	3.93701^{2}	6.21371^{-6}	1.00000^{1}	5.39957^{-6}	1.98839^{-3}	1.09361^{-2}
Feet	3.04800^{9}	3.04800^{1}	1.00000	1.20000^{1}	3.04800^{-4}	3.04800^{-1}	3.04800^{5}	1.20000^{4}	1.89394^{-4}	3.04800^{2}	1.64579^{-4}	6.06061^{-2}	3.33333^{-1}
Inches	2.54000^{8}	2.54000	8.33333^{-2}	1.00000	2.54000^{-5}	2.54000^{-2}	2.54000^{4}	1.00000^{3}	1.57828^{-5}	2.54000^{1}	1.37149^{-5}	5.05050^{-3}	2.77778^{-2}
Kilometers	1.00000^{13}	$1.00000^{[illegible]}$	3.28084^{3}	3.93701^{4}	1.00000	1.00000^{3}	1.00000^{9}	3.93701^{7}	6.21371^{-1}	1.00000^{6}	5.39957^{-1}	1.98839^{2}	1.09361^{3}
Meters	1.00000^{10}	1.00000^{2}	.28084	3.93701^{1}	1.00000^{-3}	1.00000	1.00000^{6}	3.93701^{4}	6.21371^{-4}	1.00000^{3}	5.39957^{-4}	1.98839^{-1}	1.09361
Microns	1.00000^{4}	1.00000^{-4}	3.28084^{-6}	3.93701^{-5}	1.00000^{-9}	1.00000^{-6}	1.00000	3.93701^{-2}	6.21371^{-10}	1.00000^{-3}	5.39957^{-10}	1.98839^{-7}	1.09361^{-6}
Mils	2.54000^{5}	$2.54000^{[illegible]}$	8.33333^{-5}	1.00000^{-3}	2.54000^{-8}	2.54000^{-5}	2.54000^{1}	1.00000	1.57828^{-8}	2.54000^{-2}	1.37149^{-8}	5.05050^{-6}	2.77778^{-5}
Miles ( )	1.60934^{13}	1.60[illegible]	5.28000^{3}	6.33600^{4}	1.60934	1.60934^{3}	1.60934^{9}	6.33600^{7}	1.00000	1.60934^{6}	8.68977^{-1}	3.20000^{2}	1.76000^{3}
Millimeters	1.00000^{7}	$1.00000^{[illegible]}$	3.280[illegible]$^{-3}$	3.93701^{-2}	1.00000^{-6}	1.00000^{-3}	1.00000^{3}	3.93701^{1}	6.21371^{-7}	1.00000	5.39957^{-7}	1.98839^{-4}	1.09361^{-3}
Miles(Nautical)	1.85200^{13}	1.8520[illegible]	6.0761^{3}	7.29134^{4}	1.85200	1.85200^{3}	1.85200^{9}	7.29134^{7}	1.15078	1.85200^{6}	1.00000	3.68250^{2}	2.02537^{3}
Rods	5.02920^{10}	5.02920^{2}	1.65000^{1}	1.98000^{2}	5.02920^{-3}	5.02920	5.02920^{6}	1.98000^{5}	3.12500^{-3}	5.02920^{3}	2.71555^{-3}	1.00000	5.50000
Yards	9.14400^{9}	9.14400^{1}	3.00000	3.60000^{1}	9.14400^{-4}	9.14400^{-1}	9.14400^{5}	3.60000^{4}	5.68182^{-4}	9.14400^{2}	4.93737^{-4}	1.81818^{-1}	1.00000

1. LENGTH

TABLE 2

Area (A)

	Acre	in. (Circular)	Mil (Circular)	cm^2	ft^2	$in.^2$	m^2	mm^2	Rod^2	$Yard^2$
Acre	1.00000	7.98657^{6}	7.98657^{12}	4.04686^{7}	4.35600^{4}	6.27264^{6}	4.04686^{3}	4.04686^{9}	1.60000^{2}	4.84000^{3}
Circular inch	1.25211^{-7}	1.00000	1.00000^{6}	5.06707	5.45415^{-3}	7.85398^{-1}	5.06707^{-4}	5.06707^{2}	2.00336^{-5}	6.06017^{-4}
Circular mil	1.25211^{-13}	1.00000^{-6}	1.00000	5.06707^{-6}	5.45415^{-9}	7.85398^{-7}	5.06707^{-10}	5.06707^{-4}	2.00336^{-11}	6.06017^{-10}
Centimeter^2	2.47105^{-8}	1.97353^{-1}	1.97353^{5}	1.00000	1.07639^{-3}	1.55000^{-1}	1.00000^{-4}	1.00000^{2}	3.95368^{-6}	1.19598^{-4}
Feet^2	2.29568^{-5}	1.83347^{2}	1.83347^{8}	9.29030^{2}	1.00000	1.44000^{2}	9.29030^{-2}	9.29030^{4}	3.67310^{-3}	1.11111^{-1}
Inch^2	1.59423^{-7}	1.27324	1.27324^{6}	6.45160	6.94444^{-3}	1.00000	6.45160^{-4}	6.45160^{2}	2.55076^{-5}	7.71605^{-4}
Meter^2	2.47105^{-4}	1.97353^{3}	1.97353^{9}	1.00000^{4}	1.07639^{1}	1.55000^{3}	1.00000	1.00000^{6}	3.95368^{-2}	1.19598
Millimeter^2	2.47105^{-10}	1.97353^{-3}	1.97353^{3}	1.00000^{-2}	1.07639^{-5}	1.55000^{-3}	1.00000^{-6}	1.00000	3.95368^{-8}	1.19598^{-6}
Rod^2	6.25000^{-3}	4.99161^{4}	4.99161^{10}	2.52928^{5}	2.72250^{2}	3.92040^{4}	2.52928^{1}	2.52928^{7}	1.00000	3.02500^{1}
Yard^2	2.06611^{-4}	1.65012^{3}	1.65012^{9}	8.36127^{3}	9.00000	1.29600^{3}	8.36127^{-1}	8.36127^{5}	3.30578^{-2}	1.00000

2. AREA

TABLE 3

Volume (V)

	Centimeter^3	Feet^3	Gallon	Inch^3	Liter	Meter^3	Yard^3
Centimeter^3	1.00000	3.53146^{-5}	2.64171^{-4}	6.10236^{-2}	9.99972^{-4}	1.00000^{-6}	1.30794^{-6}
Feet^3	2.83163^{4}	1.00000	7.48052	1.72800^{3}	2.83161^{1}	2.83168^{-2}	3.70370^{-2}
Gallon (U.S.)	3.78543^{3}	1.33680^{-1}	1.00000	2.31000^{2}	3.78533	3.78543^{-3}	4.95113^{-3}
Inch^3	1.63871^{1}	5.78704^{-4}	4.32900^{-3}	1.00000	1.63866^{-2}	1.63871^{-5}	2.14335^{-5}
Liter	1.00003^{3}	3.53156^{-2}	2.64178^{-1}	6.10253^{1}	1.00000	1.00003^{-3}	1.30798^{-3}
Meter^3	1.00000^{6}	3.53146^{1}	2.64171^{2}	6.10236^{4}	9.99972^{2}	1.00000	1.307[illegible]
Yard^3 (U.S.)	7.64554^{5}	2.70000^{1}	2.01974^{2}	4.66560^{4}	7.64538^{2}	7.64554^{-1}	1.00000

3. VOLUME

TABLE 4

Mass (m)

	Grain	Gram_m	Kilogram_m	Ounce (avdp)	Pound_{mass}	Slug	Ton (short)
Grain	$\underline{1.00000}$	6.47988^{-2}	6.47988^{-5}	2.28571^{-3}	1.42857^{-4}	4.44012^{-6}	7.14285^{-8}
Gram_{mass}	1.54323^{1}	$\underline{1.00000}$	$\underline{1.00000}^{-3}$	3.52739^{-2}	2.20462^{-3}	6.85216^{-5}	1.10231^{-6}
Kilogram_{mass}	1.54323^{4}	$\underline{1.00000}^{3}$	$\underline{1.00000}$	3.52739^{1}	2.20462	6.85216^{-2}	1.10231^{-3}
Ounce (avdp)	$\underline{4.37500}^{2}$	2.83495^{1}	2.83495^{-2}	$\underline{1.00000}$	6.25000^{-2}	1.94256^{-3}	3.12500^{-5}
Pound_{mass}	$\underline{7.00000}^{3}$	4.53592^{2}	4.53592^{-1}	$\underline{1.60000}^{1}$	$\underline{1.00000}$	3.10809^{-2}	$\underline{5.00000}^{-4}$
Slug	2.25218^{5}	1.45939^{4}	1.45939^{1}	5.14785^{2}	3.21740^{1}	$\underline{1.00000}$	1.60870^{-2}
Ton (short)	$\underline{1.40000}^{7}$	9.07184^{5}	9.07184^{2}	$\underline{3.20000}^{4}$	$\underline{2.00000}^{3}$	6.21618^{1}	$\underline{1.00000}$

4. MASS

TABLE 5

Time (t, τ)

	Day	Hour	Microsecond	Millisecond	Minute	Second
Day	$\underline{1.00000}$	$\underline{2.40000}^{1}$	$\underline{8.64000}^{10}$	$\underline{8.64000}^{7}$	$\underline{1.44000}^{3}$	$\underline{8.64000}^{4}$
Hour	4.16666^{-2}	$\underline{1.00000}$	$\underline{3.60000}^{9}$	$\underline{3.60000}^{6}$	$\underline{6.00000}^{1}$	$\underline{3.60000}^{3}$
Microsecond	1.15741^{-11}	2.77778^{-10}	$\underline{1.00000}$	$\underline{1.00000}^{-3}$	1.66667^{-8}	$\underline{1.00000}^{-6}$
Millisecond	1.15741^{-8}	2.77778^{-7}	$\underline{1.00000}^{3}$	$\underline{1.00000}$	1.66667^{-5}	$\underline{1.00000}^{-3}$
Minute	6.94444^{-4}	1.66667^{-2}	$\underline{6.00000}^{7}$	$\underline{6.00000}^{4}$	$\underline{1.00000}$	$\underline{6.00000}^{1}$
Second	1.15741^{-5}	2.77778^{-4}	$\underline{1.00000}^{6}$	$\underline{1.00000}^{3}$	1.66666^{-2}	$\underline{1.00000}$

5. TIME

TABLE 6

Angle (*a*)

	Degree	Minute	Quadrant (right angle)	Radians	Revolutions	Seconds
Degree	1.00000	6.00000^{1}	1.11111^{-2}	1.74533^{-2}	2.77778^{-3}	3.60000^{3}
Minute	1.66667^{-2}	1.00000	1.85185^{-4}	2.90889^{-4}	4.62963^{-5}	6.00000^{1}
Quadrants (right angle)	9.00000^{1}	5.40000^{3}	1.00000	1.57080	2.50000^{-1}	3.24000^{5}
Radians	5.72958^{1}	3.43775^{3}	6.36620^{-1}	1.00000	1.59155^{-1}	2.06265^{5}
Revolutions	3.60000^{2}	2.16000^{4}	4.00000	6.28320	1.00000	1.29600^{6}
Seconds	2.77777^{-4}	1.66667^{-2}	3.08642^{-6}	4.84815^{-6}	7.71605^{-7}	1.00000

6. ANGLE

TABLE 7

Velocity (v)

	cm/sec	ft/min	ft/sec	Kilometer/hr	Knot	Meter/min	Meter/sec	Mile/hr
Centimeter/second	1.00000	1.96850	3.28084^{-2}	3.60000^{-2}	1.94384^{-2}	6.00000^{-1}	1.00000^{-2}	2.23694^{-2}
Feet/minute	5.08000^{-1}	1.00000	1.66667^{-2}	1.82880^{-2}	9.87473^{-3}	3.04800^{-1}	5.08000^{-3}	1.13636^{-2}
Feet/second	3.04800^{1}	6.00000^{1}	1.00000	1.09728	5.92484^{-1}	1.82880^{1}	3.04800^{-1}	6.81818^{-1}
Kilometer/hour	2.77778^{1}	5.46807^{1}	9.11344^{-1}	1.00000	5.39957^{-1}	1.66667^{1}	2.77778^{-1}	6.21371^{-1}
Knot	5.14444^{1}	1.01268^{2}	1.68781	1.85200	1.00000	3.08667^{1}	5.14444^{-1}	1.15079
Meter/minute	1.66667	3.28084	5.46807^{-2}	6.00000^{-2}	3.23974^{-2}	1.00000	1.66667^{-2}	3.72823^{-2}
Meter/second	1.00000^{2}	1.96850^{2}	3.28084	3.60000	1.94384	6.00000^{1}	1.00000	2.23694
Mile/hour	4.47040^{1}	8.80000^{1}	1.46667	1.60934	8.68976^{-1}	2.68224^{1}	4.47040^{-1}	1.00000

*One knot = one nautical mile per hour

7. VELOCITY

TABLE 8

Force (f)

	Dyne	Gram force	Kilogram force	Newton	Poundal	Pound force
Dyne	$\underline{1.00000}$	1.01972^{-3}	1.01972^{-6}	$\underline{1.00000}^{-5}$	7.23301^{-5}	2.24809^{-6}
Gram force	$\underline{9.80665}^{2}$	$\underline{1.00000}$	$\underline{1.00000}^{-3}$	$\underline{9.80665}^{-3}$	7.09316^{-2}	2.20462^{-3}
Kilogram force	$\underline{9.80665}^{5}$	$\underline{1.00000}^{3}$	$\underline{1.00000}$	9.80665	7.09316^{1}	2.20462
Newton	$\underline{1.00000}^{5}$	1.01972^{2}	1.01972^{-1}	$\underline{1.00000}$	7.23301	2.24809^{-1}
Poundal	1.38255^{4}	1.40981^{1}	1.40981^{-2}	1.38255^{-1}	$\underline{1.00000}$	3.10809^{-2}
Pound force	4.44822^{5}	4.53594^{2}	4.53594^{-1}	4.44822	3.21740^{1}	$\underline{1.00000}$

8. FORCE

TABLE 9

Pressure (p)

	Standard Atmosphere	Bar**	Dynes/cm^2 (Barye)	Feet Water at 60°F*	Gm$_f$/cm^2	Inches Hg at 32°F*	Inches Water at 60°F*	Kg$_f$/cm^2	Lb$_f$/ft^2	Lb$_f$/in.^{2}	Micron Hg at 32°F	mm Hg at 32°F*
Standard Atmosphere	1.00000	1.01325	1.01325^{6}	3.39320^{1}	1.03323^{3}	2.99213^{1}	4.07184^{2}	1.03323	2.11622^{3}	1.46959^{1}	7.60000^{5}	7.60000^{2}
Bar**	9.86923^{-1}	1.00000	1.00000^{6}	3.34882^{1}	1.01972^{3}	2.95300^{1}	4.01859^{2}	1.01972	2.08854^{3}	1.45038^{1}	7.50062^{5}	7.50062^{2}
Dynes/Centimeter^2 (Barye)	9.86923^{-7}	1.00000 [illegible]	1.00000	3.34882^{-5}	1.01972^{-3}	2.95300^{-5}	4.01859^{-4}	1.01972^{-6}	2.08854^{-3}	1.45038^{-5}	7.50062^{-1}	7.50062^{-4}
Feet Water (at 60°F)*	2.94701^{-2}	2.98612^{-2}	2.98612 [illegible]	1.00000	3.04500^{1}	8.81801^{-1}	1.20000^{1}	3.04500^{-2}	6.23664^{1}	4.33100^{-1}	2.23977^{4}	2.23977^{1}
Gram$_{force}$/Centimeter^2	9.67841^{-4}	9.80665^{-4}	9.80665^{2}	3.28408^{-2}	1.00000	2.89590^{-2}	3.94089^{-1}	1.00000^{-3}	2.04816	1.42233^{-2}	7.35559^{2}	7.35559^{-1}
Inches Mercury at 32°F*	3.34211^{-2}	3.38639^{-2}	3.38639^{4}	1.13404	3.45315^{1}	1.00000	1.36085^{1}	3.45315^{-2}	7.07262^{1}	4.91154^{-1}	2.54000^{4}	2.54000^{1}
Inches Water at 60°F*	2.45589^{-3}	2.48843^{-3}	2.48843^{3}	8.33333^{-2}	2.53750	7.34834^{-2}	1.00000	2.53750^{-3}	5.19720	3.60917^{-2}	1.86648^{3}	1.86648
Kilogram$_{force}$/Centimeter^2	9.67841^{-1}	9.80669^{-1}	9.80665^{5}	3.28408^{1}	1.00000^{3}	2.89590^{1}	3.94089^{2}	1.00000	2.04816^{3}	1.42233^{1}	7.35559^{5}	7.35559^{2}
Pound$_{force}$/Feet^2	4.72541^{-4}	4.78803^{-4}	4.78803^{2}	1.60343^{-2}	4.88243^{-1}	1.41390^{-2}	1.92411^{-1}	4.88243^{-4}	1.00000	6.94444^{-3}	3.59131^{2}	3.59131^{-1}
Pound$_{force}$/Inch^2	6.80460^{-2}	6.89476^{-2}	6.89476^{4}	2.30894	7.03070^{1}	2.03602	2.77072^{1}	7.03070^{-2}	1.44000^{2}	1.00000	5.17149^{4}	5.17149^{1}
Microns Mercury at 32°F*	1.31579^{-6}	1.33322^{-6}	1.33322	4.46474^{-5}	1.35951^{-3}	3.93701^{-5}	5.35768^{-4}	1.35951^{-6}	2.78450^{-3}	1.93368^{-5}	1.00000	1.00000^{-3}
Millimeters Mercury at 32°F*	1.31579^{-3}	1.33322^{-3}	1.33322^{3}	4.46474^{-2}	1.35951	3.93701^{-2}	5.35768^{-1}	1.35951^{-3}	2.78450	1.93368^{-2}	1.00000^{3}	1.00000

*For g = 980.665 centimeters per second^2

**Some writers erroneously use the term bar for barye.

9. PRESSURE

TABLE 10

Energy, Work (e)

	Btu	I.T. Calorie	electron volt	erg (dyne-cm)	$ft\text{-}lb_f$	$gm_f\,cm$	hp hr (Mech)	ab. Joule	Kilocalorie	$kg_f\,m$	kw hr	watt hr	ft poundal
Btu	1.00000	2.51996^{2}	6.58577^{21}	1.05504^{10}	7.78158^{2}	1.07584^{7}	3.93009^{-4}	1.05504^{3}	2.51996^{-1}	1.07584^{2}	2.93067^{-4}	2.93067^{-1}	2.50365^{4}
I T Calorie	3.96832^{-3}	1.00000	2.61344^{19}	4.18674^{7}	3.08790	4.26928^{4}	1.55958^{-6}	4.18674	1.00000^{-3}	4.26928^{-1}	1.16298^{-6}	1.16298^{-3}	9.93528^{1}
electron volt	1.51842^{-22}	3.82637^{-20}	1.00000	1.60200^{-12}	1.18157^{-19}	1.63358^{-15}	5.96755^{-26}	1.60200^{-19}	3.82637^{-23}	1.63358^{-20}	4.45000^{-26}	4.45000^{-23}	3.80160^{-18}
erg (dyne-cm)	9.47831^{-11}	2.38849^{-8}	6.24220^{11}	1.00000	7.37562^{-8}	1.01972^{-3}	3.72506^{-14}	1.00000^{-7}	2.38849^{-11}	1.01972^{-8}	2.77778^{-14}	2.77778^{-11}	2.37304^{-6}
$ft\text{-}lb_f$	1.28509^{-3}	3.23836^{-1}	8.4632	1.35582^{7}	1.00000	1.38255^{4}	5.05050^{-7}	1.35582	3.23836^{-4}	1.38255^{-1}	3.76616^{-7}	3.76616^{-4}	3.21740^{1}
$gram_f\,cm$	9.29505^{-8}	2.34231^{-5}	6.121	9.80665^{2}	7.23301^{-5}	1.00000	3.65304^{-11}	9.80665^{-5}	2.34231^{-8}	1.00000^{-5}	2.72407^{-11}	2.72407^{-8}	2.32715^{-3}
hp hr (Mech)	2.54467^{3}	6.41196^{5}	1.67578^{25}	2.68452^{13}	1.98000^{6}	2.73745^{10}	1.00000	2.68452^{6}	6.41196^{2}	2.73745^{5}	7.45700^{-1}	7.45700^{2}	6.37046^{7}
ab. Joule (Watt sec)	9.47831^{-4}	2.38849^{-1}	6.24220^{18}	1.00000^{7}	7.37562^{-1}	1.01972^{4}	3.72506^{-7}	1.00000	2.38849^{-4}	1.01972^{-1}	2.77778^{-7}	2.77778^{-4}	2.37304^{1}
Kilocalorie	3.96832	1.00000^{3}	2.61344^{22}	4.18674^{10}	3.08798^{3}	4.26928^{7}	1.55958^{-3}	4.18674^{3}	1.00000	4.26928^{2}	1.16298^{-3}	1.16298	9.93528^{4}
$kg_f\,m$	9.29 [illegible]$^{-3}$	2.34231	6.121^{19}	$9.$[illegible]^{7}	7.23301	1.00000^{5}	3.65304^{-6}	9.8066	2.34231^{-3}	1.00000	2.72407^{-6}	2.72407^{-3}	2.32715^{2}
kw hr	$3.$[illegible]19^{3}	8.59858^{5}	2.24719^{25}	3.60000^{13}	2.65522^{6}	3.67098^{10}	1.34102	3.60000^{6}	8.59858^{2}	3.67098^{5}	1.00000	1.00000^{3}	8.54293^{6}
watt hr	3.41219	8.59858^{2}	2.24719^{22}	3.60000^{10}	2.65522^{3}	3.67098^{7}	1.34102^{-3}	3.60000^{3}	8.59858^{-1}	3.67098^{2}	1.00000^{-3}	1.00000	8.54293^{4}
ft poundal	3.99417^{-5}	1.00651^{-2}	2.6304^{17}	4.21401^{5}	3.10810^{-2}	4.29710^{2}	1.56974^{-8}	4.21401^{-2}	1.00651^{-5}	4.29710^{-3}	1.17056^{-8}	1.17056^{-5}	1.00000

Definitions: $\frac{1\ \text{Btu}}{\delta_F\ lb_m} = \frac{1\ \text{I.T. Cal}}{^{\circ}C\ gm}$

1 I T Cal = 1/860 int. watt hr

1 int. watt = 1.000165 ab. watt

1 Btu_{mean} = 1055.8 absolute joules

1 $Btu_{39}{}^{\circ}F$ = 1060 absolute joules

1 $Btu_{60}{}^{\circ}F$ = 1054.6 absolute joules

1 B Btu = 1055.04 absolute joules

1 B $Cal_{15}{}^{\circ}C$ = 4.1854 absolute joules

1 B Calorie = 4.18674 absolute joules

1 Cal_{mean} = 4.190 absolute joules

1 $Cal_{20}{}^{\circ}C$ = 4.181 absolute joules

1 Thermochemical Calorie = 4.1840 absolute joules

1 International Joule = 1.000165 absolute joules

1 I.T. Calorie = 1/860 international watt hour

Kilocalorie or large calorie = 1000 calories

10. ENERGY

TABLE 11

Power (P)

	Btu/hr	Btu/min	Btu/sec	I.T.Cal/hr	I.T.Cal/min	I.T.Cal/sec	erg/sec	$ft\text{-}lb_f/min$	$ft\text{-}lb_f/sec$	hp (Elec)	hp (Mech)	hp (Metric)	$kg_f m/sec$	kw	w
Btu/hr	1.00000	1.66667^{-2}	2.77778^{-4}	2.51996^{2}	4.19993	6.99988^{-2}	[illegible]	1.29693^{1}	2.16155^{-1}	3.72851^{-4}	3.93009^{-4}	3.98460^{-4}	2.98845^{-2}	2.93067^{-4}	2.93067^{-1}
Btu/min	6.00000^{1}	1.00000	1.66667^{-2}	1.51197^{4}	2.51996^{2}	4.19993	1.75840^{8}	7.78158^{2}	1.29693^{1}	2.35711^{-2}	2.35805^{-2}	2.39076^{-2}	1.79307	1.75840^{-2}	1.75840^{1}
Btu/sec	3.60000^{3}	6.0000^{1}	1.00000	9.07183^{5}	1.51197^{4}	2.51996^{2}	1.05504^{10}	4.66894^{4}	7.78158^{2}	1.41426	1.41483	1.43446	1.07584^{2}	1.05504	1.05504^{3}
I.T.Cal/hr	3.96832^{-3}	6.61387^{-5}	1.10231[illegible]	1.00000	1.66667[illegible]	2.77778^{-4}	[illegible]	5.14663^{-2}	8.57772^{-4}	1.55896^{-6}	1.55958^{-6}	1.58122^{-6}	1.18591^{-4}	1.16298^{-6}	1.16298^{-3}
I.T.Cal/min	2.38099^{-1}	3.96832^{-3}	6.61387^{-5}	6.00000^{1}	1.00000	1.66667^{-2}	[illegible]	[illegible]	5.14663^{-2}	9.35376^{-5}	[illegible]	9.48730^{-5}	7.11547^{-3}	6.97790^{-5}	6.97790^{-2}
I.T.Cal/sec	1.42859^{1}	2.38100^{-1}	3.96832[illegible]	3.60000^{3}	6.00000^{1}	1.00000	[illegible]	[illegible]	3.08798	5.61225^{-3}	5.61451^{-3}	5.69238^{-3}	4.26928^{-1}	4.18674^{-3}	4.18674
erg/sec	3.41219^{-7}	5.68699^{-9}	9.47831^{-11}	8.59898^{-5}	1.43310^{-6}	2.38849^{-8}	1.00000	4.42537^{-6}	7.37562^{-8}	1.34048^{-10}	1.34102^{-10}	1.35962^{-10}	1.01972^{-8}	1.00000^{-10}	1.00000^{-7}
$ft\text{-}lb_f/min$	7.71052^{-2}	1.28500^{-3}	2.14181[illegible]	1.94302^{1}	[illegible]	[illegible]	[illegible]	[illegible]	1.66667^{-2}	[illegible]	[illegible]	3.07233^{-5}	2.30425^{-3}	2.25970^{-5}	[illegible]
$ft\text{-}lb_f/sec$	4.62631	7.71052^{-2}	[illegible]	[illegible]	1.94302^{1}	[illegible]	[illegible]	[illegible]	1.00000	[illegible]	[illegible]	1.84340^{-3}	1.38255^{-1}	1.35582^{-3}	1.35582
hp (Elec)	2.54549^{3}	4.24249^{1}	7.07081^{-1}	6.41453^{5}	[illegible]	1.78182^{2}	[illegible]	[illegible]	5.50221^{2}	1.00000	1.00040	1.01427	7.60707^{1}	7.46000^{-1}	7.46000^{2}
hp (Mech)	2.54447^{3}	4.24079^{1}	[illegible]	6.41195^{5}	[illegible]	1.78110^{2}	[illegible]	[illegible]	5.50000^{2}	[illegible]	1.00000	1.01387	7.60402^{1}	7.45701^{-1}	7.45701^{2}
hp (Metric)	2.50966^{3}	4.18277^{1}	6.97189^{-1}	6.32424^{5}	1.05404^{4}	1.75673^{2}	[illegible]	[illegible]	5.42476^{2}	[illegible]	[illegible]	1.00000	7.50000^{1}	7.35500^{-1}	7.35500^{2}
$kg_f m/sec$	3.34622^{1}	5.57703^{-1}	9.29505^{-3}	8.43233^{3}	1.40539^{2}	2.34231	[illegible]	4.33981^{2}	7.23301	1.31456^{-2}	1.31509^{-2}	1.33333^{-2}	1.00000	9.80665^{-3}	9.80665
Kilowatt	3.41219^{3}	5.68699^{1}	9.47831^{-1}	[illegible]	[illegible]	2.38849^{2}	[illegible]	[illegible]	7.37562^{2}	[illegible]	[illegible]	1.35962	1.01972^{2}	1.00000	[illegible]
watt	3.41219	5.68699^{-2}	9.47831^{-4}	[illegible]	1.43310^{1}	2.38849^{-1}	[illegible]	4.42537^{1}	7.37562^{-1}	1.34048^{-3}	1.34102^{-3}	1.35962^{-3}	1.01972^{-1}	1.00000^{-3}	1.00000

11. POWER

TABLE 12

Mass Flow Rate ($\dot{m}$)

	Gm_m/sec	Kg_m/sec	Lb_m/hr	Lb_m/min	Lb_m/sec	Slug/sec
Gm_m/sec	1.00000	1.00000^{-3}	7.93664	1.32277^{-1}	2.20462^{-3}	6.85218^{-5}
Kg_m/sec	1.00000^{3}	1.00000	7.93664^{3}	1.32277^{2}	2.20462	6.85218^{-2}
Lb_m/hr	1.25998^{-1}	1.25998^{-4}	1.00000	1.66667^{-2}	2.77778^{-4}	8.63360^{-6}
Lb_m/min	7.55987[illegible]	7.55987^{-3}	6.00000^{1}	1.00000	1.66667^{-2}	5.18016^{-4}
Lb_m/sec	4.53592[illegible]	4.53592^{-1}	3.60000^{3}	6.00000^{1}	1.00000	3.10809^{-2}
Slug/sec	1.45939[illegible]	1.45939^{1}	1.15826^{5}	1.93044^{3}	3.21740^{1}	1.00000

12. MASS FLOW RATE

TABLE 13

Pumping Speed or Volume Flow (V)

	Cm^3/sec	Ft^3/min	Ft^3/sec	Gal/min	Liter/min	Liter/sec	M^3/hr	M^3/min
Cm^3/sec	1.00000	2.11888^{-3}	3.53147^{-5}	1.58503^{-2}	5.99983^{-2}	9.99972^{-4}	3.60000^{-3}	6.00000^{-5}
Ft^3/min	4.71947^{2}	1.00000	1.66667^{-2}	7.48052	2.83160^{1}	4.71934^{-1}	1.69901	2.83168^{-2}
Ft^3/sec	2.83168^{4}	6.00000^{1}	1.00000	4.48831^{2}	1.69896^{3}	2.83160^{1}	1.01941^{2}	1.69901
Gal/min	6.30902^{1}	1.33680^{-1}	2.22801^{-3}	1.00000	3.78530	6.30884^{-2}	2.27125^{-1}	3.78541^{-3}
Liter/min	1.66671^{1}	3.53156^{-2}	5.88594^{-4}	2.64179^{-1}	1.00000	1.66667^{-2}	6.00017^{-2}	1.00003^{-3}
Liter/sec	1.00003^{3}	2.11894	3.53156^{-2}	1.58508^{1}	6.00000^{1}	1.00000	3.60010	6.00017^{-2}
M^3/hr	2.77778^{2}	5.88578^{-1}	9.80963^{-3}	4.40287	1.66662^{1}	2.77770^{-1}	1.00000	1.66667^{-2}
M^3/min	1.66667^{4}	3.53147^{1}	5.88578^{-1}	2.64172^{2}	9.99972^{2}	1.66662^{1}	6.00000^{1}	1.00000

13. PUMPING SPEED

TABLE 14

Temperature* (T, θ)

To convert from the units below to those on the right, perform the indicated operations in order.	^{o}C	^{o}F	^{o}K	^{o}R
^{o}C	x 1	x 9/5 + 32	+ 273.15	x 9/5 + 491.67
^{o}F	- 32 x 5/9	x 1	x 5/9 + 255.372	+ 459.67
^{o}K	- 273.15	x 9/5 - 459.67	x 1	x 9/5
^{o}R	x 5/9 - 273.15	- 459.67	x 5/9	x 1

*Based on the thermodynamic temperature scale as defined by the Tenth General Conference on Weights and Measure meeting at Paris in October 1954.

Temperature of triple point of water = 273.16°K = 491.688°R = 32.018°F = 0.01°C

Temperature of ice point of water = 273.15°K = 491.67°R = 0°C = 32°F

14. TEMPERATURE

VOLUME I

TABLE OF CONTENTS
VOLUMES I, II and III.

VOLUME II

VOLUME II

GPSR Compliance
The European Union's (EU) General Product Safety Regulation (GPSR) is a set of rules that requires consumer products to be safe and our obligations to ensure this.

If you have any concerns about our products, you can contact us on

ProductSafety@springernature.com

In case Publisher is established outside the EU, the EU authorized representative is:

Springer Nature Customer Service Center GmbH
Europaplatz 3
69115 Heidelberg, Germany

www.ingramcontent.com/pod-product-compliance
Ingram Content Group UK Ltd.
Pitfield, Milton Keynes, MK11 3LW, UK
UKHW021332070726
13610UKWH00011B/45

* 9 7 8 3 6 4 2 8 2 6 7 9 5 *